JUN - 3 2009

BLAZING THE TRAIL

THE EARLY HISTORY OF SPACECRAFT AND ROCKETRY

BLAZING THE TRAIL

THE EARLY HISTORY OF SPACECRAFT AND ROCKETRY

Mike Gruntman

Published by
American Institute of Aeronautics and Astronautics, Inc.
1801Alexander Bell Drive, Suite 500, Reston VA 20191-4344

American Institute of Aeronautics and Astronautics, Inc., Reston, Virginia

2 3 4 5

Library of Congress Cataloging-in-Publication Data

Gruntman, Mike, 1954-
Blazing the trail : the early history of spacecraft and rocketry / Mike Gruntman.
p. cm.
Includes bibliographical references and index.
ISBN 1-56347-705-X (alk. paper)
1. Rocketry—History. 2. Rockets (Aeronautics)—History. 3. Space vehicles—History. I. Title.

TL781.G78 2004
621.43'56—dc22

2004006792

Cover design by Gayle Machey

Copyright © 2004 by the American Institute of Aeronautics and Astronautics, Inc. All rights reserved. Printed in the United States of America. No part of this publication may be reproduced, distributed, or transmitted, in any form or by any means, or stored in a database or retrieval system, without the prior written permission of the publisher.

FOREWORD

... A CHARIOT OF FIRE AND HORSES OF FIRE ...

Liftoff of Titan IV. Photo courtesy of NASA.

"... And I looked, and, behold, a whirlwind came out of the north, a great cloud, and a fire infolding itself, and a brightness was about it, and out of the midst thereof as the colour of amber, out of the midst of the fire.

... And ... I heard the noise of their wings, like the noise of great waters ... .

... And above the firmament ... was the likeness of a throne, as the appearance of a sapphire stone: and upon the likeness of the throne was the likeness as the appearance of a man above upon it. And I saw as the colour of amber, as the appearance of fire round about within it, from the appearance of his loins even upward, and from the appearance of his loins even downward, I saw as it were the appearance of fire, and it had brightness round about ... ."

Ezekiel, 1:4,24,26,27

Was it the first, approximately 2600 years ago, description of spaceship landing or liftoff, as some suggest?

Table of Contents

PREFACE

This book introduces the reader into the history of early rocketry and the subsequent developments that led to the space age. The exciting achievements of space exploration that began in the late 1950s are very well known to many. Not that many, however, are familiar with the beginning: How we prepared for this breakthrough into space; What happened in the early days of rocketry and spacecraft; and Who were those often unappreciated and half-forgotten visionaries, scientists, engineers, and political and military leaders who opened the way to space. Publications on these topics usually target the specialists and address in detail specific aspects of rocket history. On the other end of the spectrum are popular books that just scratch the surface and have minimal and sometimes inaccurate technical details. This book bridges the gap.

In a compressed lecture-notes style, the book presents the fascinating story of the events that paved the way to space. People of various nations and from various lands contributed to the breakthrough into space, and the book takes the reader to faraway places on five continents. It also includes many quotes to give readers a flavor of how the participants viewed the developments.

The book contains numerous technical details usually unavailable in popular publications. The details are not overbearing, and anyone interested in rocketry and space exploration will navigate through the book without difficulty. In addition, there are 340 figures, including many photographs, some appearing for the first time.

The selection of topics was guided by the desire to present the whole story in a short book, with the emphasis on the first steps in new technology and the pioneers who took them. The follow-on programs, regardless of their impact, that built upon these breakthroughs are beyond the scope of the book. Consequently, many important events and persons that would otherwise be prominently presented in a comprehensive history are omitted.

The book focuses on the early history of rocketry and the very beginning of spacecraft history. The presented story ends at the first American and Soviet launches of late 1950s. Thus, satellite description is largely confined to Sputnik, Explorer, and Vanguard, leaving out the numerous advances in spacecraft that usually incrementally built upon these firsts. However, a few selected and highly important for satellite technology developments that originated during 1950s and "spilled over" to 1960s and beyond are also briefly discussed. In particular, "opening the skies" by early reconnaissance missions that remained largely unknown until recently is presented in some detail. In addition, one chapter focuses on the very first steps of the space-faring nations other than the United States and Soviet Union.

Knowledge and appreciation of the early history of rocketry and spacecraft are important for those who are the keepers of the flame. This book is for present and future "rocket scientists" and for all those who are a part of the great space enterprise. It is for those who dream to join this endeavor in the future; for university and college instructors and students; for high-school teachers and students; and for all interested in and fascinated by space. This book could also be used as a basis for introductory and history-oriented lectures in various astronautics and space technology courses; support university classes on the history of technology and military history; and serve students as an information source for university and high-school projects.

The aerospace field traditionally uses a combination of metric and nonmetric units. Therefore, for the convenience of the reader, many of the presented parameters are given simultaneously in the most widely used units such as centimeters (cm), inches (in.), feet (ft), meters (m), miles, nautical miles (n miles), and kilometers (km) of length; pounds (lb), kilograms (kg), and tonnes (t) of mass; and pounds (lbf), newtons (N), and tonnes (t) of thrust. The unit "tonne" with the symbol "t" means "metric ton" throughout the text, that is 1 t = 1000 kg = 2204.6 lb. We will use the same symbol "t" of metric ton for both mass and weight.

ACKNOWLEDGMENTS

Many individuals and organizations kindly helped me in the preparation of this book by providing information, photographs, and publications and by offering advice and suggestions on specific issues. The support was tremendous, as you will see from the following list. I want to specially thank Bob Brodsky of Redondo Beach, California, for continuous support and interest, numerous discussions, advice, and encouragement. Bob read the draft of a large part of the manuscript, and his thoughtful comments helped to improve it significantly. Bob Sackheim of Los Angeles, California, and Huntsville, Alabama, kindly read and commented on Chapter 12. Patience and expertise of the dedicated editorial staff of the American Institute of Aeronautics and Astronautics made this publication possible.

Nothing is perfect though, and a very few organizations in the United States and a larger number abroad, concentrated in a few countries, chose not to share their contributions to the breakthrough to space. Because dis-acknowledgments are not customary yet, their names will be omitted.

I would like to thank each and all of those who helped me to write this book. I was blessed by your support. I am especially grateful to the following:

Ranney Adams, Edwards Air Force Base, California;
Aerojet-General Corporation, Sacramento, California;
Aerospace Corporation, El Segundo, California;

John Ahouse, University of Southern California, Los Angeles, California;
American Institute of Aeronautics and Astronautics;
American Institute of Mining and Metallurgical Engineers;
Orianne Arnould, Documentation Center, Centre National d'Études Spatiales, France;
Tanya Arvan, Redondo Beach, California;
Elliot Axelband, University of Southern California, Los Angeles, California;
Michael E. Baker, U.S. Army Aviation and Missile Command, Redstone Arsenal, Alabama;
Robert Louis Benson, National Security Agency;
Yurii V. Biryukov, S.P. Korolev Memorial House-Museum, Moscow, Russia;
Boeing — Rocketdyne, Canoga Park, California;
Sam Bono, National Atomic Museum, Albuquerque, New Mexico;
Karel J. Bossart, Jr., Augusta, Maine;
Newell C. Bossart, Medina, Washington;
British Interplanetary Society;
Dino Brugioni, Hartwood, Virginia;
Stephen Butler, Defense Science and Technology Organization, Australia;
Central Intelligence Agency;
Center for the Study of Intelligence, Central Intelligence Agency, Washington, D.C.;
Office of Public Affairs, Central Intelligence Agency, Washington, D.C.;
Centre National d'Études Spatiales, France;
Charles Stark Draper Laboratory, Cambridge, Massachusetts;
Arthur C. Clarke, Colombo, Sri Lanka;
Mark C. Cleary, 45th Space Wing, Patrick Air Force Base, Florida;
Pierre Louis Contreras, Educational Directorate, Centre National d'Études Spatiales, France;
E. Couzin, K.E. Tsiolkovsky Museum of Cosmonautics, Kaluga, Russia;
Vyacheslav Danilkin, State Rocket Center "Makeyev Design Bureau," Miass, Russia;
Werner Dappen, University of Southern California, Los Angeles, California;
Dwayne A. Day, Vienna, Virginia;
Defense Science and Technology Organization, Australia;
FOIA Office, Defense Intelligence Agency, Washington, D.C.;
Yurii G. Demyanko, Keldysh Research Center, Russia, Moscow;
Deutsches Museum, Munich, Germany;
Roger Easton, Canaan, New Hampshire;
Richard Eckert, Space Command, Air Force;
Jim Eckles, White Sands Missile Range, New Mexico;
Editions Flammarion, Paris, France;
Dwight D. Eisenhower Library, Abilene, Kansas;
David J. Evans, The Aerospace Corporation, El Segundo, California;
Hans Fahr, Bonn, Germany;
Bill Feldman, Los Alamos National Laboratory, Los Alamos, New Mexico;
Larisa A. Filina, S.P. Korolev Memorial House-Museum, Moscow, Russia;

Harold B. Finger, Chevy Chase, Maryland;
Public Affairs Office, Fort Bliss, Texas;
Jerold C. Frakes, University of Southern California, Los Angeles, California;
Herb Funsten, Los Alamos National Laboratory, Los Alamos, New Mexico;
Paul G. Gaffney, National Defense University, Washington, D.C.;
Giancarlo Genta, Politecnico di Torino, Italy;
Library, Getty Research Institute, Los Angeles, California;
Glenn Research Center, NASA, Cleveland, Ohio;
Stuart Grayson, Information Unit, British National Space Center, Great Britain;
Claude Gourdet, International Astronautical Federation, Paris, France;
Sergei A. Gruntman, Moscow, Russia;
Stanislaw Grzedzielski, Warsaw, Poland;
John Hargenrader, NASA Headquarters, Washington, D.C.;
Aby Har-Even, Israel Space Agency, Israel;
Jim Harford, Princeton, New Jersey;
Peter Harrington, Brown University Library, Providence, Rhode Island;
Charles B. Henderson, Leesburg, Virginia;
Henry R. Hertzfeld, George Washington University, Washington, D.C.;
Mark Hess, Goddard Space Flight Center, Greenbelt, Maryland;
Colin Hicks, British National Space Center, Great Britain;
Stephen Hood, Australian Embassy at Washington, D.C.;
K.C. (Johnny) Hsieh, Tucson, Arizona;
Hua Shih Press, Taipei, Taiwan, Republic of China;
Indian Space Research Organization, Bangalore, India;
Institute of Space and Astronautical Science, Japan;
International Astronautical Federation;
International Launch Services, McLean, Virginia;
Israel Space Agency;
Boris Katorgin, NPO Energomash, Khimki, Russia;
Sergei N. Khrushchev, Providence, Rhode Island;
Kenneth Klein, University of Southern California, Los Angeles, California;
Keith Koehler, Wallops Flight Facility, Wallops Island, Virginia;
Misha Kogan, Jerusalem, Israel;
Anatoly Koroteev, Keldysh Research Center, Moscow, Russia;
Bart Kosko, University of Southern California, Los Angeles, California;
Tom Krimigis, Applied Physics Laboratory, Laurel, Maryland;
Valerii M. Krylov, Military-Historical Museum of Artillery, Engineers and Communications, St. Peterburg (Leningrad), Russia;
Joseph A. Kunc, University of Southern California, Los Angeles, California;
Office of Public Affairs, NASA's Langley Research Center, Hampton, Virginia;
Roger Launius, NASA Headquarters, Washington, D.C.;
Susan Lemke, National Defense University, Washington, D.C.;
Lockheed Martin — Astronautics, Denver, Colorado;
Vladas V. Leonas, Sydney, Australia;
Little, Brown, and Co., Publishers;

Public Affairs Office, Los Alamos National Laboratory, Los Alamos, New Mexico;
Lowell Observatory, Flagstaff, Arizona;
Claus R. Martel, U.S. Army Aviation and Missile Command, Redstone Arsenal, Alabama;
Don McMullin, University of Southern California, Los Angeles, California, and Naval Research Laboratory, Washington, DC;
Dennis McSweeney, NASA's Liaison Office in Moscow, Russia;
Musée de l'Armée, Paris, France;
National Aeronautics and Space Administration;
National Archives and Records Administration, College Park, Maryland;
National Central Library, Taipei, Taiwan, Republic of China;
National Defense Industrial Association;
National Reconnaissance Office;
Deng Ningfeng, China Astronautics Publishing House, Beijing, China;
Oldenbourg Wissenschaftsverlag, Munich, Germany;
Lucia Padrielli, Bologna, Italy;
Simon Ramo, Beverly Hills, California;
Alan Renga, San Diego Aerospace Museum, San Diego, California;
Roswell Museum and Art Center, Roswell, New Mexico;
Franklin D. Roosevelt Library, Hyde Park, New York;
Royal United Services Institute, London, Great Britain;
Kevin Ruffner, Central Intelligence Agency;
Wesley Rusnell, Roswell Museum and Art Center, Roswell, New Mexico;
Russian Academy of Sciences;
Stefano Sandrelli, Astronomical Observatory of Brera, Milan, Italy;
Ross Scimeca, University of Southern California, Los Angeles, California;
John F. Schuessler, Mutual UFO Network, Inc., Morrison, Colorado;
Marje Schuetze-Coburn, University of Southern California, Los Angeles, California;
Gary Simmons, Defense Science and Technology Organization, Australia;
Simon and Schuster, Inc.;
Smithsonian Institution, Washington, DC;
Polina Soloviev, Moscow, Russia;
Victor Soloviev, Moscow, Russia;
Igor E. Spektor, Moscow, Russia;
Ernst Stuhlinger, Huntsville, Alabama;
Vladimir Sudakov, NPO Energomash, Khimki, Russia;
Paul F. Sullivan, Department of the Navy, Washington, D.C.;
William G. (Greg) Thalmann, University of Southern California, Los Angeles, California;
TRW, Inc.;
U.S. Army;
U.S. Air Force;
U.S. Air Force Museum, Wright-Patterson Air Force Base, Ohio;
U.S. Air Force Space Command, Peterson Air Force Base, Colorado;

U.S. Army Aviation and Missile Command, Redstone Arsenal, Alabama;
U.S. Navy;
James A. Van Allen, Iowa City, Iowa;
David Van Keuren, Naval Research Laboratory, Washington, DC;
VDI-Verlag GmbH, Verein Deutsche Ingineure, Düsseldorf, Germany;
Videocosmos, Woodbridge, Virginia;
Harry Waldron, Los Angeles Air Force Base, California;
Ruth Wallach, University of Southern California, Los Angeles, California;
Vince J. Wheelock, Boeing — Rocketdyne, Canoga Park, California;
Albert D. Wheelon, California;
Public Affairs Office, White Sands Missile Range, New Mexico;
Mike Wright, NASA Marshall Space Flight Center, Huntsville, Alabama;
Alik Zholkovsky, University of Southern California, Los Angeles, California;
V.A. Zubkov, the Russian State Library, Moscow.

The opinions expressed in this book are solely mine and not necessarily shared by the individuals and organizations that helped in preparation of the book. Needless to say, I take the responsibility for all errors.

1. HUMBLE BEGINNINGS

PRINCIPLE OF ROCKET PROPULSION

The principle of rocket propulsion was observed by the ancient Greeks. Hero (or Heron; A.D. ~65–125) of Alexandria built an aeolipile in the 1st century, which demonstrated the idea of reactive propulsion. Hero was a famous Greek geometer (Hero's formula of the area of a triangle) and inventor. He published many books on geometry, mechanics, and pneumatics.

Hero of Alexandria

The aeolipile was a hollow sphere with a few bent tubes, jets, or nozzles, pointed tangentially. The sphere could turn on a hollow axial shaft that provided steam to the sphere. The steam flowed tangentially to the sphere's surface through the outlets (nozzles) causing the sphere to revolve. This was the first known device to transform steam into rotary motion: an early reaction steam turbine.

Aeolipile

Fig. 1.1. Aeolipile. Figure courtesy of NASA, EG-1999-06-108-HQ.

The phenomenon of jet propulsion was not understood or explained at those ancient times. The Greeks considered the aeolipile a curiosity not useful for any practical purpose.

FIRST ROCKETS

Gun-powder

The earliest rockets were solid-propellant rockets (or *solid rockets*) with gunpowder (black powder) as propellant. Black powder is a mechanical, not chemical, compound of

Saltpeter and Sulfur

sulfur, charcoal, and saltpeter (potassium nitrate KNO_3). The key to the discovery of gunpowder was saltpeter. Sulfur brought the desired plasticity to the compound and improved its storage (some sort of explosive could even be produced without sulfur).

> **GUNPOWDER**
>
> Strictly speaking, the term "gunpowder" is incorrect here because the guns appeared only in the early 14th century. By that time the gunpowder (black powder) was widely used for at least several centuries.

Charcoal was obviously known even to very primitive cultures. The Western World knew saltpeter (niter) and sulfur (brimstone) since the times of the Old Testament:

> Then the Lord rained upon Sodom and upon Gomorrah brimstone and fire from the Lord out of heaven, *Genesis 19:24*;
>
> As he that taketh away a garment in cold weather, and as vinegar upon nitre, so is he that singeth songs to an heavy heart, *Proverbs, 25:20*;
>
> For though thou wash thee with nitre, and take thee much soap, yet thine inequity is marked before me, saith the Lord God, *Jeremiah, 2:22.*

Saltpeter in China and India

Though rare in Europe, saltpeter naturally occurred in soil in China and India. This substance was even called the *Indian snow* in many countries. Naturally abundant saltpeter and wide availability of sulfur were apparently critical for the discovery of gunpowder.

Many believe that gunpowder and later the rockets first appeared in China. Some would even argue that gunpowder was known for more than 2000 years. It is not inconceivable, however, that the Hindus independently discovered gunpowder, maybe even earlier than the Chinese did. It is also possible that some secrets of pyrotechny penetrated into both China and India much earlier from the ancient civilization of Egypt. The known Chinese records of fireworks simply antedate those in other countries.

Export Control of Strategic Materials

The detailed descriptions of gunpowder preparation, storage, and military use date back in Chinese sources for about 1000 years. The Chinese clearly understood military importance of gunpowder constituents. The edict of the year A.D. 1067 prohibited export of saltpeter and sulfur to foreign countries in an early example of export control of strategic materials.

Fig. 1.2. Early Chinese fire arrow-rocket. The use of a separate powder container led to the development of powder-propelled arrows, that is, early rockets. Figure courtesy of Hua Shih Press, Taipei, Taiwan, from *Wu Pei Chih* (Mao Yuan-i Chi 1991, Vol. 12, 5168).

Fire-crackers

Firing off crackers was an ancient Chinese custom since at least the seventh century. A piece of fresh bamboo thrown into fire would crack with a noise. It was hoped that the noise of cracking of the burning bamboo would frighten evil spirits. Gunpowder could substantially add to noise and was certainly used in firecrackers as early as in the 10th century. Thus, the bamboo tubes

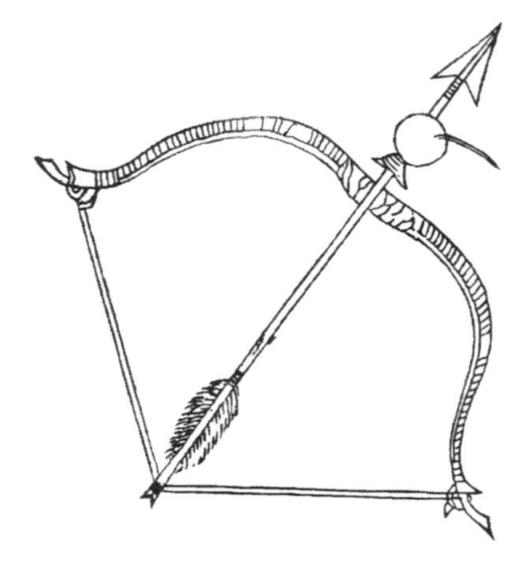

Fig. 1.3. Chinese bow-shot incendiary arrow, ca. 1621. Similar incendiary arrows were widely used by the Chinese many centuries earlier. Such arrows were often misidentified later by erroneous translations of the ancient records as rockets. Figure courtesy of Hua Shih Press, Taipei, Taiwan, from *Wu Pei Chih* (Mao Yuan-i Chi 1991, Vol. 12, 5178).

filled with saltpeter, sulfur, and charcoal were tossed into ceremonial fires during festivals.

Bamboo Tubes

Now it was only a question of time to find more practical — no offence to the entertainment industry here — applications of the new technology. Somebody undoubtedly noticed that if one side of the bamboo tube was tightly sealed with clay then a harmless firecracker turned into a small and potentially deadly projectile — a rocket. It should have been a *painful* experience! Certainly by the year A.D. 1045 gunpowder and probably some kind of rocket predecessors had been widely used by the Chinese military.

1000 Years Ago

The first primitive rockets thus appeared almost 1000 years ago. Yes, almost 1000 years ago! It was about the time when William the Conqueror landed on the shores of England, Leif Ericsson (Leif the

Fig. 1.4. Incendiary bow- and cross-bow-shot arrows were widely employed in warfare throughout the world for many centuries. This figure from a European war manual (1640) illustrates the use of incendiary arrows against enemy ships. Figure courtesy of Library, Getty Research Institute, Los Angeles, California, from *Traité de fevx Artificiels povr la Gverre, et povr la Recreation* (Malthus 1640, 19).

Lucky) reached the North American coast, and the Western (Roman) and Eastern (Orthodox) Christian churches formally split.

Fire Arrows

In the official Chinese (mandarin) language the word "rocket" is rendered as *hu-o chi-en* with the literal meaning "fire arrow." Chinese documents used this term even before A.D. 1045. It is not clear however whether they meant conventional arrows tipped with flammable substance or the rocket-propelled arrows.

Fig. 1.5. Rocket basket for launch of up to 20 arrows, ca. 1621. Rocket steel tips were smeared with poison. Figure courtesy of National Defense Industrial Association, from "Fire for the Wars of China," by T.L. Davis, *Ordnance*, Vol. 32, No. 169, 1948, p. 52.

Fire Spears

Chinese incendiary arrows were conveniently shot from bows. The arrow powder typically included saltpeter, sulfur, and willow charcoal with small additions of eggplant charcoal and white arsenic. Bamboo spears were also often tipped with small powder-filled bamboo tubes. The fire was directed forward from the spear and served as an extension of the weapon. Obviously, the fire spears were not the rockets, and the powder-produced thrust actually pushed the spear in the wrong (backward) direction.

Incendiary Weapon

Fire arrows were prominently represented in the historical records of the 11th and 12th centuries. The arrows burned enemy ships, military materials and machines, and inflicted casualties to infantry and cavalry. Only one step was left for the rockets to appear.

Fire Arrows Lead to Rockets

Incendiary gunpowder was initially attached directly to the tip of a conventional arrow. A major improvement came with the subsequent introduction of separate containers for gunpowder. This new arrangement would finally lead to a gunpowder-propelled arrow, that is, a rocket. In some designs, the gunpowder was even divided between two containers, incendiary ("warhead") and pro-

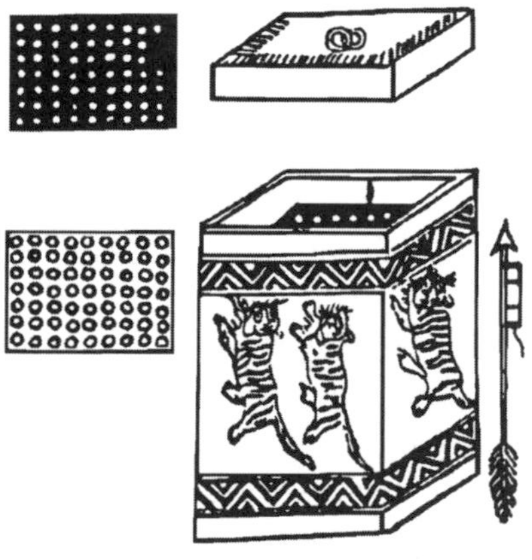

Fig. 1.6. Chinese rocket launcher for 100 arrows, ca. 1621: "the rockets ... will rush ... like 100 tigers." Figure courtesy of National Defense Industrial Association, from "Fire for the Wars of China," by T.L. Davis, *Ordnance*, Vol. 32, No. 169, 1948, p. 52.

pelling, that were separately attached to the arrow's shaft. The propulsion potential of gunpowder was thus clearly demonstrated.

Basic Rocket Design for 800 Years

The early design of a rocket has soon emerged: a gunpowder-propellant container attached to a guiding stick to stabilize the rocket flight. This basic arrangement would preserve its main features for more than 800 years until the middle of the 19th century when William Hale introduced a stickless rocket.

The nomad Mongols led by the great Genghis Khan increased military pressure on China in the beginning of the 13th century. In a pattern that would repeat itself many times in history, the more developed society of the Chinese Sung (Song) Dynasty relied on advanced technology to counter the military threat. Chinese ordnance experts introduced and perfected many types of projectiles and explosives at these times of trouble.

Advanced Technology to Counter Threat

Fig. 1.7. Rocket launcher for 40 arrows, ca. 1621. The diverging pointing of the projectiles in the launcher spread the rockets laterally over several hundred feet. Such launchers were an efficient weapon in an open country against exposed enemy troops. Figure courtesy of National Defense Industrial Association, from "Fire for the Wars of China," by T.L. Davis, *Ordnance*, Vol. 32, No. 169, 1948, p. 53.

Some historians argue that the first definite record of rockets dates to A.D. 1232, five years after the death of Genghis Khan. The Chinese rulers put a garrison of Kin Tartars to defend the city of K'ai-fung-fu, north of the Yellow River. The defenders repelled the Mongol troops in that year after a fierce battle.

First Use of Rockets

The Tartars were apparently familiar with gunpowder for some time and used explosive bombs 11 years earlier when they had attacked another Chinese city. At K'ai-fung-fu the defenders employed the *fei huo ch'iang*. The *fei huo ch'iang* that could be literally translated as "flying fire spear" ("lance") or "javelot of flying fire" is considered by many historians as a rocket. Not everybody agrees, though. It is possible that these were the fire spears that threw fire, up to 10 paces, forward.

Whether the battle of K'ai-fung-fu indeed witnessed for the first time the rockets or not, the unambiguous description of rockets could be often found since that time in numerous Chinese historical records.

Multiple Rocket Launchers

The Chinese military favored multiple rocket launchers that fired simultaneously up to 100 small fire arrow-rockets. A typical powder section of such an arrow-rocket was between 1/3 and 1/2 ft (10–15 cm) long. The bamboo arrow shafts varied from 1.5 ft (45 cm) to 2.5 ft (75 cm) long, and the striking distance reached 300–400 paces. In addition, the rocket tips were often smeared with poison. A rocket launcher could be carried and operated by a single soldier-rocketeer.

Throughout first centuries of rocketry, the Chinese were content with the size and performance of the originally introduced rockets. Their rockets remained small and inefficient until the 19th century. Significant attention was paid, however, to devising multiple launching capabilities while improvements in the performance of individual rockets were neglected.

Chinese and Indian Ways

The Chinese approach strikingly differed, as we will see later, from the rocket development in India. A few hundred years after the battle of K'ai-fung-fu, the Indians already built very big and sophisticated war rockets and employed them on an exceptionally large scale.

We do not definitively know when and where exactly the first rockets appeared. The existing Chinese records are simply the oldest found. In any event, a fortunate marriage of the military necessity and emerging technology took place some time not later than the 13th century. This marriage had started a set of technological developments that would eventually lead man to space.

2. ROCKET PROLIFERATION — THE FIRST WAVE

Mongols Learn Gunpowder and Rocketry

The nomadic Mongols were victorious in their struggle with China in the 13th century, and they quickly learned the gunpowder technology and rocketry from the conquered Chinese. The new rulers did not trust the "southern people" (that is, the Chinese), and the reliable "northern people" (Mongols) replaced many Chinese in the facilities producing gunpowder. The learning was not always smooth: inexperience of the newcomers caused a terrible explosion at the Wei Yang arsenal, which killed at least several hundred people.

Vast Empire

Genghis Khan and his sons built an enormous empire stretching from Korea in the East to Europe in the West. The movements of troops, top military commanders, civil advisors, artisans, and entire peoples led to accelerated dissemination of the knowledge and technology throughout the vast region. Many foreign adventurers, scholars, soldiers, physicians, and merchants were welcome at the court of the Mongol Khan. The foreigners from distant Byzantine, Armenia, Turkestan, Persia, and even Europe served in positions of importance, accelerating cultural and technological exchange.

Rockets Fired at Japan

The advancing imperial troops rapidly brought rocketry to many other countries, which in turn adopted the new weaponry. The stage for the first great wave of rocket proliferation was set. The Mongolian ships and ground forces showered the defenders of Japan by fire arrows and probably rockets in the unsuccessful invasion in 1274. A few years later, the Mongol Khan Kublai felt personally humiliated when the Japanese beheaded his envoys. He ordered the second, much bigger invasion of Japan in 1281. Kublai Khan sent an armada of more than 4000 vessels with 140,000 troops. Rockets were employed on a great scale. It was in that year that *kamikaze* ("divine wind"), a typhoon, destroyed most of the fleet of the invaders. The word kamikaze was later adopted by the Japanese suicide pilots during World War II.

Although the Japanese had been exposed to the modern war marvels, gunpowder and rockets, they did not learn or adopt the new technology. The Portuguese would reintroduce the gunpowder and firearms to an isolated Japan a few centuries later.

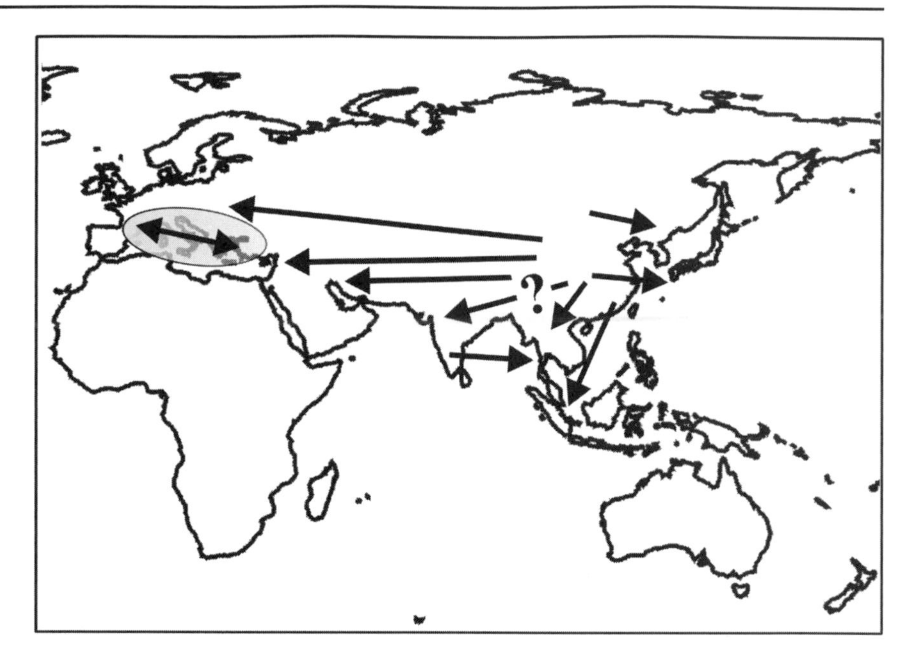

Fig. 2.1. First wave of proliferation of rocket technology. It is not clear whether rockets were independently developed in India or benefited from the Chinese experience. Europeans and Byzantines likely produced their rockets as a result of indigenous developments. Courtesy of Mike Gruntman.

In their turn, Korea and Java learned rocketry from the Mongols. Rockets also appeared in India. It is not clear however whether the Indian rockets were a local development, or brought by the invading Mongol troops, or if the technology was transferred through merchants. Both the Mongols and Indians likely carried rockets further to Southeast Asia, through projection of the military power and through trade and religious contacts, respectively.

Rockets Reach Europe

By some accounts, rockets were used for the first time in Europe in 1241. In April of that year, the advancing Mongol troops defeated the army under Prince Henry II in the Battle of Legnica (Legnitz) in Silesia (modern southwestern Poland). Again, the historical records are not clear whether the Mongols used rockets or incendiary arrows.

> **CONFUSION**
>
> For many centuries, historians and chroniclers did not distinguish and often confused hand- and machine-thrown containers with flammable (incendiary) substances, such as Greek fire, with self-propelled missiles (rockets).

Byzantine Empire

The Mongols are often credited with bringing rockets to the Near East. Fabled merchants of the Orient could have also contributed to proliferation of the knowledge of gunpowder and rocketry from China. This was not however a simple, one-way flow of the technology from China. The Byzantine Empire had independently developed and known incendiary and explosive weapons for several centuries by that time. Greek fire appeared some time in the sixth or seventh centuries, and Marcus Græcus

GREEK FIRE

Greek fire is a combustive substance that was introduced by the Byzantine Greeks in the sixth or seventh century and widely used in warfare. Greek fire was an exceptionally efficient weapon thrown from tubes and in pots. It was particularly deadly against the wooden ships, wooden military machines, and fortifications. Naphta was likely the basis of this substance that could not be extinguished by water.

When the Byzantine naval vessels destroyed the attacking Arab fleets by Greek fire in the seventh century, this terrifying weapon was known as "liquid fire" or "marine fire." Only much time later, the Crusaders coined the term "Greek fire." The deadly efficiency of Greek fire critically contributed to the long survival of the Byzantine Empire under attacks by its numerous enemies.

The composition of Greek fire was a closely guarded secret in the Byzantine Empire, and it is not known to this day. The secret of the original Greek fire was probably lost by 1204, when the Crusaders sacked Constantinople. What the Saracens, the Mongols, and the Europeans of that time knew as Greek fire was actually some less efficient derivative.

Marcus Græcus

described gunpowder-like mixtures and incendiary and explosive projectiles as early as in the ninth century.

In the 11–13th centuries, the Islamic Arab world was a home of great scholars, astronomers, mathematicians, doctors, and geographers. They were well positioned to absorb, implement, and advance novel ideas and technologies. Historical records indicate that the Arabs definitely knew about gunpowder by the end of the 13th century.

Gunpowder in the Arab World

SARACENS

Saracens were a name referring to the Islamic enemies — Turks, Arabs, Egyptians, Syrians, and others — by the Christians in the Middle Ages. The name entered the European culture and languages through the Byzantines and Crusaders.

The Saracens likely fired some rocket-propelled devices at the Crusaders of the Seventh Crusade led by King Louis IX (later to become Saint Louis) of the French. To reach and attack the enemy in the Nile river delta, the Crusaders began building a causeway across the Nile's tributary near Damietta. The construction work was protected by wooden towers. The Saracens attempted to burn the towers soon after Christmas of 1249. Jean de Joinville, a crusader from Champagne who later became friends with the king, provided an account of the event.

Damietta

The Saracens, as Joinville described, brought forward a petrary, a missile-throwing machine based on a principle of a catapult, and put Greek fire into its sling. Although Joinville wrote that the enemy threw Greek fire, his description pointed to a possible firing of something else, a kind of rocket-propelled devices. The projectiles produced a frightening effect. As this likely rocket darted toward the crusaders, in appearance it

> was like as a large tun, and its tail [of fire] was of the length of a long spear; the noise which it made was like a thunder; and it seemed a great

"... Noise ... like a thunder; ... dragon of fire ..."

> dragon of fire flying through the air, giving so great a light with its flame, that we saw in our [crusaders] camp as clearly as in broad day. ... Each time our good king St. Louis heard them make these discharges of fire, he cast himself on the ground, cried with a loud voice to our Lord, and shedding heavy tears, said: 'Good Lord God Jesus Christ, preserve thou me, and all my people.' (De Joinville 1848, 407)

The stories of new weaponry were reaching Europe and strengthening curiosity and interest in innovations. Incendiaries and explosives were not unknown to the Europeans at this time. On the contrary, many countries experimented with these technologies. For example, the same Louis IX of the Seventh Crusade established in 1227 in Paris a facility "to research in saltpeter," the key ingredient of gunpowder.

Facility "to research in saltpeter" in 1227

The rich Italian cities, Venice, Genoa, and Pisa, led the European technology development benefiting from their constant contacts, in trade and war, with the Byzantine and Islamic Mediterranean lands. These Italian cities initiated a serious work that led to perfection and wide introduction of firearms and artillery to many European lands in the early 14th century.

Firearms and Artillery in Italian Cities

Some argue that the Arabs also brought the rocket warfare to the Iberian Peninsula in 1249 and that the rockets were used there again in the siege of Valencia

Fig. 2.2. Capturing of Damietta by the Crusaders of Louis IX in 1250. The Joinville's description does not support this miniature (dated mid-14th century). The city was captured with little or no resistance. Figure from *Histoire de Saint Louis* (de Joinville 1874, 88).

WAS BAGHDAD UNDER ROCKET FIRE IN 1258?

Mongol troops led by Hülagü stormed and sacked Baghdad in 1258. They put to death the Caliph with all members of his house, thus violently ending more than five centuries of the Abbasid dynasty.

Many publications on rocket history state that the Mongol troops fired rockets on Baghdad. Although it is not excluded that the Hülagü's troops indeed used rockets, the historical sources rather point to mechanically thrown incendiary devices.

Rashid al-Din's *History of Mongols in Persia* is considered probably the most reliable primary source on the history of the Baghdad siege. Rashid al-Din wrote that the Mongol war machines threw the stones (and sometimes pieces of palm trees instead of scarce stones) and "pots of naphta." Therefore, *Baghdad under rocket fire in 1258* seems to be an apocryphal story, a myth of rocket history.

in 1288. The historical records are however uncertain, and the chroniclers likely again confused the hand- and machine-thrown incendiary containers with the gunpowder-propelled rockets.

Brave Travellers to the Orient

Brave European travellers, merchants, monks, and diplomats penetrated deep into the Orient. Rabbi Benjamin of Tudela (traveled in 1160–1173), Friar John of Pian de Carpini (1245–1247), Friar William of Rubruck (1253–1255), Marco Polo (1271–1295), and Friar Odoric (1318–1330) wrote extensive accounts about their adventures and journeys to mysterious distant countries. Their writings were widely read, especially the famous book of Marco Polo. Many Europeans thus learned about the peoples, cities, economies, and cultures of the various parts of the Mongol Empire and its successor states.

Progress in Europe

The stories about technological wonders in far away lands might have sometimes triggered the thought of European scientists and engineers. The Europeans however came up with a growing number of their own discoveries and inventions and, very importantly, developed the scientific method. Europe would soon become an unquestionable leader and dominate the world science and technology for several centuries.

It is impossible to establish who exactly pioneered the technology first, who developed it later independently, and who just copied it. One cannot state, however, that the rockets were brought from China to the Near East and later to Europe. The historical records simply do not support such a conclusion.

WORD *ROCKET*

The word *rocket* likely originates from the *rocchetta*, a diminutive of the Italian word *rocca* for distaff. A distaff is a staff for holding the bunch of flux or wool from which the thread is drawn by spinning.

Fireworks had become highly popular all over across Europe by the 16th century. Italian pyrotechnics, in particular the famous Ruggieris of Bologna, brought the new word together with their spectacular fireworks to many European capitals.

Sustainable Level of Scientific and Engineering Knowledge

The gunpowder and related technologies sporadically appeared and disappeared in various places in Europe and the Mediterranean region during many centuries. The sustainable, but often absent, level of the general scientific and engineering knowledge was needed to maintain and further develop such advanced technologies. No nation or land in the world had a monopoly on exploding and burning substances and on the various contraptions that threw liquid fire and other incendiaries.

From Mechanical Power to Projecting Power of Explosives

History gives us numerous examples of using incendiaries and explosives and the machines to project them: when the Byzantine burned the fleet of the Muscovite Prince Igor in 941; when the King of Hungary, Salomon, attacked Belgrade in 1073; when the Arabs assaulted Lisbon in 1147; when the city of Dieppe defended against the English in 1193; and many others.

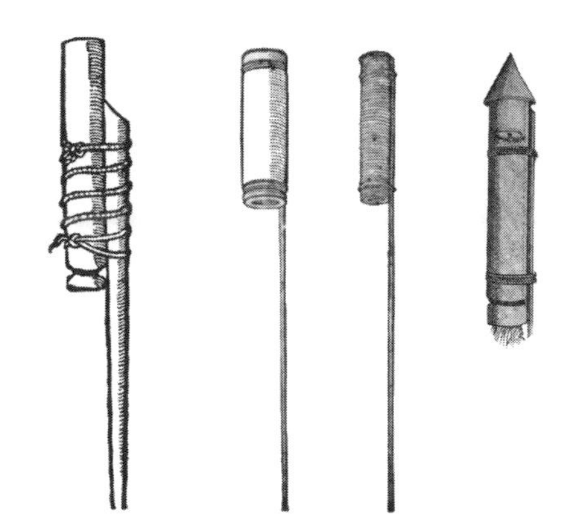

Fig. 2.3. Early European rockets from the manuals on artillery and fireworks dated (left to right) 1591, 1626, 1626, and 1620. Figures courtesy of Library, Getty Research Institute, Los Angeles, California, from (left to right) *Büchsenmeisterey, das ist, Kurtze doch eigentliche Erklerung deren Ding ...* (Brechtel and Gerlach 1591, Part 2, p. 23); *Il perfeto Bombardiero et Real Instrvttione di artiglieri ...* (Gentilini 1626, 109); *Recueil de Plusiers Machines Militaires ...* (Thybourel 1620, Livre V, p. 11).

The throwing devices gradually shifted from using mechanical power of the spring in various forms to projecting power of explosive substances. This development has ultimately led to the introduction of gun and rockets. It is also likely that the predecessors of rockets appeared independently in several places at approximately the same historic time.

Advancement of Science

What is known with certainty though is that rockets had established a foothold in Europe by the year 1300. By that time, Roger Bacon (1220–1292) in England and Albertus Magnus (1200–1280) in Germany had described preparation of black powder. Albertus discussed devices resembling in many aspects rockets, while Bacon, a major proponent of experimental science, proposed flying

Fig. 2.4. Albertus Magnus, or St. Albert the Great, 1200–1280, famous bishop and philosopher and a teacher of St. Thomas Aquinas. Albertus described preparation of black powder and discussed likely rocket-propelled devices. A papal decree of 1941 declared Albertus Magnus the patron saint of (natural) scientists. Figure from *Opera Omnia* (Albertus Magnus 1890, frontispiece).

GUNPOWDER AND SALTPETER

Production of gunpowder was a key for rocket development. Black powder consists of charcoal, sulfur, and saltpeter. It burns rapidly and produces about 40% gaseous and 60% solid products. The gunpowder rocket propellant was modified to achieve lower burning rates and, correspondingly, less pressure in rocket casings.

	"Standard" Black Powder	Rocket Propellant
Saltpeter (by weight)	75%	68%
Charcoal (by weight)	14%	19%
Sulfur (by weight)	11%	13%

Charcoal, sulfur, and saltpeter had been known since the time immemorial. Charcoal is produced by partially burning and heating wood with limited air access. Sulfur naturally occurs in sedimentary and volcanic deposits. Saltpeter that literally means *rock salt* is naturally abundant in China and India but rare in Europe. Therefore, manufacturing of saltpeter became critically important for many European countries.

Fig. 2.5. Apparatus for distillation of sulfur from Vannoccio Biringuccio's *De la pirotechnia* published in 1540. Figure courtesy of American Institute of Mining, Metallurgical, and Petroleum Engineers (AIME), from *De la pirotechnia* (Biringuccio 1942, 89).

Saltpeter forms in places protected from rain (on the ground and in stables, caves, and cellars). The surface capable of "growing" saltpeter should contain plant or animal substances with access to air. The walls containing lime and plaster also "produce" saltpeter.

In 1540, Vannoccio Biringuccio gave the following guidance for search of saltpeter. "The best and finest of all saltpeters is that made of animal manure transformed into earth in the stables, or in human latrines unused for a long time. Above all the largest quantity and the best saltpeter is extracted from pig dung. This manurial soil ... should be well transformed into a real earth and completely dried of all moisture ... Assurance that it contains goodness is gained by tasting with the tongue to find if it is biting, and how much so." (Biringuccio 1942, 405)

machines. Europe was rapidly accumulating knowledge, establishing the scientific method, and developing and learning new technologies.

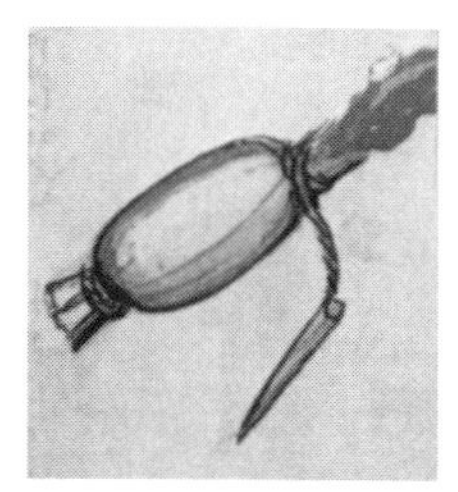

Fig. 2.6. Fragment of a manuscript *Bellifortis*, p. 102, by Conrad Kyeser showing a sky rocket and a launching trough, ca. 1400. This is one of the first drawings of a rocket in the European sources. Figure courtesy of VDI-Verlag, Verein Deutscher Ingenieure, Düsseldorf, Germany, 1967.

First European Records of Rockets

The first drawings of rockets appeared in European records around the year of 1400. A German Conrad Kyeser included a skyrocket drawing in his manuscript known under the title *Bellifortis*. About 20 years later, an Italian Giovanni Di Fontana showed a flying rocket and rocket-propelled carts in his *Bellicorum Instrumentorum Liber*. An Italian engineer, Mariano Taccola, sketched a missile in his notebook known as *De Ingenes*. The Taccola's drawing, however, most likely showed not a rocket but a hand-thrown missile with the incendiary material.

First Rockets in European Warfare

The first records of rockets in European warfare can be traced back to a war between two prominent Italian rivals, the republics of Venice and Genoa. King of Hungary and Lord of Padua allied with the Genoese against Venice in a major conflict that became known as the *Naval War of Chioggia* (1378–1381).

A Venetian chronicle by Andreæ Danduli describes, as a routine event, how the Paduans employed rockets in 1379,

> At this time our state was disturbed in many ways. It did not satisfy the King of Hungary to have seized the harbor of Dalmatia and galleys, but he sends an army in great strength to the regions of Tarvisina under the noble lord Charles of Hungary. And the Paduans very strongly attack Mestre; they capture and seize the neighboring city of St. Laurence, firing the rockets at our straw houses ... . (Muratori 1728, 448)

The other chronicle by Danielis Chinatii describes an attack by the Venetians on a strategic tower occupied by the Genoese,

"... shot numerous rockets ..."

> On the 13th of the month, the army of the Venetians went on many boats to the Bebe Tower that was not willing to surrender. An Engineer of his Lordship went to the edge of the moat with a timber cat with mortars and crossbows, breaking and piercing the tower in many places and inflicting losses on the defenders through the openings they made. And they shot numerous rockets at the top of the tower; one of them set fire to the summit and the defenders could never extinguish it. Seeing this,

CAT

A *cat* is a mobile timber penthouse used in sieges.

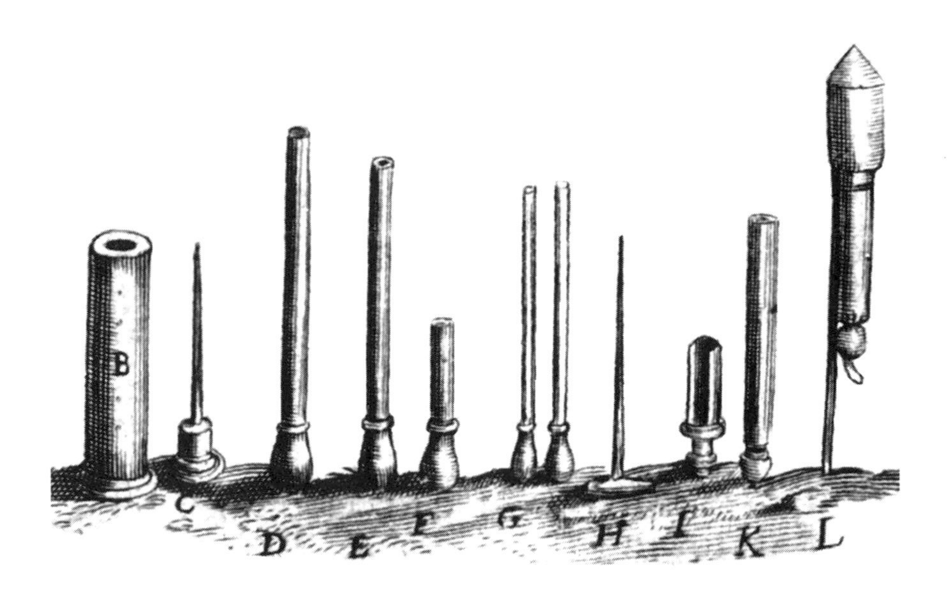

Fig. 2.7. Specialized rocket workshop ca.1620 (bottom): "To make the rockets, many things are necessary. One needs the models, [gunpowder] ramming rods, well-glued two-layer paper, ramrods, dies, mortars, sieves, mallets, and various components from which the rockets are fabricated." Various tools for rocket manufacturing, ca. 1640 (top): mold B is about 6 in. (15 cm) long. Figures courtesy of Library, Getty Research Institute, Los Angeles, California, from *Recueil de Plusiers Machines Militaires* ... (Thybourel 1620, Livre V, p. 4) and *Traité de fevx Artificiels povr la Gverre, et povr la Recreation* (Malthus 1640, 62).

the soldiers on the armed boats approached the defenses along the river and started to demolish them, in spite of those who wanted to resist them .... (Muratori 1729, 769)

Rockets in the Hundred Years War

It is fair to say that the military use of rockets peaked in the middle of the 15th century during a major European conflict, the Hundred Years War. The French rockets burned English towers and military machines at the siege of Orleans in 1428. Rockets were employed near Bray-sur-Seine and Montereau in 1437; against Pont-Audemer defended by the English in 1449; in retaking of Vire and Avranches in 1450; at the siege of Bordeaux in 1451; in the attack on Gand in 1453; etc.

The English, Dutch, and Germans adopted and advanced the new military technology, and the rocketry became common in Europe. Military use, however, quickly declined and soon practically disappeared. Rocket fireworks became and remained for several centuries a major entertainment.

Gunpowder Production

Rocket development relied on the perfection and availability of high-quality gunpowder. Firearms gained wide acceptance throughout Europe, and

Fig. 2.8. Early war rockets (1598) in Europe. Figure courtesy of Library, Getty Research Institute, Los Angeles, California, from *Recueil de Plusiers Machines Militaires* ... (Thybourel 1620, Livre V, p. 38).

"How to directly fire a rocket horizontally or in other ways ... It is necessary to assure the proper composition of the rocket [gunpowder], according to the rocket weight and size, which you want to have in order not to fail in your enterprise. Position your rocket with the appropriate [guiding] stick on smooth surface and assure that it can go by rocking and turning it... The launching surface could be mounted on a tripod with a small peg easily entering the hole in the said surface. Then aim and level the rocket in the direction you wish ... Arm the rocket and ignite: it will go directly toward the desired place, if the [gunpowder] composition is good and the distance is not too long for the fire to carry the rocket." (Thybourel 1620, 37)

GUNPOWDER'S SOUL

"Sulfur and charcoal form the body of gunpowder: saltpeter gives it the soul. It is the saltpeter that invigorates the monster ... ."

General Louis Susane (1874, 11)

black powder manufacturing became a thriving activity in the 14th century. Production of charcoal, sulfur, and saltpeter, the gunpowder ingredients, and their mixing and blending techniques were continuously perfected. A mortar and pestle were the early tools for grinding. Some time later, in the 15th century, a water-driven mechanisms were introduced in gunpowder manufacturing.

Saltpeter was rare in Europe. Consequently, securing overseas sources of saltpeter or developing its domestic production became a critically important "national security" task of European governments since the Middle Ages. Conrad Kyeser described an early saltpeter plantation around the year 1400:

Saltpeter Plantation

> Make a pit for a substance layer on a fertile flat field, with the grass completely burned out and some soil mixed with it. Then, you sprinkle the layer with any salty liquid, urine or wine, and then cover this layer with another layer of crushed lime, and pour again the above-mentioned liquid. So, a layer after layer, until the pit is sufficiently full. You water this place as often as needed to obtain the salt (substance). Always collect every 15 days what is to be collected ... .
>
> You can similarly work with the equally big clay pots, in which a layer after layer are laid and sprinkled similarly and with the prepared ingredients. You would store these pots in the soil or in a suitable storeroom ... .
>
> Cook the salt (substance) with three-day-old urine of small young boys until one third is vaporized. From here the valuable saltpeter would be obtained. (Kyeser 1967, 80)

Saltpeter, Saltpeter, Saltpeter

Fig. 2.9. "How to make the gunpowder for arquebuses and pistols," ca. 1630. Wide manufacturing of gunpowder in Europe provided an important foundation for rocket development. Figure courtesy of Library, Getty Research Institute, Los Angeles, California, from *La Pyrotechnie de Hanzelet de Lorrain, ou sont representer les plus rares ...* (Appier-Hanzelet 1630, 263).

Fig. 2.10. Arranging the fireworks (1620). "... If it is necessary to make some pleasant things, this is done on two boats, on which one erects wooden houses, or a few small castles, in order to attach at their exterior the bright rockets ... ." Figure courtesy of Library, Getty Research Institute, Los Angeles, California, from *Recueil de Plusiers Machines Militaires* ... (Thybourel 1620, Livre V, p. 36).

For many centuries saltpeter remained an important strategic material, playing a critical role in war. Special saltpeter plantations, called nitriaries, were established in France, Sweden, Germany, and other countries to supply their armies. For example, the Prussian farmers were required to build their fences with niter-forming materials. The fences were taken down after a few years and used for saltpeter extraction.

Nitriaries

In another example the English blockade of the European Continent in 1792 highlighted the critical importance of saltpeter supplies and led to a severe shortage of saltpeter in France. The 1793 decree asked the citizens of the French Republic to collect saltpeter in their cellars and stables and from old walls built of lime. Several thousand saltpeter refineries were promptly established, many by the municipalities. Soon, the country began to produce millions of pounds of saltpeter annually, making France independent of foreign supply.

AHEAD OF TIME

"The inventions too much ahead of their times stay unutilized until the moment when the level of general knowledge catches up."

Napoléon III (Susane 1874, 41)

Rocketeers developed new techniques for controlling gunpowder properties and assembling rockets. It became possible to modify the burning rates and produce additives for coloring the fire in the fireworks. The rocket cases were usually made

SCIENTIFIC FOUNDATIONS OF ROCKETRY AND SPACEFLIGHT

As the first rockets were flying in Europe, another, most important development was taking place in the Judeo–Christian world. The scientific method was gradually developing, and the foundations of modern science were being laid down. Astronomy, mechanics, physics, chemistry, and engineering were advancing, based on experimental verification of the mathematically described theories.

The Universe of Claudius Ptolemy dated back to second century A.D. The unmovable Earth was in the center, and the motion was circular. The planets moved around small circles, the epicycles. The centers of the epicycles revolved in larger circles, the deferents. The Earth was slightly displaced from the center of the deferent, and the angular rate of the epicycle center was uniform as seen from the equant, positioned symmetrically to the Earth. This very complex system rather accurately predicted planetary motion.

Nicolaus Copernicus (1473–1543) noticed that the Ptolemaic epicycles had a period of one year. The assumption of the unmovable Earth transferred its own motion in inertial space to all other celestial objects. Copernicus rearranged the Ptolemaic system by placing the Sun in the center and established the correct order of the planets. He, however, believed in holiness of the circle, and had to preserve the epicycles. This revolution dethroned the Earth from the center of the Universe.

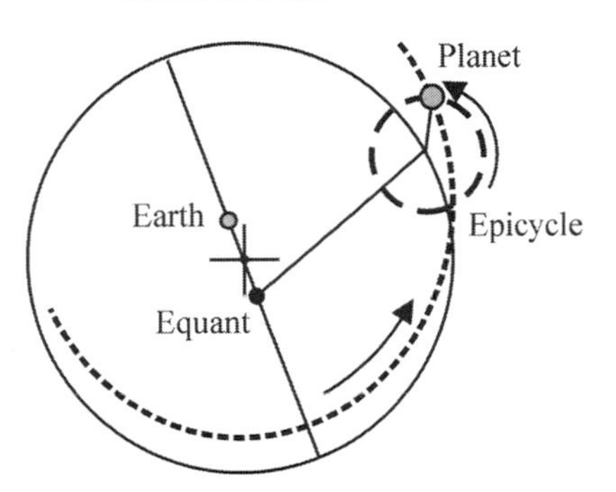

Galileo Galilei (1564–1642) further challenged the established order by applying the newly invented telescope to astronomy. Galileo discovered the phases of Venus and Mercury thus proving the correctness of the Copernican order of the planets. Observation of the four moons orbiting Jupiter proved that a center of motion can be in motion itself. Galileo's work and writings shook the world view of the Catholic Church. The contemporary "commissars" of political correctness forced him to recant, and he "abjured, cursed, and detested" his past "errors."

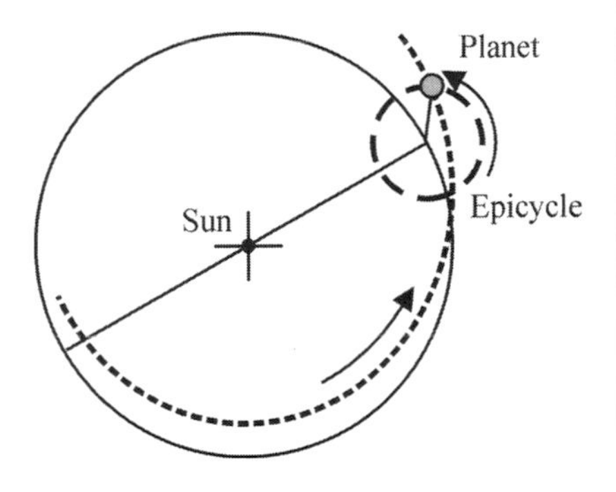

Fig. 2.11. Evolution of the understanding of the solar system, from the Ptolemaic system to Copernican. Courtesy of Mike Gruntman.

Galileo established mechanics as science. Johannes Kepler (1571–1630) derived the laws of orbital motion from the extensive observations by Tycho Brahe at the Prague Observatory. The revolution in science was culminated by Isaac Newton (1642–1727), who introduced the laws of motion and formulated the law of universal gravitation.

Nicolaus Copernicus, 1473–1543

Galileo Galilei, 1546–1642

Johannes Kepler, 1571–1630

Isaac Newton, 1642–1727

Fig. 2.12. Nicolaus Copernicus, Galileo Galilei, Johannes Kepler, and Isaac Newton laid the scientific foundation for the future science and technology advancements that would enable rocketry and spaceflight. Figures from *A History of Astronomy* (Bryant 1907, 31, 41).

Specialized Rocket Technology

from several layers of paper glued together, and various tools such as special sieves, mortars, and mallets were specifically designed for rocket fabrication. In addition, procedures for careful ramming of the gunpowder without igniting it into the cases were perfected.

Roger Bacon described in detail the making of the black powder in 1242. Later manuscripts and books added and preserved much valued information on chemistry, mechanics, and technological processes. Rocketry became widely known. By the 16th century, the military books and manuals had commonly included chapters on design, manufacturing, and use of rockets, mostly in fireworks and signaling but with some limited applications as a weapon.

Revolution in Science

This was the time when the true revolution in science was taking place. Mathematics, physics, mechanics, chemistry, and astronomy advanced at a brisk pace. Bold ideas of Nicolaus Copernicus about the solar system, introduction of a telescope by Galileo Galilei, meticulous astronomical observations by Tycho Brahe, and their brilliant analysis by Johannes Kepler changed the understanding of the world and cosmos. Isaac Newton's laws of motion and gravitation culminated this revolution in science.

The Newton's third law of motion provided a mathematical description of the jet reaction principle. (The conservation of momentum is the basis for understanding rocket acceleration.) The precious scientific and engineering knowledge was analyzed, preserved, accumulated, and nurtured, laying the scientific foundation for future technology development, including rocketry and spaceflight.

The first great wave of proliferation of rocket technology covered enormous areas stretching from Japan to the European Atlantic seaboard. Although, the European countries had mustered and advanced the rocket technology, rockets remained a curiosity. The gun ruled in Europe as a weapon of choice.

3. UNDER ROCKET FIRE IN INDIA

Events in faraway India boosted the European interest to rocketry in the late 18th century. This was the time when several Indian princes vigorously resisted European advances in the subcontinent, widely using war rockets. The reports on Indian rocketry did not pass unnoticed in Europe and would open with time a new chapter in rocket development.

War Elephants

War rockets had been known in India for several centuries. The great Timur (Tamerlane), the mighty leader of nomad warriors, faced rocket fire when he invaded India in 1398. India was famous at that time for its war elephants. A good war elephant valued as much as 500 horses was a formidable weapon, "in bulk and strength like a mountain; and in courage and ferocity like a lion" (Pant 1997, 213). The Indians often used elephants as platforms for launching war rockets.

Timur Braves War Rockets in the Battle of Delhi in 1399

Timur was known for his ruthlessness and military victories from Russia and Anatolia in the west to Syria and Iraq in the south to China in the east. His seasoned and bold troopers were not intimidated, as the history would testify, by the "tales about the strength and prowess and appearance of the [war] elephants of Hindustan" (Jackson 1906, 198).

The decisive battle took place near Delhi on 17 December 1399. Timur recounted that the opposing forces of Sultan Mahmud included "125 elephants covered with armor, most of them carrying howdahs in which were men to hurl grenades, fireworks, and rockets" (Jackson 1906, 206). The Indian army was defeated and destroyed in the ensuing battle, and Delhi was captured and reduced to ruins.

Rockets Fit Rough Terrain

By the mid-18th century, rockets had become a common weapon of warriors in India. Saltpeter was abundant. Readily available bamboo made excellent straight and light guiding sticks, an indispensable element of the rockets. Rockets perfectly fitted the warfare in the rough terrain. They did not require bullocks and elephants for transport, in contrast to expensive and heavy artillery. Whereas the technically sophisticated artillery was often commanded and serviced by experienced European officers and soldiers, the rockets could be carried and effectively fired by native warriors with only rudimentary training.

Indian states continuously warred against each other and fought with the European powers that wished to expand their control of the subcontinent. The native Indian troops widely used war rockets, both offensively and defensively.

Fig. 3.1. Elephant battery of heavy artillery along the Khyber pass at Campbellpur in 1895. Indian troops widely used war elephants for centuries in combat and later, with the advent of firearms, to haul heavy artillery. Photo courtesy of Library of Congress.

For many years, the British struggled with Mysore, where rockets were employed on an exceptionally large scale.

Hyder Ali of Mysore

Around 1761 a brave and politically shrewd Muslim leader Hyder Ali (Hydur Ali or Haidar 'Ali Khan) became a ruler of the state of Mysore. A contemporary chronicler described his qualities in a pompous and inflated language common for the time, place, and circumstances, as "... Hydur Ali, the noise of whose courage and political ability had reached the utmost corners of the earth" (Hussein Ali 1842, 66).

Hyder Ali's dominion was peopled by six million inhabitants. His unusually well-disciplined army reached almost 200,000 troops and was well equipped with more than 20,000 pieces of cannon of various caliber. Many European officers served Hyder Ali. He was however an independent ruler and introduced Persian and Tartar terms for the words of command, replacing English and French.

New Delhi
INDIA
Bombay
Calcutta
Mysore
Madras

Fig. 3.2. State of Mysore in India.

Rocketeer Force

Hyder Ali established a large rocketeer force of about 3000 men in 1767. This force had grown up to 5000 by 1786. According to one report, the regular infantry consisted of 16 cushoons, or brigades, and each of those included 200 rocketmen. The rocketmen were paid approximately the same amount as artillerymen and their detachment usually accompanied (together with cavalry, infantry, and pikemen) Hyder Ali on his trips.

Hyder Ali's rocketeers introduced a major innovation: metal cylinder cases to contain the combustion of rocket black powder. The soft iron was crude, but the strength of the container was substantially higher than the earlier paper or

MYSORE AND BANGALORE

Mysore was an Indian state in the southern part of the subcontinent. It is interesting that Bangalore, the capital of the present-day Karnataka state (formerly Mysore), is now an important center for development of rocket and space technology in India.

bamboo constructions. Rockets usually carried 2 lb (0.9 kg) of black powder and had the range up to one mile. They were either armed with an explosive or incendiary warhead or sharpened at the end.

First Metal Case Rockets

The new Indian rocketry significantly advanced the state of the art. Chinese paper rockets remained small and inefficient, while the Indians developed their rockets into large and impressive devices with mass up to 12 lb (5.5 kg) and range of 1.5 mile (2.4 km).

Not everybody was impressed by rocket performance as a weapon, though. Hyder Ali's chief of artillery, the French officer Maistre de la Tour, wrote that

> the implement is, on the whole, far more expensive than useful; which, I suppose, chiefly arises from the want of care and attention in making them up; however they have been sometimes productive of dreadful effects, by setting fire to ammunition wagons. These rockets are very well adapted for setting fire to towns and villages ... . A body of cavalry, not used to this kind of instrument, would be quickly thrown into disorder by it; for the rockets falling at the feet of horses, emit a flame, which frightens them ... . (de la Tour 1855, 158)

"... more expensive than useful"

Fig. 3.3. Hyder Ali, 1722–1782. Figure from *The History of Hyder Shah* (de la Tour 1855, iv).

Not only cavalry was scared by rocket fire but also war elephants. Fighting the elephants by fire was a common tactic for centuries. In one battle, where the Timur's army faced Indian war elephants, he "marshalled a squadron of camels; these he sent forward each bearing a load of dry grass ... . No sooner had the fight begun than fire was set to these loads of inflammable stuff when the camels all in flames did so terrify the elephants that all took to flight" (Gonzáles de Clavijo 1928, 255). The elephant units and cavalry required special training to stand fire.

Incendiary and Terror Weapon

British reports from India and local chroniclers routinely mentioned war rockets, which occasionally inflicted serious casualties or killed a prominent officer. For example in the year 1780, the troops of the British General Coote launched a night attack on Hydar Ali's Fort Selimbur (Chillambrum). "The Kiladár, or

Fig. 3.4. Elephant brigade being taught to stand fire at Moulmein, Burma, in 1853. Figure from *Illustrated London News*, Vol. 22, No. 616, p. 269, 9 April 1853.

officer in commanding that fort, whose name was Yousuf Khan, was, however, a brave soldier, and with three hundred men, defended the fort gallantly; and by continual shower of musket balls, rockets, and shells, he so effectually beat off the assailants, that between two and three hundred of the English army lost their lives, without any advantage gained" (Hussein Ali 1842, 426).

British Under Rocket Attack

The high mobility of rocketeers invited the guerilla tactic to harass regular troops. So, in the same campaign of 1780, the supply train of General Coote was attacked. "The rocketeers ... each man taking post behind the Kewra (Spikenard) trees ... fired their rockets among the followers and baggage of the English; and the poor people, frightened, were thrown into the utmost confusion" (Hussein Ali 1842, 427). Two years later in another engagement, "the troops of Nawaub [Hyder Ali] .... vigorously attacked the General's army on all sides; and by suddenly charging them, and plying them with the rockets, they carried away strength and stability from the feet of infidels" (Hussein Ali 1842, 456).

The British also attempted to use rockets against their adversaries but without much enthusiasm and success. On one occasion they fired rockets at the cavalry

SALTPETER IN INDIA

Saltpeter, an indispensable ingredient of gunpowder and rocket propellant, was widely available in India. A British scientist surveying the dominions of the Rajah of Mysore for the East India Company described in the late 18th century that "the makers of saltpeter received advances from government, and prepared the saltpeter from the earth. ... The earth seems to contain the nitre ready formed, as no potash was added to it by the makers. It is only to be found in hot season ... The soil ... is sandy and rocky, and the ways passing over ... are much frequented by men and cattle. From the 10th of January until the 10th of February the saline earth is scraped from the surface, and is lixiviated, boiled, and crystallized twice" (Buchanan 1807, 316).

INACCURATE MISSILES

The early unguided rockets were highly inaccurate and unpredictable. They could have a devastating effect on poorly trained and poorly led militia-type troops and frighten horses. In essence, the rockets were a weapon against large-area targets and against the troops they were to intimidate rather than to injure. During World War II, inexpensive and not very accurate unguided missiles produced a storm of protective fire in the skies and saturated the approaches to the navy ships and ground targets. The rockets were effective against exposed and concentrated infantry but did little damage to properly entrenched and dispersed seasoned troops. This characteristic of rockets remained until introduction of guided missiles in the 1950s.

of Hyder Ali. The cavalry was, however, "habituated to the fire by various exercises performed with paper rockets, and the horse, instead of being frightened, marched fiercely over them [the British]" (de la Tour 1855, 159).

Dangerous Weapon

Rockets were a weapon dangerous in storage and transportation. The history records a number of deadly events, including the heavy losses among the Hydar Ali's troops in fights with their perennial adversaries, the Mahrattas. In 1768 "it chanced that a shot from one of the Mahratta guns, fired at a considerable distance, fell among a string of camels carrying rockets, and threw them into disorder; and, in the tumult and crowd of men, the rockets took fire, and flying among the baggage and followers, threw them into utter confusion. To increase their misfortune, a rocket, which had taken fire, fell on one of the boxes of ammunition, and blew it up; and in black cloud of smoke, which rose up to heaven, many of Hydur's brave soldiers were carried up to a great height ... ." (Hussein Ali 1842, 195)

Tippoo Sultan of Mysore

The British–Mysore conflict had lasted for a number of years when Hyder Ali died in 1782. Tippoo (Tipu or Tippu) Sultan, a brave but not very politically sophisticated son of Hyder Ali, became a new and strongly anti-British ruler of Mysore. The stage for decisive battles that would effectively end this spectacular chapter of the Indian rocketry was set.

Fig. 3.5. Tippoo Sultan, 1749–53 – 1799. Figure from *A View of the Origin and Conduct of the War with Tippo Sultaun* (Beatson 1800, title page).

A new British Governor-General, Lord Cornwallis, was appointed in 1786. He was the same General Cornwallis who five years earlier surrendered the British Army at Yorktown, Virginia, during the American War of Independence. The Government instructed

Lord Cornwallis to refrain from aggressive wars and annexations. The British–Mysore hostilities, however, intensified.

Tippoo was taught military tactics by French officers in the employ of his father. The French wrestled with the British for control of India and were eager, as always, to twist the lion's tail and help British adversaries. In addition, numerous soldiers of fortune from many European nations served in various Indian armies at that time.

Tippoo Sultan increased his rocketeer force by some accounts to 5000 men. In 1792 the Tippoo's troops surprised the British by a large-scale use of rockets in the battle of Seringapatam. The rockets had a strong effect on the Indian native troops, in particular the cavalry that moved in great bodies. Undoubtedly, rocket noise and fire frightened horses and elephants.

Decisive Clash

The last and decisive clash, the fourth Mysore war, was triggered in 1799 by the discovery of the Tippoo's secret negotiations with France. Tippoo "privately

Fig. 3.6. View from the northwest front at the siege of Seringapatam in May 1799. "...At sunrise on the 2nd of May ... soon after the batteries opened up, a shot having struck a magazine of rockets in the fort, occasioned a dreadful explosion ... ." Figure from *A View of the Origin and Conduct of the War with Tippo Sultaun* (Beatson 1800, 122).

despatched ambassadors to the Mauritius ... to solicit the aid of 10,000 Europeans and 30,000 Negro troops" (de la Tour 1855, 298). The resistance of Mysore was finally crushed by the British, and Seringapatam, the Tippoo's capital city and the seat of Mysore rajahs since 1610, was successfully stormed. During the siege of the city, an artillery shot struck a magazine of Tippoo's rockets causing a dreadful explosion. Three days later, the victorious British troops entered Seringapatam and captured almost 10,000 Indian rockets. This decisive victory effectively put an end to the serious rocket development in India, which would reemerge only in 1960s.

10,000 Rockets Captured

SPARKLING ROCKETS ...

The rockets were so widespread in India and so deeply penetrated the culture that one song mourning Tippoo Sultan, *The Dirge of Tippoo Saltaun from Canara*, had the following lines (de la Tour 1855, 331):

His mountain-forts of living stone
Were hewn from every massy rock,
Whence bright the sparkling rockets shone,
And loud the vollied thunder spoke

Rocket Development Shifts to Europe

The reports on Indian weaponry reached Britain together with the captured rockets. Rocket lessons were also learned by the French who fought with their Indian allies against the British. The interest in new weapons grew rapidly, fueled by the imminent showdown between archenemies, Napoleon's France and Great Britain. The center of rocket development would consequently shift to Europe.

4. THE CONGREVE ROCKET

The rockets fired at Seringapatam in 1792 and 1799 "hit" an unintended target, a young British inventor William Congreve (Congreve the Younger). Congreve (1772–1828) was immensely impressed by the accounts of the Indian rockets and quickly recognized the new exciting opportunities. He set to turn this crude weapon into an efficient and deadly means of war.

William Congreve

The advantages of rockets included portability, freedom from recoil, rapid firing, and the effects of noise and fire on mounted troops. The rocket was propelled by its own impulse and did not require ordnance to project it. Thus, very large shot and shell sizes became possible.

Advantages of Rockets

Congreve particularly valued the new possibilities opened by recoilless launch. It first occurred to him in 1804 that "as the projectile force of the rocket is exerted without any reaction upon the point from which it is discharged, it might be successfully applied, both afloat and ashore, as a military engine in many cases where ... the use of artillery ... is either entirely impossible, or at best very limited" (Congreve 1827, 15).

Minimal Recoil

A common artillery cannon was usually accurately pointed at its target and discharged only once. The recoil disturbed the original aim. Accurate retargeting of the cannon was often not possible because the target had been obscured by the smoke and no longer visible by that time. So, the cannon could continue firing only in a general direction of the target. The recoilless rockets, in contrast, could be fired from the same stand in exactly the same direction, hampered only by the natural irregularities of the rocket flight.

From Boat and from Airplane

The strong and shock-proof design of the vessels was required to accommodate the artillery recoil on the ships. This feature thus effectively limited the size of naval guns and mortars, with the limits on the artillery use from small and maneuverable boats especially severe. Quite oppositely, a rocket, even the largest one, could be fired from a boat just capable of carrying this rocket. Much later, the same feature of minimal recoil made rockets attractive for use on aircraft as a large caliber weapon.

Many rockets, hundreds at a time, can be planted directly on the ground against the enemy. A devastating volley can then be fired at the same instant. A fiery train of rockets is clearly visible, with a terrifying effect. No horse can stand the hissing and sight fire of a rocket leading to the ensuing confusion of cavalry units. In addition, the rocket puts on fire everything on its trail. The rocket trajectory is also unpredictable. After bumping on an obstacle or ground, the missile could continue its flight in a new direction.

FATHER AND SON

William Congreve was born into a distinguished family. His father, William Congreve (Congreve the Elder) achieved the rank of lieutenant-general in the British Army and held important positions involving artillery design and production.

Terrifying Volleys

William Congreve began development of a family of war rockets in the early 1800s. A typical skyrocket available at that time consisted of a gunpowder grain wrapped in paper and attached to a guiding stick. The skyrockets were used for signalling and fireworks, and they did not meet military requirements: their size was small, the range was trivial, and no two rockets flew in a similar way. The earlier British experiments, conducted by General Desaguliers, to develop better war rockets were unsuccessful.

Improved Design and Fabrication

Congreve set to remedy the skyrocket flaws, with the fundamental rocket design concept intact. A new family of efficient war rockets emerged from his work. Congreve standardized gunpowder composition and developed a new, improved technology of ramming it into the rocket. Now the blows of monkey urged a wooden or metallic cylinder pressing the gunpowder into the case. (*Monkey* is a falling weight used for driving something by percussion as the falling

INVENTOR, PUBLIC SERVANT, AND COLONEL

William Congreve (Congreve the Younger) was a prominent inventor, with at least 18 patents to his credit. His contribution to military technology included such major advancements as improved manufacturing of gunpowder and a way of mounting and aiming the naval guns. Among his nonmilitary achievements were clockwork, hydropneumatic lock, and a new design of steam engine. Congreve established that detailed relief engraving in different colors provided the best security against forgery of paper currency. Subsequently, financial institutions used the latter technique for nearly 150 years.

Congreve was elected a Fellow of the Royal Society and held positions of Comptroller of the Royal Laboratory and Superintendent of Military Machines at Woolwich. In addition, he pursued a political career serving for many years as a Member of Parliament. He was also appointed Chief Equerry. The political connections included access to the Prince Regent, which helped Congreve in gaining the acceptance of his rocket system.

Sometimes one reads about Colonel Congreve. William Congreve never held a regular commission in the British Army. He however received the largely honorary rank of lieutenant-colonel in the Hannoverian Army by a peculiar twist of the royal arrangements.

CENTRAL BORE

Rocket thrust is roughly proportional to pressure of the hot gases in the rocket case. The pressure increases with the increase of the burning area. A central cylindrical cavity was therefore bored in the gunpowder of the Congreve rockets to increase thrust while protecting the rocket case from hot gases. Varying the burning area by shaping the solid rocket propellant is a common way to achieve the desired thrust variation in time in modern solid rockets.

weight of a pile driver or a drop hammer.) Introduction of a hydrostatic press for this purpose would have to wait for 40 more years.

Higher Pressure and Thrust

Sheet-iron replaced paper in rocket walls allowing higher pressure and correspondingly higher thrust for a given rocket size. A cylindrical cavity, called the bore, was made in the gunpowder along the rocket axis to increase the burning area and, consequently, the pressure of hot gases and thrust. The rocket was usually rammed solid, and the center was subsequently bored out with a bit.

The interior iron walls of rockets were lined with glued paper to prevent corrosion. This wall protection improved storage and reduced the rate of accidents. The guiding sticks were also significantly shortened. In addition, the sticks could now be assembled from several much shorter parts in a compact army version of the rocket.

Fig. 4.1. A typical rocket consisted of a warhead, cylindrical gunpowder grain with conoidal chamber in the rocket case (cross section shown in the figure), baseplate with the exhaust orifice, and guiding stick. Courtesy of Mike Gruntman.

Technology Advancement

Many rocket parts were made standard, which improved production quality and fabrication efficiency. Congreve introduced explosive and incendiary warheads with separate (from the rocket) ignition, which could now be independently timed by a separate fuse cord. New missiles were evaluated at the first real rocket test facility established at the artillery range of the Royal Laboratory at Woolwich, England.

Weapon System

In short, William Congreve has created a modern, technologically superior rocket weapon system. The smallest rocket in the family was 1 13/16 in. (46 mm) in diameter with mass 3 lb (1.3 kg). The largest rocket was 8 in. (203 mm) in diameter and 6 ft 4 in. (1.93 m) long; its mass was 300 lb (135 kg). The guiding sticks ranged from 6 ft (1.8 m) to 27 ft (8.2 m) in length. A popular 6-lb (2.7-kg) missile included a 3-lb (1.3-kg) warhead. The warhead of the largest rocket was almost 50 lb (22 kg).

In September 1805, Congreve demonstrated his rockets to Prime Minister William Pitt and War Secretary Castlereagh on Plumstead marshes. The Royal Navy clashed with the Navy of the Napoleon's France at that

ROCKET BASICS 1: THRUST AND SPECIFIC IMPULSE

In a liquid-propellant rocket, a liquid fuel and oxidizer are injected into the combustion chamber where the chemical reaction of combustion occurs. The combustion process produces hot gases that expand through the nozzle, a duct with the varying cross section.

In solid rockets, both fuel and oxidizer are combined in a solid form, called the *grain*. The grain surface burns, producing hot gases that expand in the nozzle. The rate of the gas production and, correspondingly, pressure in the rocket is roughly proportional to the burning area of the grain. Thus, the introduction of a central bore in the grain allowed an increase in burning area (compared to "end burning") and, consequently, resulted in an increase in chamber pressure and rocket thrust.

Modern nozzles are usually converging-diverging (De Laval nozzle), with the flow accelerating to supersonic exhaust velocities in the diverging part. Early rockets had only a crudely formed converging part. If the pressure in the rocket is high enough, then the exhaust from a converging nozzle would occur at sonic velocity, that is, with the Mach number equal to unity.

The rocket thrst T equals

$$T = \dot{m}U_e + (P_e - P_a)A_e = \dot{m}\left[\frac{U_e + (P_e - P_a)A_e}{\dot{m}}\right] = \dot{m}U_{\text{eq}}$$

where $\dot{m}$ is the propellant mass flow (mass of the propellant leaving the rocket each second), U_e (exhaust velocity) is the velocity with which the propellant leaves the nozzle, P_e (exit pressure) is the propellant pressure at the nozzle exit, P_a (ambient pressure) is the pressure in the surrounding air (about one atmosphere at sea level), and A_e is the area of the nozzle exit; U_{eq} is called the equivalent exhaust velocity. Rocket performance is often characterized by specific impulse I_{SP} dfined as

$$I_{\text{SP}} = \frac{U_{\text{eq}}}{g_E}$$

where $g_E = 9.81$ m/s^2 (or 32.2 ft/s^2) is the gravitational acceleration on the Earth's surface. Specific impulse is measured in the units of seconds. For chemically based rockets, the higher specific impulse I_{SP} usually means the more efficient rocket. (This characterization is not applicable directly to electric propulsion thrusters.)

Specific impulse less than 100 s was typical for Congreve rockets. Modern solid-propellant rockets achieve $I_{\text{SP}} = 250$–300 s. For comparison, a high-performing liquid-propellant rocket engine, such as the space shuttle main engine (SSME) using liquid oxygen and liquid hydrogen, would have $I_{\text{SP}} = 410$–450 s. Specific impulse and thrust usually increase with the increasing altitude as the ambient pressure decreases.

Congreve rockets had only a converging nozzle. At the exit of such a rocket, the exhaust velocity was thus typically equal to the local speed of sound, with exit pressure approximately one-half of the pressure inside the rocket, and the temperature 10–25% smaller than the gas temperature inside the rocket case.

A typical Congreve rocket would burn its propellant in a few seconds. During this time interval, the rocket thrust would accelerate the missile. After that, only forces of gravity and air drag would affect rocket flight on a ballistic trajectory.

STRENGTH THROUGH TECHNOLOGICAL SUPERIORITY

William Congreve was convinced in the necessity of "the improvement of every branch of offensive and defensive warfare." He considered technological advancement especially important for relatively small England.

"England is now at war with one half of the world, and has the other half to defend!" wrote he in 1811. "Need one say more to prove that, with a limited population, the ordinary implements of war cannot suffice — that is one of the first interests of the government to hold this forth as an era for the improvement of military mechanics, and by every liberal inducement in their power to promote cultivation of this branch of art, that we may thereby take the lead of the enemy in those mechanical aids, which are calculated to increase the powers of our navies and armies, and having got it, that we maintain it."

First Rocket Attack

time, and the rockets looked attractive as a new weapon to employ against the vessels of the French invasion flotilla being assembled in the Boulogne harbor. Specially equipped rocket boats with the trained units of the Royal Marine Artillery attempted, for the first time, to fire the new still-experimental missiles at Boulogne on 21 November 1805. The British forces carried about 5000 rockets, most armed with 3-lb incendiary warheads. Only 200 rockets had been launched before the attack had to be aborted because of the changed wind. One or two rockets did not burn on the ground and were picked up and examined by the French next day.

Boulogne Harbor

In April 1806, the Royal Navy successfully tried rockets on a small scale in the Mediterranean. The second rocket attack against Boulogne, the likely jumping

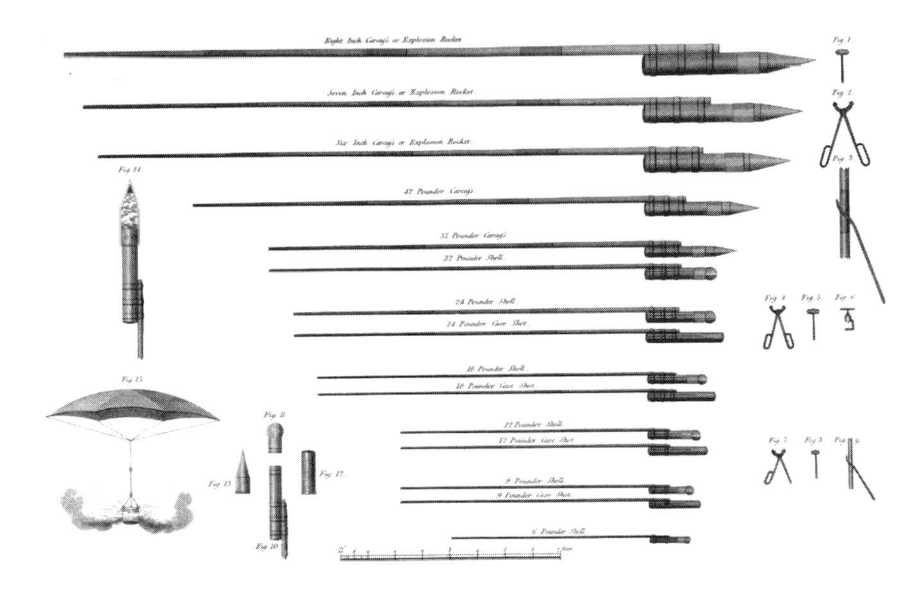

Fig. 4.2. Congreve rocket family (weapon system) with rockets from 6 lb (bottom) to 300 lb (top). Figure courtesy of the Anne S.K. Brown Military Collection, Brown University Library, Providence, Rhode Island.

off point of the French invasion of Britain, was launched on the night of 8 October 1806. Several hundred rockets were discharged and set fire to storehouses and other buildings in the town. This attack was apparently successful proving the satisfactory range of missiles and confirming practicability of their firing from small craft. Some French accounts, however, challenged reports of military effectiveness of rockets at Boulogne.

First Success

One of the British launches was sunk during the engagement at Boulogne. The French later recovered its rockets, and they were studied by the chemist Nicolas-Louis Vauquelin. The French were eager to learn the design and evaluate the prospects of the new weapon. In 1809, the retrieved rockets were examined by other famous scientists, Joseph-Louis Gay-Lussac and Joseph d'Arcet.

French Scientists

Since this second attack on Boulogne, war rockets had been accepted in Europe as means of modern war. One year later the British Army and Navy besieged the city of Copenhagen. On 2 September 1807, the bombardment began with the discharges from mortar and rocket boats. Within the first five minutes the city was on fire in several places. Many thousands of rockets were expanded during the three-day bombardment. Much of Copenhagen was burned with heavy casualties among the inhabitants and the garrison and the devastating moral effect. (Some reports disputed the effect usually attributed to rockets at Copenhagen.)

Copenhagen

Fig. 4.3. William Congreve directing discharge of his war rockets into the town of Copenhagen in 1807. The naval bombardment burned much of the city to the ground. The burning city can be seen in the middle on the left. Figure courtesy of the Anne S.K. Brown Military Collection, Brown University Library, Providence, Rhode Island.

Basque Roads

More than 1200 rockets were deployed on the fire-ships in the daring attack on the French fleet at anchorage at the Basque Roads (Ile d'Aix) near La Rochelle in April 1809. The British battleship *Caesar* waited nearby to pick up the crews of the returning fire-ships. A battleship gunner described the action, in which "shells and rockets were flying about in all directions, and the blazing light all around gave us a good view of the enemy." One of the greatest British seamen of the 19th century, Thomas Cochrane, devised and led this bold attack. Ten years later Cochrane would introduce rockets to South America.

Numerous rockets were discharged against a town of Flushing (Vlissingen) on the Walcheren Island on 13–15 August 1809. Rockets and artillery fire burnt to the ground a great part of the city, including the town hall. New missile attacks would follow.

Fig. 4.4. One of the greatest British seamen in the 19th century Thomas Cochrane (Lord Cochrane, 10th Earl of Dundonald) in 1807. Cochrane employed rockets on fire-ships against the French fleet at the Basque Roads in 1809. Ten years later, he would introduce rockets on a massive scale in Chile. Figure from *Dundonald* (Fortescue 1906, frontispiece).

RMA and RHA Rocket Units

The use of the new weapon rapidly expanded. In 1812, the Royal Navy formed rocket units in the Royal Marine Artillery (RMA) for shore operations. The Army also joined the experiments and later activated two mobile rocket troops in the Royal Horse Artillery (RHA) in January 1814. (An RHA's rocket troop detachment already distinguished itself in combat in 1813.) The extreme range of army rockets was about 2000 yards. Each rocketeer had usually one rocket ready in his hands and three rockets in a case on his back. He also carried three additional sticks that had to be attached to the rocket before launch. One rocketeer could fire four rockets a minute from a portable stand.

Fig. 4.5. 100-lb Congreve rocket in the National Air and Space Museum. The guiding stick is attached to the right side of the rocket. Photo courtesy of Mike Gruntman.

Rocket School at Woolwich

William Congreve established the first rocket school at Woolwich in January 1813. The new army and navy units were instructed in rocket equipment for six weeks and then dispatched to North America (War of 1812), Iberian Peninsula, and Germany. The Royal Marine Artillery rocket detachment under

Fig. 4.6. Firing a Congreve rocket from a launching frame on a boat. The crew is retired to the stern, while a marine artillerist discharges the rocket by a trigger-line. The mast and the sail are standing. The wet sail protects the crew from the sparks and smoke. Figure courtesy of the Anne S.K. Brown Military Collection, Brown University Library, Providence, Rhode Island.

Lieutenant Gilbert played an active role in the long siege of a large French garrison in Danzig (or *Dansic*, as it was spelled by contemporaries; modern *Gdansk* in Poland). Only during a three-day bombardment in October 1813 more than 1500 rockets were launched setting on fire numerous houses, barracks, and storage depots.

Congreve System

By the end of the Napoleonic Wars, rockets had become firmly established. The Congreve system included rockets that weighed from a few pounds to 300 lb; had various types of warheads (solid shot, shell, case shot, incendiary); and several ground and naval launching frames. The newly developed interchangeable warheads could be attached to the rocket body as required. The rocket itself became known as the Congreve.

Peninsula Campaign

Although the Royal Navy successfully employed the new weapon to project its power and even carried them to a new war in North America (War of 1812), the Army was not especially enthusiastic. The rocket reputation was not particularly high among the British generals.

The Duke of Wellington, in command of the British troops in Portugal and Spain during the Peninsular Campaign tried Congreve rockets in an engagement with the opposing French. The exhibition of rocket prowess was most discreditable.

UNCERTAIN WEAPON

... Nor is it the least appalling part of a rocket's progress [after being fired], that you see it coming, and yet know not how to avoid it. It skips and starts about from place to place in so strange a manner, that the chances are, when you are running to the right or left to get out of the way, you run directly against it; and hence the absolute rout, which a fire of ten or twelve rockets can create, provided they take effect. But it is a very uncertain weapon. It may, indeed, spread havoc among the enemy, but it may also turn back upon the people who use it, causing, like the elephant of the other days, the defeat of those whom it was designed to protect.

George Robert Gleig (1825, 298)

Whether it was strong side wind, as some claimed, or something else, many rockets turned and came back at the British. From that day forward, the Duke disliked this exotic weapon. As British officer A.C. Mercer described it, "Duke [of Wellington] ... looked upon rockets as nonsense" (Mercer 1927, 91). Rockets were still used, though on a small scale by his troops.

"... rockets as non-sense"

In one notable engagement in February 1814, an advance detachment of 600 foot guards was caught after crossing the Adour River near Bayonne in the Pyrenees by a determined counterattack of several thousand opposing French soldiers of General Soult. A party of rocketeers under Captain H.B. Lane, armed with 160 Congreves, rushed across the river and by well-directed fire stopped and put to flight the French units. A witness told that "the rocket men throwing their

Fig. 4.7. Rocket ship fires the Congreve rockets. Figure courtesy of the Anne S.K. Brown Military Collection, Brown University Library, Providence, Rhode Island.

ROCKETS AT ADOUR IN 1814

Army in Gascony

Extract of a letter from an officer in Lord Wellington's army.

In my last [letter], I informed you that the rocket brigade was to be employed in the crossing of the Adour. Forty artillerymen were accordingly ordered, and carried with them 160 rounds [rockets]; one round of which was carried by hand, and the remaining ones in the pouches for that purpose. They marched from Bidart in the evening of the 22d [February 1814], under command of Captain Lane of the artillery, and arrived, with the brigades of guards, close to the mouth of the harbor before daylight. The pontoons not having all arrived, it was impossible to throw a bridge across. About 10 or 11 o'clock, some of the light infantry of the Goldstream and 3rd Guards got across in small boats, which they continued to do, unmolested, till about half past four, or perhaps five o'clock; when the enemy [French troops] made their appearance, in number about 2000, in two columns, and a small one on their right. The rocketeers were immediately sent over, and arrived just at a proper and happy moment. The positions of two batteries of 20 rockets were directly taken up, and most part of them sent off rapidly after one another, at a distance of about 300 yards. The enemy gave way, and ran. Directly on this, the rocket party advanced on the left, in "double-quick," occasionally firing a few to accelerate their flight.

The light infantry of the guards, rapidly advancing, regained the heights they had been driven from. The rocket battery still advanced beyond their front, and fired a few more rounds. The enemy run most manfully. The force of the enemy was so greatly superior to ours, that the party of the guards at that time across the Adour, not exceeding 600 men, must have suffered very severely, and the passage for the other divisions and troops must have been much procrastinated. Thousands of troops were present on the other bank of the river, who witnessed the instantaneous effect the rockets produced on the enemy. Their reputation is now firmly established with those divisions who saw them used.

Only three companies of the guards were engaged, not more than 250 men, these and forty artillerymen, armed with rockets, put to flight 2000 men.

Had we had one hour's more daylight, it is probable the greatest part of the enemy's column would have been cut up.

I saw in one place seven Frenchmen killed and wounded by one rocket, and in another spot four. We buried 33 within our lines, and the peasants report their loss in killed and wounded to be 400. This, I think, must be exaggerated; however, they suffered severely, and we have heard of several officers being killed and wounded since the affair. Many of the men were dreadfully scorched by the fire of the rockets, and by the explosion of their cartouches as the rockets passed through the ranks.

The 32 lb shell rockets were used against the enemy's boats in the morning, to cover the flank of the 18th Portuguese brigade against the enemy's gunboats; this they performed effectually, by driving them away in the greatest consternation — one was struck, and her loss of men was great.

The Edinborough Evening Courant, No. 16,045, p. 4

Monday, 11 April 1814

diabolical engines with extraordinary precision, at the very instant when the [British] infantry fired a well-directed volley, the confusion created in the ranks of the enemy beggars all description" (Gleig 1825, 298).

Diabolical Engines

Another witness described that one rocket

> carried away the two legs of one man [French soldier], set the knapsack of another on fire, and knocked about many more, throwing them [the French troops] into great confusion and wounding several. An old [French] serjeant, whom we [the British] made prisoner said, during all his service, he had not known what fear was before; but these machines [rockets] were perfect devils, running up and down, and picking out and destroying particular victims, as it were, in one place and then another. (*Edinborough Evening Courant* 1814, No. 16,045, p. 4)

Perfect Devils

The potential of the army rockets was clearly demonstrated in a major "Battle of the Nations" that took place near Leipzig (or Leipsic, as it was spelled in the 19th century) in October 1813. Allied Austrian, Prussian, Russian, Swedish, and British forces of the Anti-French coalition decisively defeated the Napoleon's Grand Armée in this battle in the heart of Germany. The "Rocket Brigade," a detachment of the Royal Horse Artillery, was the only British unit that took part in the battle. Captain Bogue commanded the battery that had 194 officers, noncommissioned officers, and men.

Battle of the Nations

The first Congreves were fired on 16 September 1813 when the (French) Marshal Davout's detachment under General Pecheux was destroyed in a raid near Görda. The rockets contributed considerably to the success of the coalition troops in this battle where the French lost more than 1000 men. The effect of the rockets was entirely new to the French troops. Marshal

ROCKET WALTZ

A composer named Samuel Wesley, mostly known for his organ music, wrote "The Sky-rocket, a new Jubilee Waltz, for the Pianoforte" and inscribed to Colonel Congreve.

The new piece did not enjoy success however. The September 1814 issue of *The Gentleman's Magazine* caustically noted in a review of new musical publications that "it is certain that the ingenious Colonel [Congreve] has been more successful in sky rockets than the composer."

GREATEST PHYSICAL AND MORAL EFFECT

... To employ that [rocket] arm on proper occasions in salvoes projecting 20 shells and case shot or even more; and in favorable ground even double that number, a mass of fire which, when judiciously used, must be productive, as at the battle of Leipsic [Leipzig] and on the Görda, of the greatest physical and moral effect, both from the novelty of the weapon, the extraordinary and appaling noise accompanying its flight from the first moment of ignition to that of the explosion of the projectile, and from its visibility during the flight.

Memorandum to Duke Wellington on the armament and employment of the 2nd Rocket troop, May 1815
(Whinyates 1897, 134)

ROCKET BASICS 2: ROCKET DYNAMICS

Rocket acceleration is proportional to thrust of the rocket engine

$$T = M\,a$$

where T is thrust, M is the rocket mass, and a is acceleration. The calculation of rocket velocity is somewhat complicated because the rocket mass is not constant and decreases with time as the propellant being consumed.

William Moore, an instructor of the Royal Military Academy in Woolwich, England, seemed to be the first to correctly calculate the velocity and altitude of a rocket at burnout (i.e., when all propellant is consumed). Moore was in an excellent position to follow the experiments with Congreve's rockets at Woolwich and got further interested in the subject "when the Academy of Copenhagen proposed [in 1810] as a prize question, the curve that a rocket describes, when projected, in any oblique direction, in vacuo ..." (Moore 1813, iii).

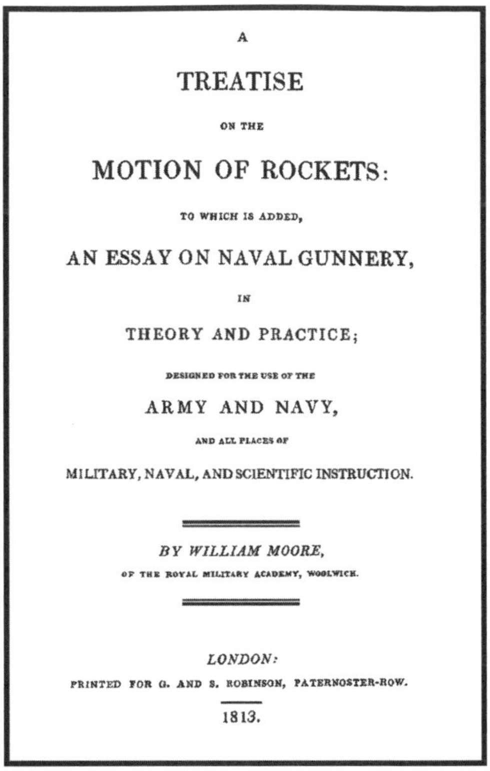

A
TREATISE
ON THE
MOTION OF ROCKETS:
TO WHICH IS ADDED,
AN ESSAY ON NAVAL GUNNERY,
IN
THEORY AND PRACTICE;
DESIGNED FOR THE USE OF THE
ARMY AND NAVY,
AND ALL PLACES OF
MILITARY, NAVAL, AND SCIENTIFIC INSTRUCTION.

BY WILLIAM MOORE,
OF THE ROYAL MILITARY ACADEMY, WOOLWICH.

LONDON:
PRINTED FOR G. AND S. ROBINSON, PATERNOSTER-ROW.
1813.

Fig. 4.8. Cover of William Moore's book on rocket dynamics published in 1813.

In 1813, Moore published a book, *A Treatise on the Motion of Rockets*, where he addressed various aspects of rocket dynamics. In fact, Moore introduced dynamics of a body with the varying mass, coming close to obtaining the rocket equation. Moore began his work with considering a vertical launch of a rocket with a constant thrust T, constant propellant mass consumption rate, and in absence of atmospheric drag. He obtained the velocity V_B and altitude H_B at burnout (in modern notaion)

$$V_B = \frac{T t_B}{M_P} \ln\left(\frac{M_0}{M_0 - M_P}\right) - g_E t_B$$

$$H_B = \frac{T t_B^2 (M_0 - M_P)}{M_P^2} \ln\left(\frac{M_0 - M_P}{M_0}\right) + \frac{T t_B^2}{M_P} - \frac{g_E t_B^2}{2}$$

where M_0 is the initial rocket mass, M_P is the total mass of the consumed propellant, and t_B is the burnout time. These formulas (usually expressed through specific impulse and the mass ratio) can be found in modern textbooks.

Bernadotte reported about this engagement that "the English artillery and rocket corps deserve the highest encomiums."

The Battle of the Nations was fought a month later on October 16–18. The British rocket unit was usually attached to the Marshal Bernadotte's bodyguard but was allowed some freedom of action during the engagements. On the 16th of October, Captain Bogue boldly deployed his rocketeers and opened well-directed devastating fire on the columns of a French brigade near the village of Paunsdorf. The panic struck, and the whole brigade was utterly dispersed. Between 2000 and 3000 enemy soldiers surrendered when Captain Bogue charged at the head of the squadron of dragoons attached for protecting his battery.

Devastating Rocket Fire

Leading a further advance of the Rocket Brigade, Bogue was shot by an enemy rifleman dead, the only British casualty in the battle. Lieutenant T. Strangways, took the command of the rocketeers and gallantly led them through the conclusion of the engagement.

A witness wrote that

> each time a rocket was fired ..., one saw whole files [of soldiers] hurled down. The scorched and battered bodies lay in great piles .... At first the French did not seem familiar with this new weapon of death and stood up against it; but when they saw what fearful destruction it wrought and in what a ghastly manner the victims died, ... there was no holding them. Whenever they saw the rocket coming, whole columns ran away and abandoned everything .... (Brett-James 1970, 188)

Devil's Own Artillery

A Russian General Wittgenstein reportedly characterized Congreve's rockets at Leipzig "as if they were made in hell, and surely are they the devil's own artillery" (*Edinborough Evening Courant*, 20 January 1814, No. 16,009, p. 2).

The successes at Görda and Leipzig led to reorganization and expansion of rocket units. In 1814, the RHA rocket detachments were reorganized into the First Rocket Troop under Captain W.G. Elliot and the Second Rocket Troop under Captain E.C. Whinyates, which were activated at Woolwich and on the European continent, respectively. The Second Rocket troop saw more action against the French in Germany, then served in Holland, and later joined the army under the Duke of Wellington in Belgium for the final major battle against Napoleon at Waterloo.

Rocket Units in Action

> **ROCKET TROOP AT LEIPSIC**
>
> "His Royal Highness the Prince Regent, in the name and on the behalf of His Majesty, has been pleased to command that the Rocket Troop of Royal Artillery, which was present at battle of Leipsic [Leiptzig], be permitted to wear the word 'Leipsic' on their appointments, in commemoration of their services on that occasion."
>
> *Order*, 16 May 1815
> (Duncan 1874, 37)

The Duke was of so low opinion of rockets since his experience in the Peninsula Campaign that at one time he ordered Whinyates to put the rockets in storage and have his troop supplied with guns instead. It was with "much difficulty and pleading" (Duncan 1874, 416) that the permission was secured to carry 12-lb rockets into the battle. The British artillery commander George Wood wrote that "the Duke's prejudice against rockets was unmistakable ... but the official reason he gave was

Duke's Prejudice

that when he had a proper proportion of artillery attached to his army ... then he would bring the Rocket Corps into play; but that he thought ... the gun was a superior weapon" (Duncan 1874, 416). Fortunately, Wood recalled later that "after the good appearance of our friend Whinyates's troop, and the plan and mode he suggested to his Grace, he [Duke of Wellington] has permitted him [Whinyates] to take into the field eight hundred rounds of rockets with his six guns" (Duncan 1874, 417).

Fig. 4.9. "Battle of Waterloo. An Ammunition Wagon on Fire & the Horses Taken Fright." Lithographic print, 1816, London. This is a unique contemporary image showing war rockets (in the sky on the left) in action. Figure courtesy of Musée de l'Armée, Paris, France.

Rockets at Waterloo

The days of the battle of Waterloo arrived. On 16 June 1815, the British began to withdraw from an important cross road town of Quatre Bras after an inconclusive engagement. The rockets flew the next day at the nearby town of Genappe in an attempt to slow down the advancing French. A British officer Captain A.C. Mercer described that the rocketeers

> had placed a little iron triangle [tripod] in the road with a rocket lying on it. The order to fire is given — port-fire applied — the fidgety missile begins to sputter out sparks and wriggle its tail for a second or so, and then darts forth straight up the chaussée. A [French] gun stands right in its way, between the wheels of which the shell in the head of the rocket bursts, the gunners fall right and left, and, those of the other guns taking to their heels, the battery is deserted in an instant. (Mercer 1827, 153)

Other rockets went astray and "some actually turned back upon ourselves [British] ... . Meanwhile the French artillerymen ... returned to their guns and opened a fire of case shot ..." (Mercer 1827, 153).

The next day, on 18 June 1815, the Whinyates rocket battery successfully covered a broken retreat of the British heavy cavalry under General Uxbridge when it was caught after the initially successful charge against French infantry.

Fig. 4.10. Congreve rockets with the center-mounted guiding stick, demonstrated in the late 1815, could be launched from tube, a significant improvement as compared to side-mounted-stick rockets. Figure from *Projectile Weapons of War* (Scoffern 1859, 172).

The rocketeers moved down the hill, discharged rockets at the opposing French formations, and then rejoined their troop on the crest of the main ridge. Fifty-two rockets were fired on that day. Three out of four officers of the rocket troop, including Captain Whinyates and the veteran of the Leipzig battle Lieutenant Strangways, were wounded at Waterloo.

Centrally Mounted Guiding Stick

War rockets were significantly improved in 1815, when William Congreve introduced a new design with centrally mounted guidesticks. A circular baseplate was attached to the rear of the rocket. The

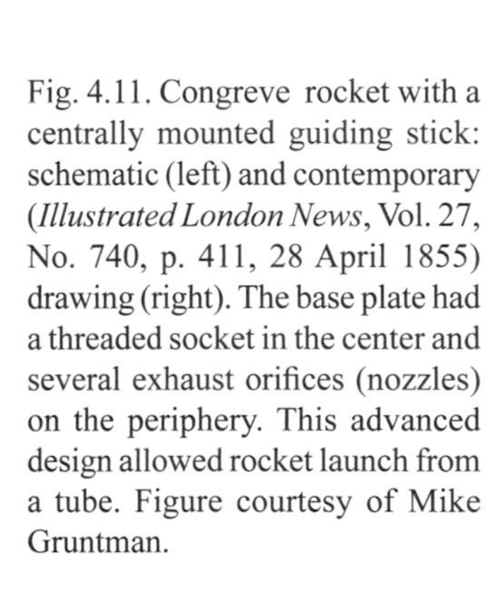

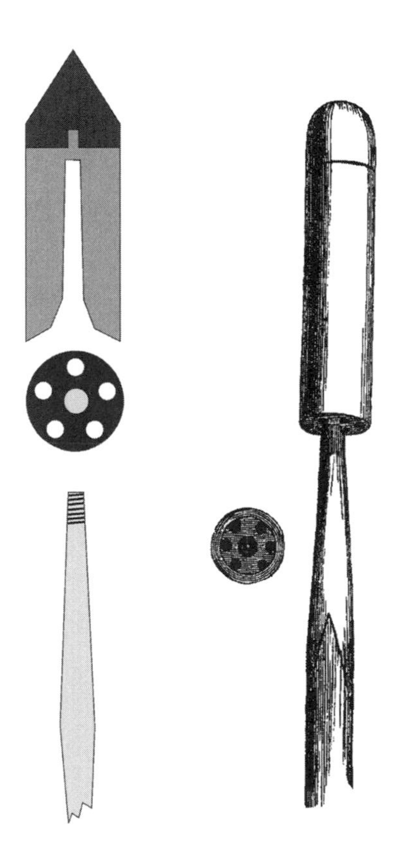

Fig. 4.11. Congreve rocket with a centrally mounted guiding stick: schematic (left) and contemporary (*Illustrated London News*, Vol. 27, No. 740, p. 411, 28 April 1855) drawing (right). The base plate had a threaded socket in the center and several exhaust orifices (nozzles) on the periphery. This advanced design allowed rocket launch from a tube. Figure courtesy of Mike Gruntman.

guidestick was screwed into a special threaded socket in the center of the baseplate. Five holes (nozzles) around the central stick served for exhaust of hot gases. The first rocket with a center-mounted stick was successfully tested on 30 December 1815. The new rockets showed convenience of service, superior accuracy, allowed firing from tubes, and were quickly adopted internationally.

The Congreve system significantly advanced the rocket technology, improved rocket performance, and returned them into the modern warfare arsenal. A number of countries witnessed the efficiency of the new weapon and embarked on the development of their own rockets. The stage was set for the second great wave of rocket proliferation.

5. ROCKETS COME TO AMERICA

The British military and colonists brought the first skyrockets to America, where fireworks were not uncommon in the colonial days. On the occasion of His Majesty's birthday in December of 1767, for example, the fireworks were conducted "after his Majesty's and many others loyal healths were drank" in New York (Barber 1851, 168). The fireworks consisted of three sets. Each set commenced with two signal rockets and included, among other fiery displays, a dozen of skyrockets. The Revolutionary War did not witness the rocket use other than for signaling. A new major conflict was needed to introduce war rockets to America.

Rockets in Colonial America

Such a war broke out between the United States and Great Britain in June 1812. The British expeditionary force brought the Congreves to North America, and they were fired across the vast land from Hampden in Maine to Upper Canada (Ontario) to New Orleans in Louisiana.

War of 1812

The British war vessels blockaded all of the principal Atlantic ports from New York to Maine, and the Royal Navy embarked on numerous operations against American coastal towns. Raiding parties not only successfully kept the fleet supplied in cattle and other provisions but also inflicted economic damage and put political pressure on the young republic.

Coastal Towns Raided

A 50-man strong Royal Marine Artillery rocket unit under Lieutenants G.E. Balchild and J.H. Stevens was trained at Woolwich and attached to two battalions of Marines of the British expeditionary force. In addition, several Royal Navy vessels, such as the *HMS Mariner*, were specially outfitted for rocket warfare. A typical Congreve rocket weighed 32 lb (14.5 kg) and was 4 in. (10 cm) in diameter and 40 in. (1 m) long; its cost was about 1£ apiece.

The mobility of rocketeers provided British raiding parties with important firepower. Rockets were often fired before the raids were launched and troops landed. On 2 May 1813, the British attacked Havre de Grace, a small town of 60 houses near the mouth of the Susquehanna River in Maryland. Hissing rockets set many wooden structures ablaze. Many other Chesapeake Bay towns were raided and suffered a similar fate.

Havre de Grace

On June 6, Congreves were fired against American forces in the failed British attack on Craney Island in Norfolk, Virginia. Later, on 24 June, 2500 British soldiers, supported by the artillery and rocket fire, captured the nearby village of Hampton. In this engagement, the British were assisted by their French prisoners of war who had enlisted as chasseur Britanniques. After a sharp engagement the troops entered Hampton. The infamous "booty and beauty" atrocities followed. The French chasseurs were conveniently blamed for the outrages and dismissed from the British service.

Hampton

North Carolina

On 13 July 1813, the British attacked two American privateer vessels at Ocracoke Inlet in North Carolina. Gunfire was exchanged for some time until rockets forced the American crews to abandon their vessels.

Northern Frontier

Fighting on the northern frontier also saw the first Congreves, when the British marines under Lieutenant Stevens supported with rocket fire the successful landing at Oswego on Lake Ontario on 5 May 1814. Two-and-a-half months later, on 25 July 1814, the British discharged rockets, without much effect, in the battle of Lundy's Lane on the shores of the Niagara river. Winfield Scott, the youngest general in the U.S. Army, commanded a brigade in this sharp engagement and was severely wounded. Many years later, in 1847, Scott would lead the

Lundy's Lane

Fig. 5.1. British raiding party leaves behind the ravaged coastal town. Figure from *The Pictorial Field-Book of the War of 1812* (Lossing 1869, 606).

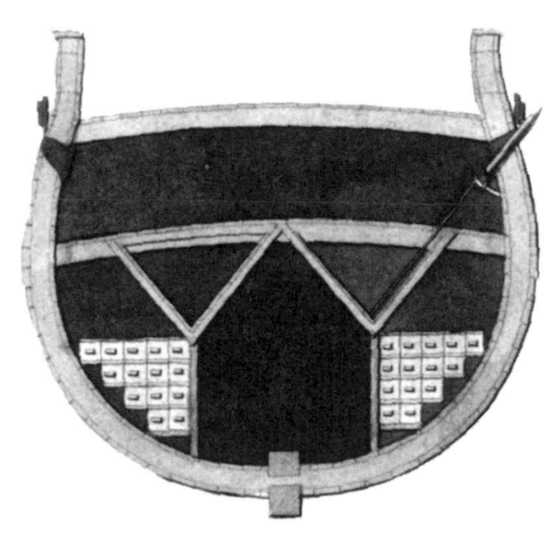

Fig. 5.2. Congreve's rocket ship. A rocket battery can be deployed at a vessel with a between-deck, in addition to the gun batteries on her spar deck. Figure courtesy of the Anne S.K. Brown Military Collection, Brown University Library, Providence, Rhode Island.

army expedition to Mexico where the Americans would fire their rockets for the first time in a major military operation.

Plattsburg

In September of 1814, the Americans repulsed the invading army at Plattsburg on Lake Champlain. The British again widely employed rockets. At one moment, as an American officer reported, "the enemy commenced a simultaneous bombardment of our works from 7 batteries, from which several hundred shells and rockets were discharged, which did us very little injury; and our artillery had nearly succeeded in silencing them all before the contest on the lake was decided" (Barber 1851, 74).

Blockade

The third British Marine battalion with the attached rocket brigade joined Vice-Admiral Alexander Cochrane's force in July. The continuing British blockade with the numerous raids on coastal towns ruined the economic life of the country. Cochrane instructed his squadrons, "You are hereby required and directed to destroy and lay waste such towns and districts upon the coast as you may find assailable" (Adams 1999, 218).

Stonington Under Rocket Attack

British ships sent a storm of bombshells and rockets on the small town of Stonington, Connecticut, for four days on 9–12 August 1814. The American militia drove back all of the attempts to land the troops, and the British attack failed. Stonington became a symbol of Yankee tenacity, and Congreve rockets were remembered in a song "Battle of Stonington":

> The bombs were thrown, the rockets flew,
> But not a man of all their crew,
> (Though every one was full in view)
> Could kill a man of Stonington

Rocket Superior Range

In 1814, British forces began their push up the Chesapeake Bay. The armed American barges commanded by Commodore Joshua Barney tried to defend the area, conducting maritime guerilla warfare, against the British flotilla. The first British Congreves were fired in a naval engagement on 1 June 1814. New rocket attacks followed on June 8 and 10. The damage from rockets was small, but their superior range allowed the British to fire at the Americans without being exposed to American shell. Commodore Barney was finally compelled to withdraw to the Patuxent river.

Rocket Ships in the Chesapeake

The main British advance up the Chesapeake began in August. The British naval forces included a specially outfitted rocket-ship *HMS Erebus*. The *Erebus* was equipped by William Congreve at Woolwich in February and March with 21 permanent breach-loading rocket launchers on its broadsides. Each launcher could fire a rocket once a minute. Congreve's design allowed the discharge of rockets from the broadside scuttles "at the highest angles for bombardment, or used at low angles, as an additional means of offense or defense against other shipping in action" (Congreve, 1827). The position of launchers on the ship protected the vessel from rocket sparks. Lieutenant Beauchant was in command of the rocketeers.

On the shore, a rocket section of the Royal Marine Artillery under Lieutenant John Lawrence, a veteran of the fighting in Canada, joined the British troops advancing on the American capital. The rockets proved an effective weapon. Lawrence's men almost immediately saw action, dispersing the harassing American mounted men on 19 and 21 August.

Battle of Bladensburg

Rockets played a prominent role in the battle of Bladensburg near Washington on 24 August 1814. By a blunder that proved disastrous, three Baltimore regiments were moved before the battle into an entirely uncovered position, within the reach of the British Congreves. Two militia regiments broke and fled in the wildest confusion under a flight of hissing rockets. British General Ross praised in his report "the well-directed discharge of the rockets throwing the Americans into confusion." Another witness Rear-Admiral Cockburn noted "with much pleasure the precision with which the rockets were thrown under the direction of First-Lieutenant Lawrence of the Marine Artillery" (Fraser and Carr-Laughton 1930, 266).

"well-directed discharge of rockets"

Fig. 5.3. Remains of the Capitol building in Washington. Figure from *The Pictorial Field-Book of the War of 1812* (Lossing 1869, 933).

Capital Lost

The battle of Bladensburg was lost, and the road to the American capital was open. Later the same day the British troops entered and occupied Washington. In those days, the city had about 900 houses scattered over three miles. The British burned the Capitol, President's house, many private and public buildings, and the great bridge across the Potomac.

HEAPS OF SMOKING RUINS

"Little, however, now remained to be done [in Washington by the British troops], because every thing marked out for destruction was already consumed. Of the Senate-house, the President's palace, the barracks, the dock-yard, &tc. nothing could be seen, except heaps of smoking ruins; and even the bridge, a noble structure upwards of a mile in length, was almost entirely demolished."

George R. Gleig (1827, 136)

Rockets on the Potomac

On 17 August, British Captain Gordon's squadron, including the rocket ship *Erebus*, started off up the Potomac. By 30 August, they reached Alexandria and captured and destroyed Fort Washington on the opposite bank of the river. During the withdrawal, the squadron was attacked by the Americans assembled on the banks of the Potomac. Rockets were flying time and again as the British fought their way through. Captain Gordon reported that "The *Erebus* while [temporarily] aground [on a shoal] fired rockets with the most decisive effect." (Fraser and Carr-Laughton 1930, 272)

Fort McHenry Under Rocket Fire

Baltimore was the next point of the British combined land and sea assault. On the land, the rocketeers under Lieutenant John Lawrence accompanied the troops of General Ross and were praised for "rendered essential service." On 13 and 14 September 1814, five British bomb vessels and the *Erebus* poured a heavy fire on Fort McHenry guarding access to Baltimore. Lieutenant Beauchant's detachment on the *Erebus* fired Congreves from the extreme range of two miles, which could be done only with the smallest, 8-lb (3.6-kg) warheads. The British warships were out of range of the Fort's guns. A large American flag, 42×30 ft (12.3×9 m), was proudly flown over the Fort.

Fig. 5.4. View of the bombardment of Fort McHenry near Baltimore. The lithograph shows only bombs fired by Royal Navy's mortars. A large number of Congreve rockets were also discharged at the fort. Figure courtesy of the Anne S.K. Brown Military Collection, Brown University Library, Providence, Rhode Island.

Fig. 5.5. Francis Scott Key, 1780–1843. Figure from *The National Cyclopædia of American Biography* (1907, Vol. 5, 498).

One bombshell damaged a 24-pounder gun in the southwest bastion, killing an officer and wounding four men. British Vice Admiral Cochrane ordered several of his bombarding ships to come a half a mile nearer Fort McHenry. The delighted Fort Commander Major George Armistead opened fire on the British with all guns. Several British ships were hit, and in half an hour they withdrew to their old anchorage. The rocket ship *Erebus* was injured by the American fire and had to be towed by small boats to safety. The bombardment continued.

With permission of President Madison, a young lawyer, Francis Scott Key, and John S. Skinner, an agent for the exchange of prisoners, went to negotiate with the British the release of a friend of Key's. The American party was detained by the British lest they could disclose the intended attack on Baltimore.

Rockets' Red Glare

Key and Skinner observed the bombshell and rocket bombardment of Fort McHenry from a cartel-ship *Minden*, under guard of the British marines. The Royal Navy artillery and rockets bombarded the Fort through the entire day of the 13th of September and the early hours of the 14th.

As dawn broke out, anxious Key saw the American flag, tattered but intact, still there flying over the rampart. It was these Congreve rockets that inspired Francis Scott Key's famous lines:

> ... And the rockets' red glare,
> the bombs bursting in air,
> Gave proof through the night
> that our flag was still there ...

Key's lyrics set to the air of a popular drinking song "To Anacreon in Heaven" later became the de facto National Anthem. The Congress officially recognized the Star Spangled Banner in 1931, immortalizing the Congreve rockets in the National Anthem.

Battle of New Orleans

The Battle of New Orleans in December 1814 and January 1815 was the last time when rockets were heavily employed by the British in North America. The British forces included a rocket brigade, 98 officers and men and 150 rockets, commanded by Captain H.B. Lane. Rockets again proved its superior mobility under conditions unfavorable for transporting artillery.

Captain Lane's rocketeers provided important fire support during the most critical day of the battle for the troops under Colonel Thornton, which crossed

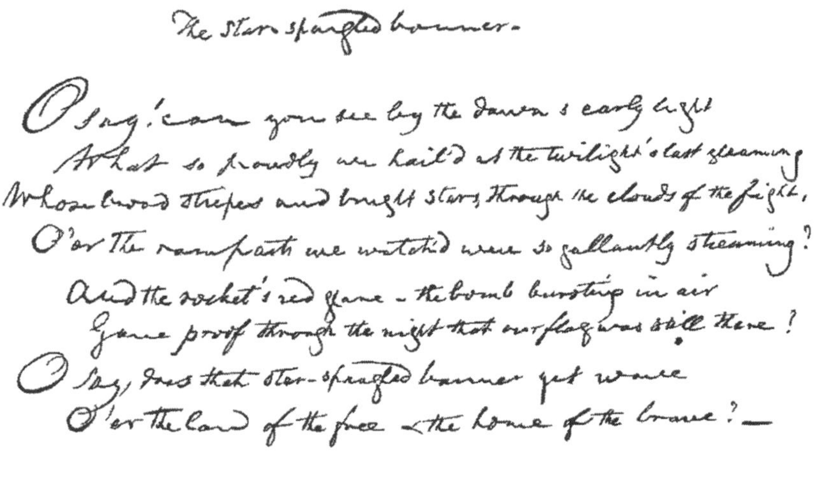

The Star spangled banner.

O say! can you see by the dawn's early light
What so proudly we hail'd at the twilight's last gleaming
Whose broad stripes and bright stars, through the clouds of the fight,
O'er the ramparts we watch'd were so gallantly streaming?
And the rocket's red glare - the bomb bursting in air
Gave proof through the night that our flag was still there?
O say, does that star-spangled banner yet wave
O'er the land of the free & the home of the brave?—

Fig. 5.6. The first stanza of Key's "Star-Spangled Banner." Note the spelling of the original version "the rocket's red glare." Figure from *The Pictorial Field-Book of the War of 1812* (Lossing 1869, 957).

the river and successfully carried the American positions on the right bank of the Mississippi. This was probably the only successful action in the otherwise disastrous, for the British, operation. Colonel Thornton later reported that "Major Michell of the Royal Artillery afforded me much assistance by his able direction of the firing of some rockets, it not having been found practicable in the first instance to bring over the artillery attached to his command" (Duncan 1874, 409).

Fig. 5.7. Andrew Jackson in the Battle of New Orleans, January 1815. Figure from *New History of the United States* (Lossing 1889, 440).

The American troops under General Andrew Jackson were showered by rockets throughout the campaign, though with little effect. Jackson's principal engineer, Arsene Latour, later observed that the British

> had great expectation from the effect of this [rocket] weapon, against the enemy who had never seen it before. They hoped that its very noise would strike terror into us; but we soon grew accustomed to it, and thought it little formidable; for in the whole course of the campaign, the rockets only wounded ten men, and blew up two caissons. (Latour 1999, 88)

Unfulfilled Expectations

Latour summarized his impression of the rocket inefficiency as "that weapon must doubtless be effectual to throw amongst squadrons of cavalry, and frighten the horses, or to set fire to houses; but from the impossibility of directing it with certainty, it will ever be a very precarious weapon to use against troops drawn up in line of battle, or behind ramparts" (Latour 1999, 89). A British officer and future author, George R. Gleig, described the response to an American night attack on the British bivouac on 23 December: "a few [Congreve] rockets were discharged which made a beautiful appearance in the air; but the rocket is at best an uncertain weapon, and these deviated too far from their object to produce even terror amongst those against whom they were directed" (Gleig 1827, 287). These descriptions of experienced officers on both sides of the conflict were not unlike those of the British in India 15 years earlier. Only the introduction of rockets with much bigger warheads (not requiring precise guidance) and precisely guided missiles in the 20th century would change rocket efficiency.

"... beautiful appearance in the air ..."

Fig. 5.8. The 100-ft (30.5-m) Chalmette Monument marks the battlefield at New Orleans, the last major military action where rockets were used against Americans on the American soil. Photo courtesy of Mike Gruntman.

The British were decisively defeated in the Battle of New Orleans with the large losses of killed and wounded. A number of Congreve rockets were captured by the American forces. Andrew Jackson reported to the Secretary of War that "some entire Congreve-rockets have been found, and a rest from which they are fired, which is my intention to forward to the seat of government whenever a proper opportunity shall offer" (Latour 1999, 268). The battle of New Orleans was the last time that the rockets were fired against Americans on the American soil.

6. FIRST AMERICAN ROCKETS

The War of 1812 brought war rockets to America. After the war the American Army began experimenting with this new weapon on a small scale. The Army Ordnance Department created by Congress in May of 1812 supervised rocket fabrication and conducted the tests. This was the beginning of the interest of the Army Ordnance in rocketry that would culminate in placing the first American satellite Explorer I in orbit by the Army's space launcher in 1958.

Army Ordnance Depart-ment

In early 1820s, Army's arsenals fabricated about 150 experimental rockets annually. It took many years to turn the experimental missiles into a weapon of war. Secretary of War Joel R. Poinsett organized a Board of Ordnance in 1837 under Brevet Brigadier General J.R. Fenwick, with J. Erving, G. Talcott, R.L. Baker, and A. Mordecai as the members. Following a proposal by Secretary Poinsett, the Board endorsed in 1838 the introduction of rockets into the Army.

Rockets in Europe and America

The Ordnance Department followed the developments in artillery and rocketry in other countries. In 1840 the officers of a special Board of Ordnance spent a year in Europe studying new designs, manufacturing, and use of ordnance. The Board included Captain Alfred Mordecai, one of the most scientifically minded Army officers. Mordecai would serve for 29 years in the Ordnance Department and direct the evaluation of the new design of war rockets and their fabrication during the Mexican War. This military operation would become the first major campaign for the U.S. Army to fire rockets in combat.

> **THE SECRET APPEARS KNOWN**
>
> "In all [European] countries visited [by officers of the Board of Ordnance], war rockets are made on a more or less limited scale. All nations make secret of the details of their manufacture, but the secret appears known to all nations."
>
> *Report to the Senate*, 1841 (Olejar 1946, 21)

The technical information and observations brought back from Europe by the American officers of the Board of Ordnance of 1840 were later credited "in a

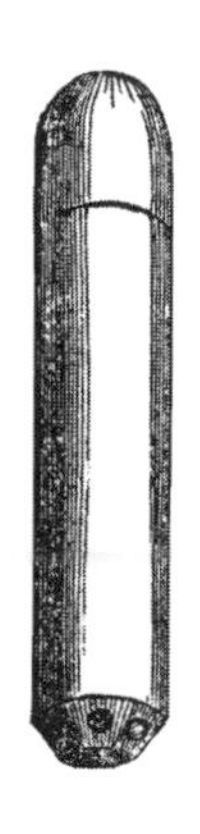

Fig. 6.1. Contemporary drawing of the Hale rocket from 1840s. One can see five oblique exits at the baseplate to spin the rocket. Figure from *Illustrated London News*, Vol. 27, No. 740, p. 411, 28 April 1855.

great measure" for that "our present [1853] system of artillery is so complete and efficient" (Benét 1880, 530). A special section of the Board's report described war rockets that were made in England, France, Prussia, and Sweden. Fabrication of Congreve-type rockets continued on a small scale in the American arsenals as well. The troops received 250 rockets in 1843, 160 in 1844, and 544 in 1845.

Invention of William Hale

Two important events in the mid-1840s brought into focus the Army's interest in war rockets, an invention and a war. In 1844 the English inventor from London, William Hale (1797–1870), introduced rocket spin to stabilize the flight trajectory and to improve accuracy without a cumbersome guiding stick.

Hale's first design was based on the main central exit (nozzle) to propel the rocket along the axis and additional five small oblique exhaust holes in the rocket baseplate. The exhaust through these small holes produced the torque that spun up the missile.

Spinning Missiles

About 10 years later, William Hale proposed a new design with the oblique holes near the rocket head. Then he patented in 1866 a new spinning system with the three curved vanes attached to the baseplate. It is possible that the curved vanes were actually invented by Hale's son, William Hale, Jr.

Most of William Hale's inventions were implemented and further developed in the United States. As the report of the American Military Commission to Europe in 1855–1856 noted, "Hale's rockets seem to have found no favor

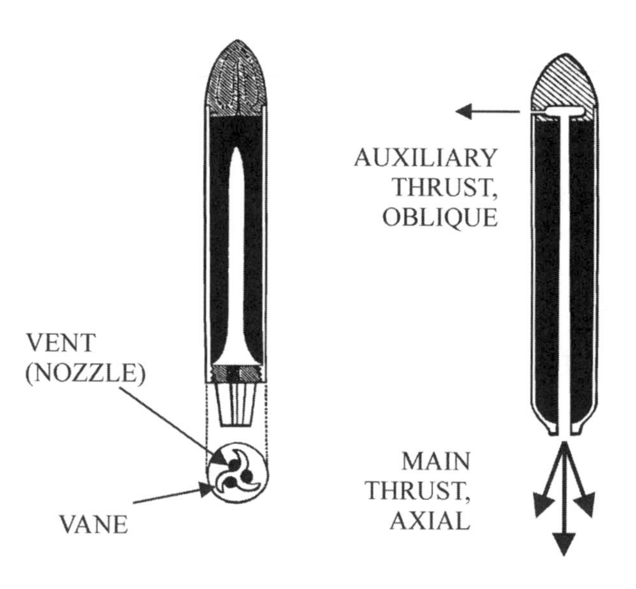

Fig. 6.2. Schematics of Hale's rockets that followed his initial design of 1844: right (introduced in 1855) — with the large central vent (nozzle) and three small vents at the base of the head; left (introduced in 1866) — with three vanes. Both small vents at the rocket head and curved vanes were used for spinning the rocket up. The figure is based on 1883 drawings (Benton 1883, 98, 99).

NAME ON THE MOON

A lunar crater is named after an English inventor William Hale in recognition of his contribution to rocketry.

with the authorities at Woolwich [artillery establishment in England]" (Mordecai 1860, 105). As a result, Hale's rockets were used only on a very limited basis by the British in the Crimean war (1853–1856).

Hale's Rockets Gain Popularity

The new rockets did not require guiding sticks and thus offered superior convenience of service. The Hales, as the rockets became known, quickly gained popularity, especially in America. Several other countries including France, Prussia, Russia, and Switzerland closely inspected and evaluated Hale's rockets. Austria purchased the invention and adopted the design for its army.

WASHINGTON ARSENAL

Washington Arsenal was a principal ordnance facility of the United States. It stood on the Greenleaf's Point, at the mouth of the Eastern branch of the Potomac and possessed every facility for receiving and shipping military stores. Its extensive buildings housed workshops with various machinery and provided storage for ordnance

Fig. 6.3. Washington Arsenal in 1852. Courtesy of the National Defense University, Washington, D.C.

equipment. Arsenals of the Army Ordnance created and maintained the tradition of in-house development of new weapon systems. The territory of Washington Arsenal is now a part of Fort McNair in the District of Columbia and occupied by the National War College of the National Defense University.

Many countries continued, however, to use Congreve-type rockets that were much simpler in manufacturing. Alternative approaches were also tried to achieve stability of rocket flight without guiding sticks and without spin. In particular, the Russian and Swedish armies experimented with rockets stabilized by fins.

MEXICAN WAR

The war broke out when the Mexican troops, in the words of President Polk, "shed American blood upon American soil" north of Rio del Norte (Rio Grande) in April 1846. The initial successes in Northern Mexico of the Army commanded by General Zachary Taylor did not lead to a peace settlement. President Polk then ordered Major General Winfield Scott to land an army at Vera Cruz, capture this seaport, and advance to Mexico City. The time was ripe for introduction of technological innovations.

Time for Technological Innovations

On 19 November 1846, the Ordnance Department proposed to General Scott to form a battery armed by mountain howitzers and rockets. The Department recently obtained "a knowledge of the great use made by the French army in Algeria of the mountain howitzer" (Benét 1880, 223) and was eager to try the new weapons. Scott favorably endorsed the proposal the next day.

The new Hale rockets were demonstrated to the U.S. Army in the fall of 1846 by Joshua B. Hyde, who purchased the invention from William Hale. Thirteen 2 3/4-in. (70-mm) rockets were successfully fired on land and water in tests conducted on 24 and 27 November 1846 at Washington Arsenal. A joint Army and Navy board issued a highly favorable report on the test results on 1 December 1846.

Successful Tests

Ten days later the agreement was made for purchase of "the full plans, instructions and drawings, requisite for making" the Hale's rockets (Benét 1880, 152). The Army and the Navy agreed to pay J.B. Hyde $2000. The agreement called for experimental manufacturing of "ten rockets of two inches and ten of three inches" (Benét 1880, 153). After satisfactory tests of the American-made rockets, an additional $18,000 was paid. The Secretary of War and Secretary of the Navy acted jointly with the departments sharing the expenses.

Speedy Procurement

The experimentally manufactured rockets successfully passed their first trials at Washington Arsenal on 5 January 1847. Fifteen 3-in. (76-mm) and 13 2-in. (51-mm) rockets were tested. The Army and Navy representatives at the tests were headed by Captain A. Mordecai and Commodore L. Warrington, respectively. Thus after a few months only, the new weapon system was ready for the introduction to the troops. By today's standards, this was an incredibly fast evaluation, prototyping, and procurement of a new weapon accomplished in cooperation of two military services.

Rocket Battery Authorized

On 3 December 1846, the Secretary of War authorized a new Howitzer and Rocket Battery to accompany General Scott's expedition against Mexico. The new unit was to be assembled at Fort Monroe near Old Point Comfort in Virginia. The battery was attached to the Regiment of Voltigeurs and Foot Riflemen, organized by the act of 11 February 1847, for and during the war with Mexico.

Both mountain howitzers and rockets were the innovations that required special technical skills of the officers and men. The rocketeers had also to develop a method of attaching the ammunition (warheads) to the rockets. The battery was manned by the Ordnance Department because, as the Department argued

(Benét 1880, 156), "the army at large" was "believed to be entirely unacquainted" with the new arms, while the ordnance officers had "the benefit of experimental firings for a number of years." Such an attitude and the role of the Ordnance Department in the Army structure and the Department's relations with the artillery corps were a sore point of bitter contention for many years.

Special Technical Skills Required

> **ROCKET AND MOUNTAIN HOWITZER BATTERIES**
>
> Wanted one hundred active, brave young men to serve with rocket and mountain howitzer batteries now preparing by the Ordnance Department for immediate departure.
>
> In pay, provisions, and clothing, this corps will be superior to any other yet raised, and, from the kind of arms, will be constantly in *the advance* where the hardest fighting may be expected.
>
> The highest character for *courage* and *physical ability* will be required for admission.
>
> Apply to ...
>
> Two dollars paid to citizens for each recruit.
>
> Army Advertisement, issued on 4 December 1846 (Benét 1880, 152)

Experienced ordnance officers, all West Point graduates, were appointed to the new battery. The first battery commander First Lieutenant George H. Talcott was in charge of a major army arsenal in Augusta, Georgia, before the war and supervised the armament of defenses of the Pensacola harbor in Florida. Talcott's subordinate officers were Brevet First Lieutenant Franklin D. Callender and Brevet Second Lieutenant Jesse Lee Reno. Before the War, Callender was in command of Detroit Arsenal, Michigan, and Reno served as an Assistant Ordnance Officer at Watervliet Arsenal in New York.

First American Rocket Battery

The Howitzer and Rocket Battery consisted of 110 men, which were gathered for training at Fort Monroe in Virginia. Fort Monroe provided the battery with 50 2-in. war rockets and six 12-pounder mountain howitzers with carriages and ammunition. Small arms including rifles, carbines, and pistols were sent from the New York Arsenal. The battery was organized as a highly mobile unit. New light-barreled mountain howitzers weighed only 500 lb (230 kg) compared to 2300 lb (1040 kg) for a similar field howitzer.

> **ARTILLERY CALIBER**
>
> Before the second half of the 19th century, the artillery caliber was measured by the weight of spherical balls fitting the gun barrel, thus, 6-pounder, 12-pounder, etc.

The rocket section of the battery was armed with the Congreves and carried the new Hale rockets for service tests. Six portable stands with special troughs were available for launching rockets. The stands weighed only 20 lb (9 kg). Lieutenant Reno was usually in charge of rocketeers.

UNCOMMON SOLDIER

Alfred Mordecai (1804–1887) played an important role in introduction of rocket weapons into the American Army. Mordecai was born to a North Carolina Jewish family and he was a most uncommon soldier. Mordecai exemplified growing professionalism and modernization of the American army and introduction of new technology. He graduated first in the West Point class (of 35 cadets) of 1823.

When the Army Ordnance Department was reestablished by the act of 5 April 1832, Mordecai was appointed one of its first 10 captains. He directed fabrication of rockets for the first American rocket battery during the Mexican War. During his military career, Mordecai commanded three of the largest Army arsenals, in Washington, D.C., Frankfort, Pennsylvania, and Watervliet, New York. He also became well known for his original research on ammunition and ballistics.

Fig. 6.4. Alfred Mordecai, ca. 1855. Photo courtesy of the Library of Congress.

As a member of the Ordnance Board, appointed in 1839, Mordecai worked on unification of artillery systems. In 1840, he visited Europe with a commission studying artillery, ordnance, and arsenals in several countries. In 1849 Mordecai published a book, *Artillery for the United States Land Service.* This influential tract became the first full and accurate description of the artillery systems.

In 1855 Secretary of War Jefferson Davis sent Mordecai to Europe with the special commission to study and report on the latest military developments during the Crimean War. (Other commission members were Major Robert Delafield and Captain George B. McClellan.)

Mordecai retired to a private life in Philadelphia at the beginning of the Civil War, being torn between his ties to the "aged mother, brothers, and sisters in the South" (Virginia and North Carolina) and his allegiance to the U.S. Army and his son, who graduated West Point and fought for the Union. After the Civil War, Mordecai worked as an engineer on the construction of a transcontinental railroad from Vera Cruz and served secretary and treasurer of the Pennsylvania Railroad Company.

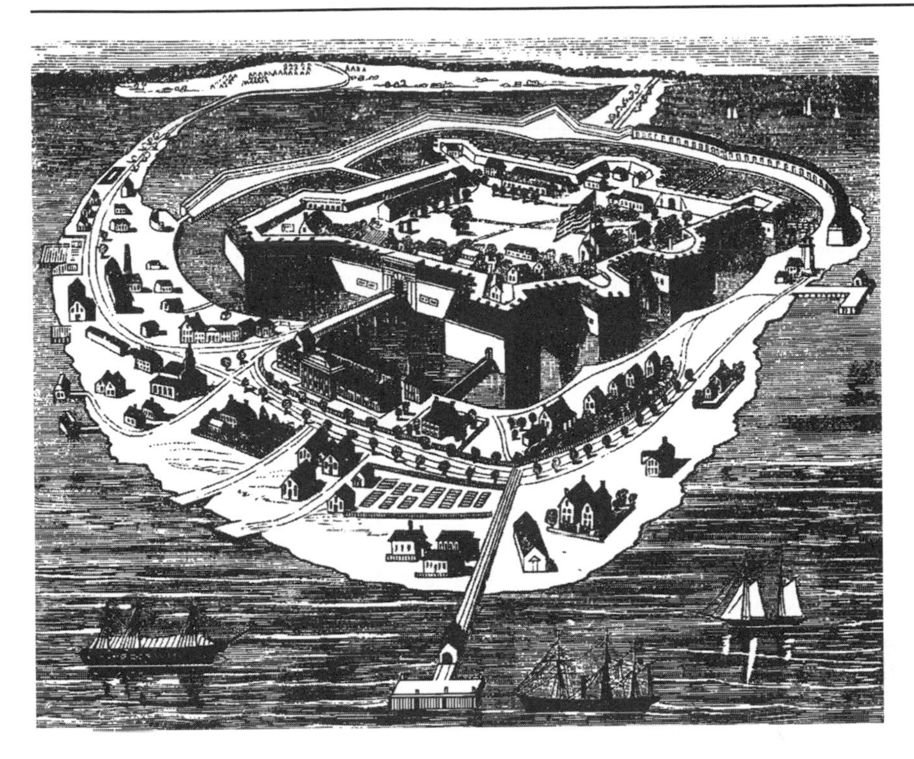

Fig.6.5. Fortress Monroe in 1861 at Old Point Comfort in Virginia, where the first rocket battery was assembled. Fort Monroe was among the largest American military posts. The fortress was established in 1819, and its construction cost two-and-a-half million dollars. It was probably the only fort with the proper range for artillery practice of any kind. The first artillery school, a "school of practice," was established in Fort Monroe in 1824. This was a school for entire units, not individuals, a new arrangement at the time. The artillery school was closed in 1835 when all of the students were sent to Florida during the Second Seminole War. Figure from *Pictorial History of the Civil War* (Lossing 1866, 499).

Rockets Rushed into Manufacturing

In the meantime the new rockets were rushed into manufacturing at Washington Arsenal under direction of Captain Mordecai. More than 2000 of 2 1/4-in. (57-mm) and 3 1/4-in. (83-mm) rockets had been made by 30 June 1847. The Ordnance Department had issued 1328 rockets to Talcott's battery by the fall of that year.

Mexican Congreves Fired First

On 1 February 1847 the Howitzer and Rocket Battery sailed from Fort Monroe for Vera Cruz with 105 men, six howitzers, rocket launchers, and the initial supply of rockets. On the evening of 9 March 1847, the battery landed near Vera Cruz with the main body of the General Winfield Scott's army. The siege of the town and the heavily fortified Castle of San Juan de Ulloa began. The Mexican troops were the *first* to fire their Congreves at the American soldiers constructing battery positions on 20 March 1847.

American Rocket Fire

Four days later, on the night of 24 March, the American rockets were fired for the first time in a major military operation. The rocket battery ran up in the darkness to a position close to Fort Santiago. A half an hour before midnight,

FIRST OFFICERS OF THE FIRST ROCKET UNIT

Three experienced officers of the Army Ordnance, all West Point graduates, were appointed to the new Howitzer and Rocket Battery. All three first American rocket officers would be brevetted for "gallantry and meritorious conduct" during the Mexican War, and all three would be wounded.

George H. Talcott, 1811(?)–1854, from New York, graduated 11th of the class (of 33 cadets) of 1831 in West Point. First Lieutenant Talcott was the first commander of the Howitzer and Rocket Battery. He was promoted to Captain of Ordnance on 3 March 1847 in the beginning of the campaign and appointed Major of Voltigeurs on 9 April after successful siege of Vera Cruz. Talcott commanded the Howitzer and Rocket Battery in the Battle of Cerro Gordo. For gallant and meritorious conduct in the Battle of Molino del Rey, where he was wounded, Talcott was breveted to Lieutenant Colonel on 8 September. After the War, Talcott commanded New York Ordnance Depot and Augusta Arsenal in Georgia. He died in 1854.

Franklin D. Callender, 1817(?)–1882, from New York, graduated 8th of the class (of 31 cadets) of 1839 in West Point. First Lieutenant of Ordnance Callender was one of the first three officers of the Howitzer and Rocket Battery. He was severely wounded (two wounds) in the Battle of Contreras, 19–20 August 1847. Callender was breveted Captain on 20 August for gallant and meritorious conduct in the Battles of Contreras and Churubusco. He fought in the Civil War and was breveted Brigadier-General of the U.S. Army in 1865. Callender died in 1882.

Jesse Lee Reno, 1823–1862, born in Virginia (appointed from Pennsylvania), graduated 8th (of 59 cadets) in the famous class of 1846 in West Point. Reno was one of the first three officers of the Howitzer and Rocket Battery and he was usually in charge of rocketeers. During the Mexican War, he received two brevets for gallant and meritorious conduct. In the Battle of Chapultepec, Reno was severely wounded. He became a major-general of U.S. Volunteers during the Civil War. In September 1862, Reno was engaged in the Battle of South Mountain near Sharpsburg, Md., three days before the Battle of Antietam. His 9th Corps repelled the attack of Confederates under General Lee, but Reno was killed "while gallantly leading his men" (Cullum 1868, Vol. 2, 146). He was 39 years old. The city of Reno in Nevada and Fort Reno (near El Reno) in Oklahoma were named after him. Reno's son, Jesse Wilford Reno, became a prominent engineer who invented a moving stairway or escalator, pontoon system for raising sunk submarines, and antisubmarine net.

Fig. 6.6. Jesse Lee Reno. Figure from *The National Cyclopædia of American Biography* (1897, Vol. 4, 103).

Fig. 6.7. Bombardment of Vera Cruz. Figure from *New History of the United States* (Lossing 1889, 491).

Stampede

about 40 hissing rockets were thrown into Vera Cruz. The rocket attack caused, as described by one of the General Scott's "indefatigable" engineers, Lieutenant George B. McClellan, "a stampede amongst the Mexicans — they fired escopettes and muskets from all parts of their walls" (McClellan 1917, 72). Both sides threw the rockets on that night. At midnight of 25 March, 10 more rockets, the Hales, were fired against Vera Cruz.

In about three weeks, the rockets were fired again, this time in the Battle of Cerro Gordo. The rocket battery was attached to the division of General Twiggs. The division marched along a rough pass in the supposedly impassable "rabbit-proof country," whereby the Mexican position could be turned. Army engineers Lieutenant Pierre G.T. Beauregard and Captain Robert E. Lee found this pass after several days of reconnaissance. On April 17, the rocket battery was deployed on the captured top of the hill La Atalaya, 700 yards in front of a key Mexican position on the opposing hill, El Telegrafo. (El Telegrafo got its name because of the semaphore that had formerly stood on the top of the hill.)

El Telegrafo, Contreras, and Chapultepek

At the break of day on April 18, artillery shell and about 30 screaming war rockets rained on the Mexican positions on El Telegrafo. The regulars of the Twiggs division dashed down the slope of La Atalaya and up El Telegrafo. The decisive American victory at Cerro Gordo opened the road to Mexico City.

MEXICAN ROCKETS CAPTURED

During the campaign of 1847, 36 Congreve rockets and 10 dozen signal rockets were captured from the Mexican Army.

On 19 August, the rockets were fired in the Battle of Contreras. Then on 12 September, the rocket battery took an active part in storming the fortress of Chapultepec, which forced the surrender of Mexico City. Lieutenants McClellan and Beauregard took temporarily charge of the rocketeers in these two engagements, respectively, when the battery officers, Callender and Reno, were wounded.

Rockets Battery Discharged from Service

In the summer of 1848 the Howitzer and Rocket Battery was discharged from the U.S. Army service. The experience with rockets in the Mexican War had not been exceedingly impressive. Rocket eccentricity in flight, instability, premature explosion, and deterioration in storage remained the main faults.

Prairie Artillery

The Army continued some experimentation, in particular with the newer Hale designs. The tests of the rocket company equipment were requested in 1849 in order "to understand practically the use" of these weapons. The Ordnance Department was pleased with the "lightness and facility of motion" of mountain howitzers and rocket units allowing their "use in countries without roads and by troops moving with any degree of celerity" (Benét 1880, 345). The establishing of "prairie artillery" was also considered, and some units carried rockets as a weapon. The records show, for example, that "a supply of light balls of

ROCKET DROOPING

Rockets were usually launched from 10-ft (3-m) troughs or tubes. The rocket moved slowly in the beginning of its flight, and it could significantly deviate from the desired direction after leaving the launcher with an insufficient velocity. Therefore, the launching tube was desired to be as along as practicably possible to reduce drooping.

It takes a certain time to build up full rocket thrust after ignition. The longer the delay is, the larger drooping. William Hale proposed a solution: a special contraption that mechanically held the rocket preventing its motion on the launching stand until thrust was sufficiently built up.

the caliber of the prairie howitzers and of Hale's rockets" was sent to the troops under command of General Brooke in Texas and New Mexico in 1850 (Benét 1880, 345).

Two sizes of Hale's rocket were in service of the U.S. Army before the Civil War, 2-in. (51-mm) and 3-in. (76-mm) rockets. The rocket size designated the exterior diameter of the cases. The missiles weighed 6 lb (2.7 kg) and 16 lb (7.3 kg), respectively. The range of rockets depended on the elevation angle of firing. The smaller 2-in. missile had a range (first graze) of 1200 yards (1100 m) when fired at 15 deg. The larger missile could fly 100–200 yards farther. At 47-deg elevation, the ranges were 1760 yards (1600 m or 1 mile) and 2200 yards (2000 m), respectively.

Army Rockets Before the Civil War

The rockets were made of sheet iron, lined with paper, or wood veneer. The case was charged solid, and the core was then bored out. The rocket head was of cast iron and could be either a solid shot, or a shell with a fuse communicating with the rocket. The recently fabricated rockets usually performed well. Storage of rockets, particularly under field conditions, remained a major problem limiting their usefulness for the Army. Rockets deteriorated and became unsafe and unfit for service after being kept for only several years.

Although rockets remained in the Army arsenals, the continuing improvements in artillery made the latter decisively superior to rockets.

CIVIL WAR

Rockets never achieved the prominence and were rarely used in the Civil War or the War Between the States of 1861–1865. Both sides of the conflict were well familiar with war rockets, and the knowledge of the basics of their performance, field use, and fabrication were required from ordnance officers. For example, a problem "Describe Hale's rocket" was typical at the examination for commission of lieutenants of ordnance in the army of the Confederate States of America. Both sides organized rocket units, but rockets were never used more than on a small scale.

Union's Rocket Battalion

The Union Army activated the Rocket Battalion in late 1861. One-and-a-half years later, it was renamed to 24th Independent Battery, New York Light Artillery, U.S. Volunteers. The battery was equipped with the launching equipment by March 1862 and went through training, including firing the rockets. After that the battery was armed with the conventional artillery pieces and never used rockets again.

The first attempt to organize a rocket battery in the Confederate Army in 1861 did not succeed. It actually led to a damaging clash between some leading Southern personalities. Another Confederate battery was organized in Texas in late 1863. The battery never received the required number of missiles nor had all the required personnel. It was finally disbanded in July 1864.

Confederate Rocket Batteries

The Confederate Field Manual (1862) for ordnance officers stated that "no regular organization of a rocket battery has been arranged" (*Field Manual* 1862, 107). The type of the rockets and the number of rockets and carriages were supposed to be determined by the character of the intended service of rocket units. During the War, the Confederate Arsenal in Richmond, Virginia, produced 3985 rockets, including signal and war rockets.

FIRST CONFEDERATE ROCKET BATTERY

After the Battle of Bull Run, General Pierre G.T. Beauregard and Ordnance Chief Josiah Gorgas took steps to organize a rocket battery in the Confederate Army. Both Beauregard and Gorgas became familiar with the rockets during the Mexican War. The rocket battery never materialized, being buried in the red tape. A bitter exchange between Beauregard and Acting Secretary of Army Judah P. Benjamin followed.

Jefferson Davis, 1808–1889

Pierre G.T. Beauregard, 1818–1893

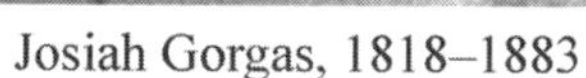

Josiah Gorgas, 1818–1883

Judah P. Benjamin, 1811–1884

Fig. 6.8. Leading personalities of the Confederacy including President Davis, Secretary of War Benjamin, Chief of Ordnance Gorgas, and General Beauregard, were involved in an ill-fated attempt to create a Confederate rocket battery. Davis, Beauregard, and Benjamin figures from *Pictorial History of the Civil War* (Lossing 1866, 252, 315, 232) and Gorgas figure from *The Long Arm of Lee* (Wise 1915, Vol. 1, p. 32).

The rocket battery incident aggravated the growing disagreements between Beauregard and President Jefferson Davis. The damage to their relations became irreparable. "The story of the first Confederate rocket battery which, while never organized, was actually more destructive in its effect than any other of the war" (Donnelly 1961, 79).

Rockets in Skirmishes

It seems that the first rockets of the war were fired in the Peninsula campaign in June and July of 1862. These were the rockets of the Horse Artillery battery of the troops under command of the Confederate General J.E.B. Stuart. The records also describe Confederate rockets employed south of the James river in the fall of 1862. Many rockets and launchers, or dischargers as the latter were called, of the Union Army were captured by the Confederates. Sometimes the Federals recaptured the rockets, as happened on 1 December 1862, near Franklin and the Blackwater river.

The rockets were fired on a number of occasions but usually small in numbers and without much effect. Some observers noted, however, that rockets were successful in a number of skirmishes, such as in chasing picket-boats and driving the enemy from the positions that could not be easily approached by artillery.

"experience ... is not ... favorable"

On the whole, the rocket performance remained unimpressive. George D. Ramsey, the U.S. Army Chief of Ordnance, wrote on 15 July 1864, to the Secretary of War that "experience with rocket batteries during this war is not at all favorable to their usefulness. The same number of men and horses can produce more effect with the improved cannon and projectiles now used. Rockets have but little range and accuracy compared to rifled projectiles, and are liable at times to premature explosions and great eccentricity of flight" (Donnelly 1961, 91).

After the Civil War, the war rockets were preserved for some time in the American army and then practically disappeared.

7. ROCKET PROLIFERATION — THE SECOND WAVE

The Royal Navy and the British Army carried Congreve rockets to all parts of the Empirc. Rockets were discharged in places such as far apart as Dalmatian coast in the Adriatic (supporting Montenegrans against French garrisons in 1813) and in Asia. Rockets flew on Borneo in 1845, at Lucknow in India in March 1858, and in China. In Africa, British missiles were fired in Sudan, Benin, and at the south of the continent. Many countries got interested in the new weapon.

British Empire

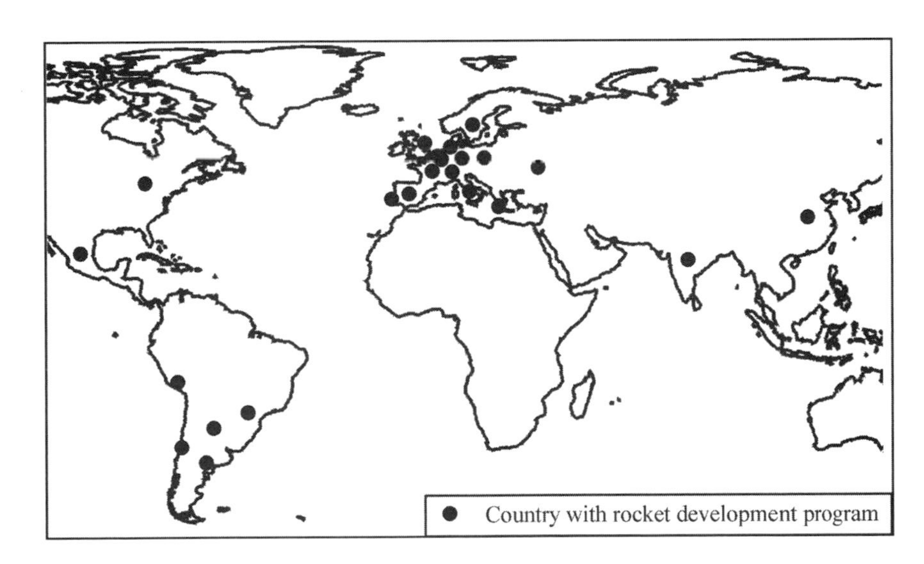

Fig. 7.1. Second wave of proliferation of rocket technology: countries with rocket establishments in the 19th century. In Europe, major rocket programs were under way in Great Britain, Austria, Denmark, France, Poland, Russia, Sweden, and Switzerland; smaller programs were conducted by Belgium, Germany, Greece, Hungary, Italy, the Netherlands, Portugal, and Spain. India and China used rockets in Asia. In Americas, besides the United States, Chile, Argentina, Brazil, Mexico, Paraguay, and Peru built or employed war rockets. Courtesy of Mike Gruntman.

ROCKETS DESTROY THE PIRATE'S NEST

On 18 August 1845 the fleet under Vice-Admiral T. Cochrane reached the Maluda Bay, at the northern extremity of Borneo. Up the shallow river was a stronghold of a notorious colony of pirates commanded by an Arab named Sheriff Osman. A small steamer, the *Vixen*, and 20 boats with 450 men went up the river to destroy the pirate's fort. There, the British were met with murderous fire. The fire was returned from the boats, and a rocket battery was quickly deployed on shore. The battery, under command of Lieutenant Paynter, threw the rockets into the fort with perfect accuracy. The marines stormed and captured the fort.

All Corners of the World

Denmark began the first experimenting with the rockets after the deadly attack on Copenhagen in 1807. France, United States, Russia, Austria, Sweden, and Switzerland were quick to join the development effort. With time, rockets were built and used by many other European countries. In South America, Chile pioneered rocketry in 1819 in a spectacular way and was followed by Argentina, Brazil, Paraguay, and Peru. The Mexican Army was armed with Congreves in 1840s. The second wave of proliferation of the rocket technology had begun.

The 19th-century rockets remained a weapon effective against large masses of poorly trained, undisciplined, and exposed troops and for incendiary purposes. They would quickly put on fire large areas and wooden buildings. Rockets could also be conveniently deployed in the areas with the limited access of artillery. These features explain the wide use of war rockets by Russia and Britain for many years, until the very end of the 19th century, in their colonial wars. Rocket

Fig. 7.2. Making war rockets by hydrostatic-driven process. Figure from *Illustrated London News*, Vol. 27, No. 740, p. 441, 28 April 1855.

FIRST ROCKETS IN SOUTH AMERICA

The Napoleonic wars ended, but the rockets were not forgotten. The British were quick to fire the Congreves already in 1816 against the pirates on the Barbary Coast of North Africa. Interestingly, the first country, after Great Britain, that fired its own rockets in a major military action was not in advanced Europe. It was Chile in South America.

In 1818, Chile was a young country fighting for independence against its former colonial master, Spain. Lord Thomas Cochrane was appointed Vice Admiral of Chile and Commander-in-Chief of the Naval Forces of the Republic. Cochrane was amongst the most dashing captains of the Royal Navy, and he was also famous for challenging his superiors and other authorities when they acted, in his opinion, without vigor and determination. After making a distinguished record in war with France, he offered his service to the young American republic.

The Spanish shipping concentrated at Callao (near Lima), Peru, and was an obvious target for the Chilean attack. Cochrane arranged the manufacture of more than 1000 Congreve rockets at the Valparaiso arsenal. Cochrane was familiar with this new weapon — it was he who devised and led the famous attack on the French fleet in 1809 at the Basque Roads, which employed 1200 rockets on fire-ships. The Chilean rockets were manufactured under supervision of the Brit Stephen Goldsack, a former assistant of William Congreve.

The Chilean squadron attacked the Spaniards at Callao on the night of 2 October 1819. Two special rocket rafts were formed of two tiers of large timber logs. Lieutenant-Colonel Charles, the commanding officer of the Chilean marines, was in charge of the rockets. At first, the war vessel *Araucano* went in and fired several rockets. After dark, the rocket rafts under Charles and Captain Hind's command were placed within 800 yards of the enemy batteries and opened fire.

The attack did not produce the desired results. Only one rocket in six functioned properly. Some rockets exploded, and some took the wrong direction because the guiding sticks were made of the knotty wood. As Cochrane later wrote (Cochrane 1859, 26, 27), "some [rockets], in consequence of the badness of the solder used, bursting ... before they left the raft, and setting fire to others." Cochrane charged that "... the fault lay in ... having ... [not] supplied [Mr. Goldsack] ... with proper workmen and material.... From scarcity and high price of spelter, ... [Mr.. Goldsack] had also been compelled to make use of inferior solder for the tubes." In addition, "the filling of the [rocket] tubes was, from motives of parsimony, entrusted [by the Chilean government] to Spanish prisoners, who ... had embraced every opportunity of inserting handfuls of sand, sawdust, and even manure, at intervals in the tubes, thus impeding the progress of combustion."

The rocket attack at Callao was attempted again on October 6. Later, the Chilean army and marines used rockets in a number of engagements on the ground. For example on 7 November 1819, the marines and soldiers landed at Pisco "for the purpose of procuring brandy for the use of the squadron." Rockets were effectively employed to disperse the opposing Spanish troops there. Rockets were also fired at the Island of Chiloe and in the valley of Mirabe in 1820 and 1821, respectively.

Argentina, Brazil, Peru, Paraguay, and Cuba experimented with rockets at some periods of their history. Argentina and Brazil fired rockets against each other at the Battle of Monte Caseros on 3 February 1852. Paraguay successfully manufactured efficient rockets during the War of Triple Alliance (1865–1870), while Mexico launched its Congreves at the American troops at Vera Cruz in 1847.

salvos in Benin in February 1897 and on the Nile river in Egypt and Sudan in April 1898 were among the last in British military operations.

Rocket Evolution

As the technology progressed, many innovations were brought into rocket design and manufacturing. The centrally mounted guiding stick was introduced by William Congreve in 1815. William Hale demonstrated rocket stabilization by rotation in 1840s and further advanced this approach during the following 20 years.

Hydrostatic Press

Rocket manufacturing has substantially improved. William Hale introduced hydrostatic pressure instead of ramming gunpowder by mallets and monkeys (falling weights). The press dramatically improved the safety of manufacturing. The old technology produced a cloud of the gunpowder dust that was injurious when inhaled. In addition, the monkey driving process occasionally ignited the rocket charges with disastrous consequences. In contrast, high hydrostatic pressure safely compressed the gunpowder in a stone-like formation. More gunpowder could thus be rammed into the rocket, and it would be less exposed to air and consequently more stable in storage and transportation.

Rocket technology widely proliferated throughout the world in the 19th century, reaching all continents, and many countries developed establishments that were manufacturing war rockets.

FRANCE

Napoleon's Commission

France was traditionally enamored with new scientific methods in the warfare and learned of the rocket military potential from the experience of its officers in India. The Congreves launched against French warships and garrisons between 1806 and 1809 prompted the action. On 8 February 1810, Napoleon created a special commission of artillerists and scientists to develop incendiary rockets. Several rockets were built and tested by the end of the year, with the largest missiles reaching distances up to 3000 m.

Fig. 7.3. Division General (Artillery) Louis Susane, 1810–1876, directed development of war rockets in France in the 19th century. Photo courtesy of Musée de l'Armée, Paris, France.

Manufacturing of the French rockets was established at first in Seville in Spain utilizing the high-quality local gunpowder. Some rockets demonstrated the range up to 4000 m. Spain was a battleground between the British and French troops at that time, and it is possible that both sides discharged rockets against each other at Cadiz in 1811. Another French Army workshop, in Toulon, also manufactured rockets.

Two years later, the French established production facilities in the German towns of Hamburg and Danzig (present day Gdansk, Poland). The workshop in Hamburg manufactured rockets based on the design of the Danish officer Andreas Schumacher. Lieutenant Schumacher of the Army Engineers became interested in rocketry after the devastating rocket bombardment of Copenhagen in 1807. With support of King Frederick, Schumacher initiated a successful rocket development in Denmark on a secluded island Hjelm. The French were not the only ones who benefited from spectacular achievements of Schumacher. The Danish rocket technology was also transferred several years later to the Austrians.

Andreas Schumacher

The French missile development came to a standstill after the Napoleon's defeat and was revived on a significant scale only in 1824. A major rocket manufacturing facility was then established in Metz on the basis of the Central School of Pyrotechnics. The Metz establishment grew with time, especially when Louis Susane took the command. Susane was probably the most important French rocketeer in the 19th century, becoming a general in 1857. Up to 20,000 rockets could be produced annually in Metz by the early 1850s, with some experimental missiles reaching distances of almost 7 km (4.4 miles).

School of Pyrotechnics in Metz

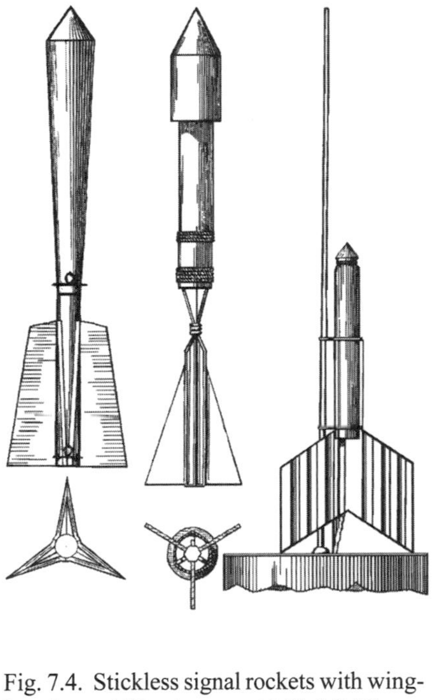

Fig. 7.4. Stickless signal rockets with winglets for stabilization, ca. 1866. Middle — French signal rocket; left and right — Russian signal rockets with wings made of cardboard material and wood, respectively. Figures from Skripchinskii (1866, 615–621).

French military campaigns in Algeria provided a test ground for the new weaponry from 1830 through late 1850s. For the first time war rockets were successfully employed in combat there at the battle of Staoueli. Many French units carried rockets in 1837–1844, but the troops were not always impressed by combat performance. Interestingly, Algeria would remain the center of testing of French missiles after World War II until mid-1960s.

French Rockets in Algeria

French war rockets reached the peak of their usage in the late 1850s and in early 1860s. The French army fired thousands of rockets during the Crimean War against the Russian Empire in 1853–1856. The army and the navy employed war rockets in combat in such faraway places as Algeria, Senegal, and China. General Susane culminated his development work by introducing a complete war rocket system that was adopted by the army.

Complete War Rocket System

French rockets went to Mexico in September of 1862, but only very few were fired in combat. Apparently the last time the French military fired a war rocket was in the battle of Guadalajara in 1864. One year later, the last war rocket

Fading Away

was discharged by the navy. Thus, this 19th-century chapter of the French rocketry was closed. The weapon would come back during World War I when *Le Prieur* missiles would be launched from French fighter airplanes at Verdun.

AUSTRIA

The manufacturing of the Austrian rockets of the Congreve type was first undertaken in 1808 under direction of the chief fireworks-master. The first missiles were tested on 24 March of that year. The development work was however soon suspended because of the war that rapidly engulfed the central Europe.

Vincenz von Augustine

Production of Austrian rockets resumed in 1815 when an enthusiastic young officer Vincenz von Augustin established the rocket works at Wiener-Neustadt not far from Vienna. Two years earlier Major Augustin was directed by his superior, Crown Prince von Schweden, to look for opportunities to learn the theoretical and practical aspects of Congreve rockets during the military campaigns of 1813 and 1814. At the end of 1814, Augustine made an acquaintance with the leading Danish rocketeer Schumacher.

With Royal Permission

The interest of the army in rockets was supported at the very top of the Austrian Empire. King of Denmark agreed to satisfy the wish of the Emperor and gave the permission to reveal rocket secrets to Augustine. Major Augustine

Fig. 7.5. Volley launch of Congreve rockets. Accuracy is a secondary consideration when the target is of considerable magnitude, such as a dense mass of infantry or cavalry. A ground volley is called for in such a situation. A large number of rockets, up to several hundred, lay on the ground at certain distances from each other. The rockets are connected by a quickmatch and fired as a volley. The first 10–150 yards the rockets proceed near the ground, then they could rise, laterally deflected, and saturate the area target. One can imagine the efficiency of such a volley against an infantry or cavalry formation. Figure from *Projectile Weapons of War* (Scoffern 1859, 174).

ROCKETS FORCE OPENING THE PARANÁ RIVER

An armed conflict between the Argentine Republic and the neighboring Banda Oriental (Uruguay) raged in 1845. The Argentinians closed navigation on the Paraná river. In November of that year however, the combined British–French squadron broke the passage through the river for merchant ships.

In the spring of 1846, the loaded merchant vessels gathered for the convoy back at a point 110 miles (177 km) up the river, beyond the reach of the Argentinian posts. In the meantime the Argentinians heavily fortified high cliffs at one area on the river bank near San Lorenzo. This formidable fort blocked the way down Paraná to the Atlantic Ocean.

Fig. 7.6. The Paraná expedition: the rockets in action at San Lorenzo on 4 June 1846. Figure courtesy of the Royal United Services Institute, London, from *The Royal Marine Artillery* (Fraser and Carr-Laughton 1930, 357).

After careful reconnoitering, a party of British rocketeers under Lieutenant Mackinnon landed under cover of darkness on a low, flat island facing the fortified cliffs at the opposite side of the river. The group carried 24-pounder and 12-pounder rockets with three launching tubes of each type. The next day the group stayed hidden on the island because an unfavorable wind prevented the convoy to start off down the river.

The weather changed for the better next day, 4 June 1846, and the rocketeers were ready: rocket stands were erected and the guiding sticks were screwed in the Congreves. When the convoy of merchant vessels approached the fortified cliffs, the battery opened up, creating panic and confusion among the Argentine troops. One of the British rockets hit an Argentine two-wheel ammunition cart setting off an explosion. At one moment the firing had to be suspended because the surrounding grass and reeds caught fire from the backfire of the rockets.

The fire from the war-steamers and the rocket battery allowed merchant vessels to pass by the enemy positions without loss. After expending its ammunition, the rocketeers withdrew to the other side of the island, embarked on their boats, and safely joined the convoy.

returned back to Vienna on 15 March 1815, to build the Austrian rocket program. In the same year two first Austrian rocket batteries were sent to participate in the siege of Huningen. When Augustin and his rocketeers joined the army, the town had already been taken.

After return from this expedition, the new machine for loading propellant was invented, and fabrication of improved rockets was started in 1816. The Emperor personally inspected rocket fabrication on 13 July 1817 and witnessed the demonstration in which rockets exceeded artillery pieces in accuracy. The impressed Emperor awarded the Order of Leopold to Major Augustin on that occasion.

Rise and Fall of the Royal Imperial Rocket Corps

The Austrian Rocket Corps quickly grew into a large establishment. On 14 November 1817, the rocketeers were reorganized into an independent unit of the Austrian Army. The Wiener–Neustadt rocket establishment became one of the largest and best equipped in Europe, and von Augustin was promoted to a general.

The Royal Imperial Rocket Corps distinguished itself in numerous military actions in Hungary and Italy in 1849. The rockets were efficient and could be conveniently deployed in the areas with limited access and poor roads. The commander of the Austrian artillery in the Hungarian Campaign, Field-Marshal-Lieutenant Franz von Hauslab, described the specifics of rocket weapon:

Weapon Against Cavalry

> Rockets cannot and should not replace the artillery pieces. Cannons are definitely superior to rockets in accuracy and force of the projectiles. However, the rockets are better in accuracy then howitzers. War rockets when employed in an open field provide a great advantage since cavalry cannot stand rocket fire. During the Hungarian Campaign [of 1849], [enemy] cavalry always fled wherever rockets were used against them. Therefore, rockets not only could often gainfully replace and complement artillery pieces, but they are very efficient against cavalry .... (*Ob Upotreblenii ...* 1854, 55)

End of a Glorious Chapter

The performance of rockets in the war with Prussia in 1866 was below expectations, and the Austrian Rocket Corps was disbanded next year.

RUSSIA

Beginnings of Russian Rocketry

The Russian Empire began its war rocket development in 1810. Four years later military engineer Kartmazov of the Military Study Committee demonstrated incendiary and explosive rockets, 2 in. and 3 1/2 in. in diameter. The range of the larger rocket exceeded 3000 yards. The successful testing of the Kartmazov's rockets led the Russian War Ministry to introduce rockets into army service in 1817.

Approximately at the same time and independently of Kartmazov, an

ADAPTATION OF TECHNOLOGY

I sought to discover how rockets might be used for incendiary purposes, and although I never had an opportunity to see, much less to obtain information as to how the English manage so to use them in war, I nonetheless thought that what they claim as such an extraordinary and important discovery is nothing other than a properly adapted conventional rocket.

Alexander Zasyadko, 1817
(Sokol'skii 1967, 24)

Fig. 7.7. Russian troops bombard Fortress Varna (modern Bulgaria) by rockets in the Russo–Turkish War in 1828. The war was sparked by Greeks' struggle for independence and led to significant Russian advances in Caucasus and on the eastern shore of the Black Sea. Figure courtesy of the Military-Historical Museum of Artillery, Engineers and Communications, St. Petersburg (Leningrad), Russia.

Fig. 7.8. Lieutenant-General Alexander D. Zasyadko, 1779–1837. Figure courtesy of the Military-Historical Museum of Artillery, Engineers and Communications, St. Petersburg (Leningrad), Russia.

artillery officer Alexander Zasyadko began development of his rockets. By the early 1817 Zasyadko had produced incendiary rockets with the diameters of 2 1/2 in. and 4 in. with the range similar to that of the Kartmazov's rockets. Both types of rockets had a laterally attached guiding stick. In 1823, the Okhtensk Gunpowder Works in St. Petersburg hired a British technician, Massingbird-Turner, to introduce manufacturing of rockets with centrally attached sticks.

Rockets in Caucasus and Russo–Turkish War

The first war rockets sent to the Caucasus for service tests in military operations proved however a failure. The rockets rapidly deteriorated after transportation on great distances. In contrast, the rockets assembled locally performed much better. The year 1827 saw the first Russian missiles fired in the battles of Ushagan, near Algoz, and against the Ardebil Fortress.

Alexander Zasyadko became a general and head of the Mikhailovskoye Artillery School. In 1827, he succeeded in establishing the first rocket company in the Russian Army.

War rockets were widely employed against fortresses in the Russo–Turkish War of 1828–1829. A total of 1191 rockets, 380 incendiary and 811 shell, were fired by the Russian army in the campaign of 1828. Russian rocketeers distinguished themselves in storming Varna (in present-day Bulgaria) and

Rockets on Flatboats

in a river battle on the Danube near Silistria. Flatboats armed with rocket launchers quickly dispersed the Turkish flotilla in the latter engagement. The rocket flatboats

Fig. 7.9. Engineer-General Karl A. Shil'der, 1785–1854. Figure courtesy of the Military-Historical Museum of Artillery, Engineers and Communications, St. Petersburg (Leningrad), Russia.

SHIL'DER'S MISSILE SUBMARINE

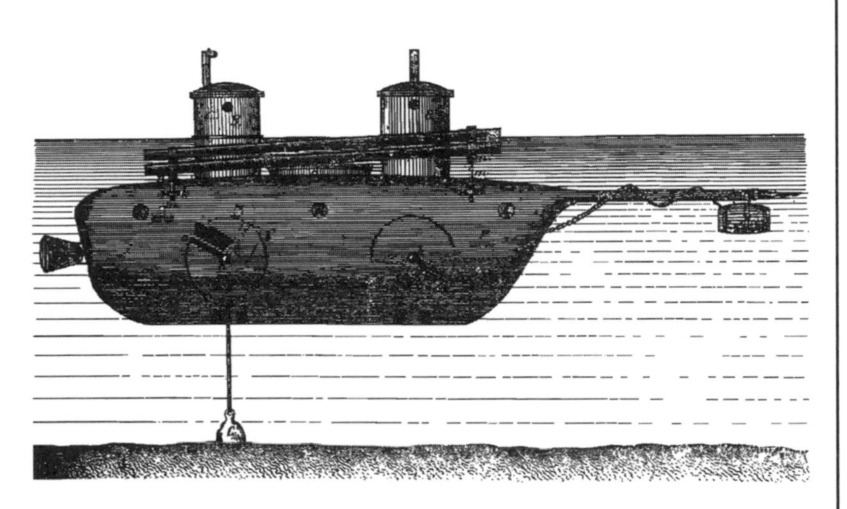

Fig. 7.10. Submarine built by Karl Shil'der in 1834. The submarine was 20 ft long, 5 ft wide (in the widest part), and 6 ft high with the 10-man crew. Two sealed tubes, 2 3/4 ft in diameter served for access to the submarine. A simple periscope allowed observation from the submerged state. Submarine could submerge down to 40 ft. Two rocket stands streched along the top submarine surface could fire three rockets each from under water. The rockets were protected from the sea water by special seals. After being galvanically ignited, the rocket pushed out the seal and left the stand. Figure courtesy of the Military-Historical Museum of Artillery, Engineers and Communications, St. Petersburg (Leningrad), Russia.

Fig. 7.11. Shil'der's missile submarine was transported by a special barge also armed with rockets. Figure courtesy of the Military-Historical Museum of Artillery, Engineers and Communications, St. Petersburg (Leningrad), Russia.

were built by General Karl Shil'der, another important contributor to the Russian rocket technology.

General Shil'der was a most innovative military engineer. In May 1834, he designed and built the first ironclad submarine equipped for rocket launching from under water. The Russian Emperor personally approved this project and followed its progress.

Rocket Submarine

The submarine was tested before the Emperor on 29 August 1834. The rockets were ignited by an electric discharge of chemical batteries, and the test program included missile launches from the submerged position. The tests were apparently successful because the Emperor awarded large monetary awards to the submarine crew four days later. The second experimental rocket submarine had been built by December of 1834. The fate of this improved vessel is not known because the development program was so secret that only very few documents were preserved.

Fig. 7.12. Lieutenant-General Konstantin I. Konstantinov, 1817 (or 1819) – 1871. Figure courtesy of the Military-Historical Museum of Artillery, Engineers and Communications, St. Petersburg (Leningrad), Russia.

Scientific Method of Rocket Design and Testing

Another outstanding individual that significantly contributed to rocketry in Russia was General Konstantin I. Konstantinov. Konstantinov introduced scientific methods in design and testing of the rockets, published almost 50 articles on various aspects of rocketry, and several books on rocket engineering. He became the leader of Russian rocketeers and significantly improved the quality of Russian war rockets.

Rockets in Colonial Wars

The Russian army widely employed rockets in the colonial wars in Turkestan (modern Kazakhstan, Kirghyzstan, Tajikistan, Turkmenistan, and Uzbekistan) until the end of the 19th century. The first experience in Turkestan showed the advantages of transporting rockets on camels, and the special rocket boxes that could be mounted on both one-hump (dromedary) and two-hump (Bactrian) camels were produced.

In addition to the campaign in Turkestan, Russian war rockets were regularly fired in the Caucasus and in the Balkans, particularly in the

DEVOTED ROCKETEER

"Commander of the Rocket Establishment [in St. Petersburg], Colonel Konstantinov, has all necessary information and personal abilities in order to advance rocket technology and he is engaged in this activity with love, the quality absolutely indispensable for achieving the goals [of the Rocket Establishment]."

Report to the Artillery Department of the Military-Scientific Committee, 1853 (*Ob Upotreblenii* ... 1854, 56)

FIRST ROCKETS NEAR TYURATAM (BAIKONUR)

It was not in the 1950s when the semidesert near the first Soviet space port, the cosmodrome, Tyuratam (Baikonur) witnessed first rockets. One hundred years earlier, in June of 1853, a rocket unit of the Russian Army under Second Lieutenant Iogansen passed by Tyuratam on the way to Ak-Mechet', 130 miles (200 km) up the Syr-Dar'ya river. The Russian Army was fighting to subdue the Kokand Khanate (the rulers of the parts of the present day Kazakhstan and Uzbekistan), and General V.A. Perovskii besieged the fortress Ak-Mechet'.

Iogansen's rocketeers fired 200 rockets during the fighting in July 1853, when the fortress was captured by the Russians. Then in December 1853, 84 Russian rockets were discharged to repel a counter attack by the Kokand troops. Ak-Mechet' was renamed *Perovsk* and later, in the Soviet times, *Kzyl-Orda* (means *Red-Camp*). Tyuratam was administratively in the Kzyl-Orda district, and for long time the cosmodrome's mailing address (P.O. Box) was *Kzyl-Orda-50.*

During the siege of Ak-Mechet', Russian missiles killed more than 100 Kokand soldiers. A military report (*Ob Upotreblenii* ... 1854, 10) described that "on one day the Kokand troopers were gathered in numbers in a mosque. A rocket with a shell warhead penetrated the roof and exploded in the crowd, killing five persons and wounding many. Generally the prisoners told [to the Russians] that when they heard the sound of a flying rocket, they tried to hide. When asked how did they like these new [rocket] projectiles that produced so much noise in flight, the Kokand prisoners replied, 'O, these were real fiery devils!'" Second Lieutenant Iogansen was awarded a military order for his performance and received one-year salary as a bonus. A few of his enlisted rocketeers were also awarded medals.

Fig. 7.13. "In [Central] Asia, everything is transported by camels; there are no [even elementary] roads in many places and therefore transport of artillery is exceptionally difficult ... . On the contrary, rockets of various sizes, together with launching stands, could be easily transported on camels ... . On the basis of the information [provided by the Commander of the Arsenal of the Orenburg Military District], the St. Petersburg Rocket Establishment began development of two special rocket boxes, for 16 rockets each, convenient in size and weight for mounting on camels" (*Ob Upotreblenii* ... 1854, 11). Figure from *Ob Upotreblenii* ... (1854, 26).

Russo–Turkish War of 1877–1878. Small weight, simplicity, and service convenience of rockets were an important asset in the warfare in difficult mountainous and desert terrains with very few roads. In many areas the artillery ordnance quickly deteriorated in harsh climates. In addition, the efficiency of rocket fire against native cavalry unaccustomed even to gunshots provided another advantage. Many fortifications in Turkestan were made of the soft material — dried clay — and were thus easily broached by the rockets. A small number of soldiers needed for servicing rockets offered another advantage in expeditions through inhospitable land, when everything, including food and water, had to be carried by the troops.

End of Another Glorious Chapter

But this 19th-century chapter of rocketry also came to an end in January 1886 when the Russian Artillery Committee ordered the production of war rockets to stop.

CRIMEAN WAR

Rocket Fire Exchange

War rockets saw the action on a large scale in the Crimean War, 1853–1856. This clash between the Russian Empire on one side and Great Britain, France, Ottoman Sultan, and Sardinia-Piedmont on the other arose from the conflict of great powers in the Middle East and Balkans. It was not the first time though that both sides of a conflict employed rockets against each other. The first time two major opposing armies, American and Mexican, fired rockets at each other was

Fig. 7.14. British attack on the town of Gheisk in the Sea of Azov on 5 November 1885, during the Crimean War. The landing parties succeeded, with the rocket support from small gunboats, to burn "the extraordinary quantity of corn, rye, hay, wood, and other supplies, so necessary for the existence of Russian armies, both in the Caucasus and the Crimea." Captain Osborn of the Royal Navy, who commanded the expedition, reported that "these vast stores should have been collected here, so close to sea, whilst we were still in the neighborhood, only can be accounted for by their supposing that they could not be reached by us" (*Illustrated London News*, Vol. 27, No. 775, p. 716, 22 December 1855). This was the second attack on Gheisk and the last engagement in the operations in the Sea of Azov. The small boats armed with rockets demonstrated that they could achieve the objectives impossible for larger ships in shallow waters. Figure from *Illustrated London News*, Vol. 27, No. 775, p. 716, 22 December 1855.

Fig. 7.15. Small boats carrying rockets became a powerful weapon in the shallow waters and among the rocks inaccessible for larger ships. A boat from *HMS Harrier* discharges rockets at Russian shipping near Nystad (Baltic Sea) on 7 July 1855, during the Crimean War. Figure from *Illustrated London News*, Vol. 27, No. 755, p. 716, 4 August 1855.

Rockets on the Land and on the Sea

at Vera Cruz in 1847. Five years later, Argentina and Brazil exchanged rocket fire in the battle of Monte Caseros in 1852.

The time had come for a conflict in Europe. France and Great Britain, now allies, landed troops in the Russian Crimea, where they were joined by the Turkish soldiers. With time, the hostilities expanded to the Baltic Sea.

Transportation and Storage

Allied troops brought numerous war rockets employed in a year-long siege of Sevastopol, the main Russian fortress at the Crimea and the home of the Russian Black Sea Fleet. The French School of Pyrotechnics manufactured rockets with the range 5–7 km (3.1–4.3 miles). The French Minister of War reported that 7000–8000 war rockets were delivered to the Crimea from Marseilles without a single accident.

The Russian Army returned rocket fire with their own rockets, though much fewer in numbers. In addition, Russian rockets suffered from deterioration during transportation and storage. Army regulations of 1854 required (Konstantinov, 1867) that "rockets stored for two years in depots be considered potentially unreliable. Before release to troops, such rockets must be tested, 10% from each set, and if a single rocket explodes at the launching stand, the whole set should be considered unreliable and the rockets should be destroyed." (To address the deterioration problem, the Russian Army invested a major effort into determining the best way of fabricating rockets. In particular, gunpowder composition and ramming the powder into rocket cases in dry and wet forms were experimentally studied. In addition, to avoid long transportation, rocket manufacturing was relocated in 1860s from St. Petersburg to a Black Sea port Nikolaev, close to the army fighting with Turkey.)

Ramming Dry and Wet Powder

Sevastopol

Most of the French rockets were launched by the army units against the Russian fortifications and troops in the besieged Sevastopol, while the Royal Navy excelled in firing missiles from small vessels in shallow waters. The companies of the British Royal Marine Artillery widely employed rockets on shore, particularly repelling Russian attacks at Yevpatoria on 14 November 1954 and 16 February 1855. The First Rocket Troop of the Royal Horse Artillery saw action in affairs at the heights of Bulganak and at Mackenzie's Farm and took part in battles of Alma, Balaklava, and Inkerman in 1854. The next year the rocket troop participated in the siege of Sevastopol.

Coastal Areas Attacked

The attacks on the coastal areas in the Sea of Azov disrupted Russian transport and communications and led to destruction of immense supplies of grain, forage, and timber accumulated to support Russian military operations. Small boats could approach close to the shore and provide fire support to landing parties. Rocket bombardment set on fire storehouses in heavily defended Taganrog on 3 June 1855. Berdyansk, Yenitchi, Mariupol, and Gheisk were attacked that year. The rocketeers bombarded Gheisk the second time on 6 November in the last landing of the Azov campaign.

Rockets on the Baltics

The conflict has eventually expanded to the northern part of Europe, where British gunboats successfully employed rockets against Russian shipping in the Baltic Sea. The Baltic campaign culminated when the British flotilla bombarded a major Russian fortress Sweaborg, covering the deep-water approaches to Helsingfors (modern Helsinki, Finland) between 9 and 11 August 1855. Rocket boats discharged numerous 24-pounder Congreves helping to expand a fire on shore that was caused by the bombardment and led to explosion of Sweaborg's main magazine.

ARTILLERY DECISIVELY WINS THE COMPETITION

Many countries experimented with the fabrication of rockets or purchased the Congreves from Britain. However, the fate of war rockets was sealed in the middle of the 19th century by the substantial improvement in accuracy, firing range, and discharge rate of artillery pieces. Several important technological advancements played a critical role in this development.

Rifled Barrels

Introduction of rifled barrels in 1860s led to significantly improved accuracy. The special spiral groves were made in the barrels that caused the artillery projectile to spin; its flight was much more stable and cancelled or mitigated many undesirable effects of imperfect projectiles.

Fig. 7.16. Rifled barrel spins up the projectile. Photo courtesy of Mike Gruntman.

Breach Loading

The practicality of breach loading was also demonstrated in 1860s. Consequently, the shooting rate of artillery pieces increased dramatically. At the same time, the new technology of fabricating stronger heat-resistant cannon barrels by the Bessemer steel process was introduced together with forging of the barrels from wrought iron and later steel. Much stronger barrels allowed a significant increase of pressure in the barrel, which correspondingly increased the range of artillery projectiles.

Bessemer Process

Steam-Powered Ironclad Ships

War rockets were especially effective against wooden ships and their sails. In another major technological development, new steam-powered ironclad ships were protected by armor that made war rockets ineffective.

Artillery decisively won the competition. And the rockets? Well, rockets were still used for signaling and battle field illumination. Other applications included entertainment (fireworks). The whaling industry was especially active in rocketry with the rocket-propelled harpoons offering a number of important advantages.

Whaling Industry

William Congreve was among the first to propose rockets for harpooning whales. He and Lieutenant of Royal Artillery James N. Colquhoun obtained a British Patent "Application of Rockets to the Destruction and Capture of Whales, &c." in 1821.

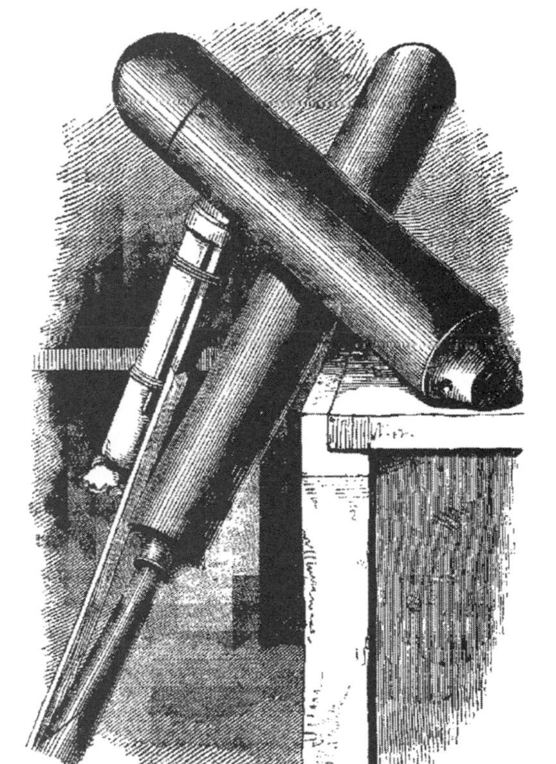

Fig. 7.17. Nineteenth-century rockets: Congreve, Hale, and skyrocket. Figure from *Projectile Weapons of War* (Scoffern 1859, 178).

Colquhoun and Congreve proposed

> using the rocket against the whale or other fish or other animal, whether it be used merely as an auxiliary to the present harpoons or other means of entanglement or capture, or whether the rocket ... be made to act as the harpoon itself, and to convey the line and fasten it to the fish by means of barbs or other such contrivance fixed to the rocket, and driven into the animal by projectile force of the weapon ... (Congreve and Colquhoun 1821, 2).

Harpoon Gun

Whaling rockets were commercially produced by a number of companies and available until the end of the 19th century. The rockets were subsequently replaced by more efficient harpoon guns introduced by the Norwegian whale master Svend Foyn. Foyn developed an efficient and convenient gun for launching harpoons with an explosive charge that killed the whales.

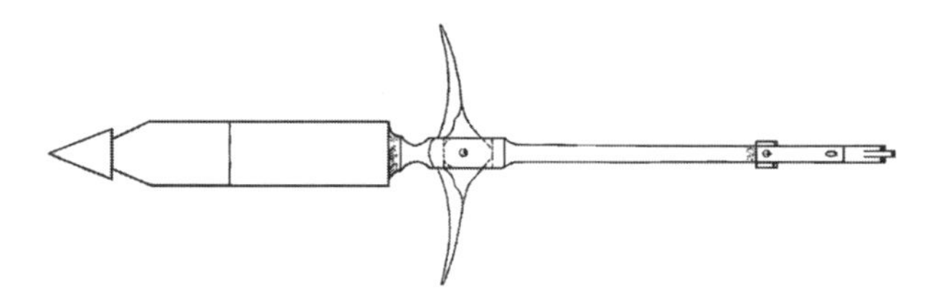

Fig. 7.18. Rocket-propelled harpoon invented by the famous American whaler Captain Thomas W. Roys in 1860–1861. The rocket harpoon was discharged from a "shoulder-gun" and its "chief use ... [was] to shoot bomb-harpoon for the purpose of killing whales" (Roys 1861, 1). Figure from Roys (1861).

Life-Saving Rockets in the 19th Century

Rockets not only hunted the whales but also saved numerous lives. The lifesaving missiles were available of both shore-to-ship and ship-to-shore types. Thus, the rockets carried the lifelines to shipwrecked seamen in the 19th century.

A few dozen of coastal stations were established along the English coast and equipped with the line-carrying rockets. The range of lifesaving rockets was improved dramatically in 1855 after General Edward M. Boxer of Royal Artillery introduced a rocket consisting of two cases. The Boxer's rocket was not exactly a two-stage rocket. The second case was ignited when the first case burned its propellant out, but the first case was not dropped off.

Thus, the fascinating chapter of the 19th-century rocket history was closed.

8. PUBLIC IMAGINATION ON FIRE

With the military interest limited to signal rockets, rocketry became confined to the whaling industry, lifesaving operations near the coast, and occasional spectacular fireworks. The men of plume replaced the men of sword as the keepers of the public interest in rocketry and spaceflight.

Fiction writers began to frequently sent their characters on journeys through space. The moon was an obvious and "highly visible" destination, but the planets were on the itineraries as well. Other worlds always attracted imagination of man. These flights of fancy were bold thoughts that required discarding a traditional point of view that the Earth was in the center of the universe. Discussing other worlds was for centuries considered heretical by many in positions of authority, and such ideas were practically inconceivable until the time of the Renaissance.

Space Travel in (Science) Fiction

Space travel could be found in fiction writings of many authors. The ancient Greek rhetorician and pamphleteer Lucian of Samosata (A.D. 120 – after 180) published the novels of travel to the Moon, *True History* and *Icaro-Menippus*. The Italian poet Ludovico Ariosto (1474–1533) also sent his main character to the Moon in *Orlando Furioso*.

Fiction Story by Johannes Kepler

Johannes Kepler (1571–1630) — yes, the same Kepler of the famous Kepler's laws of orbital mechanics — also wrote a science fiction story, *Somnium*, that was published after his death. The historian and Bishop of Hereford, Frances Godwin (1562–1633), wrote the first English space travel story *The Man in the Moone*. Godwin published this book under a pseudonym of Domingo Gonsales.

Another English Bishop and Oliver Cromwell's brother-in-law, John Wilkins (1614–1672), wrote a book *The Discovery of a World in the Moon* that inspired space enthusiasts for generations. This book played an exceptionally important role in defense of the new Copernican astronomy that dethroned the Earth and put the sun in the center. Wilkins later became the founding president of a group dedicated to physico-mathematical experimental learning. This group would be incorporated two years later as the venerable Royal Society.

JOHN WILKINS AND AMERICAN INTERPLANETARY SOCIETY

The first public activity of the just-founded (in 1930) American Interplanetary Society was a ceremony at which the famous polar explorer Captain Sir Hubert Wilkins presented the Society a copy of one of the first books on interplanetary travel written by his ancestor, John Wilkins, in the 17th century. The book was actually found and bought by the Society's president David Lasser in a secondhand bookstore.

Space Travel Fiction and New Astronomy

The French satirist and dramatist Cyrano de Bergerac (1619–1655) published stories describing journeys to the Moon and Sun. His novel *The Comical History of the States and Empires of the Moon* contributed to the debate on the new astronomy and drew on the work of Wilkins.

Many authors published (science) fiction stories about space travel. The list of those who tried the pen in this genre included the English novelist and journalist Daniel Defoe (1660–1731) and the American writer and poet Edgar Allan Poe (1809–1849).

Fig. 8.1. "The approach to Earth. The velocity became inconceivably great, but the increased rate of motion was in no way perceptible; there was nothing to disturb the equilibrium of the car in which they were making the aerial adventure... . The glowing expanse of the earth's disc ... ." Jules Verne, *Off on a Comet*. From *Works of Jules Verne* (Horne 1911, Vol. 9, 272).

Fascination with Science

Science fiction flourished in the 19th century. The seemingly unstoppable development of science, new machines and technological processes of the industrial revolution, accessibility of the printed word to wide portions of the populations, all of these factors contributed to public fascination with science fiction and especially with spaceflight.

Nobody however captured the imagination of the

Fig. 8.2. "The future express. 'Yes, gentlemen,' continued the orator, 'in spite of the opinion of certain narrow-minded people, who would shut up the human race upon this globe, as within some magic circle which it must never outstep, we shall one day travel to the moon, the planets, and the stars with the same facility, rapidity, and certainty as we now make the voyage from Liverpool to New York!'" Jules Verne, *From the Earth to the Moon*. From *Works of Jules Verne* (Horne 1911, Vol. 3, frontispiece).

Jules Verne

Fig. 8.3. Jules Verne, 1828–1904. Photo from *Les Enfants du Capitain Grant* (Verne 1906, frontispiece).

public more than the French writer Jules Verne. Verne often sent his characters on the trips around the globe and beyond. They were surrounded by the incredible, for the time, high-tech gadgetry and traveled on submarines, air balloons, and elephants. They even flew to the comets and explored the center of the Earth. The spirit of exploration and fascination with the power of science reigned. Jules Verne foresaw many scientific developments and numerous inventions, and his writings were enormously popular throughout the world and laid the foundations of the modern science fiction.

Foundations of Modern Science Fiction

Jules Verne's classic space novel *De la Terre a la Lune* (*From the Earth to the Moon*) became a seminal work on spaceflight. The book was published in 1865, when the bloody Civil War was still fought in the United States. The writer

Fig. 8.4. "Fire! An appalling, unearthly report followed instantly, such as can be compared to nothing whatever known, not even to the roar of thunder, or the blast of volcanic explosions! No words can convey the slightest idea of the terrific sound! An immense spout of fire shot up from the bowels of the earth from a crater. The earth heaved up and with great difficulty some few spectators obtained a momentary glimpse of the projectile victoriously cleaving the air in the midst of the fiery vapors!" Jules Verne, *From the Earth to the Moon*. From *Works of Jules Verne* (Horne 1911, Vol. 3, 144).

Fig. 8.5. Edward Everett Hale, 1822–1909. Photo from *The Life and Letters of Edward Everett Hale* (Hale 1917, Vol. 2, frontispiece). By permission of Little, Brown and Company, Inc.

looked beyond it and dreamed about other applications of the awesome technology perfected during the war. The book described how the civic-minded Baltimore Gun Club built a huge cannon and launched a manned projectile to the Moon.

De la Terre a la Lune

Jules Verne's *From the Earth to the Moon* ignited the imagination of the generations of future scientists and engineers. The book inspired those special teenagers that would become the visionaries and great pioneers of the space age: Konstantin Tsiolkovsky, Robert Esnault-Pelterie, Robert Goddard, and Hermann Oberth. The leading rocketeers of the next generation, Wernher von Braun, Sergei P. Korolev, Valentin P. Glushko, and many, many others were also fascinated by Jules Verne's novels. They would lead the exploration and conquering of the unknown.

Inspiration for Great Pioneers

Science fiction novels sent their characters on space travel for various reasons: satisfying curiosity, as a bet, or escaping debts. Nobody thought however about practical use of space flight as an aid to everyday needs of the people. The situation was about to change in 1870, when Edward Everett Hale, a most inventive American author, published a story entitled *The Brick Moon* in the *Atlantic Monthly*.

First Applications Satellite

In Hale's story, a huge water-powered flywheel flung an artificial satellite into orbit along the Greenwich meridian. As our natural moon allowed

Navigation Aid

> **DISTINGUISHED CARREER**
>
> Dr. Edward Everett Hale was well known for his writings and political activities. He was a prominent abolitionist. During the Civil War, Hale published a novel, *The Man Without a Country*, that did more to stimulate Northern patriotism than any other wartime writing. He authored many inspirational quotes, popular to this day among the Boy Scouts. Hale's works, in 10 volumes, appeared in 1898–1900. In 1903 Reverend Hale was named chaplain of the U.S. Senate.

> **A RING IN THE SKY**
>
> Now if any one would build a good tall tower at Greenwich, straight into sky, — say a hundred miles into the sky, — ... and could see it, we could tell how far east or west we were by measuring the apparent height of the tower above the horizon ... But nobody will build such a tower at Greenwich, or elsewhere... .
>
> ...If ... there were a ring like Saturn's which stretched round the world, above Greenwich and the meridian of Greenwich, and if it would stay above Greenwich, turning with the world, any one who wanted to measure his longitude or distance from Greenwich would look out of window and see how high this ring was above his horizon ... .
>
> So if we only had a ring like that, not round the equator of the world ... but vertical to the plane of the equator... from that ring ... we could calculate the longitude.
>
> E.E. Hale, *The Brick Moon* (Hale 1968, 32–34)

Fig. 8.6. Edward Everett Hale at the Boston Common. How many passersby know that this man was the first to write about an application satellite? Photo courtesy of Mike Gruntman.

determination of geographical latitude, the new moon was visible from Earth and helped make the determination of longitude easier for navigators.

First Space Station

This was the first novel where an artificial satellite was built for practical purposes as a navigational aid for sailors, particularly for those in danger. The new moon, argued Hale's characters, must be large "that it might be seen far away by storm-tossed navigators" (Hale 1968, 35). The agreed distance to the artificial moon was 4000 miles because this was "a good way off to see the moon even two hundred feet in diameter." It was evident for the Hale's inventors that "a smaller moon would be of no use, unless ... [they] meant to have them near the world, when there would be so many that they would be confusing and eclipsed most of time" (Hale 1968, 37).

By accident a number of people were trapped on Hale's artificial moon during its premature accidental "launch," adding a drama to the story. The inventive passengers found a way to communicating with the people on Earth, providing probably the first description of possible space communications. (The moon inhabitants jumped when observed from Earth, with the short and long leaps providing dots and dashes, respectively, of the Morse code.) Hale's novel was apparently the first description of a manned space station.

CANALI ON MARS

Extraterrestrial Spectacles

While the writers of fiction kept the public dreaming about other worlds, the imagination of the mankind was further excited by an astronomical discovery. The stars and other celestial bodies fascinated inquisitive minds since time immemorial. Occasionally, spectacular and mysterious phenomena such as comets further focused attention on the things extraterrestrial, on cosmos. This time the astronomical discovery did not come from a comet but from the Earth's close neighbor, the planet Mars.

Fig. 8.7. Fragment of the Bayeux tapestry showing the comet Halley in 1066. Such spectacles in the skies, seen by everybody, attracted attention to the things extraterrestrial. Figure from *Mémoires Biographique et Philosophique d'un Astronome* (Flammarion 1911, 413).

In 1877 the Italian astronomer Giovanni Schiaparelli observed a system of rectilinear markings on the surface of Mars. Schiaparelli was an excellent astronomer who began his career at the observatory in Milan by discovering the asteroid *Hesperia* in 1861. He became famous for establishing a link between a comet and the meteors, the discovery for which he was awarded a prestigious prize by the French Academy of Sciences in 1868. His other important contribution was detailed observations of planets. The periods of rotation of Venus and Mercury were believed to be close to that of Earth. Schiaparelli was the first to show that their rotation was much slower.

Giovanni Schiaparelli

Fig. 8.8. Giovanni Schiaparelli in the observatory. Photo courtesy of the Library of Astronomical Observatory of Brera, Milan.

Giovanni Schiaparelli was not the first to notice some structure and straight streaks while observing Mars. In fact the first blurred irregular spots on the planet surface were reported more than two centuries earlier by an astronomer in Naples, Francesco Fontana, in 1636 and 1638.

Canali Discovered

It was Schiaparelli's detailed observations, however, that brought attention of the broad public to these features and made an enigma of Mars a point of fascination by many. Schiaparelli described the straight streaks that he observed on Mars as canali in his native Italian. Canali could mean "channels" or "canals." These canali are attributed today to the optical illusions caused by accidental alignments of craters and other surface features on Mars and characteristic disturbances of the atmospheric air.

Streaks Looked Real

For Schiaparelli and the other astronomers who had noted the streaks earlier, these features looked real. The existing astronomical instruments did not allow one to reach definitive conclusions. The angular resolution of the telescopes was not high, and the human eye, complemented by sketches made by the observer, remained the most reliable and common recording technique. Astronomical photography was in its infancy at those days.

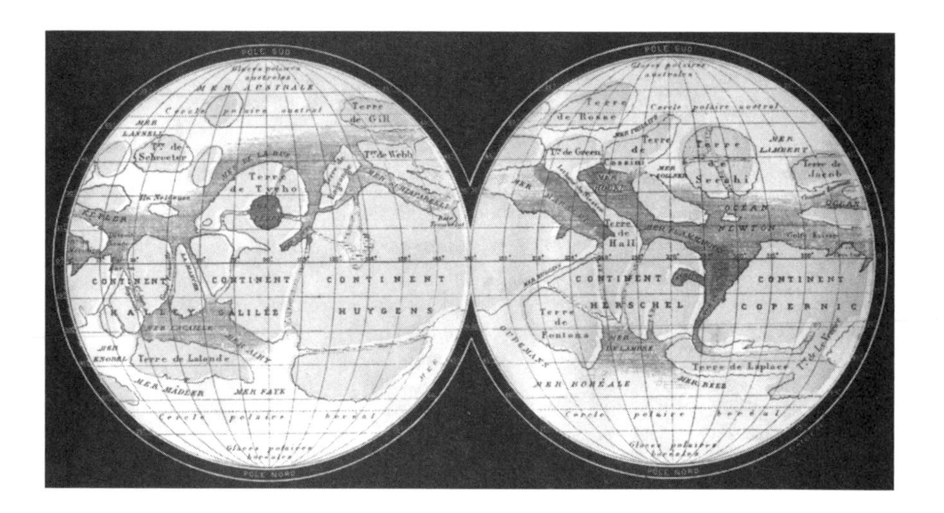

Famous Names on Maps of Mars

Fig. 8.9. Map of planet Mars. Figure from *Les Terres du Ciel* (Flammarion 1884, frontispiece).

"It was convenient to give the names of famous founders of the modern astronomy to the continents and main oceans [on Mars], and we first assigned the immortal names of Copernicus, Galilei, Tycho, Kepler, Newton, and Laplace. Naturally these names were followed by the names of astronomers who were most occupied with the study of Mars, to mention the earliest: Huygens, Fontana, Cassini, Hooke, Maraldi, Schrœter, Herschel, Mädler, and Beer; and those of our [19th century] time: Arago, Dawes, Secchi, Kaiser, Schmidt, Webb, Lockyer, Phillips, Proctor, and Terby.

Two great oceans that extended over the central region [of Mars] were named after the two immortal minds that built the theory of the world: Kepler and Newton. Four principal continents were named after Copernicus, Galilei, Huygens, and Herschel. The lands of Tycho, Laplace, Schrœter, Cassini, and Secchi followed. Beer and Mädler were associated, as during their lives, with the seas that born their names" (Flammarion 1884, 41).

Canali Become Canals

The Schiaparelli's canali were translated into other languages as canals, implying artificial waterways. Some scientists speculated that these features were the bands of vegetation, thus explaining the apparent variability of the observed streaks. Some even argued that when the life on the planet Mars was dying, the inhabitants built a set of irrigation canals to bring water from the polar caps. The hypothesis of life on Mars was also supported by the early spectroscopic observations, performed in the 1860s and 1870s, that revealed the presence of water. The new observations in 1894, with much more sophisticated instrumentation and accuracy, however, did not find any water.

Life on Mars?

Fig. 8.10. Camille Flammarion, 1842–1925, at the age of 20. Figure from *Mémoires Biographique et Philosophique d'un Astronome* (Flammarion 1911, 209).

3450-Mile Long Canal

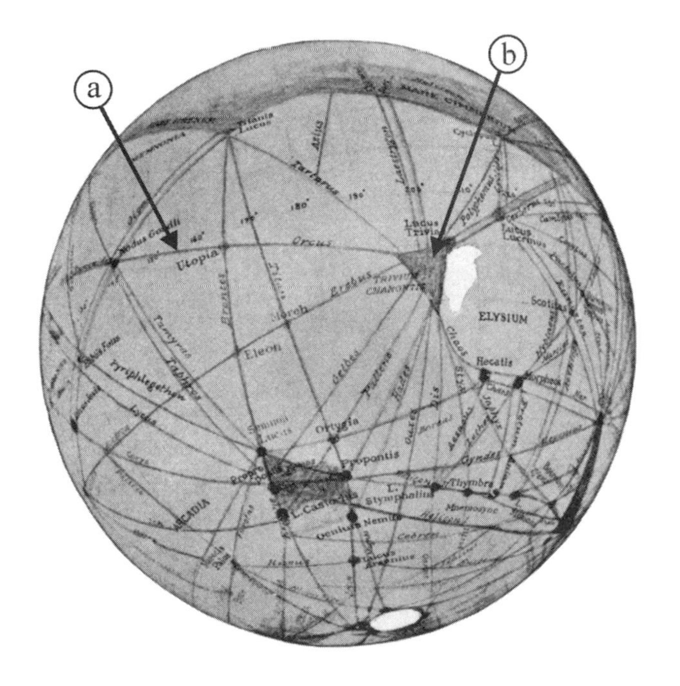

Fig. 8.11. Map of Mars, published by Percival Lowell in 1908, showing (a) a section of the huge canal Eumenides-Orcus terminating in (b) the Junction Trivium Charontis. Lowell wrote that this canal, "3450 miles [5550 km] long, for all its unswervingness to right and left, actually curves in its own plane through an arc of some 90° round the planet. It is much as if a straight line joined London to Denver, or Boston to Bering Straight" (Lowell 1910, 150). Figure from *Mars as the Abode of Life* (Lowell 1910, 150).

Fig. 8.12. "... Carefully examining the planet, an astronomer determines and sketches the continents, rivers, and islands of the geography of Mars ..." (Flammarion 1884, 5). Such pictures of awesome and sophisticated astronomical instruments brought credibility to astronomers' claims and ignited public enthusiasm about the possibility of life on Mars. Note that the astronomer records the observed image by making sketches. Figure from *Les Terres du Ciel* (Flammarion 1884, 5).

Fig. 8.13. Giovanni Schiaparelli, 1835–1910. Photo courtesy of the Library of Astronomical Observatory of Brera, Milan.

The idea of the canals on Mars was enthusiastically adopted by many astronomers and popular writers, including the French Camille Flammarion. Flammarion was not only an astronomer, but he was also a prolific writer who extensively published on astronomy, planets, and extraterrestrial life ("thc multitude of worlds"). His popular book on astronomy, published in 1879, became a best-seller. It was also Flammarion who founded, in 1887, the *French Astronomical Society* (*Société Astronomique de France*).

Camille Flammarion

Not everybody, however, embraced this fascinating explanation of the observations. Some astronomers argued that the apparent linear streaks, the "canals," were only the "summation of a complexity of details" of the irregular features on the surface of Mars. For example, thc French astronomer Eugène Antoniadi wrote, somewhat sarcastically, in 1913 that "to account for these wonderful phenomena [observed on Mars], the vast powers of Nature were found totally inadequate; and thus it was that Schiaparelli was led to enunciate the idea of the artificial origin of the canals, conceiving the larger of them to be composed of six different watercourses, whose dykes would be opened now and then by the Martian minister of agriculture" (Antoniadi 1913, 417).

"... Martian Minister of Agriculture..."

Fig. 8.14. Percival Lowell, 1855–1916. Photo courtesy of Lowell Observatory.

Lowell devoted his life to literature, travel, and astronomy. He performed detailed mathematical studies of the perturbations of the orbit of the planet Uranus and concluded that an undiscovered planet should exist at large distances from the sun. Lowell initiated search of this planet at his observatory at Flagstaff, Arizona, which he had earlier built for study of Mars. The observatory is known today as the *Lowell Observatory*. The planet search eventually succeeded when Clyde Tombaugh had discovered Pluto in 1930.

Fig. 8.15. Photographic apparatus of the Lowell Observatory devised by Carl O. Lampland and used in getting the photographs of Mars. Figure from *Mars as the Abode of Life* (Lowell 1910, 154).

Fig. 8.16. Dome of the Percival Lowell's observatory in Tacubaya, near Mexico City, Mexico, where Mars was observed during the winter of 1896–1897. Lowell carried out numerous observation of Mars from Flagstaff, Arizona, and other locations. Figure from *Mars as the Abode of Life* (Lowell 1910, 208).

Fig. 8.17. Percival Lowell in the library. Photo courtesy of Lowell Observatory.

For many years the pioneers of this discovery of another world had their revelations strictly to themselves, decried as baseless views and visions by the telescopically blind. So easily are men the dupes of their own prejudice. But in 1901 attempts began to be made at Flagstaff to make them tell their own story to the world, writing it by self-registration on a photographic plate. It was long before they could be compelled to do so. The first attempts showed nothing; the next, two years later, did better, evoking faint forms to the initiate, but to them alone; but two years later still, success crowned the long endeavor. At last these strange geometricisms have stood successfully for their pictures. The photographic feat of making them keep still sufficiently long — or, what with heavenly objects is as near as man may come to his practice with human subjects, the catching of the air-waves still long enough to secure impression of them upon a photographic plate — has been accomplished by Mr. Lampland. After great study, patience, and skill he has succeeded in this remarkable performance, of which Schiaparelli wrote [to P. Lowell] in wonder ...: "I should never have believed it possible."

Percival Lowell, 1908 (Lowell 1910, 153)

The American astronomer Percival Lowell also became a champion of the idea of life on Mars. Lowell, a member of the prominent Massachusetts family, was interested in astronomy since his boyhood. The book by Camille Flammarion *La Planète Mars* triggered Lowell to make a drastic turn in his life and devote himself to the astronomical study of a possibility of life on Mars.

Fascination by Mars

Fig. 8.18. The sun sets on the canals of Mars. Figure from *Les Terres du Ciel* (Flammarion 1884, 65).

"... most astounding objects to be viewed in the heavens ..."

Lowell was fascinated by the canals on Mars. He wrote in 1908:

> In 1877, however, a remarkable observer made still more remarkable discovery; for in that year Schiaparelli, in scanning these continents [on Mars], chanced upon long, narrow markings in them which have since become famous as the canals on Mars. Surprising as they seemed when first imperfectly made out, they have grown only more wonderful with study. It is certainly no exaggeration to say that they are the most astounding objects to be viewed in the heavens. There are celestial sights more dazzling, spectacles that inspire more awe, but to the thoughtful observer who is privileged to see them well there is nothing in the sky so profoundly impressive as these canals on Mars. Fine lines and little gossamer filaments only,

> cobwebbing the face of Martian disk, but threads to draw one's mind after them across the millions of miles of intervening world. (Lowell 1910, 146)

Observatory at Flagstaff, Arizona

Percival Lowell initiated extensive astronomical observations of Mars and founded in 1894 and built an observatory in Flagstaff, Arizona. This observatory bears the Lowell's name today. As Lowell explained in 1916,

> Flagstaff was specially selected as a site because of the excellence of its planetary images. Far from the smoke of men it enjoys a climate which renders its air steadier than any so far found in the United States. This permits not only the detection of detail invisible elsewhere, but a precision of measurement of the detail detected, necessary to any accurate deductions from it. In consequence the art of observation has there been studied carefully, to the revealing of atmospheric and of optical conditions which, if not duly reckoned with, are pitfalls to precise results. (Lowell 1916, 421)

Continents, Oceans, and Rivers on Maps of Mars

The maps of Mars surface, obtained by painstaking observations with imperfect instruments, included continents, oceans, rivers, and other features, not unlike those on Earth:

> Odd as is the look of the individual canal, it is nothing to the impression forced on the observer by their number and still more by their articulation.

Fig. 8.19. Mariner 4 image of the Martian surface from 12,600 km (7830 miles). The image covers the area 250 by 254 km (155 by 158 miles). The 151-km (94-mile) *Mariner* crater is in the center. Photo courtesy of NASA.

> When Schiaparelli finished his life-work, he had detected 113 canals; this figure has now been increased to 437 by those since added [at the observatory] at Flagstaff. As with the discovery of the asteroids, the later found are as a rule smaller and in consequence less evident than the earlier. (Lowell 1910, 151)

Percival Lowell carried out numerous astronomical observations of Mars at Flagstaff and at other locations, including Mexico. He introduced photographic registration of the Martian surface features to convince the world in the reality of his findings.

Eugène Antoniadi and *Grand Lunette*

The astronomical world was not unanimous in embracing the Lowell's hypothesis, and the results of the observations were not unambiguously convincing. A breakthrough came in 1909 when the French astronomer Eugène Antoniadi (1870–1944) performed a series of observations with the *Grand Lunette*, the 33-in. (84-cm) refractor at the Meudon Observatory near Paris. Antoniadi clearly saw that what appeared as regular lines were in fact rather irregular borders and sequences of rugged shadings, knotted spots, and irregular streaks. The canals apparently were just fragmentary optical illusions, seen by glimpses. Gradually, the fascinating story of the canals on Mars and its appeal to the public were fading away.

***Mariner 4* Flyby of Mars**

The theory of the canals on Mars was finally put to rest by the flyby of the planet by the first spacecraft on 14 July 1965. The American *Mariner 4* flew by Mars and sent back 22 photographs. The images revealed numerous surface features, including craters, but no canals.

9. GREAT PIONEERS

The second half of the 19th century brought the realization that until the rocket was perfected there would be no trips through outer space, no landing on the Moon, and no visits to other planets. The practical work had to begin to develop the science and engineering of powerful rockets and sophisticated spacecraft. And this work had begun.

Practical Work on Rocketry and Spaceflight

A long period followed when isolated visionaries and thinkers, including amateurs, sketched out the sinews of the spaceflight concept. Various technical details had precarious credibility and were usually dismissed as ridiculous by elitist self-confident intellectuals of the day and assorted "competent authorities."

> "Where there is no vision, the people perish ... ."
>
> *Proverbs 29:18*

Many outstanding individuals laid the foundations of the rocket technology and spaceflight. Four visionaries in four countries working under very different conditions became the great pioneers of the space age: Konstantin E. Tsiolkovsky, Robert Esnault-Pelterie, Robert H. Goddard, and Hermann Oberth. They were followed by many more enthusiasts and emerging rocket and spaceflight societies.

KONSTANTIN E. TSIOLKOVSKY (1857–1935)

Dreamer from Kaluga

Konstantin E. Tsiolkovsky was a high school mathematics teacher in a provincial Russian town of Kaluga, about 140 km (90 miles) south of Moscow. Konstantin became nearly deaf at the age of 11 after scarlet fever. This misfortune prevented Tsiolkovsky from studying in institutions of higher learning that simply

> **JULES VERNE'S INFLUENCE**
>
> "My aspiration for space travel was seeded by the known French dreamer J. Verne ... ."
>
> K.E. Tsiolkovsky 1926
> (Tsiolkovsky 1954, 103)

MANKIND ... WILL CONQUER ALL THE SPACE

"Mankind will not remain on the earth forever, but in pursuit of light and space it will at first timidly penetrate beyond the limits of the atmosphere, and then conquer all the space around the sun."

K.E. Tsiolkovsky, 12 August 1911 (Tsiolkovsky 1968, 14, 331)

would not admit him. He however partly overcame his disability through self-education.

Self-Education

At the age of 16, Konstantin came to Moscow where he lived for three years spending most of the time in the Rumyantsev Library. (This was the main Russian library, later known as the Lenin State Library. Presently it is called the Russian State Library; the library is similar in many of its functions to the Library of Congress in the United States.) Near-deafness made Konstantin an isolated, withdrawn, and lonely person. These qualities however made it easier for Tsiolkovsky to pursue his dreams of spaceflight, surrounded by the wall of indifference and often ridicule.

Some of Tsiolkovsky's thoughts on space travel can be traced to his 1883 manuscript *Free Space*. This work, described by Sergei P. Korolev as a kind of scientific diary, was written when Tsiolkovsky served as a teacher for 12 years in a high school in a small town of Borovsk in the Kaluga region. Two works on the possibilities of space travel, *Dreams of the Earth and the Sky* and *The Effects of Universal Gravity*, followed in 1895, when Tsiolkovsky settled permanently in Kaluga.

Fig. 9.1. Konstantin E. Tsiolkovsky in 1909. Photo courtesy of K.E. Tsiolkovsky Museum of Cosmonautics, Kaluga.

"Research into Interplanetary Space by Means of Rocket Power"

Tsiolkovsky's first major "serious" article, "Research into Interplanetary Space by Means of Rocket Power," appeared in 1903 in the magazine *Nauchnoe Obozrenie* (*Scientific Review*). He was clearly in awe of the potential offered by the rocket technology. Tsiolkovsky wrote that

"... appealing and significant promises"

> instead of using them [cannons] or air balloons [to achieve high altitudes] ... I propose a reactive device, that is a kind of rocket, but one of enormous dimensions and specially designed. The idea is not new, but the relevant calculations have yielded such remarkable results that it is unacceptable to keep silent about them. This my work by far neither examines all aspects of the problem nor it resolves the practical issues of its feasibility; however in the far distant nebulous future one perceives such appealing and significant promises as hardly anyone today dreams of. (Tsiolkovsky 1954, 29; Tsiolkovsky 1968, 55)

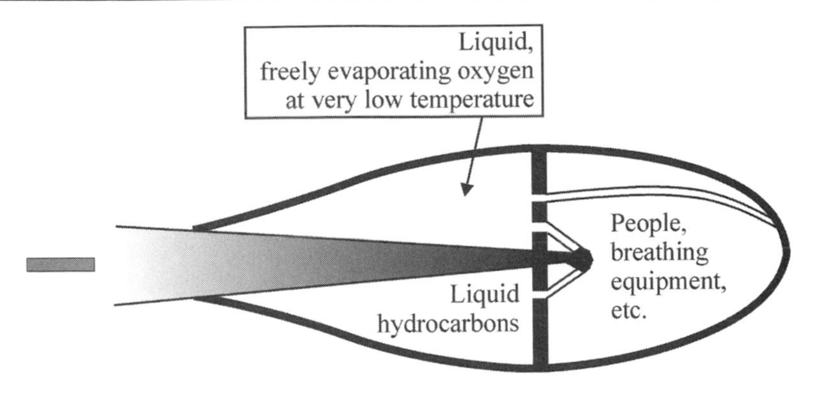

Fig. 9.2. Schematic of Tsiolkovsky's hydrocarbon-oxygen rocket proposed in 1903. The thrust vector control is performed by a rudder in the exhaust flow. The central expanding tube (nozzle) is surrounded by a jacket with the circulating liquid metal or a liquid with high thermal conductivity.

Rocket Equation

In this famous publication Tsiolkovsky obtained the *rocket equation*, which is an equation describing the dynamics of bodies with variable mass for a given propellant exhaust velocity. This important equation is called *formula Tsiolkovskogo*, or the *Tsiolkovsky formula,* in the Soviet Union.

Advantages of Liquid Rocket Propulsion

Tsiolkovsky pointed out that the rocket was the only means of propelling vehicles in space. He realized energy limitations of solid-propellant rockets and advocated liquid fuels, particularly liquid oxygen and hydrocarbons. Liquid-propellant rockets, he argued, would make it possible to overcome the limitations of solid rockets, provide higher exhaust velocities, and allow control of thrust as well as starting and stopping the engine at will.

Fig. 9.3. Konstantin E. Tsiolkovsky in 1930. Photo courtesy of K.E. Tsiolkovsky Museum of Cosmonautics, Kaluga.

TSIOLKOVSKY'S NAME SPELLING

Tsiolkovsky's name is sometimes spelled as *Ziolkowski* or *Ziolkovsky*. His original Russian name is not in the Latin alphabet, but in Cyrillic, and its spelling depends whether it is transliterated or phonetically rendered in a Latin-based language. *Ziolkowski* is a common phonetical rendering of his name in German.

ROCKET BASICS 3: ROCKET EQUATION

The equation that describes rocket acceleration is of such fundamental importance for rocket flight that it is called *the rocket equation*. The equation relates the velocity acquired by a rocket with the equivalent exhaust velocity, U_{eq} (Rocket Basics 1 and 2; Chapter 4), and the amount of the consumed propellant. The latter quantity is conveniently expressed through the dimensionless mass ratio

$$R = \frac{M_0}{M_B}$$

where M_0 is the initial rocket mass; $M_B = M_0 - M_P$ is the rocket mass at burnout; and M_P is the mass of the consumed propellant. According to the rocket equation, the rocket velocity increment ΔV would be

$$\Delta V = U_{eq} \ln R$$

for constant U_{eq} in the absence of gravity and air drag.

The rocket equation combines dynamics of a body with the varying mass and the relation between the accelerating force (thrust) and the propellant exhaust velocity. In 1813, William Moore described the relevant dynamics for constant thrust and constant propellant consumption rate acting on a rocket with the varying mass. Moore however did not relate thrust and the exhaust velocity and, therefore, did not relate the rocket velocity increment and the exhaust velocity of the propellant flow.

Derivation of the rocket equation is rather elementary. By the middle of the 19th century, problems related to rocket flight (requiring derivation of the rocket equation) had been given to university students as a standard exercise in particle dynamics (Tait and Steele 1856, 255). Many researchers would independently obtain this simple equation again and again throughout many years. Konstantin Tsiolkovsky described it in 1903. (Another Russian, Ivan V. Meshchersky, 1859–1935, obtained in 1897 a differential equation describing dynamics of a point with a variable mass.) Esnault-Pelterie, Goddard, Oberth, and many others independently derived the equation later.

Consider, for example, a Congreve or Hale rocket with specific impulse $I_{SP} = 80$ s and propellant constituting one-third of the total rocket mass, $M_P = M_0/3$ and correspondingly the burnout mass $M_B = M_0 - M_P = 2M_0/3$. The equivalent exhaust velocity would be $U_{eq} = I_{SP} \times g_E = 80 \times 9.81 \approx 785$ m/s, and the mass ratio $R = M_0/M_B = 1.5$. When all rocket propellant is consumed, the rocket would achieve the velocity

$$\Delta V = U_{eq} \ln (R) = 785 \times \ln (1.5) \approx 318 \text{ m/s}$$

in the absence of gravity and air drag. For a typical 3-s propellant burning time, Moore's equations (Rocket Basics 2, Chapter 4) for vertical launch (with gravity but disregarding air drag) would give the velocity and altitude at burnout 289 m/s and 401 m, respectively. Such a rocket would have reached an altitude of 4.65 km (2.89 miles) in the absence of air drag.

A similar solid-propellant rocket with a modern propellant and nozzle design could have $I_{SP} = 260$ s. In the absence of gravity and air drag, such a rocket with the same mass ratio would achieve the velocity 1034 m/s. In a vertical launch and in the absence of drag, the rocket would reach the velocity and altitude at burnout 1004 m/s and 1402 m, respectively, and the total altitude 52.8 km (32.8 miles). This example demonstrates how superior specific impulse dramatically improves rocket performance.

PEREL'MAN AND TSIOLKOVSKY

"The greatest contribution to the promotion of the idea of space travel in the USSR was made by Ya.I. Perel'man by his beautifully written our country's first popular book on astronautics, *Interplanetary Voyages*, with 10 printings in 20 years (1915–1935), other books, and numerous articles and public lectures he delivered since 1915. Thanks to the active popularization by Perel'man, the writings of K.E. Tsiolkovsky became widely known."

Valentin P. Glushko (*Odnazhdy i Navsegda* ... 1998, 33)

Tsiolkovsky's Writings Become Known

Based on heat of formation, Tsiolkovsky also pointed out superior characteristics of a combination of hydrogen and oxygen for rocket propulsion. "Liquid oxygen and liquid hydrogen," he wrote, "supplied ... to the narrow section of the pipe [nozzle] would combine ... giving excellent explosive [propulsion] material. The water vapor at awfully high temperature, obtained by combination of these liquids, would expand, toward the end ... of the tube [nozzle] until it cools off to such a degree as to transform into a liquid, dashing in a form of fine fog along the tube toward the exit" (Tsiolkovsky 1954, 35; Tsiolkovsky 1968, 63).

Fig. 9.4. Monument to Konstantin E. Tsiolkovsky in Moscow, Russia. The inscription reads, "Tsiolkovsky, the Founder of Cosmonautics." Photo courtesy of Mike Gruntman.

Liquid Oxygen and Liquid Hydrogen

Tsiolkovsky gave the maximum exhaust velocity 5700 m/s for the hydrogen-oxygen propellant combination. He calculated that the mass ratio 193 would achieve the rocket velocity equal to that of the Earth's orbital motion around the sun (~30 km/s). Tsiolkovsky was optimistic that even higher energies, and correspondingly higher exhaust velocities, could be obtained in the future.

Experiments

Tsiolkovsky's work was mostly of theoretical nature. He tried though, in the late 1870s, to experimentally study the effects of increased loads on living organisms, expected during the rocket launch and reentry. He experimented with cockroaches and chickens on a small centrifuge, where the chickens

BOLSHEVIK PARTY AND SOVIET STATE

"I bequeath all my works on aviation, rocket flight and interplanetary communications to the Bolshevik [Communist] Party and the Soviet State, the true leaders of cultural progress."

K.E. Tsiolkovsky, 17 September 1935 (two days before his death) (Tsiolkovsky 1968, 331)

"The scientific legacy of Tsiolkovsky transferred to the Bolshevik Party and the Soviet State ... is being creatively developed and successfully continued by Soviet scientists."

Sergei P. Korolev, 1957 (Tsiolkovsky 1968, 24)

survived up to the 10-*g* loads. Tsiolkovsky also studied aerodynamics of various bodies in a simple wind tunnel that he built in 1890s. The dirigibles, in particular the idea of the all-metal dirigible, inspired him throughout all his life since his youth. He studied the dirigibles theoretically and built models to explore the concept in detail.

Father of Cosmonautics

Tsiolkovsky, who was commonly called the *Father of Cosmonautics* in the USSR, never built a rocket. His work and his writings, however, inspired a generation of Soviet rocket enthusiasts, including Sergei Korolev and Valentin Glushko. These scientists and engineers would establish and lead the Soviet rocket program in the 1930s and the space program after World War II.

Konstantin E. Tsiolkovsky wrote in 1903 (Tsiolkovsky 1968, 84), "I shall be happy if my work induces others to further effort." He has succeeded beyond his wildest dreams.

ROBERT ESNAULT-PELTERIE (1881–1957)

Robert Esnault-Pelterie, also known as REP by his initials, occupies a unique place among the great space pioneers. He was also a great aviation pioneer with remarkable contributions to the development of airplanes. The dark-haired Frenchman graduated in engineering at the Sorbonne. He was an accomplished inventor and engineer and a sculptor. The French Academy of Sciences elected him a member in 1936.

Great Aviation Pioneer

Fig. 9.5. Pioneer aviator Robert Esnault-Pelterie in his airplane REP-2. Photo courtesy of Musée de l'Air and Centre National d'Études Spatiales, France.

Robert Esnault-Pelterie was the fourth person in France who obtained the pilot's license. He started with building gliders. On 19 October 1907, he flew his first all-metal monoplane, known as the R.E.P. This was the first airplane with a completely enclosed fuselage. Esnault-Pelterie invented the aileron, a movable part at the trailing edge of the wing, allowing aircraft roll along the longitudinal axis. He also introduced a "joystick" for airplane control, four-bladed propeller, fan-shaped radial engine, safety belt, speed indicators, and numerous other now-standard devices.

Fig. 9.6. Robert Esnault-Pelterie was the fourth person in France to obtain the pilot's license (1908). Photo courtesy of Musée de l'Air and Centre National d'Études Spatiales, France.

An unfortunate crash ended the pilot carrier of Esnault-Pelterie on 18 June 1908. His airplane hit the ground at full speed, and Esnault-Pelterie was thrown from the airplane in spite of an elastic "safety" band, hit against the fuel tank and was badly injured. He was found unconscious by a nearby farmer who witnessed the crash. The farmer brought Esnault-Pelterie back to life by a stiff shot of cognac. The injuries afflicted Esnault-Pelterie for many years, and he never flew as a pilot again. However his unbounded energy and vision had already been exploring, since 1907, a new challenge — the rockets.

"Consideration on the Results of the Unlimited Lightening of Motors"

On 15 November 1912, Robert Esnault-Pelterie delivered a lecture "Consideration on the Results of the Unlimited Lightening of Motors" to the French Physical Society. As a rocket consumes propellant, it becomes lighter, thus the "lightening of the motors." This presentation was a repeat of his lecture eight months earlier in St. Petersburg, Russia.

Passive Thermal Control of Spacecraft

Esnault-Pelterie discussed the acceleration of a rocket in his lecture and derived the rocket equation. He considered the energetic properties of guncotton, hydrogen-oxygen mixture, and radium as propellants. Then he provided estimates of the required velocity increments and flight times for travel to the Moon, Venus, and Mars and requirements to the propellant for such missions. Considering interplanetary flight of humans, Esnault-Pelterie proposed passive spacecraft temperature control, "... A vehicle built in such a way that one half of its surface would be of polished metal ... . The other half of the surface ... would be ... a black surface. If the polished surface would face the sun, the temperature would decrease. In the opposite position, the temperature would increase" (Esnault-Pelterie 1913, 228).

The topic of the Esnault-Pelterie's lecture was more from the realm of science fiction than from the field of "serious" research at those days. Certainly, his fame as an aviation pioneer helped Esnault-Pelterie to gain acceptance by a serious scientific audience of the French Physical Society. It is never easy to break the barriers. The abbreviated form of his presentation was published next year in the respected *Journal de Physique*.

Article in *Journal de Physique*

This was the time when the foundations of astronautics were being laid down. Actually, the word *astronautics* did not exist yet. It was Esnault-Pelterie who later, in 1927, would name the new field of science and engineering and make it widely accepted.

On 8 June 1927, Esnault-Pelterie delivered a lecture "Rocket Exploration of the Very High Atmosphere and the Possibility of Interplanetary Travel" to the general meeting of a leading learned group, the French Astronomical Society. This work was published by the Society the next year. At the same time Esnault-Pelterie and his enthusiastic supporter André Louis Hirsch established what became known as the REP-Hirsch Award (Prix REP-Hirsch) to promote scientific research in space travel. The award played an important role during the next decade

REP-Hirsch Award

PRIX REP-HIRSCH (REP-HIRSCH AWARD)

1929
Hermann Oberth
Rumania/Germany

1931
Pierre Montagne
France

1933
(Prize of Encouragement)
Ary Sternfeld
Poland/France/USSR

1935
(Prize of Encouragement)
Louis Damblanc
France

1936
American Rocket Society
and Alfred Africano
USA

1939
Frank Malina and
Nathan Carver
USA

REP-HIRSCH PRIZE (AWARD)

In 1928 Robert Esnault-Pelterie and André Louis Hirsch provided funds (5000 francs annually) to the French Astronomical Society to establish the REP-Hirsch International Astronautics Prize. The prize was to be given to recognize the "best work of scientific origin, theoretical or experimental, capable of advancing one of the areas on which interstellar navigation depends or augments the knowledge in one of the areas relevant to the astronautical science" (Esnault-Pelterie 1930, 243)

A special Committee on Astronautics was formed to make the awards. The committee included a number of prominent scientists that helped to establish the respectability of the new science. The applicants had to submit their work in open competition, written in one of the following languages: French, English, German, Spanish, Italian, or Esperanto.

The first REP-Hirsch Award was given to Hermann Oberth who struggled to find support for his work in Germany. The prizes were awarded until 1939, with the last prize bestowed on Frank Malina, a young collaborator of Theodore von Kármán at the California Institute of Technology, and Nathan Carver of New York.

ASTRONAUTICS

It was 26 December 1927. The guests gathered at the house of the mother of André Louis Hirsch. Earlier that year, Robert Esnault-Pelterie presented his work on rocketry and interplanetary flight at the meeting of the French Astronomical Society. The dinner was arranged by Esnault-Pelterie and his friend and supporter André Louis Hirsch to form a committee to promote the work in the field of space travel. In addition to Esnault-Pelterie and Hirsch, the guests at the dinner included famous physicists Jean Perrin (a Noble Prize winner) and Charles Fabry, astronomers Ernest Esclangon and Henri Chrétien (a famous telescope builder), president of the French Astronomical Society E. Fichot, General Gustave Ferrié (who pioneered many military applications of radio telegraphy), and science-fiction writer J.H. Rosny aîné (the elder).

The group worked out the plan to establish an annual award that will become known as the REP-Hirsch Prize to promote research in space travel. Then the guests discussed the name that should be given to the new science. Esnault-Pelterie proposed *sideration* (structured similarly to *aviation*), but it did not appeal. Somebody suggested *cosmonautique* (cosmonautics), but it did not appeal either. Then, Rosny proposed the word *astronautique* (*astronautics* in English). This was it! The word was adopted at once.

Fig. 9.7. Robert Esnault-Pelterie with his book *L'Astronautique*, ca. 1950. Photo courtesy of Centre National d'Études Spatiales, France.

The Esnault-Pelterie's lecture at the meeting of the French Astronomical Society (given in 1927) was published in 1928. The publication was much bigger than the original lecture (an hour and a half to read), and Esnault-Pelterie used the new word, astronautics, several times in this publication. Two years later, in 1930, Esnault-Pelterie published his comprehensive treatise on rocketry and space travel under the title *L'Astronautique*. A new field in science and engineering has thus been named.

The word *astronaut* was used in the science fiction literature as early as by Percy Greg in his *Across the Zodiac* published in 1880. (*Astronaut* was the name of the spaceship.) What is important for us is that the word *astronautique* was not considered by the French scientists in the late 1920s as a word of science fiction literature, but it was the word coined by them for the new science field. The word became known to the world in the intended meaning, *the art or science of designing, building, and operating space vehicles*, and it entered many languages. The Russian language uses a different word, *kosmonavtika* (*cosmonautics*), with the same meaning.

in stimulating development of rocketry and building links among the rocketeers in various countries.

The year of 1930 was another major milestone in astronautics. Esnault-Pelterie published the book entitled *L'Astronautique* (*Astronautics*). He presented the new science in a consistent and comprehensive way: discussed in detail rocket motion in vacuum and air; considered gas flows in converging-diverging nozzles; applied thermodynamics to the combustion processes of various fuel-oxidizer combinations; and pointed out the exceptional properties of atomic hydrogen as propellant. Esnault-Pelterie outlined possible rocket applications including studies of the aurora borealis and the upper atmosphere, missions to the Moon and to the planets. He suggested reaction wheels for spacecraft attitude control and discussed the effects of space flight on humans.

L'Astronautique

ARY STERNFELD

Ary Sternfeld, 1905–1980, was born and studied in Poland. He continued his education and worked in France and was awarded the REP-Hirsch prize in 1933. Sternfeld moved to the communist USSR in 1935 where he published extensively on astronautics.

JEAN-JACQUES BARRÉ

A young artillery officer and graduate of the Ecole Polytechnique, Jean-Jacques Barré, attended the Esnault-Pelterie's lecture in 1927. After the lecture, Barré wrote to Esnault-Pelterie a letter, and the latter promptly responded. Within the next several years, they exchanged more than 300 letters, discussing the problems of combustion, exit velocity, and other aspects of rocketry.

Barré began practical rocket development for the French army in mid-1930s. After the humiliating Armistice of 1940, Barré concentrated on the first French liquid-propellant (gasoline and liquid oxygen) rocket EA-41. The French Vichy Government arranged the work of the Barré's group at Fort de Vancia near Lyon, without knowledge of the Germans. The allied invasion of the French North Africa in 1942 interrupted the preparations of the first rocket launch in Algeria, and all practical work ceased with the German occupation of the entire France. Barré's rocket was launched for the first time on 15 March 1945 from the Establishment of Technical Experiments at La Renardière near Toulon. The rocket exploded after 5 s.

Fig. 9.8. Jean Jacques Barré (to the right from the rocket, in jacket) during the assembly of the rocket EA-41 at Fort de Vancia near Lyon in 1942. Photo courtesy of Centre National d'Études Spatiales, France.

> **PHYSICAL REQUIREMENTS FOR INTERSTELLAR FLIGHT**
>
> "Numerous authors made a man travelling from star to star a subject for fiction ... . No one has ever thought to seek the physical requirements and the orders of magnitude of the relevant phenomena necessary for realization of this idea ... . This is the only aim of the present study."
>
> Robert Esnault-Pelterie (1913, 218)

Esnault-Pelterie was mostly interested in theoretical astronautics. He, however, initiated practical development of liquid rocket engines with support of the French army. An army officer Jean-Jacques Barré joined Esnault-Pelterie, but was recalled a year later because the army was unable to accept that the "study of rockets could absorb the activity of an officer." Barré would play an important role in development of the French rocketry during and after the war.

In October 1931, Esnault-Pelterie lost the ends of four fingers on his left hand in experiments with gasoline and tetranitromethane as rocket propellants. This accident led to selection of less-dangerous liquid oxygen as oxidizer for the future French rockets.

In 1931, Esnault-Pelterie demonstrated a liquid rocket engine working on gasoline and liquid oxygen. In December 1936, his engine achieved 1230 N (275 lbf) thrust, and later 2940 N (660 lbf), for 60 s. The propellant exhaust velocity was as high as 2400 m/s (7870 ft/s), corresponding to specific impulse of 250 s. The world war brought an end to Esnault-Pelterie's work in 1939. After the war, Esnault-Pelterie settled in Switzerland. He passed away in 1957.

ROBERT H. GODDARD (1882–1945)

First Liquid-Propellant Rocket

At 2:30 p.m. on 16 March 1926, Robert H. Goddard launched the first liquid-propellant rocket ever to fly. He ignited the rocket in a snowy pasture of Aunt Effie's (Ms. Effie M. Ward) farm at Auburn, Massachusetts. The rocket flew up to the altitude of 41 ft (12.5 m) and landed in a cabbage patch 184 ft (56 m) from the launch site, about the same distance as the first Wright Brothers flight. The first liquid rocket weighed 10.45 lb (4.74 kg) fully fueled, including 3.7 lb (1.7 kg) of liquid oxygen and 0.75 lb (0.34 kg) of gasoline. The whole flight (time in air) took 2 1/2 s.

> **CLARK, PRINCETON, AND ROSWELL**
>
> After completing the doctorate, Robert H. Goddard began to work as a research fellow at Clark University and one year later at the Palmer Physical Laboratory at Princeton University. He returned to Clark University in 1914, where he ultimately became professor and director of the physical laboratories. In 1930, Goddard moved to Roswell, New Mexico.

Robert Goddard began his experiments in rocketry while studying for his doctorate at Clark University in Worcester, Massachusetts. Goddard's doctoral dissertation "On the Conduction of Electricity at Contacts of Dissimilar Solids," 1911, was not related to rocketry, but was rather in the "mainstream" physics. Goddard experimentally studied anisotropic changes in the electrical resistance of loosely powdered substances, particularly barium sulphide. The emerging radio technology relied on such materials for receiving the signals. Today we would call this area of research experimental solid-state physics.

Experimental Solid-State Physics

His Ph.D. degree attained, Goddard actively embarked on research in rocketry, when he joined the faculty of Clark University in 1914. His basic research and development of new technology would achieve many rocket "firsts" and bring him 214 patents. He would often be called the Father of Modern Rocketry.

Fig. 9.9. Robert H. Goddard with the first liquid-propellant (gasoline and liquid oxygen) rocket ever to fly, 16 March 1926. The thrust chamber assembly at the top is supported by two vertical propellant pipes. The gasoline tank is in the center at the very bottom. The liquid-oxygen tank is immediately on the top of the gasoline tank, with the cap protecting the tanks from rocket exhaust. Goddard placed his hand on the launching frame. Photo courtesy of NASA.

Fig. 9.10. Robert H. Goddard with parts of the experimental setup built at Clark University to study rocket performance in vacuum, ca. 1916. Photo courtesy of NASA.

Early Work on Solid-Propellant Rockets

Goddard concentrated first on the study of solid-propellant gunpowder rockets and improving their efficiency. The term "efficiency," introduced by Goddard, meant "the ratio of the kinetic energy of the expelled gases to the heat energy of the powder." He increased the average velocity of ejection of gases from ordinary rockets from 1000 ft/s (300 m/s) to slightly under 8000 ft/s (2450 m/s). "These velocities were proved to be real velocities, and not merely effects due to reaction against the air, by firing the same steel chambers *in vacuo*, and observing the recoil." (Goddard, 1970) Goddard's experiments in vacuum clearly exposed a common and persisting misconception that a rocket needed some ambient medium to push against. Not everybody would listen, though.

Fig. 9.11. Robert H. Goddard loading a rocket projectile into a compact launcher on Mount Wilson, California, in 1918. Photo courtesy of NASA.

ROCKET BASICS 4: ROCKET IN VACUUM

(COMMON MISCONCEPTION)

A common misconception for many, many years emphasized the importance of ambient air for rocket propulsion. It was claimed that the rocket needs some medium to "push against." The fundamentals of rocket performance, particularly in vacuum, were not understood until the 20th century.

For example, William Moore wrote in his 1813 treatise on rocket dynamics:

> If the rocket burns in a medium, then, as there is a body reacting against the fluid [propellant] that rushes from the rocket, there is not so instantaneous a dissipation of the force of the latter the moment after it is generated; but a time of its action upon the rocket which is greater or less according to the surrounding medium is more or less dense and elastic. In this case, therefore, more motion is communicated to the body than in the former, and but for the resistance of the forepart of the rocket it would move farther in a medium than in vacuum. (Moore 1813, 25)

A standard textbook on ordnance and gunnery prepared for cadets of the U.S. Military Academy in West Point in the second half of the 19th century instructed:

> A rocket is set in motion by the reaction of a rapid stream of gas escaping through its vents. If it be surrounded by a resisting medium, the atmosphere for instance, the particles of gas, as they issue from the vent, will impinge against and set in motion certain particles of air, and the force expended on the inertia of these particles will react and increase the propelling force of the rocket. It follows, therefore, that, though a rocket will move *in vacuo*, its propelling force will be increased by the presence of a resisting medium. (Benton 1883, 95)

Konstantin E. Tsiolkovsky wrote in his 1903 seminal publication on rocketry:

> The effect of the atmosphere [ambient air] on an explosion [propulsion] is not completely clear: on the one hand, because the exploding substances [propellant] have some support in the ambient material medium, which they involve in their motion and thus contribute to the increase of the rocket velocity; but on the other hand, the same atmosphere [ambient air] inhibits [exhaust] gas expansion beyond a certain limit because of atmosphere's density and elasticity, with the result that the explosives [propellant] does not achieve the velocity they could have had expanding in vacuum. (Tsiolkovsky 1954, 32)

The rocket principle is based on conservation of momentum. A rocket does not need air to push against, and in reality, the ambient air decreases thrust. The rocket thrust contains (Rocket Basics 1, Chapter 4) an item $(P_e\text{-}P_a)\,A_e$, where P_e and P_a are the exit and ambient pressure, respectively, and A_e is the exit area of the nozzle. Thus, the reduction of the ambient pressure P_a results in increase of thrust in addition to decrease in air drag on a missile.

Although the converging-diverging (De Laval) nozzles significantly improved the performance of gunpowder rockets, Goddard became convinced that reaching high altitudes and ultimately escaping the Earth's gravity would require more powerful liquid-propellant rockets.

Early Bazooka

During World War I, Goddard experimented with solid rockets for the U.S. Army Signal Corps. Together with C.N. Hickman and H.S. Parker, he demonstrated a small rocket that could be launched from a tube and traveled straight distances of 60–80 ft (18–24 m). The rocket tests were conducted in summer 1918 at Mount Wilson Observatory in California. The development of this early prototype of bazooka had come to an end with the Armistice signed in Europe.

"A Method of Reaching Extreme Altitudes"

Robert Goddard presented the results of his early rocket work in the *Smithsonian Miscellaneous Collections*, v. 71, N.2, 1919, a publication of the respectable Smithsonian Institution. This famous treatise of Goddard, entitled "A Method of Reaching Extreme Altitudes," outlined his ideas on rocketry and included detailed calculations of rocket dynamics and results of his various tests.

Goddard's paper included a section, "Calculation of Minimum Mass Required to Raise One Pound to an 'Infinite' Altitude." Goddard presented calculations of the initial, starting mass of a rocket capable of sending 1 lb on the "parabolic" velocity, or what we would commonly call today the "escape" velocity. He was also concerned with the experimental proof that the rocket would indeed escape. Goddard wrote,

> It is of interest to speculate upon the possibility of proving that such extreme altitudes had been reached even if they actually were attained. In general, the proving would be a difficult matter. Thus, even a mass of flash powder,

All the Wisdom That's Fit to Bestow

TOPIC OF THE TIMES

... After the rocket quits our air and really starts on its longer journey [to the moon], its flight would be neither accelerated not maintained by the [proposed by Goddard solid rocket based on] explosion of the charges ... . To claim that it would be is to deny a fundamental law of dynamics, and only Dr. Einstein and his chosen dozen, so few and fit, are licensed to do that.

... That Professor Goddard with his "chair" in Clark College and the countenancing of the Smithsonian Institution, does not know the relation of action and reaction, and of the need to have something better than a vacuum against which to react — to say that would be absurd. Of course he only seems to lack the knowledge ladled out daily in high schools.

... As it happens, Jules Verne, who also knew a thing or two in assorted sciences ... deliberately seemed to make the same mistake that Professor Goddard seems to make. For the Frenchman, having get his travellers to or toward the moon into the desperate fix of riding a tiny satellite of the satellite, saved them from circling it forever by means of explosion, rocket fashion, where an explosion would not have had in the slightest degree the effect of releasing them from their dreadful slavery. That was one of Verne's few scientific slips, or else it was a deliberate step aside from scientific accuracy, pardonable enough in him as a romancer, but its like is not so easily explained when made by a savant who isn't writing a novel of adventure.

Editorial comments, *The New York Times*, 13 January 1920

> arranged to be ignited automatically after a long interval of time, were projected vertically upward, the light would at best be very faint, and it would be difficult to foretell, even approximately, the direction in which it would be most likely to appear.
>
> The only reliable procedure would be to send the smallest mass of flash powder possible to the dark surface of the moon when in conjunction (i.e. the new "moon"), in such a way that it would be ignited on impact. The light would be visible in a powerful telescope .... (Goddard 1970, p. 393)

Fig. 9.12. Robert H. Goddard with Harry F. Guggenheim (left) and Charles A. Lindbergh (right) near Roswell, New Mexico; September 1935. Photo courtesy of NASA.

"Moon Rocket"

The "moon part" of this highly technical report caught the attention of newspapers. As Goddard later observed, "from that day, the whole thing was summed up, in the public mind, in the words 'moon rocket'" (Goddard 1970, 24). The *New York Times* went much further ridiculing the idea that rocket propulsion would work in vacuum and questioning the integrity and professionalism of the author. The sensationalism of newspapers and merciless ridicule left a profound impression on Robert Goddard who became secretive about his work, shied publicity, and avoided contacts with the media.

DANIEL AND HARRY GUGGENHEIM

Philanthropists Daniel (1856–1930) and Harry (1890–1971) Guggenheim, a father and son, made a difference in development of aeronautics in the United States in 1920s. The Guggenheims provided much-needed seed money to promote critical areas of this fledgling field of technology. They used their great wealth to sponsor aeronautical meteorology, instrument flying, speed reduction at landing, experimental "model airway" between Los Angeles and San Francisco, the U.S. Library of Congress aeronautical collection, and marking the rooftops of more than 8000 American towns with the town names to aid in navigation and locating the airfields.

Through the Daniel Guggenheim Fund for the Promotion of Aeronautics (1926–1930) and the Daniel and Florence Guggenheim Foundation (founded in 1924), they provided funds to establish the first school of aeronautics at a major American university (New York University). Later, they funded the aeronautical engineering schools at the California Institute of Technology, Stanford University, Massachusetts Institute of Technology, Georgia Institute of Technology, and many others.

No More Launches in Massachusetts

Goddard continued development of various rocket components and conducted tests in Worcester. On 17 July 1929, he tried to fly a rocket with several instruments: an aneroid barometer, a thermometer, and a camera to record the readings. The launch failed and caused fire. When the neighbors complained to the state fire marshal, Goddard was enjoined from further launches in Massachusetts.

Fig. 9.13. Robert H. Goddard, ca. 1932. Photo courtesy of NASA.

Charles Lindbergh Steps in

The monoplane *Spirit of St. Louis* carried Charles A. Lindbergh solo across the Atlantic and to the world fame in 1927. Two years later, Lindbergh met with Goddard and became his lifelong supporter. It was Lindbergh who convinced the philanthropist Daniel Guggenheim to support Goddard's rocket work. In 1930, the Daniel Guggenheim Foundation for the Promotion of Aeronautics granted Robert Goddard $100,000 for four years. The new funding allowed Goddard to transfer to a new location in the West and devote his full time to rocket research and development.

Fig. 9.14. Robert H. Goddard tows a rocket to a launching site near Roswell, New Mexico, some time in 1930–1932. Photo courtesy of NASA.

Fig. 9.15. Robert H. Goddard (left) works with his crew of three on a new rocket model at Roswell, New Mexico. The rocket was completed in July 1938 and flew the next month to an altitude of 3294 ft (1004 m) and safely descended by parachute. The altitude was recorded by a small barometer. This was the last successful flight test of Goddard's rockets. Photo courtesy of NASA.

Move to Roswell, New Mexico

Goddard moved to Roswell, New Mexico, where he settled at Mescalero Ranch. The location was definitely different from the familiar New England. This was the land where Billy the Kid roamed 40 years earlier and where an unidentified flying object (UFO) would "crash" 17 years later, followed by numerous "sightings" of alien spaceships.

In Roswell, Robert Goddard established a small shop with a crew of assistants. On 30 December 1930, Goddard fired a liquid-propellant rocket to a 2000-ft (600-m) altitude. In the rocket shot on 19 April 1932, he demonstrated rocket steering with vanes in the exhaust flow, controlled by a gyroscope.

The Great Depression terminated Guggenheim's

NO DIRECT LINE FROM GODDARD

There is no direct line from Goddard to present-day rocketry. He is on a branch that died. He was an inventive man and had a good scientific foundation, but he was not a creator of science, and he took himself too seriously. If he had taken others into his confidence, I think he would have developed workable high-altitude rockets and his achievements would have been greater than they were. But not listening to, or communicating with, other qualified people hindered his accomplishments.

Theodore von Kármán, early 1960s
(von Kármán 1967, 242)

funding in 1932, and Goddard had to return to a full-time teaching at Clark University. The Smithsonian Institution provided modest funds to continue his research on rocketry though without actual rocket flights.

Lindbergh and Guggen-heim Help Again

Two years later, in 1934, Charles Lindbergh again proved critical in convincing Harry Guggenheim to resume financial support for Goddard's work. The new funds were secured this time from the Daniel and Florence Guggenheim Foundation. The old team was reassembled in Roswell and Goddard resumed his work.

The next seven years witnessed development of numerous techniques in propulsion, control, and rocket design. The gyro stabilization was improved, the control vanes strengthened, gimbal-mounted engines tried, more efficient

Fig. 9.16. One of Goddard's rockets in a launching tower in Roswell, New Mexico. Photo courtesy of NASA.

Fig. 9.17. Reconstructed Goddard's launch tower with a rocket at the Roswell Museum and Art Center in Roswell, New Mexico. The author of this book stands next to the commemorating plaque on the stone (top, right). Photo courtesy of Mike Gruntman.

Many Firsts

propellant pumps and lightweight tanks were developed, and cooling of combustion chamber and nozzle studied, with numerous static tests improving the design of engines. Robert Goddard invented, developed, and achieved many firsts, but the results of his work, because of self-imposed secrecy, remained largely unknown to other scientists and engineers. American rocketry was developing independently of Goddard without, unfortunately, benefits of his accomplishments.

Record Altitudes

In 1935 the first liquid-propellant rocket accelerated to a speed faster than the speed of sound. Two years later, on 26 March 1937, the Goddard's rocket reached an altitude 8000–9000 ft (2400–2700 m). All of this work was performed by Robert Goddard practically alone, with a few assistants. History would demonstrate, in a few years only, that the time of such individual effort has gone. The development of a modern powerful rocket would require a concerted effort

LATE RECOGNITION

Goddard earned 214 rocket patents by the time he died in 1945. Many years later, the U.S. government acknowledged use of Goddard's patents and paid $1,000,000 in settlement.

of hundreds and thousands of scientists and engineers backed by the vast resources; the task possible only with the support of a mighty state.

World War

With World War II blazing in Europe, Goddard tried to interest the military to support his work. Rockets for jet-assisted takeoff of airplanes were of the immediate interest. Under the U.S. Navy's contract, Goddard transferred his developmental work to Annapolis, Maryland, where tests of his liquid-propellant engines for takeoff of flying boats were not entirely successful. The Navy continued with another contract to Goddard, focused mostly on improvement of the various rocket-based devices developed by other contractors.

Goddard Space Flight Center

Robert H. Goddard died on 10 August 1945, in Baltimore a few days after World War II ended. The recognition of Goddard's work came only long after his death. In 1959 the U.S. Congress honored Robert H. Goddard, and NASA named after him one of its leading field centers, *Goddard Space Flight Center*, in Greenbelt, Maryland, on May 1.

HERMANN OBERTH (1894–1989)

Hermann Oberth was born on 25 June 1894, to a German family in Transylvania, a region of Austria–Hungary and now a part of Rumania. When he was 11 years old, Hermann had become fascinated by spaceflight after reading Jules Verne's *From the Earth to the Moon.*

Fig. 9.18. Hermann Oberth in 1955–1958. Photo courtesy of NASA.

First Rocket Proposal

Oberth's study of medicine in Munich was interrupted by World War I. He was called into the Austro–Hungarian Army and wounded. He found time to explore his ideas of spaceflight and developed a proposal for a long-range liquid-propellant rocket. The proposal was rejected by the German War Ministry with a brief answer, "According to experience these rockets do not fly farther than seven kilometers ... it cannot be expected that this distance can be surpassed considerably" (Oberth 1985, 131).

The war was now over, and Oberth decided to continue his education in Germany. As a Rumanian citizen and a foreigner in Germany, he was subjected to many restrictions.

Oberth studied first in Munich, then in Göttingen, and finally at Heidelberg University. Here he sought a Ph.D. degree in astronomy or physics. His dissertation, based on his rocket design, was rejected in 1922. It was considered neither a work on astronomy by astronomers nor a work on physics by physicists. Oberth had to return to Rumania. He quickly graduated at a local university and began to work as an instructor in a gymnasium (high school).

Ph.D. Disserta-tion Rejected

A famous Heidelberg astronomer Max Wolf suggested that Oberth publish his work as a book. After a long search for the publisher, the book appeared in 1923 under the title *Die Rakete zu den Planetenräumen* (*The Rocket into Interplanetary Space*). Oberth had to pay the publication expenses using the savings of his wife.

Die Rakete zu den Planeten-räumen

In his book, Hermann Oberth focused on rocket dynamics. He considered the powered vertical flight through the atmosphere and introduced a concept of the optimal velocity ("most favorable velocity," as he called it) that minimizes propellant consumption. He showed how a rocket could achieve the velocity allowing it to escape Earth's gravity. He described the details of liquid-propellant engines and calculated propellant exhaust velocities. Oberth introduced a number of ideas such as rocket staging, film cooling of the engine walls, and strengthening the structure by pressurizing propellant tanks.

Oberth's Model B

The book presented a detailed design of a two-stage rocket with extensive supporting engineering calculations. The rocket used liquid oxygen as an oxidizer. The fuel was liquid hydrogen for the second-

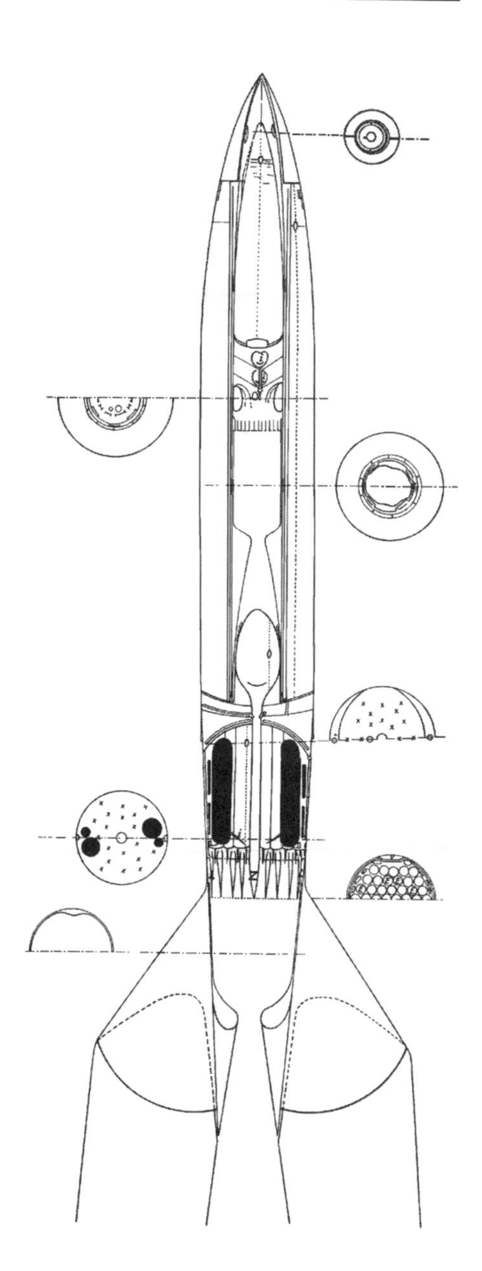

Fig. 9.19. *Model B* liquid-propellant two-stage rocket designed by Hermann Oberth, 1923. The upper stage used liquid oxygen and liquid hydrogen as propellant; the lower stage used liquid oxygen and the mixture of water and alcohol. Figure courtesy of Oldenbourg Wissenschaftsverlag, Munich, Germany, from *Die Rakete zu den Planetenräumen* (Oberth 1923, Tafel I).

(upper) stage engine and the alcohol-water mixture for the first- (lower) stage engine. The latter combination, liquid oxygen with the alcohol-water mixture, would be used in the famous German A-4 (V-2) ballistic missile in 1940s.

The Oberth's rocket, called Model B, was 5 m (16.4 ft) long, 55.6 cm (22 in.) wide, and weighed 554 kg (1221 lb). Oberth designed his rocket for study of the Earth's atmosphere, capable of reaching the altitude of 2000 km (1250 km), when launched with the assistance of an additional booster stage.

Observation Station in Orbit

Oberth's book also discussed possible applications of large rockets permanently orbiting the Earth. He called them the *observation stations*. The orbiting stations could be used for communications (Oberth thought about light signals) and for studying the Earth from space. The suggested examples included observing and photographing the unexplored lands and unknown peoples in Tibet.

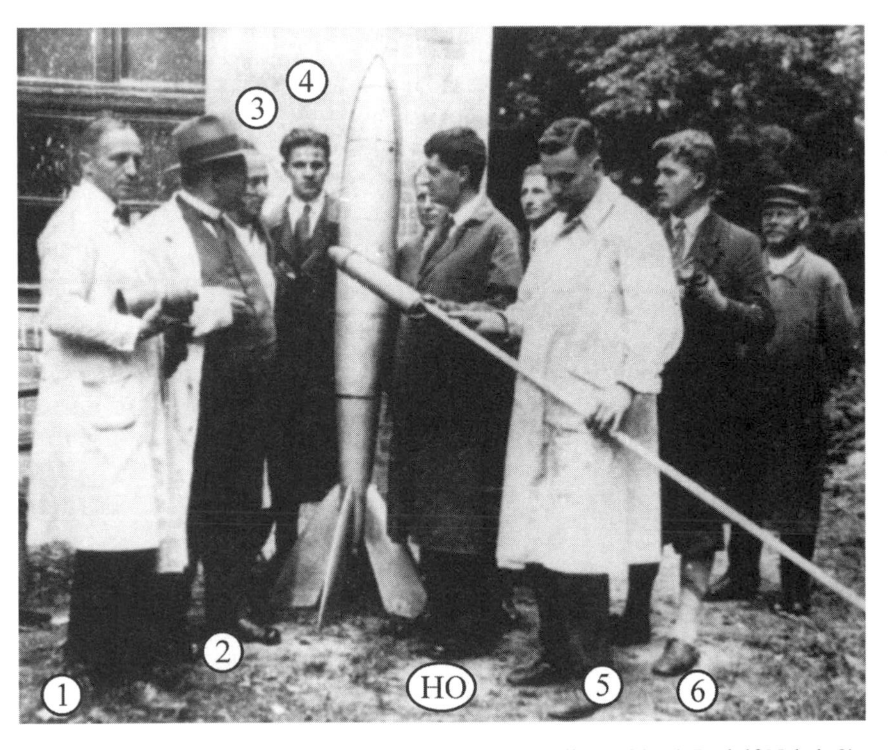

Fig. 9.20. Hermann Oberth (HO) with his team in Berlin, 1930: 1) Rudolf Nebel, 2) Ritter, 3) Barmuller, 4) Kurt Heinisch, 5) Klaus Riedel, 6) Wernher von Braun. Photo courtesy of NASA; identification by F.I. Ordway III and M.R. Sharpe (1979, 1982).

Giant Mirrors

Hermann Oberth also emphasized the particular strategic value of space stations for surveying the theaters of war operations. In addition, spotting icebergs from space and relaying the warning to the ships could have prevented the disaster of Titanic, he argued. More futuristic applications envisioned by Oberth included placing large mirrors, up to 100 km (62 mile) in size, in orbit. The mirrors would concentrate the solar light and melt the ice in polar regions, keeping the Northern harbors free of ice.

Work in Isolation

Hermann Oberth did his work on rocket dynamics independently of other researchers. He became familiar with the work of Robert Goddard and Konstantin Tsiolkovsky in 1922 and 1925, respectively. Oberth was not unique among the space pioneers accustomed to working in isolation. It was a common situation because no regular channels of communication existed among space enthusiasts at the beginning of the space age. The work of space pioneers was often met with indifference by the scientific and cultural elite, and the doors of respected journals and societies were firmly closed for them. The lectures by Robert Esnault-Pelterie at the French Physical Society and the French Astronomical Society were truly exceptional and helped to "legitimize" the new field of research.

First REP-Hirsch Award

The next Oberth's book, *Wege zur Raumschiffahrt* (*Road to Space Travel*) published in 1929, described large rockets for manned spaceflight and suggested an ion rocket (electric propulsion). This work brought Oberth the first REP-Hirsch Award. The French Committee on Astronautics was so impressed by the Oberth's treatise that it doubled the prize to 10,000 francs.

The award money allowed Oberth to finance the research on liquid-propellant rocket engines. German UFA Film Company also provided him some support to build a rocket for the movie *Frau im Mond* (*Woman in the Moon*) filmed by director Fritz Lange. The company conceived a publicity stunt to time the firing of the Oberth's rocket with the first showing of the film. It was soon realized that no rocket launch worthy of the desired publicity could take place, and the project was cancelled.

Kegeldüse

Oberth joined the German *Society for Space Travel* and became its president in 1929. The first successful static tests of his conical rocket engine, *Kegeldüse*, were performed in 1930 near Berlin. It was for this project that a young assistant, Wernher von Braun, joined Oberth's team. The rocket engine used gasoline and liquid oxygen and produced 68.5 N (15.4 lbf) of thrust for 90 s. No funding to continue the work was forthcoming, and Oberth returned to his teaching post in Rumania.

German Citizen and Contribution to War Effort

Oberth finally relocated to Germany during World War II and became a German citizen in 1940. He joined von Braun in 1941 as an advisor at the Peenemünde rocket center and was subsequently transferred to development of solid-propellant antiaircraft rockets elsewhere in 1943.

After spending some time in Switzerland after the war, Oberth became involved in work on solid-propellant antiaircraft missiles for the Italian Navy. He also got interested in unidentified flying objects, or flying saucers. In 1955, Oberth moved to Huntsville, Alabama, and joined Wernher von Braun, this time in the

FLYING SAUCERS

It is my thesis that flying saucers are real and that they are space ships from another solar system. I think that they are possibly manned by intelligent observers who are members of a race that may have been investigating our earth for centuries.

... I have examined all of the arguments supporting the existence of flying saucers and denying it, and it is my conclusion that the "Unidentified Flying Objects" do exist, are very real and are visitors from outer space.

...Flying saucers are space ships coming here — one way or another — from outside of our solar system.

Hermann Oberth (1954, 4)

United States, in the Army ballistic missile program. Von Braun always considered Oberth his mentor, and von Braun's group of German rocketeers highly respected him.

Fig. 9.21. Hermann Oberth (left), Wernher von Braun (center), and Chief of the Physics and Astrophysics Section, Research Projects Laboratory, Army Ballistic Missile Agency, Charles A. Lundquist, in Huntsville, Alabama, 28 June 1958. Photo courtesy of U.S. Army.

Oberth wanted to retire in the United States. His employment with the Army in Huntsville was, however, too short for an adequate American pension and he had to return to Germany to be eligible for his German pension. So, Oberth went back to West Germany in 1958 where he continued his theoretical work and publications, including a philosophical book in 1959. Oberth passed away in Nürnberg in 1989.

Hermann Oberth played an important role in practical development of rocketry in Germany in the 1930s and provided inspiration for a generation of European spaceflight enthusiasts.

ROCKET AND SPACEFLIGHT SOCIETIES

The great space pioneers were not alone. A growing number of enthusiasts in various countries explored the possibilities of spaceflight. A German architect from Essen, Dr. Walter Hohmann, studied various orbits and maneuvers in his

WORLD WAR I

Lack of recoil allowed firing large-caliber rockets from small boats in the 19th century. The same characteristic feature of rocket launch made rockets attractive for airplanes. The other advantage of the air-launched rocket included a possibility to discharge the weapon at increased distances from the target, often beyond the enemy air defense fire.

During World War I, the French were the first to fire rockets from a Nieuport Scout airplane in the battle of Verdun on 22 May 1916. The rocket became know as *Le Prieur* after a French naval officer, Lieutenant Y.P.G. Le Prieur, who was credited with the suggesting the use of rockets. The rockets were usually employed against the hydrogen-filled observation balloons.

Rocket and Spaceflight Concepts Advanced

1925 book *Die Erreichbarkeit der Himmelskörper* (*The Attainability of the Celestial Bodies*). He suggested aerobraking for spacecraft returning to the Earth and identified the most energy-efficient orbital transfer maneuver, which would bear his name, the *Hohmann transfer*.

Orbital Station and GEO

An Austrian–Slovenian engineer Hermann Potočnik (1892–1929) discussed in detail the concept of Earth-orbiting manned space stations in his 1929 book *Das Problem der Befahrung des Weltsraums. Der Raketen-Motor* (*The Problem of Space Travel. The Rocket Motor*). Potočnik was born in the Austro-Hungarian Hapsburg Empire, educated in German, and after the disintegration of the Empire settled in Austria. He was apparently of Slovenian and possibly Czech origin. His book was published under pseudonym Hermann Noordung. In the book, Potočnik-Noordung also suggested space station attitude control by three orthogonal reaction wheels, addressed many problems of propulsion and communications, described the geostationary orbit, provided analysis of travel to the moon and planets, and even considered some requirements to interstellar flight.

GEOSTATIONARY ORBIT (GEO)

A geostationary orbit (GEO) is an equatorial circular orbit with an orbital period equal to that of the Earth's rotation. A satellite in GEO would be seen as stationary in the sky by an observer on the ground. This orbit is probably the most valuable piece of "real estate" in space for placing communications satellites.

Spaceflight and Rocket Enthusiasts

The growing number of scientists and engineers enthusiastically discussed space flight and rocketry in the late 1920s and early 1930s. They worked on various problems of astronautics in Germany (Karl Debus, Willy Ley, H. Lorenz, Rudolf Nebel, Klaus Riedel, Alexander Scherschewsky, Max Valier, Johannes Winkler),

IT WILL EXPLODE ...

"A good rule for rocket experimenters to follow is this: always assume that it will explode."

Astronautics, No. 38, p. 8, October 1937

GEOSYNCHRONOUS SATELLITES

World War II was over. A 27-year old man has published an article "Extra-Terrestrial Relays" in the15 October 1945, issue of the British journal *Wireless World.* At that time he still served in the Royal Air Force as a radar instructor and technician, and his name was Arthur C. Clarke. He would later become president of the British Interplanetary Society, a famous and prolific science fiction writer (the reader might remember Stanley Kubrick's 1968 film *2001: A Space Odyssey* based on the Clarke's 1951 short story *The Sentinel*), and expert skin diver and photographer.

Fig.9.22. Arthur C. Clarke ca. 1945. Photo courtesy of Arthur C. Clarke.

In his 1946 article Clarke noted that a satellite in an orbit with a radius about 42,000 km, would "have a period of exactly 24 hours. A body in such an orbit, if its plane coincided with that of the earth's equator, would revolve with the earth and would thus be stationary above the same spot on the planet. It would remain fixed in the sky of a whole hemisphere and unlike all other heavenly bodies would neither rise nor set."

Clarke then suggested that three satellites in such an orbit would ensure complete coverage of the globe. His calculations showed that "small parabolas about a foot in diameter would be used for receiving at the earth end and would give a very good signal/noise ratio."

Clarke went through the calculations of the required power and proposed to use solar engines that employed mirrors to concentrate sunlight on a boiler and have already been devised for terrestrial use. Practical solar cells did not exist at that time, and Clarke observed that "thermo-electric and photo-electric developments may make it possible to utilize the solar energy more directly." He identified the importance of solar eclipses for geosynchronous satellites and correctly calculated their durations and the seasons of the eclipses.

Clarke's concept of geosynchronous satellite-based communications was supported by detailed engineering calculations. It was a solid work that would be called today a "feasibility study." He advanced his concept in 1945 when the most powerful rocket (V-2) flew only a few hundred miles, solar cells did not exist, spacecraft navigation and attitude control were rudimentary, and space environment was poorly understood. "It may be argued," noted Clarke, "that we have as yet no direct evidence of radio waves passing between the surface of the earth and outer space [which is necessary for communications via satellite]; all we can say with certainty is that shorter wavelengths are not reflected back to earth." He clearly realized that "many may consider solution proposed [in his article] ... too far-fetched to be taken seriously." Well, only 17 years later the first geosynchronous communication satellite was in orbit.

SPACE WARFARE AND DETERRENCE IN 1929

There is no such a thing as a purely civilian, nonmilitary space technology. Hermann Oberth suggested space-based mirrors to melt the ice in order to keep strategic Northern harbors ice free. Hermann Potočnik-Noordung went further describing in 1929 the potential military applications of mirrors in space. Potočnik-Noordung wrote,

> But like any other technical achievements the space mirror could also be employed for military purposes and, furthermore, it would be a most horrible weapon, far surpassing all previous weapons.
>
> ... Now, if we visualize that the observer in the space station using his powerful telescope can see the entire combat area ... including the staging areas and the enemy's hinterland ... .
>
> It would be easy to detonate the enemy's munitions dumps, to ignite his war material storage areas, to melt cannons, tank turrets, iron bridges, the tracks of important train stations, and similar metal objects. Moving trains, important war factories, entire industrial areas and large cities could be set ablaze. Marching troops or ones in camp would simply be charred when the beams of this concentrated solar light were passed over them. And nothing would be able to protect the enemy's ships from being destroyed or burned out ... .
>
> However, all of these horrible things may never happen, because a power would hardly dare to start a war with a country that controls weapons of this dreadful nature. (Noordung 1995, 122, 123)

France (Alexandre Ananoff, Jean-Jacques Barré, Louis Damblanc, Pierre Montagne, Henri F. Melot), Austria (Eugen Sänger, Friedrich Schmiedl, Franz von Hoefft, Guido von Pirquet, Franz Ulinski), the United States (David Lasser, G. Edward Pendray), the Soviet Union (Yurii Kondratyuk, Yakov Perel'man, Nikolai Rynin, Nikolai Tikhomirov, Fridrikh Tsander), Italy (Giulio Constanzi, Gaetano Arturo Crocco), Czechoslovakia (Ludvík Očenášek), Japan (Tsunendo Obara), and many others.

Rocket Societies

Small but enthusiastic groups began to coalesce in 1920s forming associations and societies to promote the rocket research. *The Austrian Society for High-Altitude Research* (*Wissenschaftliche Gesellschaft für Hohenforschung*) was founded in 1926 by Franz von Hoefft. Von Hoefft became the Society's first president, and Guido von Pirquet was named Secretary. Later Rudolf Zwerina and von Pirquet established *the Austrian Society for Rocket Technology* (*Österreichische Gesellschaft für Raketentechnik*) in 1931.

Die Rakete

The German *Society for Space Travel (Verein fur Raumschiffahrt – VfR)* was founded in Breslau (present Wroclaw in Poland) in 1927. VfR started publication of the magazine *Die Rakete* devoted to spaceflight. VfR had rapidly grown to more than 1000 members by 1930.

American Rocket Society

Then, the *American Interplanetary Society* (renamed four years later to the *American Rocket Society*, or ARS) was established in New York City in April 1930. Several other smaller American groups were also formed in the 1930s: *Cleveland Rocket Society* in Cleveland, Ohio, in 1933 by Edward L. Hanna and E. Loebell; *Peoria Rocket Association* in Peoria, Illinois, in 1934

> **AMERICAN ROCKET SOCIETY**
>
> The annual meeting of the [American Interplanetary] Society will be held this year at 8 o'clock, Friday, April 6th, at the American Museum of Natural History. Two important items of business are up for action: election of a Board of Directors, and a vote upon whether the name of the Society should be changed to "The American Rocket Society." In the opinion of many members, adoption of the more conservative name, while in no way implying that we have abandoned the interplanetary idea, would attract able members repelled by the present name . All active members are entitled to vote ... .
>
> *Astronautics,* No. 28, p. 7, March 1934

by Ted S. Cunningham; *American Institute for Rocket Research* in 1936 in Chicago, Illinois, by C.W. McNash.

British Interplanetary Society

The *British Interplanetary Society* (BIS) was founded in October 1933 by Phillip E. Cleator in Liverpool. In 1937 BIS was formally transferred to London. The practical work of British rocketeers was severely hampered by the law, the Explosives Act of 1875, forbidding private experimentation and manufacturing of various explosives, including rockets.

Then sixteen-year-old Eric Burgess established the *Manchester Interplanetary Society* in 1936. Then he and Trevor Cusack, founded the *Manchester Astronautical Association* in 1937. BIS suspended its activities in the late 1939 with the beginning of World War II and was reestablished when the war ended.

World-wide Interest

The *Dutch Rocket Society* (*Nederlandse Rakettenbouw*) was formed by Gerard A.G. Thoolen in The Hague in 1934. The *Australian Rocket Society* was founded in 1936 in Brisbane, Queensland, by Alan H. Young and Noel S. Morrison.

Fig. 9.23. (Left to right) Max Krauss, Nathan Schachner, Bernard Smith, and Alfred Africano were among the first members of the American Rocket Society, summer 1933. Photo from *Astronautics*, No. 26, p. 13, May 1933.

AMERICAN ROCKET SOCIETY (ARS), INSTITUTE OF AEROSPACE SCIENCES (IAS), AND AMERICAN INSTITUTE OF AERONAUTICS AND ASTRONAUTICS (AIAA)

In 1930, a group of space flight enthusiasts used to gather at the apartment of the *New York Herald Tribune* reporter G. Edward Pendray. David Lasser, an editor of science fiction publications, was the heart of the group. An excellent speakeasy operated at the basement of the apartment building, helping to keep the spirit of the discussions high.

On an evening of 4 April 1930, 12 enthusiasts established the *American Interplanetary Society*. In addition to Lasser and Pendray, the Society's first president and vice president, the founding members included Mrs. Pendray, Charles W. Van Devander, Adolf L. Fierst, Warren Fitzgerald, William Lemkin, Everett Long, Laurence E. Manning, Charles P. Manson, Fletcher Pratt, and Nathan Schachner.

Most of the founders were of a literary, nontechnical type: journalists and writers interested in science fiction. Lemkin was the only Ph.D. in the group. The society was successfully "launched" and soon had more than 100 members, including the growing number of scientists and engineers. These technically oriented members, H. Franklin Pierce, Bernard Smith, John Shesta, Lovell Lawrence, Alfred Africano, Roy Healy, James Wyld, and others, initiated and led an energetic experimental program and began to dominate the group. Robert Goddard also joined the society at this time.

The society was renamed the *American Rocket Society* (ARS) in 1934. World War II led to the explosive growth of rocket development in the United States. ARS was mostly an East Coast group, while new rocket centers were emerging in the West. At first, the establishing of an independent professional rocket society on the West Coast was considered. This idea was finally abandoned, and the "Westerners" began joining ARS. The Society's periodical publication, *Astronautics. Journal of the American Rocket Society*, played an important role in information exchange among the rocketeers.

The *Institute of Aeronautical Sciences (IAS)* was organized in 1932 "to promote the application of science in the development of aircraft." The Institute became the main American scientific and engineering society in the field of aeronautics. Many leading scientists, engineers, and aircraft and aircraft subsystem designers joined IAS, which had 408 and 655 members on 1 January of 1933 and 1934, respectively. The members of the Institute of Aeronautical Sciences included such well-known airplane designers and specialists in aeronautics as Joseph S. Ames, Lyman J. Briggs, Charles Lawrance, James H. Doolittle, Donald W. Douglas, Hugh L. Dryden, Jerome C. Hunsaker, Alexander Klemin, Glenn L. Martin, Lessiter C. Milburn, Clark B. Millikan, John K. Northrop, Igor I. Sikorsky, Theodore von Kármán, Albert F. Zahm, and many others. A large number of foreign scientists and engineers were also registered as Institute members.

IAS was renamed to the *Institute of Aerospace Sciences* in 1960, reflecting the rapidly growing importance of research and development in the emerging rocket-related areas, beyond the traditional realm of aeronautics.

On 1 February 1963, ARS merged with IAS to form the *American Institute of Aeronautics and Astronautics* (AIAA). At that time, ARS had more than 20,000 members — a spectacular transformation of an original group of a dozen spaceflight dreamers in New York. Today AIAA is the world's largest professional society of aerospace scientists and engineers with more than 31,000 members.

Fig. 9.24. American rocket pioneers John Shesta (left) and G. Edward Pendray with experimental rocket No. 4, summer 1934. Photo from *Astronautics*, No. 29, p. 1, September 1934.

Another *Australian Rocket Society* was formed in 1941 in Melbourne, Victoria. In France, the French Astronomical Society was the center of activities in spaceflight and astronautics, with the Society's *Astronautical Section* (*Section Astronautique*) established by Alexandre Ananoff in 1938.

Soviet State Steps In

The Communist government of the post-revolutionary Soviet Union strongly supported rocket-related research, with the military playing the most important role since 1921. A section on *Interplanetary Communications [Travel]* was established at the Red Air Force Academy in 1924. Later the same year, the section was reorganized as the *Society for the Study of Interplanetary Communications* (*Obshchestvo po Izucheniyu Mezhplanetnykh Soobshchenii*, or *OIMS*). The OIMS membership grew to almost 200 members. The society organized the First World Exhibition of Interplanetary Machines and Mechanisms in Moscow in 1927.

OIMS and other groups of space enthusiasts were absorbed within a few years into a consolidated state-sponsored military effort in rocketry. Nothing could be independent of the state in a totalitarian socialist society, and spaceflight was no exception.

WAR AND THE BRITISH INTERPLANETARY SOCIETY

What war means to rocketors is rather dramatically revealed in recent correspondence with British experimenters. For example, this letter from Arthur C. Clarke, Treasurer of the British Interplanetary Society, to [ARS's] Mr. [Alfred] Africano:

> "Owing to the war, it will be impossible for the B.I.S. to carry on an active existence, and arrangements have been made to put it in cold storage for the duration.
>
> In order that our work will not be entirely lost, whatever happens to us, I am sending you a few copies of our last two "Journals". These contain merely the general outlines of the theoretical work we have done, but they may assist others to follow up the same lines of research.
>
> After the War any surviving members of the Council will attempt to start things going again, but until then — whenever that may be — it's no good making any plans for the future .... .
>
> If the worst comes to the worst — which I don't for a moment think it will — we hope that the ARS will be able to see that our work has not been entirely wasted."

Astronautics, No. 44, p. 10, November, 1939

World War II and Astronautics

Fig. 9.25. Alexandre Ananoff, 1907–1992, was the key organizer of the First International Astronautical Congress. Photo courtesy of the British Interplanetary Society and International Astronautical Federation.

World War II led to the suspension of most of astronautical activities. After the war a number of national organization devoted to spaceflight and rocketry were gradually reestablished, reborn, or formed. The *Society for Study of Space* (*Gesellschaft für Weltraumforschung — GfW*) in Stuttgart, West Germany, suggested in 1949 to other astronautical groups, especially the British Interplanetary Society, that an international astronautical meeting be organized.

Premier Congrès International d'Astronautique

Alexandre Ananoff, then Director of the *French Astronautical Group of the French Aero Club* (*Groupement Astronautique Français, Aero-Club de France*) volunteered to organize such a meeting in Paris. Ananoff who called the meeting the *First International Astronautical Congress* (*Premier Congrès International d'Astronautique*) spectacularly succeeded with this difficult task.

The Congress opened on 30 September 1950, with the first day session attended by more than 1000 participants. The next two days were the closed meetings of 19 representatives from eight astronautical societies. With an exception of one South American, the Argentinian T.M. Tabanera, all other meeting participants were from the European societies.

The Second International Astronautical Congress was opened in September of the next year in London. The American Rocket Society pledged support for the

Fig. 9.26. Working (private) meeting of the First International Astronautical Congress on 1 October 1950. Left-to-right: H.J. Rückert, T. Mur, L. Hansen, T.M. Tabanera, A. Ananoff, G. Loesser, H.H. Koelle, W. Brugel, C. Oesterwinter, F.K. Jungklaass, and J. Humphries. Photo courtesy of International Astronautical Federation.

international astronautical organization and sent its representatives Andrew G. Haley and Frederick C. Durant, III, to the gathering. In addition to ARS, three other American groups were among the founding member: Pacific Rocket Society, Detroit Rocket Society, and Reaction Research Society.

International Astronautical Federation (IAF)

On 4 September 1951, the *International Astronautical Federation (IAF)* was born. Eugen Sänger became the first president and G. Loeser and A.G. Haley vice presidents. The Second Congress also featured the first astronautical scientific conference on 6 and 7 September, the *Symposium on Artificial Satellites*. Today, more than 150 members — government organizations, industrial companies, and professional and learned societies — from 45 countries belong to the IAF.

10. THE FIRST MODERN ROCKET

Rocket Surpassing Artillery Range

The year 1932 started a set of events that led to a quantum leap in rocket technology. That year, the *Ordnance Department* (*Heereswaffenamt*) of the *German Army* (*Reichswehr*) decided to develop a liquid-propellant rocket with the range surpassing that of artillery. Two years later, the Army initiated a major program that would lead to long-range ballistic missiles.

German Army Steps in

It was clear that only the utmost secrecy would fully exploit the subsequent surprise deployment of rockets. Therefore, the Army Ordnance decided on in-house development and put Captain Walter R. Dornberger in charge of the new program at the Kummersdorf Artillery Range 17 miles (27 km) south of Berlin. The Army's Chief of Ballistics and Munitions Colonel (later General) Karl Becker provided resources for the rocket development at the Army's facilities.

Fig. 10.1. German foremost rocket expert General Walter R. Dornberger, 1895–1980. Dornberger was at the origins of the German rocket program in 1930 and became the program's enthusiastic leader and promoter and outstanding administrator. For many years he commanded the Army's Peenemünde center and led rocket development until the very end of World War II. Photo courtesy of National Air and Space Museum, Smithsonian Institution (SI 2003-4827), Washington, DC.

Kummersdorf Artillery Range

Twenty-year-old Wernher von Braun reported for work at Kummersdorf on 1 November 1932. A new chapter in rocket history had begun. Captain Dornberger would

HERMANN OBERTH AND WERNHER VON BRAUN

Wernher von Braun read Hermann Oberth's book *Die Rakete zu den Planetenraumen* in 1925. It became a turning point in von Braun's life. Eighteen-year-old Wernher met Oberth in the spring of 1930. Von Braun, Rudolf Nebel, and Klaus Riedel assisted Oberth in the tests of his conical rocket engine, the *Kegeldüse*. This was the beginning of the friendship between von Braun and Oberth lasting almost 50 years.

become Lieutenant General and the rocket program's energetic leader and capable administrator, while technical development would be directed by von Braun.

First Disappointment and Success

The first liquid-propellant rocket designed by von Braun, the *Aggregat-1* (A-1), exploded. (The German word *das Aggregat* can be translated as *the assembly*.) The second model, the A-2, successfully flew and reached an altitude of 1.5 mile (2.4 km) in December 1934. The rocket used liquid oxygen as the oxidizer and a 75% blend of ethyl alcohol (H_2O_5OH) with 25% of water as the fuel. The addition of water to alcohol reduced the temperature in the combustion chamber, and subsequently the heat flow to the chamber walls, with only a slight loss in performance. Enormous heat flows generated by combustion challenge to the present day the designers of rocket engines.

A-3

The next, much larger model, the A-3, was more than 21 ft (6.4 m) long, weighed 1650 lb (748 kg) at launch, and its engine produced thrust of 3300 lbf (14.7 kN). A sophisticated three-axis gyroscope system was designed to control the rocket using molybdenum vanes in the engine's exhaust flow.

Fig. 10.2. Wernher von Braun, 1912–1977. Photo (ca. 1930) courtesy of U.S. Army.

Engine for A-4

Simultaneously with the building and testing the A-3, Dornberger and von Braun proceeded with the design of an operational war rocket, the A-4, capable of delivering a one-tonne warhead to a distance of 160 miles (257 km). The A-4 rocket, also known later as the V-2, required a new powerful engine with thrust 56,000 lbf (249 kN) firing for 60 s. Dr. Walter Thiel, a combustion specialist, joined the Army team in the fall of 1936 and took charge of the development of this awesome engine.

A-5 Test Vehicle

Two test launches of the A-3 failed in December of 1937, and the malfunction was traced to the guidance system. A new rocket, the A-5, had to be specially built to study and correct the uncovered problems. The A-5 was similar to the A-3 but had an improved guidance control system. Two years later the A-5 successfully flew, reaching

Fig. 10.3. Technical Director of the *Heers Anstalt Peenemünde* (Army Establishment Peenemünde) Wernher von Braun in his office in *Haus 4* at Peenemünde, 1937. Photo courtesy of U.S. Army.

altitudes exceeding five miles. The new guidance system worked well. In addition, the molybdenum vanes in the exhaust flow were replaced by graphite, which significantly improved the response time of the vanes as well as their strength. The road to the A-4 had been finally open.

Peenemünde

The rivalry between the Army and the *Luftwaffe* (the German *Air Force)* brought a significant increase in the rocket program funding in 1935. The Kummersdorf Range near Berlin was too small for a major missile development, and the rocket establishment was transferred to a new much bigger and secluded site on the island of Usedom. Usedom was one of the islands separating the Bay of Stettin (now Szczecin, Poland) and the Baltic Sea. A small fishing village,

ROCKETS AND SUPERSONIC AERODYNAMICS

Aerodynamic design of the A-3 was experimentally validated by Dr. Rudolf Hermann in the supersonic wind tunnel at the Technical University of Aachen. This work had clearly demonstrated the importance of supersonic aerodynamics for rocket development.

Hermann joined the Peenemünde team and built a new, much bigger supersonic wind tunnel with a 16×16 in. (40×40 cm) working area. This new facility provided Mach numbers up to 5.3 and played a critical role in the aerodynamic design of the A-5 and A-4 rockets and of the surface-to-air supersonic missile *Wasserfall.*

A-4 (V-2) STORY

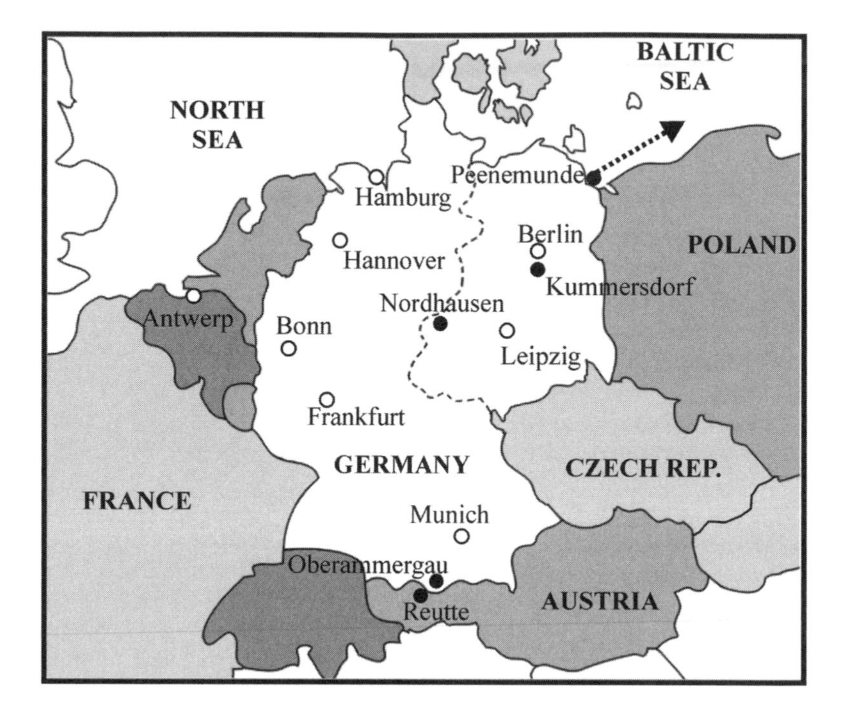

Fig. 10.4. Major locations (•) of the A-4 (V-2) story: Kummersdorf, Peenemünde, Nordhausen (near the Mittelwerk), Oberammergau and Reutte. The port and city of Antwerp were most hard hit on the continent by the V-2s.

The - - - arrow shows the direction of the A-4 test launches from Peenemünde. The eastward trajectory of vehicles passed along 250 miles (400 km) of the German-controlled Pomeranian coast with tracking and observation posts. Such trajectories preserved the rocket development secret. Only one ballistic missile deviated from this secure path during testing of the radio control system and landed in Sweden. The debris of this rocket, together with the simultaneously obtained information on another rocket (salvaged by the Polish resistance underground *Armija Krajowa* — the *Home Army* — near the German military training base in the southeastern Poland) allowed British intelligence to accurately reconstruct main V-2 characteristics in the summer of 1944.

The map shows modern country borders; ---- marks the border between the former Western and Eastern Germanies.

Figure courtesy of Mike Gruntman.

Peenemünde, was near the estuary of the Peene river crossing the island. (The village name, *Peenemünde*, can be translated as "mouth of the Peene river.")

The construction of the new facility, *Heers Versuchsstelle Peenemünde* (*Army Experimental Station Peenemünde*), began in 1935. Most of the research staff had relocated to Peenemünde by April 1937. Von Braun was appointed technical director of the Army Station, later renamed *Heers Anstalt Peenemünde* (*Army Establishment Peenemünde*). The Army part of the facility was also called *Peenemünde East*, while the Luftwaffe occupied *Peenemünde West*. The winged surface-to-surface missile Fi-103, also known as the V-1, would be developed at this Luftwaffe center.

Army Establishment Peenemünde

Fig. 10.5. Soviet Foreign Minister Vyacheslav Molotov signs the German-Soviet Nonaggression Pact in 1939. Kliment Voroshilov, Joachim von Ribbentrop, and Joseph Stalin stand behind. The pact temporarily settled the differences between the two totalitarian states and opened the way for the Germany's attack on Poland a few days later. World War II would greatly accelerate the development of rocket technology. Photo courtesy of National Archives and Records Administration.

World War

A nonaggression pact between the national-socialist Germany and the communist Soviet Union was concluded in the late August 1939. A few days later Germany attacked Poland, starting a new world war. Both dictators, Adolf Hitler and Joseph Stalin, had become enamored with rocketry. Hitler was the first to rush development, although after some delay, and the A-4 program would become a major part of the German war effort consuming enormous scientific and technological resources. Stalin did not make long-range ballistic missiles a high priority until after the war.

Army, Industry, and Academia

The German technical personnel at Peenemünde numbered about 100 in May of 1937. It had gradually grown to a few thousand by the fall of 1942. More than 10,000 people worked at the rocket center in 1944 when the new weapon system was being prepared for operational deployment.

Von Braun brought in a large number of university faculty to assist in the development of the A-4. The professors worked on integrating accelerometers, pump impellers, gyroscope bearings, radio wave propagation, Doppler tracking,

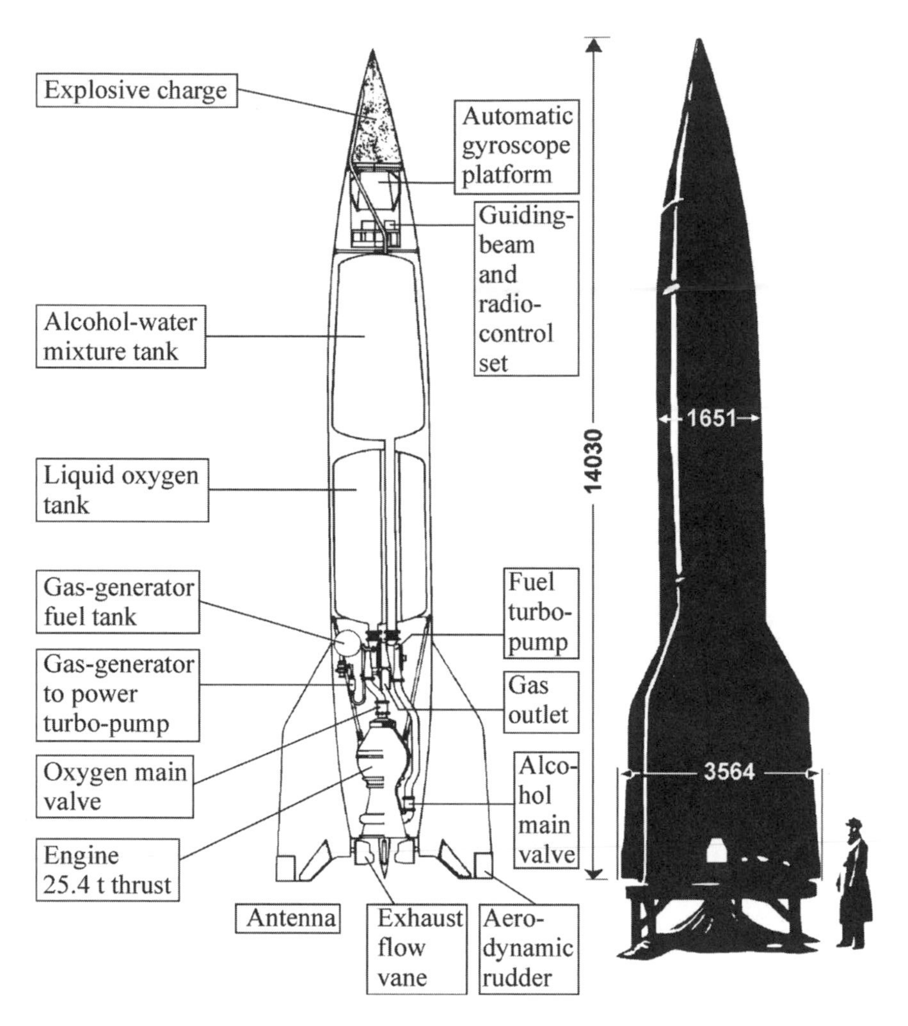

Fig. 10.6. German A-4 (V-2) rocket with the cutout view; the dimensions are in millimeters. The lower section of aerodynamic rudders, the fins, was made of plastic. The lower edge of the plastic parts was framed by copper, which served as an antenna. Original figures courtesy of NASA. Rendering of the callouts by Mike Gruntman.

antenna patterns, computing machines, and many other critically important problems. The German war rocket program combined the expertise and resources of the armed forces, industry, and academia.

Success on 3 October 1942

The first three tests of the A-4 at Peenemünde failed. 3 October 1942 had become a turning point, when the rocket was, for the first time, launched successfully. The A-4 reached an altitude of 60 miles (96 km) and hit an area 120 miles (193 km) from the launching site in the Baltic Sea. Twenty-three months of intensive testing and development followed to make the missile ready for combat.

The Führer Converted into Rocket Supporter

Germany's Minister of Armaments and War Production Albert Speer had become an enthusiastic supporter of the Peenemünde rocketeers. At Hitler's request, Speer brought Dornberger and von Braun to the Wolfsschanze headquarters in East Prussia on 7 July 1943.

PROPULSION PRACTICABLE FOR SPACE TRAVEL AND THE ROCKET AS A WEAPON

... We have invaded space with our [A-4] rocket [successfully launched today] and for the first time — mark this well — have used space as a bridge between two points on the earth; we have proved rocket propulsion practicable for space travel .... So long as the war lasts, our most urgent task can only be the rapid perfecting of the rocket as a weapon. The development of possibilities we cannot yet envisage will be a peacetime task.

Walter Dornberger, 3 October 1942 (Dornberger 1954, 17)

The Führer, skeptical for many years, was greatly impressed by the progress in rocket development and gave his full support to the A-4. Speer "proposed to Hitler that [31 year old] von Braun be appointed professor. 'Yes, arrange that at once ...,' Hitler said impulsively. 'I'll even sign the document in person'" (Speer 1970, 368).

Technological Marvel

The A-4 rocket was a technological marvel that introduced and demonstrated a number of design concepts, technologies, and materials that would be common in rocketry for the next decade and some much longer. Dornberger later summarized, that "essential prerequisites [for efficient rocket development] were the large-scale production of aluminum alloys; the ability to produce, and store, liquid oxygen in quantity ....; and finally the development of electrical and mechanical precision instruments" (Dornberger 1954, 18).

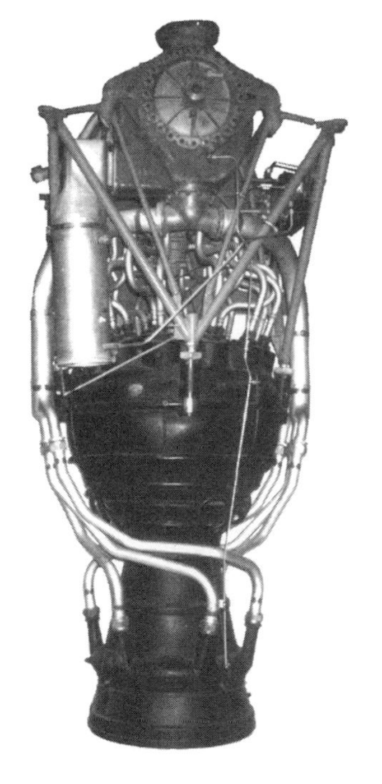

Fig. 10.7. A-4 (V-2) engine. Photo courtesy of Mike Gruntman.

TYPICAL PARAMETERS OF A-4

length: 14,030 mm
mass (fueled): 12,700 kg
warhead mass: 1000 kg
thrust at sea level: 56,000 lbf (25 t)
thrust at engine cutoff (altitude 15 miles): 70,000 lbf
acceleration at liftoff: 1 *g*
acceleration at engine cutoff (burnout): 6 *g*
burning time: 60–65 s
combustion chamber pressure: 15 atm
combustion chamber temperature: 2300°C
throat diameter: 400 mm
nozzle exit diameter: 740 mm
propellant exhaust velocity: 2050 m/s
specific impulse at sea level: 210 s
range: up to 180 miles (300 km)
rocket velocity at engine cutoff: 1500 m/s
impact velocity: 800 m/s
flight duration: 320 s

The upper end of the A-4 rocket carried a warhead with the control section immediately below. The fuel and oxidizer tanks filled the rocket midsection. The engines, vanes in the exhaust flow, and aerodynamic control surfaces were at the bottom end.

A-4 Engine

The rocket engine developed thrust of 25 t (56,000 lbf) and consumed 72 kg (159 lb) and 58 kg (128 lb) each second of the oxidizer (oxygen) and fuel (75% mixture of ethyl alcohol with water), respectively. The propellant was injected and atomized through 2160 oxidizer and 1224 fuel injectors. The temperature in the combustion chamber reached 2300°C (2600 K) and pressure almost 15 atm (215 psi). The hot gas expanded in the converging-diverging (de Laval) nozzle to a supersonic velocity at the exit.

Fig. 10.8. Launch preparations of the A-4 (V-2) rocket, ca. 1944. Rocket paint pattern facilitates visual observation of rocket roll in flight. Photo courtesy of NASA.

ALCOHOL AS ROCKET FUEL

During the A-4 development phase, use of alcohol as a fuel provided a benefit of independence from the scarce supply of oil in the wartime. Alcohol was obtained from plentiful potatoes harvested all over Germany, especially in the east. When the A-4 was moved into series production, however, the availability of alcohol in the desired quantities became a critical factor restricting massive use of the A-4s. The insufficient supply of liquid oxygen further limited rocket deployment after major underground liquefaction plants in Saar had fallen into the Allied hands.

Cooling Techniques

Alcohol cooled the engine by circulating around the combustion chamber wall before injection. This regenerative cooling technique reduced the temperature of the chamber walls. In addition, extra alcohol was injected near the walls in the combustion chamber and in the nozzle. The combustion of such an excessive-fuel mixture (also called fuel-rich mixture) produced gas with lower temperature in these areas. This latter technique, called film cooling, further reduced the wall temperature. The lightweight aluminum-magnesium propellant tanks contained 9200 lb (4170 kg) and 12,200 lb (5530 kg) of the fuel and oxidizer, respectively. The oxidizer, liquid oxygen, was cryogenic, which presented additional challenges to the rocket design.

LIQUID OXYGEN

Liquid oxygen is stored at cryogenic temperatures. The boiling temperature of liquid oxygen at pressure of 1 atmosphere is –186°C (–303°F).

Gas Generator

Both the fuel and oxidizer were pumped with turbopumps, powered by a turbine. A special gas generator produced hot gas to power the turbine. The hot gas was obtained by catalytic decomposition of 82%-strong hydrogen peroxide (H_2O_2). Hydrogen peroxide was mixed with a catalyst, a 27% solution of sodium permanganate ($NaMnO_4$), in a special chamber producing the 385 °C gas that drove the pump turbine. The turbine spun at 3800 rpm and developed 460 hp. (1 rpm = 1 revolution per minute)

Flight Profile

The rocket-powered flight lasted about 60 s. The fully fueled typical rocket weighed 12,700 kg (25,000 lb) at the launch pad, and it lifted off with an acceleration of 1 *g*. After 4 s of vertical flight, the rocket began to smoothly pitch forward (reaching and surpassing the speed of sound on the 25th second) until it reached 47-deg angle by the 43rd second of the flight. It continued further acceleration until the engine was cut off at the desired velocity. The rocket became lighter and lighter with the propellant consumed while the engine thrust remained nearly constant. Thus, the rocket gradually increased its acceleration to about 6 *g* (pronounced "six gees") by engine cutoff, which typically occurred at an altitude of 14 to 16 miles (22.5 to 26 km).

ACCELERATION

It is customary to express the acceleration of rockets in the units of *g*, or the free-body acceleration in the Earth's gravitational field at sea level.

$1\,g = 9.81\ \text{m/s}^2$, or $32.2\ \text{ft/s}^2$

The rocket velocity at cutoff determined the distance at which the rocket would hit the ground target. On most rockets, the onboard integrating accelerometers measured velocity with a 0.1% accuracy. An alternative technique based on the ground Doppler control radio station was rarely used.

Inertial Guidance and Riding the Beam

The rocket attitude control system employed gyroscopes spinning at 30,000 rpm. On some models, the azimuth (flight) direction was controlled by a ground radio beam. Such rockets "rode" the beam that was continuously transmitted from the launch site in the desired direction. About one-fifth of the rockets launched against England were beam guided.

Fig. 10.9. Wernher von Braun with German military brass during rocket tests at Peenemünde, ca. 1943. Photo courtesy of National Air and Space Museum, Smithsonian Institution (SI 78-5935), Washington, DC.

After engine cutoff, the rocket flew as a projectile in a ballistic phase of the trajectory. On operational flights, the A-4 reached an altitude of 50 miles (80 km), and its range exceeded 160 miles (260 km). The impact velocity, 800 m/s, was supersonic. Thus in contrast with the subsonic V-1, nobody could hear the sound of the approaching rocket. The warhead weighed 2200 lb (1000 kg) and contained 1650 lb (750 kg) of Amatol, a high explosive. (Amatol is a mixture of 60% of TNT and 40% of ammonium nitrate.) The warhead fuse was extremely reliable: only two unexploded warheads (out of 2000 successfully launched) were found during the war.

Reliable Warhead

The long-range rocket development was a closely

ATOMIC ENERGY FOR ROCKET PROPULSION

"After 1943 we had approached [director of the Kaiser Wilhelm Institute for Physics] Professor [Werner] Heisenberg for information about the practical possibilities [of use of atomic energy for rocket propulsion]. He could give us no firm promises of any description."

Walter Dornberger (1954, 252)

guarded military secret in Germany. Nevertheless references to this activity began to appear in British intelligence reports as early as 1939. An unusually detailed document, the so-called "Oslo Report," reached London in November of that year and identified Peenemünde as a development site of rocket weapons. It took, however, a few years and mounting evidence before the magnitude of the threat posed by these new weapons was fully realized.

"Oslo" Report

Aerial reconnaissance Mosquito planes were sent on missions over Peenemünde in May and June of 1943. A Royal Air Force report stated on 11 June 1943 that "the Germans are pressing on as quickly as possible with the development

Aerial Reconnaissance

V-1 — *VERGELTUNGSWAFFE EINS*

V-1 Winged Missiles

The Fi-103, or FZG 76, winged surface-to-surface missile was developed by the Luftwaffe at Peenemünde West. The missile was designed by Robert Lusser, chief engineer of *Fieseler Flugzeugbau*, Kassel.

The 27-ft (8.2-m)-long Fi-103 was equipped with a pulse-jet engine operating on a low-grade gasoline. The missiles were launched from a special ramp with the firing catapult (operated on hydrogen peroxide) accelerating the projectiles to 200 miles per hour (320 km/hr). At this speed, the pulse-jet would begin to operate efficiently and carry the missile to distances up to 200 miles (320 km). The operational flight altitude was between one and one-and-a-half miles and the cruise velocity 350 miles (560 km) per hour.

Fig. 10.10. Fi-103 (V-1) missile in the Imperial War Museum, London, England. Photo courtesy of Mike Gruntman.

The Fi-103 missile became known as the V-1 (*Vergeltungswaffe eins*, which stood for the *vengeance weapon one*) and carried a warhead of nearly 2000 lb (900 kg) of high explosives, approximately the same as the A-4 (V-2). The cost of manufacturing the V-1 was about one-tenth of that of the V-2.

Allied aerial reconnaissance had detected construction of the numerous launching ramps, all pointed at London. The ramps, dubbed the *ski sites*, were subsequently heavily bombed. About 10,000 V-1 missiles were fired, mostly against London. With time, countermeasures significantly reduced the effectiveness of this weapon.

of the long-range rocket at the experimental establishment at Peenemünde, and that frequent firings are taking place" (Churchill 1953, Vol. 5, 203). The new weapon, A-4, appeared on aerial photographs. Interestingly, Peenemünde was not linked at this time to the development of the Fi-103, and the western part of the establishment was not subsequently attacked.

Menace of New Weapons

The new menace to London, the most likely target for an inaccurate weapon, was taken by the British very seriously. Plans were drafted for an evacuation "of a hundred thousand persons in priority classes, such as schoolchildren and pregnant mothers, at the rate of ten thousand a day," should an attack occur. Thirty-thousand additional table stretchers were brought to the city in July of 1943.

> **BITE AND BARK**
>
> "It naturally pays the Germans to spread talk of new weapon to encourage their troops, their satellites, and neutrals, and it may well be that their bite will be found less bad than their bark."
>
> Winston Churchill, in the letter to President Roosevelt, 25 October 1943, on the threat of mysterious (V-1 and V-2) weapons (Churchill 1953, Vol. 5, 210)

RAF Bomber Command Strikes

On the night of 17 August 1943, Air Chief Marshal Sir Arthur Harris, the Commander-in-Chief of the *Royal Air Force* (RAF) *Bomber Command*, struck Peenemünde with 596 bombers, the Lancasters (324), Halifaxes (218), and Stirlings (54). To conceal the highly sensitive information about rockets, the airmen were told that "a new form of highly specialized R.D.F. [Radio Direction Finding, i.e. radar] equipment, which promises to improve greatly the German night air defense organization, is being developed and made at Peenemünde" (Middlebrook 1988, 235). The RAF operational order stressed the critical importance of the raid for the war effort: "The destruction of this [Peenemünde] experimental station, the large factory workshops and killing of the scientific and technical experts would retard the production of this new equipment [being developed at Peenemünde] and contribute largely to increasing the effectiveness of the bomber offensive." The order added an unusual statement that "it is clearly advantageous to ... [effectively damage the target] in one attack, otherwise the enemy is likely considerably to increase the existing defenses, but if necessary the attack will have to be repeated until the requisite degree of destruction has been achieved" (Middlebrook 1988, 235).

"... the attack will have to be repeated"

In the Bright Moonlight

The target was too far from the United Kingdom to be supported by the radio navigation aids and outside of the protection range of British fighter planes. The raid on Peenemünde was probably the only operation during the war when the entire force of the RAF Bomber Command was sent against such a small-area target. Therefore, this precise attack had to be carried out by moonlight and, consequently, at the mercy of Luftwaffe night fighter planes.

Eight British Mosquito planes were sent as a decoy to Berlin. The diversion was successful and draw German night fighters for some time. Finally, the Germans realized the deception and caught the bombers on their way home. Forty British bombers were shot down in the bright moonlight, and 290 men were lost, including 245 killed.

About 1800 tonnes of incendiary and high-explosive bombs fell on the Peenemünde center. Almost 750 people were killed on the ground in this raid between 1:10 a.m. and 2:07 a.m. The primary target of the attack was the settlement that housed German scientists and engineers. Two senior members of the rocket center staff, chief engineer of the manufacturing works Dr. Erich Walther and leader of the propulsion group Dr. Walter Thiel, perished in the raid. The latter was killed with his entire family by a direct hit on their shelter.

Air Raid Destruction

First aiming bombing markers were released two miles south from the planned position. Master Bomber Group Captain John H. Searby quickly spotted the error and radioed corrections to the incoming waves of bombers. Many bombers, however, overshot their intended targets and hit the barracks of foreign prisoners and laborers, killing more than 550 of them.

Fig. 10.11. Leader of the German rocket program General Walter R. Dornberger, ca. 1943. Photo courtesy of Deutsches Museum, Munich, Germany.

Dispersal of Rocket Manufacturing

The damage to Peenemünde was heavy, but the supersonic wind tunnel and many other critically important facilities survived the raid. By that time, the ballistic missile program had already been in transition to mass production. After the air attack the manufacturing facilities were dispersed throughout the country and moved underground. Some research programs were also transferred elsewhere, but most research and testing operations remained at Peenemünde. The rocket research center remained in the bombsights of the Allies: three American daytime raids hit Peenemünde in the summer of 1944 with more than 3000 tons of bombs.

The Mittelwerk (The Central Works)

The key underground manufacturing facility for ballistic rocket production, the *Mittelwerk* (Central Works), was established under Kohnstein Mountain in the Harz Mountains. The plant was next to the village of Niedersachswerfen near

Fig. 10.12. Entrance to the main tunnel of the Mittelwerk with the railroad tracks leading into the tunnel. Photo from Aircraft Division Industry Report, United States Strategic Bombing Survey, 1945.

the town of Nordhausen in Thuringia. The underground site consisted of two parallel main tunnels connected by the 57 evenly spaced cross tunnels.

The main tunnels were 1.7 km (1.05 miles) long, 35 ft (11 m) wide, and 25 ft (7.5 m) high. The cross tunnels were slightly smaller in cross section (30×22 ft or 9.1×6.7 m) and about 460 ft (140 m) long. The cross tunnels served as an area for machining and assembly of parts and components. The final assembly was done in the main tunnels. Standard-gauge railroad tracks led directly into the tunnels, where the assembled rockets could be loaded into railroad cars underground.

Tunnels in Kohnstein Mountain

Construction of the tunnels in Kohnstein Mountain began around 1917 for obtaining gypsum. The tunnels were expanded with time and used for storage of critically important supplies such as mineral oils and gasoline. In 1943, the facility

IN THE MITTELWERK

... I cannot forget a professor of the Pasteur Institute in Paris who testified as a witness at the Nurnberg Trial. He too was in the Mittelwerk ... . Objectively, without any dramatics, he explained the inhumane conditions in this inhumane factory ... . The conditions ... were in fact barbarious ... the sanitary conditions were inadequate, disease rampant; the prisoners were quartered right there in the damp caves, and as a result the mortality among them was extraordinary high ... .

Albert Speer (1970, 370)

was further expanded and converted to manufacturing of the A-4 (V-2), Fi-103 (V-1), and Junkers aircraft engines and turbojet units. The usable plant area had reached 1,200,000 sq. ft (110,000 sq. m) by the end of the war.

Savage Conditions

Concentration camp labor, mostly Soviets, French, Poles, Germans, and Italians, was widely used for construction of the Mittelwerk. The prisoners worked underground in the most savage conditions in the clouds of dust of anhydrous ammonia. Up to 6% of laborers died every month at the peak of construction work. A special concentration camp, Dora, with its own crematorium was established nearby to supply the work force. A few other subcamps provided more workers.

Fig. 10.13. Liberated prisoners of the camp *Lager Nordhausen* in the vicinity of the Mittelwerk, 12 April 1945. Photo by James E. Myers, Technician, Fourth Class. Photo courtesy of National Archives and Records Administration.

German Technicians

After construction, a large number of qualified German civilian workers and technicians were brought in for rocket manufacturing. Over 8000 people worked on the A-4 in the middle of 1944, including 5000 prisoners. The number of prisoners had decreased to 2000 by the end of the war, and their living and working conditions somewhat improved. The death rate dropped to about 1% per month.

SS Joins Rocket Program

The SS organization gradually increased its role in the German rocket program. Reichsführer Heinrich Himmler appointed SS General Hans Kammler the Commissioner General for the A-4 Program in August 1943. The new plans called for production of 900 rockets per month. The first rockets were assembled and tested by New Year's eve, with mass production starting in January 1944. In the beginning, it took about 15,000 man-hours to build one rocket. With time, improved efficiency reduced this number to 8000 man-hours per rocket.

V-2 Mass Production

As the A-4 manufacturing was gearing up, the Peenemünde center frantically worked on testing and evaluating the rockets produced in the Mittelwerk, providing critical engine calibration and quality control. Many Peenemünde specialists were transferred to the Mittelwerk to facilitate mass production. Others, including von Braun, regularly visited the manufacturing facilities to address specific problems.

Fig. 10.14. A cross tunnel for the assembly of the V-2 rocket engines at the Mittelwerk. Photo from Aircraft Division Industry Report, United States Strategic Bombing Survey, 1945.

Sabotage Suppressed

German technicians performed most of the rocket assembly in the Mittelwerk, with several thousand concentration camp laborers supporting their work. Attempts of sabotage by prisoners were ruthlessly suppressed by the German SD security service. For example, 162 prisoners, mostly Soviets and Poles, were hanged in March 1945. The analysis of the rocket failure rates did not show a statistically significant difference between the rockets assembled in Peenemünde by the purely German technicians and those manufactured in the Mittelwerk. The SD has apparently been successful.

Fig. 10.15. The new winged missiles, Fi-103, began an assault on London on 13 June 1944, exactly one week after the successful landing of the Allied troops in Normandy. The national-socialist propaganda chief Dr. Joseph Goebbels (right on the photograph together with Adolf Hitler on the left and Hermann Göring in the center) dubbed the missiles the Vengeance (*Vergeltung*) Weapon 1, or shortly V-1. The more powerful "miracle" weapon, the Vengeance Weapon 2, or V-2, was coming. Photo courtesy of Franklin D. Roosevelt Library, Hyde Park, New York.

Another important task of the Peenemünde rocket center was to increase the missile's range. The increased range had become especially important with the gradual advance of the Allied troops, which required A-4 launches from more distant sites. The effective range of many rockets increased to 220 miles (350 km) by 1945. Walter Dornberger claimed that some experimental models with the enlarged propellant tanks reached distances of almost 300 miles (480 km).

Range Increased

RECORD VERTICAL LAUNCH

The German military crews trained launching the A-4s in the southeastern Poland. These numerous launches revealed that some rockets disintegrated before striking the target. The phenomenon became known as *Luftzerleger*, or *air (reentry) burst*. A number of missiles were launched vertically at Peenemünde to examine and correct this undesired effect. In one such launch, the rocket reached an altitude of 117 miles (189 km).

Training of the military units to employ the new highly sophisticated weapon was also stepped up. A special Army training command was activated in April 1942, with the main launch site at Heidelager on the SS training ground near Blizna, Poland. The A-4 impact area was in a desolate region of the Pripet Marshes in Byelorussia 200 miles (320 km) to the east. The initial missile output of the Mittelwerk was used for training of the military units and extensive testing at Peenemünde.

Rocket Units Train

The field launch of the A-4 was a complex operation involving some 30 vehicles and trailers. The machinery included the trailers transporting the rocket and launch platform, the trucks carrying alcohol, liquid oxygen, and hydrogen peroxide, and a number of command and control vehicles. All of the support elements of the A-4 weapon system were mobile. The rocket was erected vertically

A-4 Field Launch

Fig. 10.16. The V-2 rocket is pulled out of the shelter. Photo courtesy of NASA.

on the launching table, and it took between four to six hours to prepare and fire the A-4.

V-2 Strikes

The Vengeance (*Vergeltung*) Weapon 2, V-2, struck the Allied targets in September 1944. This new mysterious weapon, the A-4 rocket, was successfully fired for the first time in the morning hours on 8 September 1944. Liberated Paris was the target. Two more rockets were launched later the same day at London. The V-2 was a supersonic missile, and therefore, in contrast to the V-1, there was no warning sound coming from the approaching rocket. People felt helpless in front of this silent, essentially impersonal and indiscriminate threat.

TEAM EFFORT

Neither the V-2 nor the V-1, nor any other great technological invention of recent decades, can be associated with the name of any one man. The days of the lonely creative genius are over. Such achievements can only be the fruit of an anonymous team of research specialists working selflessly, soberly, and in harmony.

Walter Dornberger (1954, 273)

From the military point of view, however, many questioned the effectiveness of the V-2 and argued that the development and manufacturing programs wasted enormous German scientific and industrial resources. The same resources spent on manufacturing a single V-2 rocket could have produced a half dozen fighter planes or 10 to 20 V-1 missiles. During the war, all V-2s delivered only several thousand tons of explosives at the targets. In comparison, a single major Allied air raid dropped 10,000 tons of explosives on a target within a few hours.

Peenemünde Evacuated

On the day of the last test launch at Peenemünde, 17 February 1945, the first trucks and railroad trains began evacuation of the documents, rocket parts, and equipment. The Red Army was rapidly approaching from the east. The research personnel, headed by von Braun, moved to the town of Bleicherode near Nordhausen, close to the Mittelwerk. It took more than 1000 trucks to haul the equipment from Peenemünde to central Germany.

V-2 DISPOSITION

Almost 5800 A-4 (V-2) rockets had been manufactured. About 600 rockets were used for testing and training, and almost 2000 fired operationally against England and the continental targets.

The first V-2 ballistic missiles were launched against the liberated Paris and against London on 8 September 1944. More than 1100 rockets reached England. On the continent, Antwerp, a major logistical center of the Allied Forces in Belgium, was especially hard hit. During the three weeks beginning in the middle of December 1944, 100 rockets were launched against this city weekly. Almost 100 rockets hit Liège, 65 were fired against Brussels, 15 against Paris, and 5 against Luxemburg. Eleven rockets were launched against a tactical target, the Remagen bridge over the Rhine river, on 17 March 1945.

About 1000 rockets were found in transport and field storage in the American, British, and French zones of Germany. The Red Army captured 1100 rockets, and 250 more were found in the Mittelwerk. The Mittelwerk facilities were captured by the Americans, but a month later this area of Germany was transferred to the Red Army.

MILITARY EFFECTIVENESS OF V-2s

Countering the threat of V-weapons diverted a significant fraction of the American and British air power from the strategic bombing campaign against Germany and from the tactical support of military operations. Before the landing in Normandy, from December 1943 to June 1944, the U.S. Eighth Air Force and the Ninth Air Force lost 462 and 148 men, respectively, in operations against the V-weapon targets. For two months after the landing in June 1944, about 20% of the Allied air power was directed against V-targets. This costly air campaign was considered by a number of Allied military leaders as ill conceived.

It was fortunate that the Germans spent so much effort on rockets instead of on bombers. Even our [British bombers] Mosquitoes, each of which was probably no dearer than a rocket, dropped on the average 125 tons of bombs per aircraft within one mile of the target during their life, whereas the rocket dropped one ton only, and that with an average error of fifteen miles.

Winston Churchill (1953, Vol. 6, 48)

... From the end of July 1943 on tremendous industrial capacity was diverted to the huge missile later known as V-2 ... Hitler wanted to have nine hundred of these produced monthly ... .

The whole notion was absurd. The fleets of ... [Allied] bombers in 1944 were dropping on average of three thousand tons of bombs a day ... And Hitler wanted to retaliate with thirty rockets that would have carried twenty-four tons of explosives to England daily. That was an equivalent of the bomb load of only twelve Flying Fortresses.

... [Germany] would have done much better to focus our efforts on manufacturing a ground-to-air defensive missile. It had already been developed in 1942, under the code name Wasserfall, to such a point that mass production would soon have been possible, had we utilized the talents of those technicians and scientists busy with rocket development at Peenemünde under Wernher von Braun ... .

As late as January 1, 1945, there were 2210 scientists and engineers working [at Peenemünde] on the long-rage rockets A-4 and A-9, whereas only 220 had been assigned to Wasserfall, and 135 to another antiaircraft rocket project, Taifun.

Albert Speer (1970, 365)

The threadbare argument that our [German] A-4 was too costly in comparison with the heavy bomber became more and more difficult to uphold in the light of [German] experience over England. If, as accurate statistics showed, a bomber was shot down after an average of five or six flights over England, if it could carry only a total of six to eight tons of bombs during its active existence, and if the total loss of a bomber, including the cost of training the crew, were estimated at about thirty times the price of an A-4 (38,000 marks), then it was obvious that the A-4 came off best.

Walter Dornberger (1954, 71)

Fig. 10.17. German rocketeers after they surrendered to the U.S. troops in Bavaria.

Left to right: Major General Walter Dornberger, Commander of the Army Peenemünde Center; Lieutenant Colonel Herbert Axter, a former Berlin patent lawyer and chief of the military staff at Peenemünde; Wernher von Braun (von Braun broke his arm in a car accident in March when his driver fell asleep at the wheel and the car crashed); and Hans Lindenberg, a combustion chamber specialist in Peenemünde and later in the Mittelwerk.

Photo by Technician Fifth Class Louis Weintraub, U.S. Army, Austria, 3 May 1945. Photo courtesy of National Archives and Records Administration.

In early April 1945, SS General Kammler ordered von Braun and his selected team to move south to Oberammergau, a town 40 miles southwest of Munich in the Bavarian Alps. On the orders of von Braun, Dieter Huzel and Bernhard Tessmann hid the collection of the Peenemünde technical documents in an abandoned mine in the Harz Mountains before their departure for Bavaria.

Surrender in Bavaria

Three weeks later, Hitler's suicide accelerated the crumbling of the remaining structure of the national-socialist Reich. Dornberger, von Braun, his brother Magnus, and four other close collaborators (Herbert Axter, Dieter Huzel, Hans Lindenberg, and Bernhard Tessmann) surrendered on May 2 to the soldiers of the 324th Infantry Regiment, 44th U.S. Infantry Division. The Germans were transported to the nearby town Reutte in Austria for interrogation by the Counter Intelligence Corps (CIC).

Technical Intelligence in Europe

The advancing Allied armies had deployed special technical intelligence units tasked to search for the new German military technology and key research personnel. The areas of atomic energy and rocketry were among the top priorities. The U.S. Army Ordnance Corps contracted *General Electric Company* in November 1944 for research and development of the long-range rockets. Dr. Richard W. Porter scouted Germany for this project, named *Hermes*, which included the evaluation of the V-2 missile program.

General Electric Company and Project *Hermes*

> **NEW WEAPON**
>
> "There is no quicker way to stimulate interest in a new weapon than to discover it in use by the enemy."
>
> Holger N. Toftoy, 1956
> (*Army Ordnance Satellite Program* 1958, 22)

Capturing German rocketeers and V-2 rockets was a high-priority task for Colonel Holger N. Toftoy, chief of the Army Ordnance Technical Intelligence in Europe, and his assistant Major James P. Hamill. Two key ordnance intelligence men in the field were Major Robert B. Staver (chief of the *Jet Propulsion Section*) and Major William Bromley (head of the *Special Mission V-2*).

Operation Paperclip

Several programs have been emerging at that time in various branches of the government to locate the leading German specialists in the areas critically important for military science and technology, with the objective of "importing" those willing to come the United States. With time these programs would merge into a joint operation called *Project Paperclip*.

In April 1945, the American army captured the Mittelwerk, and Richard Porter and Robert Staver rounded up 1000 former Peenemünders. Practically all A-4

> ***OVERCAST* AND *PAPERCLIP***
>
> Operation *Overcast* covered the early importation of German specialists to the United States. The name Overcast had been compromised by 1946 and was substituted by Project Paperclip on 13 March 1946.
>
> The name Paperclip related to the paperclips used for marking the files of those German rocketeers, in custody of the Army Ordnance officers in 1945, who presented interest for future employment.

OPERATION PAPERCLIP

The U.S. troops advancing into Germany were accompanied by the special teams of scientists and officers. The Navy, Army, Army Air Forces, Manhattan Project, and other branches of the government established technical intelligence groups to find, evaluate, and learn about the German wonder weapons.

The war with Japan was still in full swing, and the Armed Forces were eager to increase war-making capacity for operations in the Pacific. The Germans were providing the Japanese with important technical information on weapon development until the last days of the Third Reich. It was imperative to reveal what surprises Japan might have in store for the Americans closing on their homeland. Many military leaders became convinced of the critical importance of science in warfare and were also concerned with the postwar military research.

It became obvious that "exploitation" (in the military sense of the word) of the German scientists and engineers could save a few years of research by the American research institutions and industry in some important areas. An arrangement allowing willing German specialists, "chosen, rare minds whose continuing intellectual productivity we wish to use," to work in the United States became clearly desirable. The first specialists came through a number of programs under various arrangements.

The Navy seemed to be the first to "import" the Germans. Dr. Herbert Wagner of the *Henschel Aircraft Company* headed the development of the radio-controlled anti-aircraft rocket *Schmetterling* (butterfly). Wagner was captured with his assistants in Oberammergau by the Navy team on 1 May 1945. Eighteen days later, Wagner's group was hastily brought to the United States to help in development of countermeasures to his missile.

Several other German electronics experts fell into the Navy hands two days earlier. The German submarine U-234 was on the way to Japan carrying a team of technical experts to help the Japanese war effort. Following the cease-hostilities order by Admiral Dönitz, the boat surrendered in the Atlantic and was escorted to a Navy yard in Maine. (Two Japanese navy officers onboard committed suicide.) The German specialists, headed by Dr. Heinz Schlicke, were soon helping with their knowledge the Bureau of Ships.

"Importation of specialists" was supplemented by recording the captured technical documentation. In one year, 1946, almost 4,000,000 pages of the various technical documents were microfilmed and made available in the United States.

The growing political problems with the aggressive Soviet policy prompted the actions to ensure the "denial" of the German specialists, forestalling their utilization by the Soviet Union in the military important areas. The urgency of the situation had become evident when the Soviets rounded up and sent to the USSR many thousands of German specialists and members of their families. The French and to a lesser degree the British also actively recruited German scientists and engineers.

The Project Paperclip (initially known as operation Overcast) dealing with importation of German specialists was formalized in the spring 1946. President Truman approved the program six months later.

Arrival of the more than 100 German rocketeers of the von Braun's team got most of the publicity. The scale of the program was much larger and diverse, however. A total of 642 alien specialists were brought to the United States through Paperclip between May 1945 and December 1952.

rockets found by the American troops were incomplete with critical parts missing, particularly the precise components of the control sections. Two weeks later Dornberger, Von Braun, and their closest associates surrendered in Bavaria. Von Braun and Dornberger apparently did not disclose to the Americans what happened to the rocket technical documentation, but it did not matter much. Major Staver had located the documents hidden in the mine.

Fig. 10.18. Colonel Holger N. Toftoy disables a mine in 1944. Chief of the Army Ordnance Technical Intelligence in Europe, Toftoy was a key officer responsible for locating and transferring German rocketeers to the United States. Photo courtesy of U.S. Army.

The Mittelwerk Captured

The part of the Harz Mountains that included the Mittelwerk had to be transferred to the Red Army according to the Yalta agreements. The Americans, racing against time, brought 300 railroad cars of rocket parts to the port of Antwerp. In a great hurry the found boxes with the A-4 documentation had been removed from the mine and transported west to the American zone of occupation, just before the advancing Red Army units arrived to the Mittelwerk. Sixteen Liberty ships carried the rocket hardware to New Orleans, Louisiana, from where they headed to their final destination at the White Sands Proving Ground in New Mexico.

From Peenemünde to Texas

In 1945, the U.S. Army screened a number of German rocketeers and offered them contracts for work in the United States. It was specifically required that "scientists must not be listed as war criminals and must be volunteers" (*Army Ordnance Satellite Program* 1958, Appendix 1). Colonel Toftoy assembled von Braun and his group, a total of 127 specialists, at Fort Bliss, Texas, under the command of Major, later Colonel, Hamill. Transfer of German specialists to the United States raised a number of tricky administrative and legal questions.

OPERATION *BACKFIRE*

The British initiated an operation *Backfire* in May 1945 to assemble and fire, for the evaluation purposes, several V-2s. Under British observation the German prisoners of war prepared and launched three V-2s in October 1945 from the Krupp Naval Testing Ground near Cuxhaven on the North Sea. The original plan to launch as many as 30 rockets had to be abandoned because most of the captured rockets missed important parts, especially critical components of the control compartment. The assembly of the captured rockets for launch proved to be a difficult task, as would be later confirmed by the American experience at White Sands. The first rocket was launched on 2 October 1945 and the last took off on 15 October.

Finally, the Army assumed a custody of the "resident aliens" and brought the Germans to the United States without visas, but with the knowledge of the President.

FIRST GERMAN ROCKETEERS IN THE UNITED STATES

The first seven German rocketeers arrived at Fort Strong, New York, on 20 September 1945 and were transferred to the Aberdeen Proving Ground in Maryland. There Wernher von Braun, Erich W. Neubert, Theodor A. Poppel, August Schultze, Eberhard Rees, Wilhelm Jungert, and Walter Schwidetzky helped to process and sort out German documents on guided missiles captured by the Army. The German rocketeers joined their 120 other colleagues at Fort Bliss, Texas, in late 1945.

German Rocketeers at Fort Bliss, Texas

The specialists at Fort Bliss were assigned several tasks that included sorting and refurbishing the rocket parts that had arrived from Germany, supporting the General Electric Company's Hermes Program, and working on other projects involving various advanced rocket subsystems. The General Electric's personnel prepared and launched V-2 rockets from nearby White Sands Proving Ground, New Mexico. The Germans initially assisted in this task. By the spring 1947, they had been replaced with the Americans. Sixty-seven rockets had been fired by 1951 when Project Hermes ended.

Other victorious Allied powers were also keenly interested in German rocket technology and the specialists. The least interested was Great Britain, and only several German engineers and scientists joined British research and development efforts in the late 1940s. In contrast, a large number of Germans from Peenemünde, Kochel, and other research and industrial organizations came to work to France.

Soviet Large-Scale Effort

Most of the manufacturing machinery in the Mittelwerk fell in the hands of the Soviets. The Red Army quickly rounded up many V-2 technicians and later located a number of engineers and research men. The Soviet Union mounted a large-scale effort to learn the advanced German military technology. A special commission that was established for this purpose included 284 specialists in the field of rocketry alone. The Soviets began to gear up their research effort in long-range rocketry in 1944 when the Red Army captured first missile components at the German training and testing sites in Poland.

The Soviet approach to utilize the German achievements was essentially different from American. The USSR opened special Soviet–German research and development institutes and centers in Germany in order to collect and process existing technical documentation and to study military technologies. The institutes were directed by Soviet specialists, and many Soviet rocketeers spent from a few months to almost two years

INDUSTRY, AGRICULTURE, AND SCENERY IN GERMANY

"England received the German industry, Russia its agriculture, and the United States the scenery."

This was a common joke about the zones of occupation in the defeated Germany. In addition to the undeniable scenery, the mountainous regions of Bavaria were also the hideout of many leading German scientists and engineers that were captured by the American troops.

in occupied Germany. The list of their names was a who's who of the early Soviet rocket and space programs: Sergei Korolev, Valentin Glushko, Vladimir Barmin, Boris Chertok, Aleksei Isaev, Vassilii Mishin, Mikhail Ryazansky, and Nikolai Pilyugin. Actually many leading Soviet specialists met each other for the first time on this assignment. Their professional ties formed in Germany would later provide an organizational backbone of the powerful Soviet rocket and space establishment. As Korolev wrote in 1962, "The most valuable [outcome] of what was in Germany [in 1945–1946] was not [captured] hardware but that we [the leading Soviet specialists] united and created the foundations of the [rocket] community" (Raushenbakh and Vetrov 1998, 383).

"... the foundations of the [Soviet rocket] community"

Rocket research centers had been established in a number of German cities. Among the most prominent was the *Rabe Institute* (*Rabe* stands for the German *Raketenbau und Entwicklung*, or *rocket building and design*) in the vicinity of Bleicherode, near Nordhausen. The Institute was expanded in 1946 into a larger establishment, the *Institute Nordhausen*, headed by a high-level Communist party functionary, General Lev M. Gaidukov, and Sergei P. Korolev.

Rabe Institute and Institute Nordhausen

German scientists and engineers worked under the supervision of Soviet rocketeers. Production and testing of the V-2s was reestablished, and a number of various projects on advanced rocket technology were initiated. The research centers grew to employ a few thousand people. Many Germans from the western zones of occupation were induced to move east and join the Institute Nordhausen by offers of continuation of their professional work, a highly attractive proposition for a scientist or for an engineer. In addition, good salaries and food rations were precious items in devastated postwar Germany.

By May of 1946, the Soviet government had produced a comprehensive plan for rocket development in the USSR, a plan in which German specialists had an important role to play. Consequently, the leading German rocketeers, including Helmut Gröttrup, were forcibly moved to work in the Soviet Union in October

GERMAN ROCKETEERS AND FRANCE

The French Ministry of Armed Forces sent a team, headed by Henri Moureu, to gather information on German rocketry. Moureu reported the results of his inspection and drafted an ambitious proposal to establish a rocket development center.

The French were relatively slow to engage in rocket research with the top priority given at that time to atomic energy. The Moureu's proposal was not accepted, and instead a smaller Center for the Study of Self-Propelled Projectiles (*Centre d'Étude des Projectiles Autopropulsés*) was founded. Nevertheless a group of German rocketeers, including Wolfgang Pilz and Heinz Bringer, were employed by the French when the Ballistic and Aerodynamic Research Laboratory (*Laboratoire de Recherches Balistique et Aérodynamiques — LRBA*) was opened at Vernon on 17 May 1946 and initiated development of liquid-propellant rockets. In total, more than 120 rocketeers from Peenemünde and aerodynamicists from Kochel worked in France in the late 1940s and early 1950s. In addition, a large number of German scientists and engineers were employed at the *Aeronautical Arsenal* at Châtillon near Paris, *ONERA* (*Office National d'Études et de Réalisations Aéronautiques*), the *Franco-German Institute* at Saint-Louis in Alsace, and several other research and development organizations.

GERMAN ROCKETEERS IN THE SOVIET UNION

In 1946 the Soviet leaders decided to transfer a number of German specialists for work in the Soviet Union. General of State Security Ivan A. Serov instructed Soviet rocketeers in the Institute Nordhausen to secretly select the Germans to be sent east. The scientists and engineers would be transferred regardless of their wishes. A simultaneous operation was planned for those employed elsewhere in the Soviet part of Germany. In addition to rocketeers, many specialists in aircraft design, jet engines, air defense, and submarines would be sent to the USSR.

Boris E. Chertok was an organizer and leader of the Rabe Institute. He would become a prominent specialist in control and space navigation and a trusted deputy of Sergei P. Korolev for many years to come. Chertok recalled the events of the fateful night in his book of memoirs published in 1994:

> In the evening of 22 October [1946], a lavish banquet was organized in the restaurant "Japan" with the unlimited booze for the Germans and strict prohibition of getting drunk for the Soviet specialists, who played hosts. The banquet was announced as a celebration of the assembly of the first dozen of [V-2] rockets. In total more than two hundred people "had fun." However, only the Germans enjoyed themselves. All Russians were in a bad mood because of the prohibition to drink when such great food was available. The party ended around 1 a.m.
>
> At 4 a.m. hundreds of military Studebakers [automobiles with the Soviet security officers] noisily drove into the quiet sleeping town. Each operative surveyed earlier the house that he had to approach. ... Dazzled Germans could not understand why they had to move to the Soviet Union at 4 o'clock in the morning, with all their family and belongings. However, being brought up in the spirit of discipline, order, and following the authorities without questions, the spirit that the German people lived for many dozens of years, worked. ... There was not a single serious accident ... . (Chertok 1994, 178)

More than 100 German scientists, engineers, and technicians with their families were thus forcibly loaded into prepared trains and shipped to Russia. The group included a leading German engineer from Peenemünde Helmut Gröttrup, who, according to Chertok, had voluntarily left the American zone and joined the Rabe Institute a year earlier.

Most of the Germans were formally assigned to the *Scientific Research Institute 88* (NII-88) in Podlipki near Moscow. After some time they were assembled on the island of Udomlya in the middle of Lake Seliger, about 200 miles (320 km) north from Moscow. The secluded island settlement was behind the barbed wire and armed guards. It was like a concentration camp with the living conditions superior to those of ordinary Soviet people outside. Occasionally, the Germans were allowed making trips to the nearby town of Ostashkov.

More than 500 Germans were confined to Udomlya where the research and testing facilities were erected. An experienced Soviet rocketeer, Yurii A. Pobedonostsev, supervised the work on the island. The German specialists were never allowed integration into the Soviet rocket effort. They were squeezed for all of their knowledge and eventually, after a "cooling-off" period, sent back home. The first group of Germans was allowed to return back to the Eastern Germany in December 1951. The last group left the USSR in the end of 1953.

1946. This importation of the scientists and engineers was performed in a way very different from the contracts by which the United States had arranged arrival of von Braun and his team to Ft. Bliss in Texas. On 22 October 1946 the Soviet state security operatives simply loaded selected German specialists and their families in the guarded trains and shipped them to Moscow. The remaining parts of the Institute Nordhausen in the occupied Germany were shut down.

22 October 1946

> **USE OF GERMAN SPECIALISTS**
>
> ... Until this day, the German specialists participate in fabrication [of rockets] only as observers, without any responsibility for the quality of the work.
>
> The experience of the assembly of the first article [i.e., A-4 rocket] shows that, in order to improve the quality of the work and to accelerate training of the Soviet specialists, it is necessary that the German specialists directly participate in fabrication and bear personal responsibility for the quality of the work ... .
>
> Sergei P. Korolev, Memo to top management of rocket plant, 11 April 1947
> (Raushenbakh and Vetrov 1998, 119)

Work and Life in Isolation

In the Soviet Union, the imported Germans lived in isolation from the general population. This isolation had even become stricter with time. Helmut Gröttrup and his team worked on a variety of rocket subsystems, some in cooperation with Soviet technical groups and some independently. They designed a new rocket with a projected range of 3000 km (1900 miles), which had an unusual conical body with a separable warhead. The project never went beyond the paper study and was soon abandoned.

Fig. 10.19. "Paperclip" German rocketeers at Fort Bliss, Texas, in November 1946. Photo courtesy of U.S. Army.

"Hitlerite Scientists and Engineers"

The Soviet propaganda commonly called von Braun's team working in the United States the "Hitlerite scientists and engineers." The "Soviet Germans" were obviously fundamentally different and toiled for world peace and progressive mankind. They were never allowed, however, integration into the Soviet rocket program and society and were eventually, after being drained of their knowledge and cooling-off, sent back to the Eastern Germany.

Fig. 10.20. General Holger N. Toftoy organized and supervised the Army rocket development program at the Redstone Arsenal in the 1950s. Von Braun would serve a technical director of the Guiding Missile Development Group (later Division) reporting to Toftoy. Photo (mid-1950s) courtesy of U.S. Army.

German Rocketeers in America

The situation of the German rocketeers was quite different in the United States. The initial restrictions on the Germans assembled at Fort Bliss were gradually eased up. The renewed contracts were for longer terms and allowed bringing the families. The specialists and their families more and more participated in a normal American life.

In the late 1940s the legal status of the Germans was finally resolved. They needed to formally enter the United States and establish their immigration status. Dieter Huzel described (in 1962) this rather unusual procedure:

> In early 1950, the first concrete steps were taken to initiate legal status for the [German] group members. Extensive form filling ensued; then the legal entry followed. For this purpose, we were all brought [from Ft. Bliss] across the [Mexican] border, to the U.S. Consulate in Ciudad Juarez, Mexico. Following processing of our forms and issuance of visas, we returned, this time processed in a routine fashion at the border. Because of the proximity of two cities, Juarez and El Paso, the entire procedure took no more than a few hours; even left time for a first, excellent Mexican meal in a small border restaurant. (Huzel 1962, 226)

From Texas to Alabama

In 1950, most of the "American" Peenemünders were moved from Fort Bliss to the U.S. Army Redstone Arsenal in Huntsville, Alabama, where they worked in various Army rocket programs. General Toftoy remained closely associated with and supervised the work of the Paperclip rocketeers for the Army. In

1960, von Braun and his team were transferred to the recently established National Aeronautics and Space Administration (NASA). They formed the nucleus of the new NASA's *Marshall Space Flight Center* in Huntsville. Some Peenemünders also joined the American rocket and space industries.

Contributions to Science and Technology

Many Paperclip rocketeers fully integrated into American life and naturalized. Thirty-nine Redstone Arsenal's German specialists were sworn as U.S. citizens on 11 November 1954. Then, a larger group of 103 scientists, engineers, and members of their families were naturalized on 14 April 1955. The latter group included Wernher von Braun who took the citizenship oath in a ceremony at Huntsville High School. Huntsville's Peenemünders actively participated in the development of rockets Redstone, Jupiter, Juno, Pershing, Saturn I, IB, and V and launching of the first U.S. satellite Explorer I on 31 January 1958. They published numerous technical articles and books and achieved prominence in various areas of rocketry and space technology.

Fig. 10.21. Eberhard Rees from von Braun's group of German rocketeers served in 1970–1973 as director of NASA's Marshall Space Flight Center, Huntsville, Alabama. Photo courtesy of NASA.

The leader of the German rocket program, Lieutenant General Walter Dornberger, also came to the United States. After the war ended, Dornberger was first held by the British for two years and released in 1947. Dornberger then moved to the United States where he initially consulted the Air Force. He joined the *Bell Aircraft Corporation* in Buffalo, New York, in May 1950 and served as Bell's vice president in 1959–1965. Dornberger actively participated in development of the early reusable space launching systems.

Leading Positions in Space Establishment

Several former Peenemünders reached high management positions in the space establishment. Von Braun became technical director of the Army ballistic-weapons program in 1952 and director of NASA's Marshall Space Flight Center in 1960. Eberhard Rees succeeded him as the center director in 1970. Kurt Debus was responsible for Redstone's launch operations on Cape Canaveral in 1950s and became the first director of NASA's Kennedy Space Center. A number of German specialists worked in American industry where they participated in major programs and made important contributions to rocket and space technology.

Another Peenemünder, Arthur Rudolf, served as program manager for the Saturn V rocket. His fate though turned very differently, reflecting the origin and role of the Paperclip specialists in the war effort and crimes of the national-socialist Germany. Rudolf left the United States in 1984 and gave up American citizenship because of his role in the wartime Mittelwerk.

11. JATO AND BEYOND

While Robert H. Goddard was shooting his rockets higher and higher in isolated Roswell, New Mexico, in 1930s, two other centers of the American rocketry emerged: one on the East Coast and the other on the West.

Two New Centers of American Rocketry

The American Rocket Society steadily grew in numbers, and its programs attracted new rocket enthusiasts, including many with engineering education and technical backgrounds. The original founding members of the Society were soon joined by Alfred Africano, Carl Ahrens, Alfred H. Best, Peter van Dresser, Goodman, Max Krauss, Laurence Manning, H. Franklin Pierce, John Shesta, Bernard Smith, James H. Wyld, and others. They designed, built, and tested various rockets.

In December 1938, ARS's James Wyld built and successfully tested the first American liquid-propellant rocket engine with regenerative cooling. By summer of 1941, the new Wyld engine, with 100-lbf (445-N) thrust, was reliably fired for up to 40 s. This development of the regenerative cooling technology was a major advance opening a way for practical liquid-propellant rocket engines.

> **REGENERATIVE COOLING**
>
> In regeneratively cooled engines, the propellant flows in a special jacket around the combustion chamber and the nozzle keeping the wall temperature within the acceptable limits. After passing along the hot walls, the heated propellant enters the combustion chamber thus returning a significant fraction of the absorbed heat (that would have been otherwise wasted) to the rocket engine. Regenerative cooling prevents wall burnout, reduces loss to the walls of the chemical energy released in combustion, and enables operation of liquid-propellant rocket engines for long periods of time.

Regeneratively Cooled Engine

Independent of ARS, Harry W. Bull worked on nozzle cooling at Syracuse University in New York, and midshipman Robert C. Truax experimented with small liquid-propellant rockets at the U.S. Naval Academy. At the same time, Captain Leslie A. Skinner designed solid-propellant missiles at the Army's Aberdeen Proving Ground in Maryland. Local rocket societies emerged

Fig. 11.1. ARS rocket test in 1934 near Stockton, New Jersey. Ms. Lee M. Gregory at the shelter trench and H. Franklin Pierce and David Lasser near the rocket launching rack. Photo from *Astronautics*, No. 39, p. 4, January 1938.

in such places as Cleveland, Ohio; New York, New York; New Haven, Connecticut; and Peoria, Illinois. The rocketeers built and flew models, conducted various experiments, studied related theoretical problems, and educated and excited the public.

GALCIT

Around the same time, a new major rocket center emerged in Southern California. A prominent aerodynamicist, Dr. Theodore von Kármán, was appointed director of the *Guggenheim Aeronautical Laboratory* of the *California Institute of Technology* (GALCIT) in 1930. GALCIT quickly became a center of excellence in fundamental aerodynamic research. A group of young rocket enthusiasts formed around von Kármán some time in 1935–1936. Frank J. Malina, Weld Arnold, William Bollay, John W. Parsons, Edward S. Forman, Apollo Milton Olin ("Amo") Smith, and Hsue-Shen Tsien initiated the development of high-altitude sounding rockets.

Arroyo Seco Canyon

An analysis of the current technology quickly led the California group to the conclusion that the key to a successful rocket was efficient propulsion — a practical engine with a reasonable specific impulse. Therefore, the

Fig. 11.2. Bernard Smith prepares ARS rocket No. 3 in 1934. Photo from *Astronautics*, No. 30, p. 5, October 1934.

THEODORE VON KÁRMÁN

Theodore von Kármán (1881–1963), a Hungarian-born American scientist and engineer, played a key role in development of rocket technology in the United States starting in the mid-1930s. As a young man, von Kármán worked for several years as an assistant to Professor Ludwig Prandtl at the Göttingen University in Germany. Von Kármán became famous for his work in aerodynamics (for example, the *Kármán's Vortex Street* helped later to explain the collapse of the Tacoma Narrows Bridge, Washington, in 1940) and for developing and establishing wind tunnels for basic research.

Fig. 11.3. Theodore von Kármán in 1953. Photo courtesy of NASA.

Robert A. Millikan, President of the California Institute of Technology (Caltech), persuaded von Kármán to leave his position at the famous Aeronautical Institute in Aachen, Germany, and join Caltech as director of the Guggenheim Aeronautical Laboratory (GALCIT). The rapid progress of national socialism in Germany helped von Kármán make up his mind. He then had to fight an uphill battle to convince his mother and sister to emigrate with him to America; a land that was, as they believed, a "country of gangsters and misfits." This "enlightened" attitude amazingly survives to this day in Europe in spite of the vagaries of the 20th century history.

At Caltech, Theodore von Kármán's laboratory initiated the study of the fundamentals of rocket propulsion. Among their first outstanding achievements were the significantly improved composite solid propellants and jet-assisted takeoff of heavy airplanes. The rocket group of GALCIT was subsequently reorganized as Jet Propulsion Laboratory and became a part of the Army and later of NASA. Von Kármán was among the founders of the Aerojet Engineering Corporation, a leading American manufacturer of rocket engines.

Von Kármán played a crucial role in building the research program of the U.S. Air Force during and after World War II. On request of General Henry H. Arnold, commander of the U.S. Army Air Forces, he organized and headed the Air Corps Scientific Advisory Group in 1944 and later the Air Force Science Advisory Board.

Von Kármán became a U.S. citizen in 1936. He received the highest medals and awards that can be bestowed on a civilian for his wartime contributions. When President John F. Kennedy presented Theodore von Kármán with the first National Medal of Science in 1963, von Kármán responded pledging "his brain as long as it lasted" to his country.

Fig. 11.4. GALCIT rocket propulsion proving stand in 1938. Figure from *Astronautics*, No. 41, p. 3, July 1938.

GALCIT group initiated a study of the fundamentals of rocket propulsion. This focus on basic science was essentially different from the traditional approach of empirical rocket enthusiasts eager to shoot rocket models.

Interest in Jet-Assisted Takeoff

Some experiments of the California rocketeers were soon moved off the Caltech campus, where the group quickly became known as the "suicide squad" as a result of a few accidents. The new test ground was located in the Arroyo Seco, a dry canyon wash in Pasadena, California. Two years later, Caltech with the help of the U.S. Army would acquire this area from the city of Pasadena and lease it for GALCIT rocket experiments. Eventually, the Jet Propulsion Laboratory (JPL) would be established at the nearby site.

A practical interest to rocketry appeared from an unexpected direction. In August 1938, Reuben Fleet, President of Consolidated Aircraft Co., San Diego, California, contacted GALCIT with the idea of rocket-assisted aircraft takeoff. In December of the same year, the U.S. Army Air Corps also got interested in using rockets for assisting the takeoff of heavily loaded airplanes.

Air Force Champions Science and Technology

Chief of the Air Corps General Henry H. "Hap"

NO OTHER MAN ...

No other man has had so great an impact on the developments of aeronautical science [in the United States] ... . Hundreds of young men became his students and scientific collaborators and were inspired to greater creative effort. They were taught by him to analyze technical problems in terms of fundamental physical concepts and to apply mathematical analysis, striving for simplicity but retaining adequate accuracy for engineering purposes. Many of these men are now leaders of aeronautical science and engineering.

Congratulations by the Institute of the Aeronautical Sciences on the occasion of von Kármán's 75th birthday (*Journal of the Aeronautical Sciences*, Vol. 23, No. 5, 1956, p. 403)

Arnold was a champion of bringing the cutting-edge science and technology to the air forces. He cultivated close links with the scientific community in the 1930s, establishing a high-tech tradition prominent in the U.S. Air Force to this day.

First Air Force Rocket Development Contract

In 1938, Arnold asked the National Academy of Sciences to evaluate the scientific developments needed for the modern air forces. As a result of the study, a government contract was awarded in 1939 to the GALCIT Rocket Research Group to study the jet-assisted takeoff (JATO) problem. This was the first U.S. Army Air Corps rocket research contract.

Fig. 11.5. Commanding General of the Army Air Forces General Henry H. "Hap" Arnold (right) with Army Chief of Staff General George C. Marshall in September 1944. General Arnold strongly supported advanced science and technology and vigorously argued for rapid rocket development after the war. The Air Force established the *Arnold Engineering Research Center* in 1951 in Tennessee. The NASA center in Huntsville, Alabama, was named after George C. Marshall in 1960. Photo courtesy of Franklin D. Roosevelt Library, Hyde Park, New York.

Solid-Propellant JATO

In 1940, GALCIT rocketeers demonstrated a stable, long-duration, solid-propellant rocket motor. This success accelerated the work on practical JATO. The development culminated on 12 August 1941 by the first aircraft takeoff assisted by rockets. Six rocket motors were packed together to provide the desired thrust. Each rocket motor was made of stainless steel and contained "three pounds of amide powder mixed in corn starch and Le Page's glue" (Boushey 1993, 129). The motor end-burning grain generated 28 lbf (125 N) of thrust.

Rocket-Assisted Aircraft Takeoff

A light Army airplane, Ercoupe, piloted by Homer A. Boushey, got in the air using only 300 ft (91 m) of the runway instead of ordinary 580 ft (176 m). The takeoff time was reduced from 13.1 to 7.5 s. This first demonstration would be followed by the development of liquid-propellant rocket JATOs by both the West Coast and East Coast groups.

Soon after the successful demonstration of the rocket-assisted takeoff, the first aircraft takeoff by rocket power alone followed. On August 23, the Ercoupe, with the propeller removed, was towed by a truck to a speed of 25 miles per hour, when pilot Boushey released the rope and fired 12 rocket motors. The airplane took off the ground, though for 10 ft only, and performed, as Boushey recalled, "marginally-controlled flight for perhaps a distance of one mile" (Boushey 1993, 134).

Composite Grain

The gunpowder-based rockets were not stable and deteriorated in storage. A major breakthrough in solid propellants was achieved when the GALCIT group proved a new solid-propellant concept, the composite grain. The first composite grain, the "asphalt propellant," was produced by embedding the crystals of the potassium perchlorate oxidizer into the roofing tar serving as the binder and fuel. The new technology made it possible to directly cast the propellant mixture into rocket casings. In addition, JATOs and other rockets could now be stored for long periods of time under various temperatures and conditions. By 1944 the new Thiokol liquid polymer had replaced the tar as a binder and fuel in composite grains. This superior combination laid the foundations for the development of large solid-propellant motors, a key element for the future long-range ballistic missiles Polaris and Minuteman.

New Thiokol Liquid Polymer

In 1941 Robert C. Truax began work on a powerful liquid-propellant JATO at the Navy's Engineering Experiment Station in Annapolis, Maryland. Liquid oxygen could not be conveniently stored and thus considered by Truax an unacceptable oxidizer for practical applications in the field. Therefore, gasoline and red-fuming nitric acid (RFNA) were selected as propellant. Truax's group battled with the problem of timely ignition of the mixture in the combustion chamber, when in 1942 Ensign Ray C. Stiff discovered that aniline and RFNA would ignite

Hypergolic Propellants

Fig. 11.6. Takeoff of the first American rocket-assisted airplane from March Field, California, on 12 August 1941. Each solid-propellant rocket in the JATO booster produced 28 lbf of thrust for 12 s. Left to right are an Air Force technician, Fred Miller, and Frank Malina. Photo courtesy of NASA.

Fig. 11.7. Several minutes before the flight of an Ercoupe aircraft on rocket power alone by Captain Homer A. Boushey (right). Others in photo, left to right: Clark Millikan, Martin Summerfield, Theodore von Kármán, and Frank J. Malina. March Field, California, 23 August 1941. Photo courtesy Aerojet-General Corporation.

COMPOSITE AND DOUBLE-BASE PROPELLANTS

Modern solid propellants can be broadly divided into two main categories: the composite and double-base propellants. The composite propellant is a heterogeneous mechanical mixture of oxidizing crystals in an organic plastic-like binder. The binders are usually made of synthetic rubbers and serve as a fuel. In addition, metal powder — such as aluminum — can be added to the mixture to increase specific impulse. The GALCIT's "asphalt propellant" pioneered this new technology.

A double-base propellant is a homogeneous substance, usually on the basis of nitrocellulose dissolved in nitroglycerin. Double-base propellants are practically explosives and require special care in handling. Safety rating puts double-base propellants into the category of potentially detonable materials (hazard classification Class 1.1 or Class 7) in contrast to composite propellants classified as nondetonable (Class 1.3 or Class 2). The technology of double-base propellants was significantly advanced during the war under the sponsorship of the Office of Scientific Research and Development.

Modern composite propellants often imbed in the binder some double-base ingredients and other powerful explosives, such as RDX and HMX. RDX and HMX are, in effect, a balanced chemical combination of fuel and oxidizer. Other minor additives are used to improve processing properties of the propellants during casting and curing, to achieve reliable bonding to the motor case, and to reduce chemical deterioration with time.

> **THIOKOL**
>
> Chemist Dr. Joseph C. Patrick discovered synthetic rubber in Kansas City, Missouri, in 1926. He named this polysulfide polymer *Thiokol* from the Greek word roots for sulfur and glue. Patrick and Bevis Longstreth formed a new company, *Thiokol Corporation*, to produce the polymer.
>
> Liquid polysulfide polymer solidified and formed synthetic rubber, resistant to common solvents. Thiokol proved to be an excellent sealant for fuel tanks in wartime and was also widely used for sealing fuselages, air ducts, and gun turrets. Then Jet Propulsion Laboratory demonstrated another important application of the Thiokol polymer: it was a superior binder for rocket solid propellants.
>
> The war had ended. The only two commercial rocket manufacturers, *Aerojet* and *Hercules*, were not interested in Thiokol. To preserve the promising technology and with the help of the U.S. Army, *Thiokol Chemical Corporation* plunged into solid-propellant rocket motor manufacturing in 1947. The company became a prime developer of rocket motors ranging from tactical missiles to intercontinental ballistic missiles to space shuttle solid-rocket boosters.

on contact. Such propellant combinations are called hypergolic. By 1943 Truax had an experimental hypergolic JATO unit producing 1500-lbf (6.7 kN) thrust.

First Private Rocket Enterprise

The Navy's interest in and support of JATO development led to a remarkable event. Following the fine American tradition of private enterprise, four members of the ARS — H. Franklin Pierce, John Shesta, Lovell Lawrence, Jr., and James H. Wyld — founded the first private company dedicated to rocket development, *Reaction Motors, Inc. (RMI)*. The company with the name coined in analogy with *General Motors* was formed in a tiny room in a garage in North Arlington, New Jersey. RMI was incorporated in December of 1941, and Lawrence Jr. became its first president.

Fig. 11.8. H. Franklin Pierce was President of ARS in 1940–1942 and one of the founders of the first rocket corporation Reaction Motors, Inc. Figure from *Astronautics*, No. 46, p. 2, July 1940.

Bureau of Aeronautics' Contract

RMI focused on applications of the liquid-propellant regeneratively cooled engine demonstrated by James Wyld. For several months Navy's *Bureau of Aeronautics* (*BuAer*) followed with interest the work on the engine. BuAer signed a contract with RMI in December 1941 after the Pearl Harbor bombing. RMI was to develop, demonstrate, and deliver within five months a 100-lbf (445-N) thrust engine followed later by a 1000-lbf (4.45-kN) engine operating on aviation gasoline and liquid oxygen.

The little company had grown from the original four founding members to 55 and 473 employees by 1945 and 1947, respectively. In its first year of operation, in 1942, the company posted a profit of

HYPERGOLIC ROCKET PROPELLANTS

Red-fuming nitric acid (RFNA) and aniline ignite on contact, forming a hypergolic propellant combination.

RFNA is a nitric acid (HNO_3) that contains 5–20% of the dissolved nitrogen dioxide (NO_2). Its red-brown, as the name suggests, fumes are poisonous. RFNA is more energetic, more stable, and less corrosive than a concentrated nitric acid. The latter is often called *white-fuming nitric acid* (WFNA) and contains less than 2% of water and impurities. The so-called *inhibited RFNA*, IRFNA, also served as an oxidizer. RFNA is a storable oxidizer, and it was widely used in rocket propulsion from 1940s through 1960s.

Aniline ($C_6H_5NH_2$) is a highly poisonous, oily, and colorless substance, first obtained from indigo. Aniline is used to make dyes, drugs, and explosives.

High-Tech Venture Investing

$643. After the war the financially struggling company was rescued by Laurence S. Rockefeller. Rockefeller was a savvy early "high-tech" venture investor who helped to launch or preserve such innovative companies as *McDonnell Aircraft Company*, *Piasecki Helicopter Company*, Eddie Rickenbacker's *Eastern Air Lines,* and *Itek*.

RMI's founders, Pierce, Lawrence, and Shesta, left thc company in 1947, 1951, and 1953, respectively. Lawrence and Shesta dropped technical work all together. Pierce went to California to a citrus business, failed, and eventually joined the Douglas Aircraft Company. James Wyld continued to actively work at RMI until 1953 when he died at the early age of 41.

Fig. 11.9. A liquid-propellant rocket engine, built by Robert H. Goddard and his team, assists the takeoff of the Navy's flying boat PBY on Severn River near Annapolis, Maryland, in September 1942. Photo courtesy of the Roswell Museum and Art Center, Roswell, New Mexico.

With the Rockefeller's help in 1947 and 1948 and with improved management, RMI became profitable again. The company grew and boasted more than 1600 employees by 1958. On 30 April 1958, RMI merged with the *Thiokol Chemical Corporation*. Thiokol, the leading manufacturer of the synthetic rubber, entered the rocket propulsion field soon after the war and became a nation's premier manufacturer of solid-propellant motors.

RMI Merges with Thiokol

RMI and Truax were not the only developers of liquid-propellant JATOs in the early 1940s. Robert H. Goddard's Roswell operations were transferred to the Navy's Engineering Experiment Station in Annapolis, Maryland. Goddard brought his liquid-propellant engine built under the contract with the Navy. The program culminated with the demonstration of the takeoff assist unit on a flying boat PBY on 23 September 1942. After several tests of taxiing the plane on Severn River, the PBY took off assisted by the roaring rocket engine. The second attempt of takeoff failed when the rocket malfunctioned and put the already airborne airplane on fire. The crew was fortunately rescued to safety.

Goddard's JATO

Martin Summerfield was in charge of the GALCIT liquid-propellant JATO effort. His selected propellant combination, RFNA and a hydrocarbon fuel, showed combustion instabilities, as in Truax's work. Discovery of hypergolic propellants

Liquid-Propellant JATO at GALCIT

Fig. 11.10. Specialists of the Aerojet Engineering Corporation at Muroc Air Force Base, California, in January 1943: (left to right) Theodore C. Coleman, John W. Parsons, Edward S. Forman, Paul H. Dane (pilot of the Douglas Havoc A-20, in the background), Andrew G. Haley (Aerojet's President from 26 August 1942 till 7 August 1945), Theodore von Kármán, Frank J. Malina, Martin Summerfield, and T. Edward Beehan. Photo courtesy of Aerojet-General Corporation.

Fig. 11.11. Early rocket firing at Aerojet Engineering Corporation, Azusa, California. Photo courtesy of Aerojet-General Corporation.

by the Navy's group brought a glimmer of hope. The GALCIT team quickly adopted aniline and RFNA propellants, which worked remarkably well.

Two 1000-lbf thrust liquid rockets, developed by GALCIT, were installed on the Douglas Havoc A-20 bomber. Von Kármán described the first flight test of these JATOs in the spring of 1942 at the Muroc Field (now Edwards Air Force Base), California, "Some fifty people were on hand as [Edward] Forman and [pilot Paul] Dane climbed into the craft and took their positions. The engines were revved up, the rockets ignited, and suddenly the bomber took off with a roar, almost straight up. As smoke billowed forth, the A-20 continued to rise as though scooped upward by a sudden draft" (Von Kármán 1967, 254).

Aerojet Engineering Corporation

Incorporation of the first private rocket company, RMI, was followed in a few months by another company, this time in California. The successes of the GALCIT rocketeers with JATOs led to the formation, on 19 March 1942, of the *Aerojet Engineering Corporation*. The founders of Aerojet, von Kármán, Malina, Summerfield, Parsons, and Forman, each contributed $200 to launch this new enterprise.

Doing Business in Washington

Aerojet got off to a good start landing the contracts from Navy's Bureau of Aeronautics and the Air Force's Aircraft Laboratories at Wright Field for development of new solid- and liquid-propellant JATOs, respectively. A few months later the funding was not renewed, however. The Aerojet's president, von Kármán, was bluntly told by the military brass, as he later recalled, that he was liked "but only in cap and gown to advise us [the military] what to do in science. The derby hat of the businessman does not fit you .... Find somebody who knows something about doing business with Washington ..." (Von Kármán 1967, 259).

Andrew G. Haley, a Washington lawyer, was promptly recruited as a new president on 26 August 1942. (In 1950s, Haley became a prominent specialist in

BAZOOKA

Clarence N. Hickman, Robert H. Goddard, and H.S. Parker designed a Bazooka-type 3-in.-diam (76-mm) rocket with a leg-supported launcher tube in 1918. The hostilities of the world war had ended before the missile could be fully developed.

Hickman later became research director at Bell Telephone Laboratories and headed the Rocket Ordnance Section (H) of the National Defense Research Committee (NDRC). In June 1940, Hickman wrote a letter to the president of the National Academy of Sciences, Frank B. Jewett, calling attention to the importance of Goddard's rocket work.

From his position, Hickman supported and guided rocket development conducted in various departments and organizations. These efforts led to the undoubtedly best known American wartime rocket, Bazooka. Basic development of the new weapon, a rocket-propelled grenade, was performed in 1941 by Colonel Leslie A. Skinner of the Army Ordnance Department. Burning of the bazooka rocket was entirely completed inside the launcher tube. The flight of the rocket outside the launcher was similar to that of a gun projectile. The projectile had diameter 2.36 in. (60 mm) and reached velocity 200 miles per hour (320 km/h). Bazooka's maximum range exceeded 600 m.

Fig. 11.12. World War II Bazooka and 3.5-lb (1.6-kg) solid-propellant rocket projectile. The missile carried 0.5-lb (0.23 kg) of a high explosive. The explosion of a specially shaped charge could penetrate up to 5 in. (127 mm) of armor. Photo courtesy of Mike Gruntman.

The bazooka was successfully demonstrated against tanks in May 1942 at the Aberdeen Proving Ground. Formal demonstration to the Departments of War and Navy followed at Camp Sims in June. A production contract was issued shortly afterwards to General Electric, and the bazookas reached the troops in less than a year. In March 1943, they were already in use in Tunisia during the campaign in North Africa. A number of bazooka variations were introduced as the war progressed.

Bazooka was named after a musical instrument it resembled in appearance. This instrument, a pipe trumpet, was invented and played by a comedian Bob Burns in the late 1930s and early 1940s.

ROCKETS FOR ANTI-AIRCRAFT DEFENSE IN 1940–1941

... Combinations of a hundred guns would have been excellent [against the enemy aircraft], if the guns could have been produced and the batteries manned and all put in the right place at the right time. This was beyond human power to achieve. But a very simple, cheap alternative was available in the rocket .... The 3-inch rocket carried a much more powerful warhead than a 3-inch shell. It was not so accurate. On the other hand, rocket projectors [launchers] had the inestimable advantage that they could be made very quickly and easily in enormous numbers without burdening our hard-driven factories. Thousands of these ... [rocket launchers] were made, and some millions of rounds of ammunition .... .

When the ...[rockets] did come into action [in the middle of 1941] the number of rounds needed to bring down an aircraft was little more than that required by the enormously more costly and scanty [antiaircraft] A.A. guns, of which we were so short. The rockets were good in themselves, and also an addition to our other means of defense.

Winston Churchill (1953, Vol. 2, 347)

space law. He served as president of the International Astronautical Federation in 1958 and 1959 and was a founder of the *International Institute of Space Law*.) Aerojet rapidly grew, delivering 2000 JATO units by the end of 1943. At first located in Pasadena, California, the company moved to nearby Azusa, California, where it further expanded and built a rocket testing facility in an adjacent gravel pit. The company eventually grew to employ more than 30,000 people and became a premier designer and manufacturer of rocket propulsion systems in the United States.

Aerojet Takes Off

The World War greatly accelerated the pace of development of various rocket systems. The belligerent countries introduced many types of war rockets, including all major categories: surface-to-surface, surface-to-air, air-to-surface, and air-to-air. Rockets were fired on the land, in the air, and on the sea. Most of these missiles were unguided and inexpensive, with the numbers making up for lack of accuracy. In addition to GALCIT, Aerojet, and Navy's Annapolis establishment, major solid-propellant rocket development and fabrication were organized

Return of War Rockets

AMERICAN V-1

Germany attacked England with the V-1 missiles in June 1944. Parts of this new weapon were soon brought to the Wright Field, Ohio, of the Materiel Command of the U.S. Army Air Forces. The race to reconstruct the new weapon has begun. Within two weeks the V-1's pulse jet engine was reproduced. *Republic Aviation Corpoartion* had the first copy of the complete V-1 ready for tests by September 8. The Navy developed launching sleds, or ramps, to accelerate the missile. The first American V-1, called JB-2 (*JB* stood for *jet bomb*), flew at Eglin Air Force Base in Florida in October 1944.

Almost 1400 JB-2s had been manufactured by the Willys-Overland Company (airframe) and Ford Motor Company (engines) by the program termination in September 1945. The JB-2 did not see operational use although 200 missiles were test launched.

AMERICAN MISSILES AND SOVIET ESPIONAGE

Accomplishments of Soviet spies in pilfering atomic secrets of their wartime American ally are well publicized. The American aviation industry was also under a watchful eye of the Soviet intelligence agents and their leftist sympathizers and collaborators who dutifully reported to Moscow first steps in rocketry and missiles, particularly a development of the American copy of the V-1, called the JB-2.

The U.S. Army's *Signal Intelligence Service* and its successor the *National Security Agency* (NSA) cracked some of Soviet wartime intelligence messages to and from the United States. In a message on 15 September 1944, the KGB New York station described to the Moscow center that it was

> attempted to arrange for the robots [as the V-1s were sometimes called at that time] to be produced within a period of 60 days. The firms which were developing ... [jet systems] did not want to arrange for production within such a short time limit. Republic [Aviation Corp.] agreed to turn out the robots and started production 55 days later. 10 [units of JB-2s] were turned out for the first time on 2nd September.

Alexander Feklisov explains in his memoirs that he was, when stationed in the United States, a wartime KGB controller of Julius Rosenberg and his spy ring. Feklisov describes the enormous amount of technical material passed to the Soviets during and after World War II by a member of the ring William Perl, an aeronautical engineer at NACA's Cleveland Center. Feklisov credits Perl with obtaining "the complete blueprint of the first American jet fighter, the Lockheed P-80 Shooting Star" (Feklisov and Kostin 2001, 146). In an interesting twist, Perl provided in 1945, according to Feklisov, "two ultrasecret reports written in Russian and full of incomprehensible [for KGB officers in New York] formulas regarding rockets." Moscow believed that the reports "had been obtained in the USSR by British or American intelligence," but Perl could not find out how "these documents had landed in the offices of NACA."

The first exploratory work on space projects also got noticed after the war. Soviet spies reported on a "sky platform," obviously a satellite, and an "atomic airplane" using "nuclear fission to propel airplane engines." A leading FBI counterintelligence specialist Robert J. Lamphere wrote that in 1950 "we [the FBI] were on the verge of tieing [William] Perl more closely to the theft of data about the 'atomic airplane' and were concurrently meeting with the [Atomic Energy Commission] AEC and NACA on that subject. We were also exploring the possibility that Perl, while an assistant to von Kármán, had removed secret files from von Kármán's office, copied them and turned the copies over to Rosenberg [who passed it to his KGB controller]" (Lamphere 1995, 215).

Lamphere described that a jailhouse informer reported that "according to Rosenberg [as the latter explained to the informer], Perl had come to New York from Cleveland, gone up to Columbia University [on the Fourth of July weekend in 1949] and removed some files containing secret data from a safe; then, together with Rosenberg and two other men, ... had photographed the secret data. ... Records showed that Perl had visited the Columbia laboratory and office of his former boss, Theodore von Kármán, and von Kármán told us [the FBI] that Perl knew the combination to his locked safe and could well have removed documents from it. We discovered that after this weekend ... Perl made an unusual deposit to his bank account ... ." The FBI "had only circumstantial evidence, not enough to stand in court to prove Perl to be a spy. So in 1953 Perl was tried and convicted of perjury — having lied to the grand jury about knowing Rosenberg and [another convicted spy Morton] Sobell — and was sentenced to serve five years" (Lamphere 1995, 249, 250).

Fig. 11.13. Five-inch (127-mm) rockets being loaded under the wing of F4U Corsair. Before takeoff, the safety pins were removed. These high-velocity aircraft rockets (HVAR), nicknamed *Holy Moses*, were 5.7 ft (1.7 m) long with mass 134 lb (61 kg), and had range of up to 3 miles (4.8 km). About one million missiles of this type were manufactured during World War II. These rockets proved so reliable and effective that they were produced until 1955. Photo by Lt. David. D. Duncan, U.S. Marine Corps, Okinawa, June 1945. Photo courtesy of National Archives and Records Administration.

at the Naval Powder Factory (reorganized into the Naval Ordnance Station after the war) in Indian Head, Maryland, and at the Hercules Powder Company.

Fig. 11.14. Corsair fighter launches rockets against the Japanese stronghold on Okinawa in June 1945. Photo by Lt. David. D. Duncan, U.S. Marine Corps. Photo courtesy of National Archives and Records Administration.

Barrage Rockets

The U.S. Marine Corps' 4.5-in. (114-mm) beach barrage rockets (BBR) pounded enemy defenses as a close support weapon (with range 0.6 miles or 1 km) in a short time intervals between the end of the naval and air bombardment and actual landing of the troops. The rockets, called *Old Faithful*, were crash developed just in time to support the landing near Casablanca in the North Africa on 8 November 1942. Rocket-outfitted landing craft, LCM, could carry up to 88 barrage rockets. More than 1.6 million of such rockets had been produced by the war's end. The U.S. Army's 4.5-in (114-mm) M8 rockets with the range up to 2.8 miles (4.5 km) were usually fired from 8-tube (*Xylophone*) or 60-tube (*Calliope*) launchers.

Fig. 11.15. Rows of rocket launchers loaded with deadly missiles aboard an LSM(R), Pacific theater, 1945. Photo courtesy of National Archives and Records Administration.

Fig. 11.16. LSM(R) launches rockets, Pacific theater, 1945. Photo courtesy of National Archives and Records Administration.

The British manufactured 3- and 5-in. (76- and 127-mm) barrage rockets. These missiles were under development since 1935 and used cordite (50% nitroglycerin, 41% nitrocellulose, 9% carbonite) as the propellant. The rockets were routinely fired from 60- and 120-tube launchers on ships against shore targets in landing operations. Members of the Home Guard manned the special launching ramps, called the projectors, of antiaircraft rockets. These rockets that were often called in official British documents *unrotated projectiles*, or UPs, were rushed into service in 1941.

Missiles on the Land, in the Air, on the Sea

The British *Hedgehog* was an antisubmarine rocket system developed for the Royal Navy and also employed by the U.S. Navy. The Hedgehog fired 24 58-lb (26-kg) rockets. Great Britain also initiated development of guided missiles (*Brakemine*, *Little Ben/Longshot*, *Stooge*, and *LOPGAP*), but no practical weapon system was ready by the time the war had ended.

> **ROCKETS AGAINST SUBMARINES AND RADARS**
>
> ... In the Bay of Biscay however the Anglo-American air offensive [in the early 1943] was soon to make the life of [the German] U-boats in transit almost unbearable. The rocket now fired from aircraft was so damaging that the enemy started sending the U-boats in groups on the surface, fighting off the aircraft with gunfire in daylight. This desperate experiment was vain ... .
>
> Winston Churchill (1953, Vol. 5, 8)
>
> From Calais to Guernsey the Germans had [in 1944] no fewer than one hundred and twenty major pieces of Radar equipment for finding our convoys and directing the fire of their shore batteries. We discovered them all, and attacked them so successfully with rocket-firing aircraft that on the night before D Day not one in six was working ... .
>
> Winston Churchill (1953, Vol. 6, 9)

The German Army used the Nebelwerfer series of missile systems, originally designed for firing smoke bombs. The rocket diameter varied from 15 to 32 cm (from 6 to 12.5 in.). For example a 21-cm (8 1/4-in.) rocket had a 4-mile (6.4-km) range and carried 22 lb (10 kg) of explosives. Following the success of the American Bazooka, the German Army introduced its own antitank rockets, the Panzerfaust and Panzerschreck.

Antiaircraft Missiles

Barrage surface-to-air rockets were widely used in World War II. More than 4.5 million of such unguided 2-in. (51-mm) rockets were ordered by the British Royal Navy. Much more complex antiaircraft guided missiles were being developed in Germany to counteract the destructive American and British air raids. This high-priority program led to the demonstration of several promising missiles, the best known being *Wasserfall*, *Schmetterling*, and *Rheintochter*.

The Wasserfall was simpler and much smaller (one-third) than the A-4 (V-2), and its accurate guidance systems relied on two radars tracking the missile and the target. The supersonic missile was designed to destroy bombers at altitudes up to 10 miles (16 km) and at the range of 30 miles (48 km). Thirty-five test firings were performed in 1944 and 1945. The war ended before this antiaircraft wonder weapon could be deployed.

In 1945, the U.S. Navy also initiated a guided surface-to-air missile development program to counter the increasing danger of Kamikaze planes. This effort led to the *Lark* program in the late 1940s, with most test firings performed from the Naval Ordnance Air Missile Test Center at Point Mugu, California.

Katyusha

The Soviet Army widely used its Katyusha solid-propellant barrage rockets during the war. The design proved so efficient and undemanding in service that the Islamic and Palestinian militants used Katyusha's derivatives until late 1990s for bombarding civilian targets in northern Israel near the Lebanese border.

The Japanese also produced a few short-range rockets. Small antitank rockets were 3.5 in. (76 mm) in diameter. Various larger rocket ranged in diameter from 3 to 18 in. For example, an 8-in. (203-mm) rocket weighed 180–200 lb (82–91 kg) and had a range up to 2 miles (3.2 km).

The belligerent countries, Germany, Great Britain, the United States, and the USSR, widely employed air-to-surface missiles against submarines, surface ships, tanks, and other targets. The U.S. Navy alone expended about 100 million dollars on rockets each month in 1945.

Fig. 11.17. General Henry H. Arnold emerged as a major proponent of development of long-range guided ballistic missiles. Photo courtesy of 45th Space Wing History Office Archives, Patrick Air Force Base, Florida.

Struggle for the Future

World War II was essentially over by the middle of the summer 1945. A new battle had begun at that time, however: the battle over the future of the American rocketry and the American satellite. Their fate was anything but certain.

A number of top military leaders in the Army Ordnance, Army Air Forces, and Navy saw the promise of long-range rockets, ballistic and winged. The long-range missiles seemed technically possible for many and clearly critical for the nation's security. Intercontinental warfare thus moved high on the priority list.

Missiles and Nuclear Weapons

The successful development of the atomic bomb made rapid advancement in long-range guided rocketry crucial for all of the three military services. (The Air Force would become formally independent in September 1947.) The three services embarked on exploratory programs in rocketry, setting the stage for intense interservice rivalry and competition in a struggle for control of the emerging weapon systems.

A rocket capable of delivering a nuclear bomb for a thousand of miles and further across the oceans with the desired precision would drastically change the balance of power in the world. The exceptional explosive power of nuclear weapons has also significantly reduced the requirements to the guidance of long-range missiles. The guidance precision would remain, however, a critical technical challenge for several years.

Antiaircraft guided missiles were the other rocket area of vital importance for national defense. Some military leaders and scientists looked even further, into a possibility of Earth-orbiting satellites. The new rocket and space vision was not, however, universally shared in Washington, including by many in position of knowledge, influence, and power. This was a formidable opposition.

Science Advisory Group

The Army Air Forces aggressively worked to bring science into its planning and development. As early as September 1944, General Arnold asked von Kármán to gather a group of scientists "to develop a blueprint for air research." The von Kármán's group became known as the *Science Advisory Group* (SAG).

In December 1945, SAG issued the report, "Toward New Horizons," which covered the research areas of importance for the future Air Force. A number of prominent scientists and engineers contributed to this document, including Hugh L. Dryden, Lee A. DuBridge, George Gamow, Ivan A. Getting, William H. Pickering, Edward M. Purcell, George S. Schairer, Hsue-Shen Tsien, and Vladimir K. Zworykin.

"Toward New Horizons"

> 7 November 1944
>
> Memorandum for Dr. von Karman
>
> I believe the security of the United States of America will continue to rest in part in developments instituted by our educational and professional scientists. I am anxious that Air Forces postwar and next-war research and development programs be placed on a sound and continuing basis. In addition, I am desirous that these programs be in such form and contain such well thought out, long range thinking that, in addition to guaranteeing the security of our nation and serving as a guide for the next 10-20 year period, that the recommended programs can be used as a basis for adequate Congressional appropriations.
>
> Henry H. Arnold
> Commanding General, Army Air Forces
> (*Towards New Horizons* 1992, vii)

Considering long-range rocketry, the SAG report noted that hydrogen propellant "will give the greatest exhaust velocity .... It is estimated that if we are able to produce sufficiently high temperatures and high pressures [in the combustion chamber], the thrust produced per unit weight of the consumed material could be made about six times the present value. This would increase more than thirty times the [150-mile] range of V-2 type rockets using chemical propellants and would make rocket navigation possible up to the highest altitude beyond the stratosphere. The 'satellite' is a definite possibility" (*Towards New Horizons* 1992, 29).

"The 'satellite' is a definite possibility."

Deputy Chief of Air Staff for R&D

The Air Forces' emphasis on the advanced science and technology led to establishment of a new deputy chief of staff responsible for this important area. Major General Curtis LeMay became the first Deputy Chief of Air Staff to guide research and development. The permanent Air Forces' *Science Advisory Board*, chaired by von Kármán, was formed later.

The question of what role the science in general, and the Science Advisory Board in particular, should play in the military affairs was not simple and remained highly controversial for some time. The new Deputy Chief of Staff wisely

HSUE-SHEN TSIEN (QIAN XUESEN)

Fig. 11.18. Chinese Communist leader Chairman Mao Zedong (left) and Hsue-Shen Tsien, ca. 1956. Photo courtesy of China Astronautics Publishing House.

Hsue-Shen Tsien, a former student of von Kármán and an early member of the GALCIT group, was considered to be "an undisputed genius whose work was providing enormous impetus to advances in high-speed aerodynamics and jet propulsion" (von Kármán 1967, 308). Tsien accompanied von Kármán on a trip to Europe in 1945 to assess the status of German aeronautical developments, and he contributed to the work of the Science Advisory Group of the U.S. Air Forces.

In July 1950, Tsien was accused of being an alien Communist and, consequently, a security risk. His security clearance was promptly revoked. Because Tsien actively worked in highly classified areas, he was permitted to leave the United States for Communist China only in 1955. As von Kármán noted (1967, 308), "the United States ... gave Red China one of our most brilliant rocket experts for no really good reasons." In Tsien's words, "the great Chairman Mao saved [him, i.e., Tsien] from the difficult environment in a foreign country [United States] and personally led [him] to the road of revolution and liberated [him] from the bondage of old traditional views" (Tsien 1976).

Communist China put development of rocketry and nuclear weapons on the top of its priorities. These were prodigious tasks for an economically backward country. Tsien left the United States for his "great ancestral socialist motherland" in September 1955. Already in February 1956 he sent a proposal "Some Suggestions on Establishing the Defense Aeronautical Industry in China" to the Central Committee of the Chinese Communist Party. In April 1956 Tsien addressed the Central Military Commission, chaired by Premier Zhou Enlai, with the missile program proposal. On 8 October 1956, Tsien was appointed president of China's first missile research establishment, *Research Academy No.5* of the Ministry of National Defense.

Communist China greatly expanded its rocket and space programs, culminating in the launch of the first satellite on its own rocket in 1970. Tsien played the leading role in these developments. His writings and scientific prestige also provided important support for formulating and promoting the disastrous *Great Leap Forward* policy of Chairman Mao. Tsien became an integral part of the ruling Communist elite. He reportedly publicly supported the crackdown on Tienanmen Square in 1989. "I got where I am today," wrote Tsien, "because of what Chairman Mao and the Communist party gave me" (Tsien 1976). He is revered today by many in China, where the scientists are exhorted to follow the example of Comrade Tsien and be loyal to the Communist Party and to the socialist motherland.

WADE-GILES AND PINYIN

Two systems, the Wade-Giles and Pinyin, are commonly used for transliteration of Chinese (Mandarin) words and names. The former system originated in 1859, while the People's Republic of China introduced the latter in 1958.

The Wade-Giles and Pinyin render the name of *Tsien* as *Hsue-Shen Tsien* (*Tsien Hsue-Shen*) and *Qian Xuesen*, respectively. Chairman Mao's name is respectively spelled as *Mao Tse-tung* and *Mao Zedong*.

maintained a balance between the valuable scientific input and often justified resistance by leading operational officers. As von Kármán wrote, "General Curtis LeMay kept a foot in each camp. The General supported good research proposals but hit the ceiling whenever, as he put it, a 'longhair' had an idea that he felt was basically in the realm of operation" (von Kármán 1967, 302).

Navy Satellite

The Navy got keenly interested in the new opportunities offered by satellites. Its Bureau of Aeronautics established the *Committee for Evaluating the Feasibility of Space Rocketry* (CEFSR). In October 1945, the Committee, chaired by the Navy physicist Commander Harvey Hall, recommended BuAer to devise and launch an Earth-orbiting satellite using a single-stage rocket propelled by liquid hydrogen and liquid oxygen (H_2/LOX). This propellant combination was known to the prewar rocket pioneers but considered impractical because of the extremely low temperature and low specific density of liquid hydrogen. (Hydrogen becomes liquid at temperature 20 K at 1 atm pressure with the density 14 times smaller than that of water.) The bold concept proposed by the Navy became known as the *high-altitude test vehicle* (HATV).

H_2/LOX Propulsion

Fig. 11.19. General Curtis LeMay in 1956. LeMay became the first Deputy Chief of Air Staff for Research and Development when this office was created on 1 December 1945. He was instrumental in safeguarding the independent status of Project RAND and commissioned its first report evaluating the possibility of a "world-circling spaceship." Photo courtesy of the U.S. Air Force Museum, Wright-Patterson Air Force Base, Ohio.

Commander Hall and several other Navy scientists enthusiastically supported the satellite that promised advantages of worldwide communications and navigation. GALCIT, Aerojet, Glenn L. Martin Company, North American Aviation, and Douglas Aircraft Company, were contracted to study various technical aspects of the HATV and the satellite. The exploratory discussions that led to the establishment of CEFSR and the following Navy contracts initiated practical development of the H_2/LOX rocket propulsion technology at Aerojet and, a few years later, at Jet Propulsion Laboratory. A large hydrogen liquefier was subsequently constructed at Aerojet. Independently, the Army Air Forces organized studies of fundamental properties of liquid hydrogen and its propulsion applications by Herrick L. Johnston at the Ohio State University in Columbus. These early research and development efforts laid the foundation for the future highly efficient H_2/LOX rocket engines.

Facing the prohibitively high price tag for its satellite, the Navy approached the Army Air Forces in the early 1946 in an attempt to start a joint satellite program. The Air Forces were not especially delighted in the Navy's invasion of the area considered as their own province. In addition, the Army Air Forces had already initiated their own feasibility study of the satellite at the newly established RAND. General LeMay declined the Navy joint satellite proposal in late March 1946.

Project RAND

On 2 March 1946, the Army Air Forces signed a contract establishing its primary advisory organization, *Project RAND*. The focus of the planned research effort was specifically on intercontinental warfare. Project RAND was first a part of the Douglas Aircraft Company in Santa Monica, California, and reported to the chief of the Air Forces research and development, General LeMay.

> **PROJECT RAND**
>
> ... An order is hereby placed with you to the furnishing to the Government of the following services:
>
> Item 1 — Study and research on the broad subject of intercontinental warfare, other than surface, with the object of recommending to the Army Air Forces preferred techniques and instrumentalities for this purposes ... .
>
> Letter contract establishing Project RAND, 2 March 1946

RAND Corporation

RAND had emerged from the wartime experience of General Arnold and special consultant in the War Department Dr. Edward L. Bowles. The Douglas Aircraft Company's President Donald Douglas, Sr., Chief Engineer Arthur Raymond, and Franklin Collbohm, helped to work out the concept of the new think tank. Project RAND separated from the Douglas Aircraft Company in 1948 and became an independent nonprofit organization known as the *Rand Corporation*. General LeMay was instrumental in shaping and wisely preserving an independent character of RAND, an important feature of the organization.

Report No. SM-11827

After formation of RAND, LeMay commissioned a report on a possibility of the artificial satellite. This first RAND's report No. SM-11827 "Preliminary Design of an Experimental World-Circling Spaceship" was issued by the Douglas Aircraft Company on 2 May 1946. The report presented "an engineering analysis of the possibilities of designing a man-made satellite" and

> **SPACESHIP UTILITY**
>
> ... We can see no more clearly all the utility and implications of spaceships than the Wright brothers could see the fleets of B-29s bombing Japan and air transports circling the globe ... .
>
> *Preliminary Design* 1946, 1

concluded "that modern technology has advanced to a point where it now appears feasible to undertake the design of a satellite vehicle." In addition to the feasibility study of a rocket capable of deploying a satellite, the report considered various problems of orbital evolution, attitude control, communications, thermal control, and even satellite vehicle return to Earth and landing. The problems of the man-carrying vehicle were also discussed.

Three to Five Years to Launch a Satellite

The RAND report estimated the time needed for launching the satellite on an alcohol-oxygen rocket as "at least six months [of] additional research and preliminary design" followed by a "three to five year [development] program." It also concluded that "estimates on the hydrogen-oxygen [engine] units are virtually impossible at this time due to the unknown factors in the use of liquid hydrogen" (*Preliminary Design* 1946, 213).

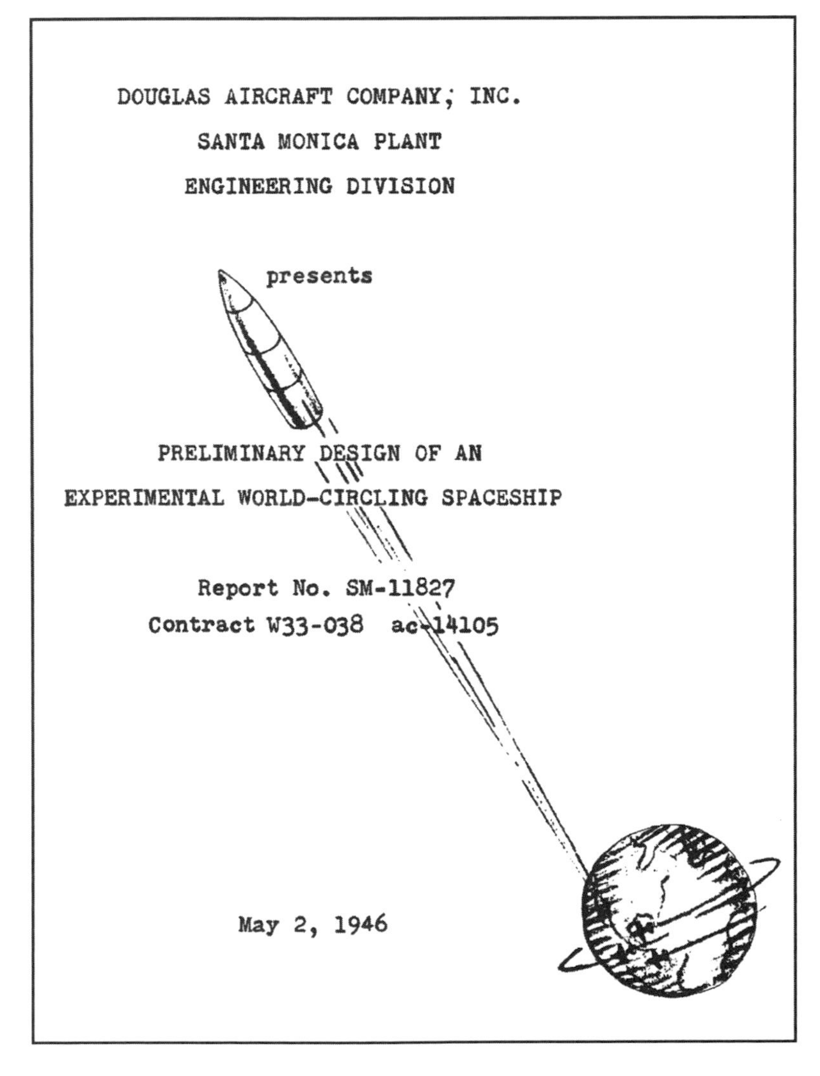

DOUGLAS AIRCRAFT COMPANY, INC.
SANTA MONICA PLANT
ENGINEERING DIVISION

presents

PRELIMINARY DESIGN OF AN
EXPERIMENTAL WORLD-CIRCLING SPACESHIP

Report No. SM-11827
Contract W33-038 ac-14105

May 2, 1946

Fig. 11.20. Cover of the RAND Report "Preliminary Design of an Experimental World-Circling Spaceship," 2 May 1946. Courtesy of U.S. Air Force.

Engineering Appraisal

"Payload of 500 lbs. and 20 cuft."

The RAND report was a thorough engineering appraisal of the required technologies for the rocket and the spacecraft. It stated clearly that the satellite "vehicle will undoubtedly prove to be of great military value. However, the present study was centered around a vehicle to be used in obtaining much desired scientific information on cosmic rays, gravitation, geophysics, terrestrial magnetism, astronomy, meteorology, and properties of the upper atmosphere. For this purpose, the payload of 500 lbs. [230 kg] and 20 cuft. [0.57 m^3] was selected as reasonable estimate of the requirements for scientific apparatus capable of obtaining results sufficiently far-reaching to make the undertaking worthwhile" (*Preliminary Design* 1946, ii).

CRYSTAL BALL

... Though the crystal ball is cloudy, two things seem clear:

1. A satellite vehicle with appropriate instrumentation can be expected to be one of the most potent scientific tools of the Twentieth Century.
2. The achievement of a satellite craft by the United States would inflame the imagination of mankind, and would probably produce repercussions in the world comparable to the explosion of the atomic bomb.

Preliminary Design (1946, 1)

Utility Had Not Yet Been Established

Following the rejection of the joint satellite proposal in March 1946, both the Army Air Forces and the Navy continued for some time independent feasibility studies of the satellites and required launchers. This preliminary satellite design identified several areas such as guidance, navigation, attitude control, power sources, and communications electronics, where significant research and development were needed. Neither transistor nor practical solar cells existed at that time. A special committee under Clark Millikan reviewed the satellite programs and concluded in March 1948 that a scientific or military utility of a satellite vehicle had not yet been established.

POSSIBILITY OF HUMAN SPACEFLIGHT

Throughout the present design study of a satellite vehicle, it has been assumed that it would be used primarily as an uninhibited scientific laboratory. Later developments could alter its capabilities for use as an instrument of warfare.

However, it must be confessed that in the back of many of the minds of the men working on this study there lingered the hope that our impartial engineering analysis would bring forth a vehicle not unsuited to human transportation.

... Next we consider the safety and welfare of the man after the vehicle has been established on the orbit. Popular fiction writers have devoted considerable thought and ingenuity to means of furnishing him with air, food and water. The most ingenious of these solutions is that of the balanced vivarium in which plants and man completely supply each other needs. Leaving these problems to the inventors, we ask ourselves the engineering questions of whether we can provide livable-temperatures and reasonable protection against meteors. ... We have seen that the answers are tentatively in the affirmative.

Preliminary Design (1946, 211, 212)

SOVIET PROPAGANDA

The Soviet propaganda vociferously attacked those Americans, in uniform and without, who advocated "the development of long-range guided missiles and of 'intercontinental warfare.'" Moscow thus energized its fellow travellers and other leftist sympathizers for action on its behalf.

The Soviet magazine *New Times*, published specifically for Western consumption, ridiculed intercontinental missiles as "wild utopia" from the military standpoint. The concept of "artificial satellites of the earth as military bases" especially enraged the Kremlin and the Americans supporting such ideas were branded "disciples of the Nazi scientists" (Rubinstein 1947). This rhetoric did not prevent the Soviet Union to concentrate, at the same time, enormous resources of the totalitarian socialist state on an all-out effort to develop the nuclear and rocket capabilities.

The budget realities and other priorities shifted the interests elsewhere. The wisdom of investing significant resources in the development of long-range rockets and a satellite was not shared by many. The Navy abandoned its studies of the satellite in 1948, while the Air Force continued some low-key work.

Opponents of Long-Range Rockets

The feasibility of practical long-range ballistic missiles was also questioned in earnest at high levels of the government. Vannevar Bush has emerged as a

VANNEVAR BUSH

Vannevar Bush (1890–1974) was a prominent scientist and administrator. He became famous for his work on the early computers. In 1931, Bush developed a differential analyzer, which was the first general equation solver capable of handling sixth-order differential equations. The other Bush's work (in 1945) was the first theoretical prototype, as some suggest, of a hypertext language. In 1932, Bush was appointed the first Dean of Engineering at the Massachusetts Institute of Technology.

In 1939–1941, Vannevar Bush headed the National Advisory Committee on Aeronautics, the predecessor of NASA. Then he chaired the National Defense Research Committee (1940) and directed the new Office of Scientific Research and Development in 1941, coordinating the science war effort. His 1945 essay, "Science, the Endless Frontier," influenced the organization and direction of the federally funded academic research for many years.

Fig. 11.21. Vannevar Bush, visiting the Langley Research Center a few months before becoming a chairman of the National Advisory Committee for Aeronautics, October 1938. Photo courtesy of NASA.

formidable opponent of rocket enthusiasts in the armed forces. During World War II, Bush coordinated a significant fraction of the nation's science war effort as a head of the *Office of Scientific Research and Development* (OSRD). He was exceptionally influential in matters of science and defense, chaired the powerful *Joint Army and Navy Research and Development Board* in 1946, and his advice was routinely sought by the government and by the Congress.

U.S. AIR FORCE

The National Security Act of 1947 established an independent United States Air Force on 18 September 1947. Development of long-range missiles had to compete for funding against the top priority of the Air Force at that time — the Strategic Air Command. In addition, ground-to-ground ballistic missiles were lower on the priority list than air-to-air and air-to-ground missiles, the latter enhancing capabilities of manned aircraft.

Bitter Debate

The debates on the prospects of long-range ballistic rockets flared up in 1945, were bitter, and lasted for several years. Bush did not pull punches. For example, he caustically wrote the following in 1949 about Hap Arnold, other Air Force leaders, and their supporting science advisory group headed by von Kármán:

> We are hence decidedly interested in the question of whether there are soon to be high-trajectory guided missiles ... spanning thousands of miles and precisely hitting chosen targets. The question is particularly pertinent because some eminent military men, exhilarated perhaps by a short immersion in matters scientific, have publicly asserted that there are. We have been regaled to scary articles ... implying that we are thus to be exterminated, or that we are to employ these devilish devices to exterminate someone else. We even have the exposition of missiles fired so fast that they leave the earth and proceed about it indefinitely as satellites, like the moon, for some vaguely specified military purposes ... . (Bush 1949, 84)

Vannevar Bush was not the only one sceptical of the rocket promise. Chair of the National Advisory Committee for Aeronautics (NACA) Jerome Hunsaker addressed the National Academy of Sciences on 23 April 1946 and noted that "the German V-2 rocket marks the highest state of the art of rocket design." Hunsaker then added, "One is tempted to speculate about the possibilities of an improved rocket of this type. An engineer cannot see much prospect for an improved propellant nor for much better material of construction ... . For a missile to be projected on an intercontinental adventure, the rocket does not yet offer much of a threat."

MISSILE PROJECTS

From 1945 to 1953, the three services, the Army, Navy, and Air Force, started more than 110 separate missile-related projects. About 30 of those survived as hardware and study projects by the end of 1953. The Navy devoted more funds to missile development in 1945–1950 than the other services.

"Satellite ... Possible"

For many other experts, however, the guided long-range rocket and the satellite looked technically feasible. In January 1948, the Air Force concluded that "progress in guided missile research and development by the Air Force, the Navy and other agencies is now at a point where the actual design, construction, and launching of an Earth Satellite Vehicle is technically, although not necessarily, possible" (Logsdon 1996, 272). Vice Chief of

ROCKET DEBATE

The debate on the technical possibility of long-range rockets and advisability of government support of such development was intense and sometimes bitter. General Hap Arnold of the Army Air Forces led the charge for rocket development. Dr. Vannevar Bush, a distinguished scientist and director of the Office of Scientific Research and Development, strongly opposed.

Hearings before the Special Committee on Atomic Energy, U.S. Senate, 3 December 1945

Dr. Bush. ... We don't have to believe in miracles. I can say this, that this situation is serious enough without going back into Jules Verne and Buck Rogers. I have been annoyed in recent weeks in reading some of the discussion about these possibilities [offered by guided missiles and rockets] ... .

... Let me say this: there has been a great deal said about a 3,000-mile high-angle rocket. In my opinion, such a thing is impossible today and will be impossible for many years ... .

... I can say with confidence that I would not know how to do it today, and I do not think that any man in the world, today, would. I am speaking about a 3,000-mile rocket ... .

Senator [Millard E.] Tydings [Maryland]. But there is no reason why the atomic bomb could not be put in that [V-1 type, buzz-bomb] device?

Dr. Bush. No, there is no reason why you could not put an atomic bomb in such a device. ... But the people who have been writing these things that annoy me haven't been talking about that. They have been talking about a 3,000-mile high-angle rocket, shot from one continent to another, carrying an atomic bomb and so directed as to be precise weapon which would land exactly on a certain target, such as a city.

I say, technically, I don't think anybody in the world knows how to do such a thing, and I feel confident it will not be done for a very long period of times to come... .

[Committee] Chairman [Senator Brien McMahon, Connecticut]. Well, Dr. Bush, I read in this week's Collier's magazine an article by General Carl Spaatz of the Air Forces. I would like to have you read that article.

Dr. Bush. That would not worry me in the slightest degree. I have just been criticizing the report of General Arnold of the Army Air Forces ... . If you were talking about 400 miles or 500 miles, I would say by all means. ... I would say yes, even with 2,500 miles. But 3,000 miles? That is not just a little step beyond, it is a vastly different thing, gentlemen. I think we can leave that out of our thinking. I wish the American public would leave that [3,000-mile rocket] out of their thinking.

Hearings before the Preparedness Investigating Subcommittee of the Committee on Armed Services, U.S. Senate, 16 December 1957

[Senator Stuart] Symington [Missouri]. ... I do not in any way criticize the sincerity of that [Dr. Bush's] statement [at the hearings in December of 1945, above], but it was made against the best thinking of probably the greatest airman who ever lived, Gen. H.H. Arnold; and also against the thinking of Dr. Theodore von Kármán, who, in my opinion, is generally recognized as the world's leading aeronautical scientist.

Where would we be today if, when it came to weapons systems, like [long-range rockets] Polaris and Atlas, we had followed the suggestions of [Dr. Bush,] the head of [the Office of Scientific Research and Development] OSRD as far back as 1945?

"... Logical Responsibility for the Satellite"

Staff, Air Force, General Hoyt S. Vandenberg stated that "the USAF, as the Service dealing primarily with air weapons — especially strategic — has logical responsibility for the Satellite" (Logsdon 1996, 272). Other services did not necessarily concur with this claim, and both the Army and the Navy pursued their satellite programs independently.

Unresolved Disputes Over the Roles and Missions

The competition for control of the emerging guided missile weapons reflected the much broader problem of the unsettled division of the roles and missions of military services in the nuclear warfare. The technology and capabilities of new weapon systems blurred the boundaries between the traditional functions of the Army and the Navy. The establishment of an independent Air Force further complicated the matters of strategic warfare, especially at the time of shrinking budgets. The *Key West Agreement* of the Joint Chiefs in 1948 and the follow up *Newport Conference* settled the disputes and disagreements only temporarily. The struggle for control of the ballistic missiles would intensify and be finally resolved in 1950s, with the research programs being transformed into a full-scale development, production, and deployment.

Lean Years

Political and financial obstacles have prevented establishment major rocket and satellite development programs in 1946. Nevertheless, the Army Ordnance, the Air Force, and the Navy, embarked on a number of separate smaller and less ambitious projects to develop various missiles. The sometimes bitter rivalry among the services would continue for many years and NASA would join the fray later. The funding for missiles, however, began to quickly shrink, and a number of the programs had to be abandoned next year. The lean years set in.

V-2 FOR SCIENTIFIC RESEARCH

The application of the V-2 for scientific study of the upper atmosphere was first considered by Erich Regener (1881–1955), the former director of the Physical Institute of the Stuttgart Technical University (*Technische Hochschule*). Regener was not favored by the National Socialist regime, and he lost his job in Stuttgart in 1937. With the help of the German Air Force (Luftwaffe), he succeeded in establishing the Research Station for Physics of the Stratosphere (*Forschungsstelle für Physik der Sratosphäre*).

Regener was contracted by Wernher von Braun in 1942 to build a set of scientific instruments for the A-4. Von Braun's immediate needs included the knowledge of the altitude profiles of atmospheric pressure, temperature, and density necessary for optimization of the A-4 design. The delays in instrument development and other pressing wartime needs prevented launch of the scientific package.

The war ended. The main German nationwide scientific establishment, the Kaiser Wilhelm Society (*Kaiser-Wilhelm-Gesellschaft*), was restored in West Germany as the Max Planck Society (*Max-Planck-Gesellschaft*), with Regener as its vice president. The old Regener's Research Station was reorganized in 1952 into the *Max-Planck-Institute for Physics of the Stratosphere*, known today as the *Max-Planck-Institute for Aeronomy* (*Max-Planck-Institut für Aeronomie* — MPAE), located in a small town Lindau, 25 miles (40 km) from Göttingen. Today, MPAE is one of the leading German research institutions focused on space physics and planetary exploration.

Advancement of rocket technology thus continued but at a much slower pace. The growing global threat to the free world by totalitarian communist states would change the situation only by 1949, with the blockade of Berlin, Soviet Union detonating the atomic bomb, and the communists taking over China. Then North Korea unleashed a massive attack on the Republic of Korea in the south in June 1950. All of a sudden, rocketry was rapidly becoming critically important for national security.

Practical work on ballistic missiles in the United States accelerated greatly immediately after the war with the arrival of the captured V-2s. The experiments with the V-2s began in 1945 and brought valuable scientific, engineering, and operational experience to the military personnel and industrial contractors.

Operation Sandy

The Navy, in *Operation Sandy*, studied whether a rocket could be fueled and launched from a ship underway. The V-2, although not designed to counteract the ship's motion, successfully lifted off the deck of the USS *Midway* in September 1947. The rocket malfunctioned during the flight and exploded several miles from the aircraft carrier. The *Midway* resumed normal operations immediately after the rocket launch.

Fig. 11.22. Operation Sandy. The V-2 is being prepared for launch aboard the USS *Midway* (CVB-41) in the Atlantic, September 1947. Photo courtesy of National Archives and Records Administration.

Fig. 11.23. Operation Sandy. The V-2s aboard the USS *Midway* in the Atlantic, September 1947. Photo courtesy of National Archives and Records Administration.

Operation Pushover

In another experimental *Operation Pushover*, a fully fueled rocket was toppled aboard a ship mock-up erected at White Sands Proving Ground. The extent of damage from such a simulated accident had a profound effect on Navy's willingness to deploy ballistic missiles, using liquid oxygen, on ships. The submariners have understandably developed a lasting bias toward solid-propellant rockets.

SHIP LAUNCH

Operation Sandy was a predecessor of satellite launches from ship. As time went by, the technology matured further with launches of Aerobee sounding rockets from the USS *Norton Sound.* In 1948, the Philadelphia Navy Yard made numerous alterations and added a special 100-ft (30-m) launch tower to the fore-deck of this former seaplane tender that had been commissioned and joined the Pacific Fleet in 1945. The new missile ship operated out of Port Hueneme near Point Mugu, California. Launches of sounding rockets from ships provided a way of probing the Earth space environment at various geomagnetic latitudes.

Space launches followed much later. The U.S. satellite *Explorer-42* was launched from an Italy's San Marco platform anchored off the Kenya coast on 12 December 1970. The satellite, named *Uhuru*, or *freedom* in Swahili, significantly advanced the emerging x-ray astronomy. Today, commercial satellites are placed into orbit from an ocean-going platform by the *SeaLaunch Company*. A satellite could even be hurled into orbit by a rocket launched from a submarine. First launches of the *Volna*, a modified Russian SLBM SS-N-18, were attempted in 2001 and 2002.

V-2s at White Sands

The Army Ordnance Corps continued its Project Hermes with the General Electric's personnel preparing and launching V-2s from the White Sands Proving Ground (WSPG), New Mexico. The German Peenemünde specialists were housed in the nearby Fort Bliss in Texas, and initially assisted in this task. They were replaced with the Americans by the spring 1947.

The test area in New Mexico, the White Sands Missile Range (WSMR), as it is called today, has emerged as a primary American test grounds. WSMR

Fig. 11.24. Preparation of the V-2 rocket for launch on 10 May 1946. This was the first successful V-2 launch at White Sands Proving Ground. Photo courtesy of White Sands Missile Range.

WHITE SANDS MISSILE RANGE (WSMR)

The White Sands Missile Range (WSMR, pronounced *wiz-mer*) is the largest land-area military reservation in the United States. Its area is almost 3200 square miles, stretching for 100 miles in the north-south direction. The range has about 9000 military and civilian employees and supports missile development and test programs of various military services, government agencies, and industry.

WSMR is located in the Tularossa Basin of south-central New Mexico, with the Sacramento Mountains on the east and the San Andres and Organ Mountains on the west. The main post of the range is near its southern border, 25 miles from Las Cruces, New Mexico. The other major town, El Paso, Texas, is about 40 miles away.

Fig. 11.25. The headquarters area, White Sands Proving Ground, 1947. Photo courtesy of White Sands Missile Range.

The range was formally established on 9 July 1945, as *White Sands Proving Ground* (WSPG), and renamed to WSMR in 1958. The site was selected after a group of officers of the War Department and the Corps of Engineers surveyed the few possible areas allowing missile test-firing over land and after-impact recovering. The northern part of WSPG was used as a bombing range during the war. There, at Trinity Site, the first atomic bomb was detonated on 16 July 1945.

The New Mexico testing site offered a number of advantages. Most of the land was owned by the government. The weather was clear almost all of the time during the year, and visibility was unlimited. In addition, the desert was sparsely populated with minimal safety problems and easy recovery of the spent rocket parts and missiles.

The first hot firing of a rocket engine at the new missile range took place on 26 September 1945, when a Tiny Tim booster was fired. The test was followed by two launches, on 11 and 12 October 1945, of WAC Corporal missiles with the Tiny Tim boosters. The missiles reached an altitude of 44 miles. Today, sounding (research) rockets are routinely launched from WSMR to altitudes up to 200 miles, limited by the safety considerations.

In addition to missile development programs, White Sands is a major test site for such futuristic weapons as high-energy lasers. WSMR is also a communication hub for the NASA's Tracking and Data Relay Satellite System (TDRSS), providing continuous communications with low-Earth-orbit satellites through spacecraft in geostationary orbit. In addition, WSMR maintains an emergency landing site for Space Shuttle Orbiters.

witnessed the first successfully launched V-2 on 10 May 1946. In total, 67 V-2 rockets had been fired by 1951 when Project Hermes ended.

Many V-2 rockets were significantly modified to perform various technological experiments. A number of rockets carried scientific instruments studying the upper atmosphere, ionosphere, solar radiation, cosmic rays, and micrometeoroids. Even a monkey got a rocket ride.

Fig. 11.26. V-2 launch at White Sands Missile Range, ca. 1946. Photo courtesy of White Sands Missile Range.

Upper Atmosphere Rocket Research Panel (UARRP)

The *Upper Atmosphere Rocket Research Panel* (UARRP) coordinated research activities on sounding rockets. UARRP was established in 1946 as the *V-2 Panel* and was afterwards called for some time the *V-2 Upper Atmosphere Panel*. Ernest E. Krause of the Naval Research Laboratory (NRL) became the first chair of the panel. At that time, Krause led a group of enthusiastic scientists and engineers at the NRL that had established the Rocket-Sonde Research Branch on 17 December 1945, for the specific purpose of studying the upper atmosphere with rockets.

Birth of Experimental Space Science

Many leaders of the emerging American space science such as Richard Tousey, James Van Allen, Fred Whipple, and Herbert Friedman, worked on the early V-2 rocket experiments. Space science has thus been born, aided by and in a unique collaboration with the military.

Emerging Field

A growing number of American research institutions and industrial contractors designed and built scientific instrumentation, telemetry systems, and participated in modification of rocket subsystems. As time went by, new sounding rockets, particularly the Aerobee and Viking, were added to the inventory of research missiles launched at White Sands.

Fig. 11.27. Missile Park at White Sands Missile Range demonstrates the range's contribution to the development of the American rocket technology. Photo (ca. 1990) courtesy of White Sands Missile Range.

Applied Physics Laboratory (APL), Naval Research Laboratory, Air Force Cambridge Research Laboratory (AFCRL), Jet Propulsion Laboratory, and numerous other research and development institutions, universities (mostly through the Air Force contracts), and industrial contractors contributed to various scientific and technological aspects of this early program. Many new research and development groups appeared, introducing a large number of scientists and engineers to the exciting emerging field. These institutions and specialists would play an important role in the rapid growth of the rocket and space technology in 1950s and afterwards.

12. BUILDING THE FOUNDATION

The U.S. Army, Navy, and Air Force entered the emerging field of guided missiles in 1940s. The three services brought with them different traditions, interests, and visions.

Army Ordnance

The Army's effort was led by the Army Ordnance, which recognized the importance of rockets and initiated their development in the early 1940s. In 1942, Ordnance's Chief of Research and Development Major General Gladeon M. Barnes established a farsighted plan to support Army's programs by obtaining technical information on enemy weapons and equipment from various theaters of operations. The resulting Ordnance's active role in Operation Paperclip, V-2 test launches at White Sands Proving Ground, and support of the German rocketeers at Fort Bliss were, however, only a part of the Army's energetic program in long-range missiles.

Fig. 12.1. Chief of the Research and Development Service of the Army Ordnance Maj. Gen. Gladeon M. Barnes, June 1944. General Barnes initiated Army rocket programs during the war, building the foundation for the advancement of rocket technology in 1940s and 1950s. Photo courtesy of National Archives and Records Administration.

ORDCIT

As early as in May 1944, General Barnes placed a large $3,300,000 *Ordnance Contract to California Institute of Technology* (ORDCIT). The contract supported von Kármán's group of GALCIT rocketeers in comprehensive research on guided missiles, propulsion, supersonic aerodynamics, materials, and remote control. This growing group had become known by that time as the *Jet Propulsion Laboratory* (JPL). This laboratory name first appeared in November 1943 in a report on the potential of long-range missiles prepared for the Materiel Command of the Army Air Forces. Frank Malina directed JPL in 1944–1946 and was followed in the director position by Louis Dunn (1946–1954) and by William H. Pickering (1954–1976).

Jet Propulsion Laboratory (JPL)

The Army Ordnance soon followed its first major missile contract ORDCIT by a new *Project Hermes*, contracted to the General Electric Corporation. In addition, the Bell Telephone Laboratories, Inc., was tasked in February 1945 to develop an antiaircraft guided missile system, the *Nike*. The Army's rapidly expanding research facilities included a new wind tunnel at the Aberdeen Proving Ground in Maryland and the White Sands Proving Ground in New Mexico. In addition, the first guided missile battalion was activated in October 1945 in Fort Bliss in Texas.

Beginning of Antiaircraft Missile Systems

JPL was rapidly expanding its research in propulsion, aerodynamics, control, radio telemetry, materials, and electronics. This expertise would become especially important later, in 1957–1958, when JPL significantly contributed to the success of a crash project to launch the first American satellite Explorer I. The laboratory had quickly outgrown its original operational structure. On 19 October 1945, Caltech deeded to the United States the 31.5-acre land tract (at the nominal cost of $164 per acre), making the U.S. Government the owner of JPL's land and facilities. At that time JPL already had a staff of 350, which had grown to 500 in 1948 and to more than 1000 in 1954.

Expansion of JPL

The Army-funded JPL concentrated on the *Private* and *Corporal* missiles. The laboratory planned to progressively increase its missiles in size and complexity, starting from the Private and continuing with the Corporal and Sergeant. The small unguided missile Private was tested in December 1944 at Camp Irwin near Barstow, California, and in April 1945 at Hueco Range, Fort Bliss, Texas. The larger missile, Corporal, was first launched in May 1947 at White Sands Proving Ground. The liquid-propellant Corporal

Private and Corporal

Fig. 12.2. The Hermes A-1 missile on display at the U.S. Space and Rocket Center in Huntsville, Alabama. Development of the surface-to-air missile Hermes, based on the German Wasserfall, was initiated in 1946 as part of the Hermes Program contracted to the General Electric Corporation by the Army Ordnance Corps. The Hermes was among the first American test missile systems developed in the late 1940s and contributed to the growing rocket expertise of the industry. Photo courtesy of Mike Gruntman.

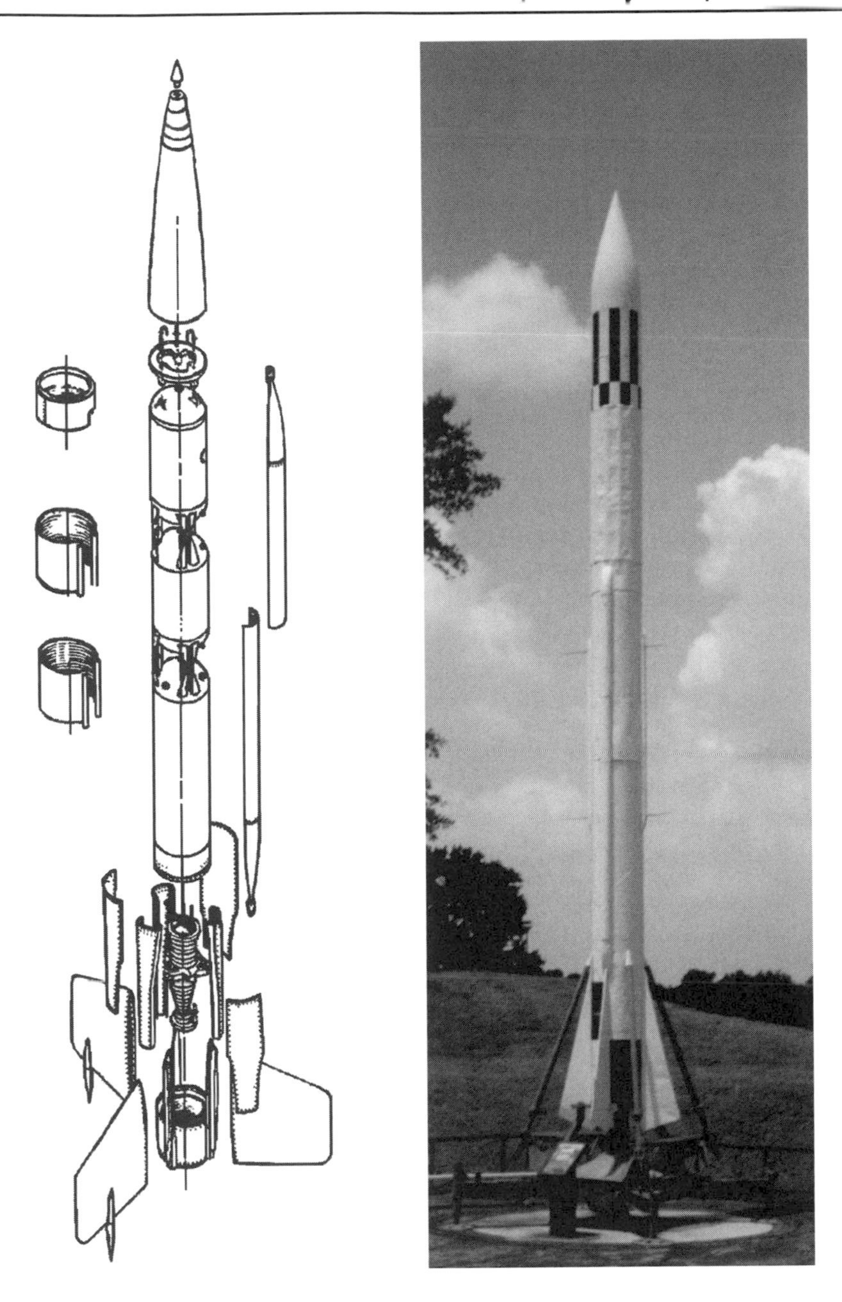

Fig. 12.3. Left: exploded view of JPL's WAC Corporal (figure courtesy of U.S. Army), first launched at White Sands in September 1945. The rocket was 12 ft 6 in. (3.81 m) long and weighed 660 lb (300 kg); the propellants (red-fuming nitric acid and aniline containing 20% furfuryl alcohol) weighed 360 lb (163 kg). Right: the liquid-propellant (monoethyaniline and nitric acid) surface-to-surface missile Corporal with the range of 75 miles (120 km) on display at the U.S. Space and Rocket Center in Huntsville, Alabama (photo courtesy of Mike Gruntman). The JPL's Corporal was based on experience derived from the Private and WAC Corporal missiles.

became the first American tactical missile approved as an atomic weapons carrier in 1950.

First American Atomic Missile *Corporal*

Von Kármán recalled later that "as we [JPL] continued to improve on the early [missile] models, and to give each new [missile] version a rank in the military order [starting with Private and Corporal], General Barnes [of the Army Ordnance] asked me how far up in rank we intended to go. 'Certainly not over colonel,' I said and then added with a smile. 'This is the highest rank that works.' The General laughed" (von Kármán 1967, 265).

Fig. 12.4. Test launch of the Army's Nike antiaircraft guided missile at White Sands in 1950s. The Nike, one of the very first American missile systems, began in January 1945 with the study by the Bell Telephone Laboratories under contract to the Army Ordnance. The program culminated when the Nike I missile successfully intercepted a remotely controlled QB-17 bomber over White Sands in April 1952. The first Nike antiaircraft unit became operational at Fort Meade, Maryland, in May 1954. Photo courtesy of White Sands Missile Range.

Missiles Rise at White Sands

The first rockets to rise from White Sands were not the captured German V-2s but the missiles built by the von Kármán team. JPL test fired 17 Private F rockets during the first months of 1945 (before WSPG was activated in July), followed by hot tests of booster rockets. On 11 October 1945, another JPL rocket, the unguided liquid-propellant WAC Corporal was launched, aided by the Tiny Tim booster. The missile reached an altitude of 235,000 ft (72 km). The success with the WAC Corporal led to the *Bumper*, a two-stage, or "step-stage" as it was called at the time, rocket. The V-2 was used as the first stage to "bump" the WAC Corporal to higher altitudes where the latter was fired. On its first launch at White Sands on 13 May 1948, the Bumper reached an altitude of 80 miles (130 km). Several more launches followed in 1948 and 1949 before the tests were moved to Cape Canaveral.

Redstone Arsenal

As JPL was steadily building up its expertise, the Army Ordnance continued to support the work of the von Braun's team at Fort Bliss. In early 1950s, the Germans were transferred from Texas to a new Army rocket center opened at the Redstone Arsenal in Huntsville, Alabama, where the development of a new single-stage ballistic missile was initiated. During the war, the arsenal provided

Fig. 12.5. Redstone's liquid-propellant (liquid oxygen and a mixture of alcohol with water) engine on display at the U.S. Space and Rocket Center in Huntsville, Alabama. The power plant was originally developed by the North American Aviation for the booster of the Navaho cruise missile and was modified for the Redstone. The engine was in fact a redesigned and improved version of the German A-4 (V-2) engine and demonstrated higher specific impulse and thrust. Photo courtesy of Mike Gruntman.

facilities for the Army's chemical program. A small north Alabama town, Huntsville, had only 16,000 inhabitants in 1950. Rapid growth of the arsenal and associated contractor establishments trebled the population in six years.

Navaho's Heritage

The new Army's ballistic rocket was christened after the arsenal, the *Redstone*, in April 1952. The rocket used a modified version of the power plant originally developed by the North American Aviation as a booster for the *Navaho* intercontinental cruise missile. The engine operated on liquid oxygen and the mixture of alcohol with water and was, in fact, a redesigned and improved version of the A-4 (V-2) engine. The new engine achieved specific impulse 218 s (specific impulse of the V-2 was 210 s) and thrust 78,000 lbf (347 kN). The Redstone was the last American major rocket system based on alcohol as fuel. The improving engines would rely on kerosene and liquid oxygen as a propellant combination of choice in mid-1950s.

Ballistic Missile *Redstone*

The Redstone was equipped with the inertial guidance system built by the Ford Instrument Company, Division of Sperry Rand Corporation. The rocket could carry a conventional or nuclear warhead for 175 n miles (325 km); it was 69 ft (21 m) tall and 70 in. (178 cm) in diameter and had mass 61,350 lb (28,800 kg) at ignition. The original Army desire to build missiles in-house soon ran into insurmountable difficulties with the long lead time for development and procurement of many components. It became clear that subcontracting of major subsystems would be unavoidable. In addition, the well-prepared prime contractor was essential for rocket mass production.

Fig. 12.6. The Army's Redstone ballistic missile lifts off in a test launch at White Sands Missile Range on 20 January 1959. Photo courtesy of White Sands Missile Range.

Chrysler Corporation

The prime contractor for the new Army's ballistic missile was sought among locomotive and automobile industries. Aircraft companies usually worked with the rival Air Force and were thus excluded from the search. Finally, the *Chrysler Corporation* became the missile prime contractor. Both the Redstone Arsenal and the prime contractor manufactured the Redstones. Chrysler temporarily shared floor space with jet-engine production at the *Naval Industrial Reserve Aircraft Plant* (NIRAP) in Detroit, Michigan. In October 1957, NIRAP became a permanent rocket manufacturing site under the name of the *Michigan Ordnance Production Plant*. The North American's Rocketdyne Division in California supplied Redstone's engines. The first Redstone was launched at Cape Canaveral in August 1953, and the ballistic missile system entered service in 1958.

Redstone's Extraordinary Service

Redstone's life as an Army tactical missile was extraordinary but short-lived: it had been deactivated by the end of June 1964. In July and August 1958, two Redstone rockets lifted from Johnston Island in the Pacific and detonated nuclear warheads at altitude 75 and 35 km, respectively, in experiments — called *Operation Hardtack* — studying the effects of nuclear explosions at high altitudes. The modified rocket would also serve as the first stage for the Juno 1, a launcher that placed the first American satellite, Explorer I, into orbit in 1958. In addition, the Redstones would carry Alan B. Shepard, Jr. and Virgil I. Grissom on first American suborbital flights in 1961.

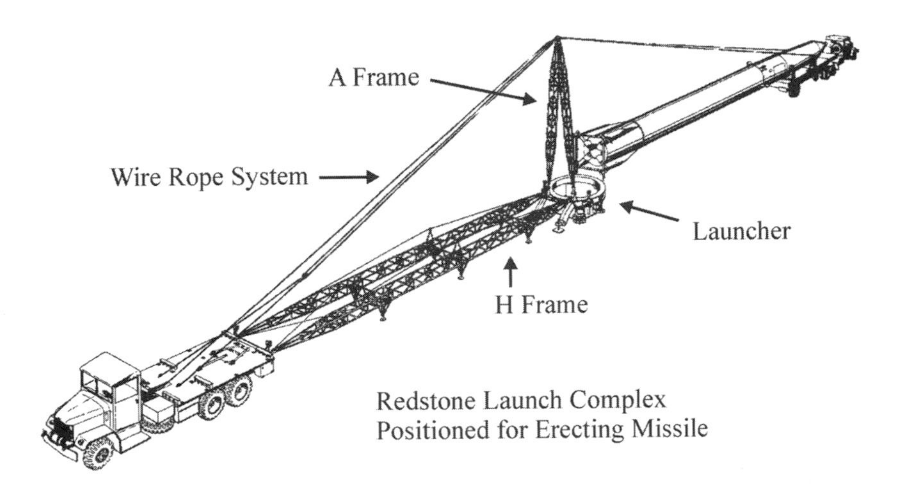

Fig. 12.7. The Redstone missile was designed as a highly mobile system that could be assembled and deployed in the field. Based on figure courtesy of U.S. Army.

Consolidation of Army's Effort

Army Ordnance's energetic steps in development of guided ballistic missiles and antiaircraft rocketry set a stage for the protracted and intense interservice rivalry over the emerging weapon systems. In 1950, the Army consolidated its mushrooming missile programs at the Redstone Arsenal. General Toftoy, who directed rounding up of German rocketeers in 1945 and their subsequent transfer to Fort Bliss, was put in charge of the arsenal.

Jupiter

The liquid-propellant Redstone was a forerunner of the America's intermediate-range ballistic missile (IRBM) *Jupiter*, also developed at the arsenal. The Jupiter program began first as a joint Army and Navy project to develop a liquid-propellant IRBM. The Navy then switched its attention to the all-solid-propellant submarine-launched *Polaris*. As a result, the Army was left alone to fight with the Air Force for control of the land-based ballistic missiles.

The Department of Defense gave the green light to the Army to proceed with the development of the 1500-mile (2400-km) IRBM Jupiter, also called IRBM

Fig. 12.8. Leading Army Ordnance rocketeers General John B. Medaris (left), Wernher von Braun (center), and General Holger N. Toftoy (right). Medaris became the first Commanding General of the U.S. Army Ballistic Missile Agency activated at the Redstone Arsenal on 1 February 1956. He was appointed in 1958 chief of the Army Ordnance Missile Command, with Toftoy as his deputy. Toftoy guided the work of von Braun's team for several years after the war. After 1952, Toftoy directed the Ordnance Missile Laboratories and later the Redstone Arsenal, supervising Army's rocket development programs. Photo courtesy of U.S. Army.

Army Ballistic Missile Agency (ABMA)

Nr. 2, in November 1955. This decision covered the organization of all of the nation's long-range ballistic missile programs and included Air Force's two intercontinental ballistic missile (ICBM) programs, *Atlas* and *Titan*, and IRBM (Nr. 1) *Thor*. It was specifically emphasized that achieving the Air Force's and Army's IRBMs should not interfere with the ICBM development. To organize the Army effort, the *Army Ballistic Missile Agency* (ABMA) was promptly activated at Redstone on 1 February 1956. The first commanding general of ABMA John B. Medaris would direct Army's effort in ballistic missiles, particularly the high-priority Jupiter.

Kerosene Fuel

The development of the Army IRBM heavily relied on the well-advanced Redstone program. Many Redstone components and subsystems were modified for and adopted by the Jupiter. The propulsion system, however, underwent a major change switching to a new superior propellant combination of kerosene and liquid oxygen. North American Aviation's engine NAA-150-200-S-3D provided 150,000 lbf (667 kN) of thrust. The technology of storage and transporting liquid oxygen had been so perfected by late 1950s that the inherent limitations of this cryogenic oxidizer did not significantly interfere any more with the concept of mobility of Army's ballistic missiles. About 1% of liquid oxygen evaporated and was lost per day in storage at rest, which corresponded to a loss of half of oxygen in 69 days. When being on the move, about 3% of oxygen was usually lost each day.

Jupiter Development

A large number of test Redstone missiles were flown under the names of Jupiter A and Jupiter C as supporting elements of the Jupiter program. Twenty-five modified Redstones rockets, designated Jupiter A, were launched between September 1955 and June 1958 with the specific goal of testing new designs and components being developed for the new IRBM. In addition, three Redstones, designated Jupiter C, were flown in support of the nose cone development. The first test launch of the Jupiter IRBM took place on 1 March 1957, at Cape Canaveral. Chrysler continued its role of the industrial contractor for series production of the Army missile. Forty-five Jupiters were deployed in Italy and Turkey in the early 1960s. After a brief service, the Jupiter IRBM was phased out in 1963.

Fig. 12.9. Jupiter being prepared for launch. Photo courtesy of U.S. Army.

Army Ordnance Missile Command (AOMC)

In the following reorganization of the Army rocket effort, ABMA became a part of the newly created *Army Ordnance Missile Command* (AOMC) on 31 March 1958. In addition to ABMA, AOMC included Redstone Arsenal, Jet Propulsion Laboratory, and White Sands Proving Ground. General Medaris was appointed commanding general of AOMC with the responsibility for all programs of the Army Ordnance in guided missiles, in ballistic rockets, and in space field.

Army's In-House Development

The Army Ordnance had a long and rich tradition, dating back to the first half of the 19th century, of experimentation and weapon development at its arsenals and testing grounds. Correspondingly, a significant part of the Army's rocket program was organized as an in-house effort, supported by industrial contractors and directed and coordinated by AOMC.

The former German rocketeers constituted an important part of the technical team at Huntsville. When the development of the Jupiter IRBM was assigned to von Braun's *Guided*

Missile Development Division of ABMA, the division employed 1600 people including 500 scientists and engineers. Among the latter almost 100 were former German "Paperclip" specialists. The significant presence of the German rocketeers created additional problems for the Army in the intense interservice rivalry reflecting, in the words of Charles A. Lindbergh, "the antagonism still existing toward von Braun and his co-workers because of their service on the German side of World War II."

"Germans" in Huntsville

In contrast to the Army, the Air Force, a much younger service, followed a different philosophy in weapons development and relied primarily on industrial contractors for its rocket program. By the end of the war, many leading industrial companies, particularly airframe manufacturers, had formed guided missile divisions or sections and initiated rocket-related research and development. The list of the early Army Air Forces contractors in guided missiles included Bell Aircraft, Bendix Aviation, Consolidated Vultee Aircraft, Curtiss-Wright, Douglas Aircraft, General Electric, Glenn L. Martin, Goodyear Aircraft, Hughes Aircraft, McDonnell Aircraft, Marquardt Aviation, M. W. Kellog, North American Aviation, Northrop Aircraft, Pratt and Whitney, and Republic Aviation. As a result, industry's expertise and capabilities were rapidly growing.

Air Force and Industrial Contractors

One of the programs supported by the Air Force was initiated by the Consolidated Vultee Aircraft Corporation. The company, also known as *Convair*, was formed in 1943 by a merger of the Consolidated Aircraft Corporation with Vultee Aircraft. This Air Force program would eventually lead to an especially important future development, the first American intercontinental ballistic missile Atlas. In the spring of 1946, Convair's Vultee Field Division in Downey, California, was awarded a contract to develop a prototype test vehicle MX-774. The MX-774 was a bold attempt to demonstrate guided missiles capable of delivering a 5000-lb (2270-kg) warhead for 5000 miles (8000 km) within 5000 ft (1520 m) of the target.

Convair's MX-774

Fig. 12.10. Karel J. "Charlie" Bossart, 1903–1975, led the development of the MX-774 test missile in late 1940s and the first American ICBM Atlas in 1950s. Photo courtesy of Karel J. Bossart, Jr.

Charlie Bossart

Convair engineer Karel J. "Charlie" Bossart became the leader of the MX-774 development. Bossart obtained a degree in mining engineering in his native Belgium and came to the United States in 1925 on a scholarship to study aeronautics. He specialized in structures and in time became chief of structures at Convair's plant in Downey, California. Bossart would subsequently earn the name of the "father of the Atlas ICBM."

ARMY'S ARSENAL SYSTEM, THE GERMANS, AND THE AIR FORCE

... At the close of the war the Army was responsible for bringing to this country a considerable group of the top-notch German scientists who had been responsible for the development and fielding of the German V-2. They were taken to White Sands Proving Ground as headquarters ... .

[In early 1950s] the present Redstone Arsenal site was selected as the base for all Army activities in the development and production of missiles. The group headed by Dr. von Braun was moved there ... .

... [The Redstone Arsenal] has an in-house capability to do prototype development and prototype tests and carry through to actual production of the missile; whereas in the balance of the missile work it is done by industry ... .

[By "in-house capability"] I mean in the [Army] Ballistic Missile Agency [ABMA], with all the scientists we have, we can do the experimental, initial experimental and theoretical calculation, the construction and assembly of the early hardware, the prototype hardware, finally the flight testing of the missile, which is done at the Air Force missile test center in Florida, but under our own firing crew who are part of our laboratories, and finally we can produce in limited quantities those missiles, as we did initially for the Redstone. It (Redstone) was then phased over to industry where engineering development was complete.

What I mean by an in-house capability, this can be done by our close group in one place, all of whom are government employees.

Maj. Gen. John B. Medaris
Commander, Army Ballistic Missile Agency
Senate Hearings, 14 December 1957 (*Hearings* 1958, 539–542)

... They [the Army] have a very competent group [of specialists headed by Dr. von Braun] there [in the Redstone Arsenal in Huntsville, Alabama]. I think that the fact that they have a very competent group there, however, is sort of an accident.

In other words, I don't go along with the arsenal philosophy of doing development and then turning it over to industry for production.

I think the Air Force philosophy of having industry do development and having the capability of planning for production simultaneously is a much better way of doing it.

The Air Force had quite a number of German scientists right after the war at Wright Field, and made, deliberately made, the decision not to try to retain that group of scientists as a group, similar to what they have done at Redstone, and they have been, most of them have gone into American industry and a lot of them are in industry today. They are at Convair [Astronautics Division of General Dynamics], they are at Bell [Aircraft], and a number of other companies, and although this is a matter of opinion, my feeling is that these people distributed to American industry, are doing equally as good a job for the United States as this one small group [under Dr. von Braun] that are still assembled at the Redstone Arsenal.

Maj. Gen. Bernard A. Schriever, USAF
Commander, Air Force Ballistic Missile Division
Senate Hearings, 9 January 1958 (*Hearings* 1958, 1637, 1638)

Fig. 12.11. MX-774 (RTV-A-2) missile on a trailer. Photo courtesy of Karel J. Bossart, Jr. and U.S. Air Force.

RTV-A-2 Hiroc

The MX-774 program was cancelled in 1947 because of budgetary pressures, but the Air Force allowed Convair to use the already allocated but unexpended funds to complete and fly three single-stage test vehicles, designated RTV-A-2. The test missiles were also known as the *Hiroc* (high-altitude rocket). At the same time, the Downey team was relocated to Convair's facilities in San Diego, California.

The starting point for the MX-774 development was the only long-range ballistic rocket, the German V-2. Convair's team introduced several major innovations in its missile, particularly in the areas of vehicle structural design and flight control. The new features — integral propellant tank structure, pressurized body, swiveling engines, and nose cone separation — led to remarkable advances in missile performance.

MX-774's Innovations

Bossart, with his experience in structural engineering, clearly saw that the weight of the V-2 airframe, or the structural weight, could be substantially reduced. Two separate propellant tanks were mounted inside the V-2 body. Similar to modern airplanes, the new MX-774 did not have separate tanks but used the

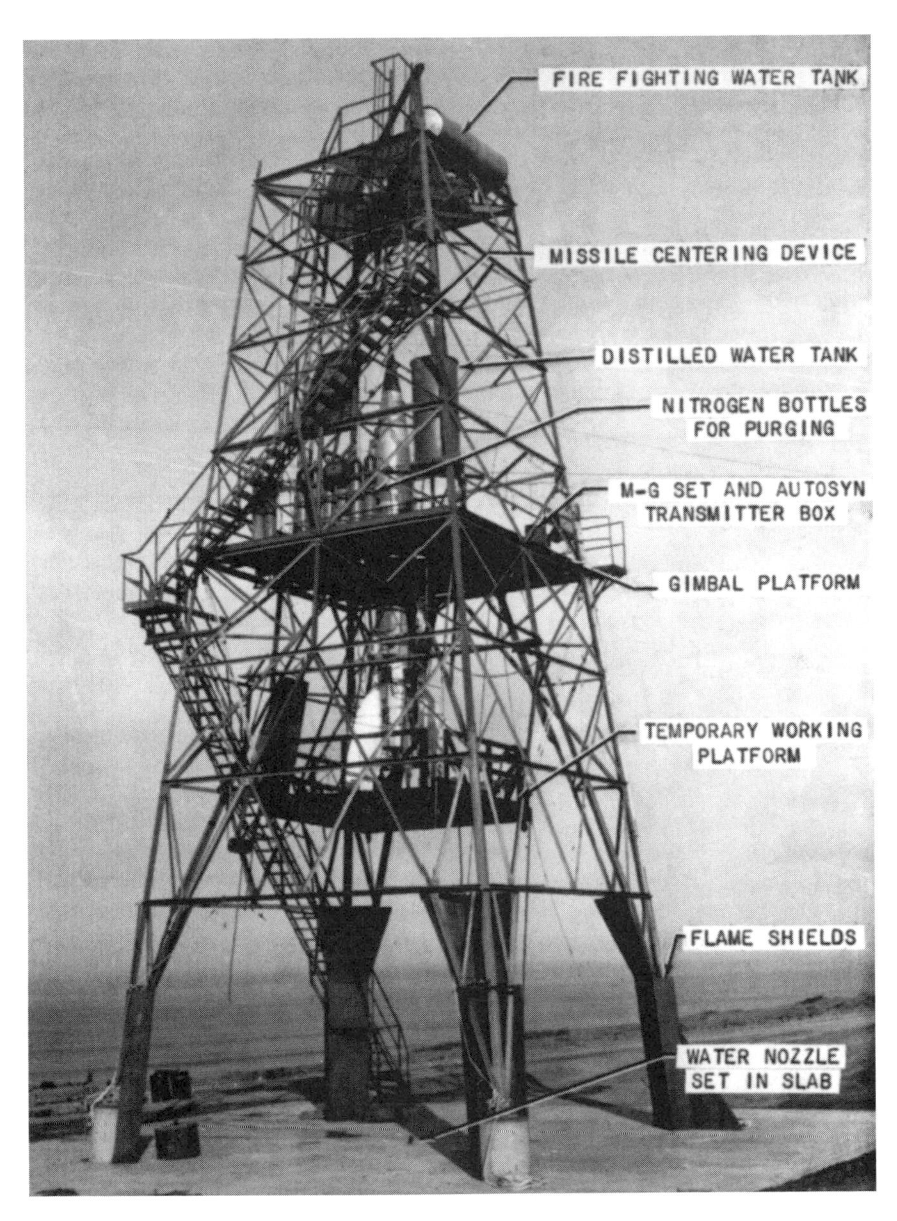

Fig. 12.12. MX-774 static firing test stand at secluded Point Loma in San Diego, California. Photo courtesy of Karel J. Bossart, Jr. and U.S. Air Force.

rocket body to directly contain the fuel and oxidizer separated by a bulkhead. The structural weight was further reduced by using pressure of the nitrogen gas inside the propellant tanks to support the rocket frame. The missile thus resembled a flying propellant tank with thin walls made of an aluminum alloy.

Integral Propellant Tank

The V-2 had been built to survive the reentry of the atmosphere, which increased the rocket's structural weight. Bossart introduced separation of the nose cone, which directly led to an additional reduction of the airframe weight. Only the warhead would now have to survive the reentry. As the result of the innovations, the fraction of structures in the overall missile weight was lowered significantly.

Separating Nose Cone

Another important new feature introduced by Convair was swiveling engines for flight control. The V-2 controlled its thrust vector by movable graphite vanes inserted in the exhaust flow. These vanes reduced specific impulse of the propulsion system by up to 15%. The new much more efficient swiveling engines brought, however, numerous mechanical and control problems.

A specially built test stand had a test missile platform that was moved by water jets installed underneath. The jets were controlled by the actual rocket servomechanism, thus simulating accelerations of the missile and response of its flight control system. Many hours of simulated flight time optimized and thoroughly tested the new thrust vector control system.

Swiveling Engines for Thrust Vector Control

Three RTV-A-2 test rockets were built and launched on 13 July, 27 September, and 2 December 1947. The missile was 31.5 ft (9.6 m) long, weighed fully fueled 4090 lb (1855 kg), and was designed to reach 422,000 ft (129 km) altitude with a 500-lb (227-kg) payload.

The missile's propulsion unit consisted of four engines, each of which could independently swivel plus or minus 10 deg from the neutral position. The engines were built by Reaction Motors, Inc., and produced 2000-lbf (9-kN) thrust each. Decomposition of hydrogen peroxide powered the turbopumps that supplied the

RMI's Engines

Fig. 12.13. Four swiveling rocket engines of the MX-774 (RTV-A-2) rocket enabled flight control (thrust vector control) making aerodynamic control surfaces (fins) obsolete. Photo courtesy of Karel J. Bossart, Jr. and U.S. Air Force.

Fig. 12.14. Launch of the first MX-774 (RTV-A-2) on 13 July 1948 at White Sands Proving Ground. Photo courtesy of Karel J. Bossart, Jr. and U.S. Air Force.

alcohol fuel and liquid oxygen oxidizer into the combustion chamber. The engines developed specific impulse 210 s at sea level, comparable to that of the V-2.

MX-774 Test Flights

The launches of the MX-774 were not entirely successful, but the engines worked up to one minute and the new lightened structure was successfully demonstrated. The swiveling engines perfectly controlled rocket attitude and flight path. In short, the basic missile concepts have been validated. During the first flight, for example, the vehicle deviation from the nominal path did not exceed 11 ft (3.3 m). The three test vehicles reached altitudes 1.2, 24, and 30 miles (1.9, 39, and 48 km), respectively.

Convair Carries On

After cancellation of the MX-774 program by the Air Force, Convair continued some of its work on long-range ballistic missiles, mostly further "paper" studies of the concept.

INDEPENDENT DEVELOPMENT OR ESPIONAGE?

In March 1947, the leading Soviet rocket designer Sergei P. Korolev gathered his engineers and outlined for them a new approach to the design of ballistic missiles. One of the engineers, Georgii S. Vetrov, recalled (Vetrov 1986) that Korolev suggested the use of two new features in future rockets, separation of the nose cone and making propellant tanks supporting elements of rocket structure.

It is possible that these Korolev's ideas grew out, logically and independently, of the Soviet work focused at that time on reproduction of the German V-2.

It is not excluded, however, that Korolev gleaned some insight from the MX-774 program in San Diego. The Soviet missile espionage in the United States likely followed the same pattern as the highly successful theft of nuclear secrets. The KGB and the military intelligence GRU provided the obtained information only to the very few top Soviet scientists and engineers. These trusted specialists subsequently introduced American discoveries and designs into their projects without revealing the sources to their subordinates. In case of the atomic bomb, for example, the head of the Soviet nuclear program Igor V. Kurchatov reportedly played such a role. Assuming the same pattern in guided missiles, Korolev would have most likely been among the very few persons briefed about the details and achievements of secret American programs.

So, were the innovations proposed by Korolev a logical development of the Soviet missile work or results of the espionage on the MX-774? The answer is likely preserved in the Russian archives.

This internally funded development, combined with the MX-774 achievements, laid the foundation for the program resurrection in early 1950s. The MX-774 significantly advanced missile performance and demonstrated what would be called today the enabling technologies for the first American ICBM, the Atlas.

NAVAL RESEARCH LABORATORY

The Naval Research Laboratory, the first modern research institution within the U.S. Navy, was established on 2 July 1923. Today, NRL with its 3000 employees is the Navy's corporate laboratory engaged in broadly based multidisciplinary scientific research and technological development. NRL actively advanced space technology, beginning with the early upper atmopsheric research, rockets, and satellites.

NRL's Viking

The Navy considered V-2 launches so important for studies of the upper atmosphere that in 1946 it conceived the *Viking* rocket as a replacement of the dwindling supply of captured German V-2s. The Naval Research Laboratory in Washington, D.C., already active in the upper atmospheric research, supervised development of this research missile. NRL's Milton Rosen led the Viking development and later would serve as technical director the successor space launcher program, the Vanguard.

Fig. 12.15. Milton W. Rosen, b.1915, joined the Naval Research Laboratory in 1940. He led development of the Viking rocket and was later technical director of its successor program, the Vanguard. In early 1960s, Rosen served at the NASA Headquarters as Director of Launch Vehicles and Propulsion in the Office of Manned Space Flight. Photo courtesy of NASA.

Glenn L. Martin Company

The NRL chose the Glenn L. Martin Company of the nearby Baltimore, Maryland, as the prime contractor. The selection was influenced in part by Martin's proximity to NRL. The Viking program was the beginning of the distinguished contribution of Martin to ballistic missiles that would include in the future the Titan family of ICBMs and space launchers.

Viking's Innovations

Altogether, 14 Viking rockets were built and launched from 1949 until 1957. The new rocket development brought a number of innovations with the missile characteristics and capabilities gradually evolving as the program progressed. For example, on the third Viking (launched in February 1950) the separate oxygen tank was eliminated, with the rocket skin directly holding the oxidizer. The rocket became somewhat shorter, and its diameter was significantly increased, from 32 to 45 in. (from 81 to 114 cm), beginning with the eight's missile (launched in June 1952).

Fig. 12.16. Viking rises at White Sands Missile Range. Photo courtesy of Naval Research Laboratory.

"Slim Pencil-Like" Viking

Similar to Convair's MX-774, the Viking widely relied on aluminum alloys in the airframe, achieving significant reduction of structural weight. The media liked to describe earlier Vikings that were 45.2–48.6 ft (13.8–14.8 m) tall as "slim pencil-like" rockets. The main reason for the rocket's small 32-in. (81-cm) diameter was the readily available 100-in. (2.54-m) wide aluminum sheets that could be rolled into cylinders.

Gimballed Engines for Thrust Vector Control

Reaction Motors, Inc., built the Viking engine based, similarly to the V-2, on alcohol and liquid oxygen; specific impulse was modest, around 200 s, and thrust 20,450 lbf (91 kN). In a major innovation, the gimballed engine provided thrust vector control, being even more technically complex than the MX-774's swiveled engines. (A gimbal is essentially a universal joint while a swivel permits rotation about one axis only.)

Fig. 12.17. The first photograph of a hurricane and a tropical storm obtained by the Viking from a high altitude. Photo courtesy of Naval Research Laboratory.

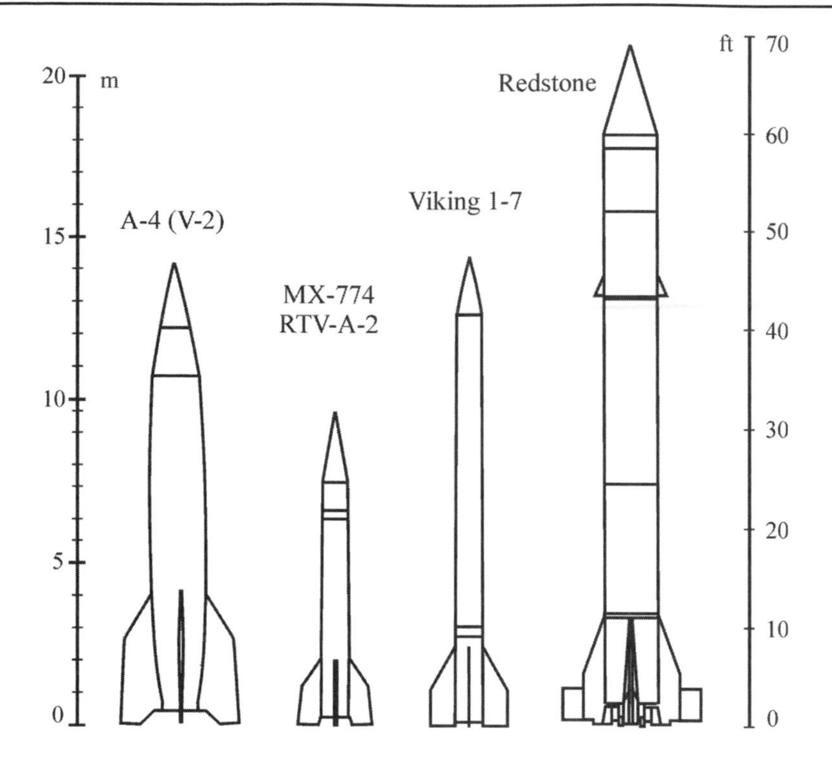

Fig. 12.18. American high-altitude rockets of the late 1940s: Air Force's MX-774 and first Navy's Viking (the follow on Vikings were somewhat shorter and larger in diameter). Also shown are the German A-4 (left) of the early 1940s and the U.S. Army's Redstone (right) of the early 1950s. Figure courtesy of Mike Gruntman.

Platform for Scientific Measurements

The first Viking flew at White Sands in May 1949 to an altitude of 50 miles (80 km). A year later, launched from the USS *Norton Sound*, the Viking reached 106 miles (170 km) on 11 May 1950; another missile reached an altitude of 158 miles (254 km) in May 1954. Before the program was terminated, the Vikings provided a platform for numerous scientific measurements above the atmosphere and obtained the breathtaking color photographs of the Earth from altitudes

APPLIED PHYSICS LABORATORY

The Navy organized the *Applied Physics Laboratory* through the Johns Hopkins University in 1942. The primary goal of the new research center was development of a proximity fuze, which has been successfully accomplished. The proximity fuze triggered explosion of an artillery shell or a bomb in proximity of the target. The device was among the most valuable technologies developed during the war.

Since the late 1940s, APL became an active participant in the emerging field of space research and applications and pioneered a number of important technologies. Today, APL is a large research and development center in Laurel, Maryland, concentrating on applied science and technology, including space science and space systems. The Laboratory has designed and built almost 60 spacecraft.

higher than 100 miles (160 km). Two last Vikings were launched in support of the development of the Vanguard space launcher, when the latter was initiated in 1955.

James Van Allen and Aerobee

Another Navy's initiative, spearheaded by the Applied Physics Laboratory, led to the most popular high-altitude research (sounding) rocket family, the *Aerobee*. James Van Allen (who discovered the radiation belts bearing his name in the late 1950s) was the head of the High Altitude Research Group at APL. During his visit to Aerojet in 1946, Van Allen learned about development of an advanced liquid-propellant engine (RFNA as oxidizer and a mixture of furfuryl alcohol and aniline as fuel) for the Army's *Nike Ajax* surface-to-air missile.

Aerojet's Engine

Van Allen convinced the Navy to initiate development of a new family of sounding rockets at Aerojet based on this new engine. The rocket was called the Aerobee, a contraction from *Aerojet* and *Bumblebee*; the latter being the name of the APL research program with the sounding rocket part managed by Van Allen. [The Bumblebee was "a major program on jet-propelled antiaircraft missiles" initiated at APL in December 1944 (Meyer and Anderson 1982, 173). The program led to the subsequent development of such missiles as Talos, Triton, Typhon, and Terrier.]

Bumblebee

Fig. 12.19. Aerojet's Aerobee-150 leaves the launching tower. The Aerobee was an unguided, fixed-fin-stabilized, guiding-tower-launched two-stage rocket. The first, booster stage was a solid rocket, and the second stage used a liquid-propellant engine. Two stages were started simultaneously, with the first stage consuming its propellant in 3 s and then jettisoned. The Aerobee became a workhorse of the upper-atmosphere research. More than 1000 Aerobees of various modifications were launched from 1947 to January 1985. Photo courtesy of Aerojet-General Corporation.

Workhorse of Upper-Atmosphere Research

The Aerobee was an unguided missile, with the later versions capable of achieving altitudes up to 350 miles (560 km). The Aerobee first flew from White Sands in November 1947. It became one of the most widely used sounding rockets in the United States. It is interesting that neither Aerobee nor Viking have been developed for military purposes,

ROLES AND MISSIONS

4. Air Force Tactical Support of the Army.

The Army will continue its development of surface-to-surface missiles for close support of Army field operations with the following limitations:

a) That such missiles be designed and programmed for use against tactical targets within the zone of operations, defined as extending not more than 100 miles beyond the front lines. As such missiles would presumably be deployed within combat zones normally extending back of the front lines about 100 miles, this places a range limitation of about 200 miles on the design criteria for such weapons ... .

5. Intermediate Range Ballistic Missile (IRBM).

In regard to the Intermediate Range Ballistic Missiles:

a. Operational employment of the land-based Intermediate Range Ballistic Missile system will be the sole responsibility of the U.S. Air Force.

b. Operational employment of the ship-based Intermediate Range Ballistic Missile system will be the sole responsibility of the U.S. Navy.

c. The U.S. Army will not plan at this time for the operational employment of the Intermediate Range Ballistic Missile or for any other missiles with ranges beyond 200 miles ... .

(The Intercontinental Ballistic Missile has previously been assigned for operational employment to the U.S. Air Force.)

Charles Wilson, The Secretary of Defense,
Memorandum, 26 November 1956

Roles and Missions

in contrast with the USSR where military missiles were converted for use as research rockets.

The Army, the Air Force, and the Navy were carrying out essentially independent development programs in guided missiles, with some overlap and bitter rivalry. On 26 November 1956, the Secretary of Defense Charles E. Wilson attempted to address the duplication problem and clarify the roles of the services in rocketry. Wilson's "roles and missions" memorandum, issued in November 1956, reduced the Army's role in the intercontinental warfare by restricting its operational responsibilities to relatively short-range missiles.

Army Loses Rocket Fight

In the words of the ABMA Commander, General Medaris, this was "a crushing victory for the Air Force. ... The maximum range for an Army missile was set at 200 miles. Everything beyond that became Air Force responsibility" (Medaris and Gordon 1960, 125). Wilson's memorandum did not obviously end the turf battle. Only the formation of the National Aeronautics and Space Administration in 1958 and the following

DUPLICATION, RIVALRY, AND COMPETITION

Science, as well as any kind of technical work, thrives on friendly competition, on the fostering of different points of view, and on the exchange of ideas developed in various surroundings. It is only too easy for a single group to become fascinated by some aspects of a development and to neglect other hopeful approaches.

Edward Teller (1955, 274)

"ENGINE CHARLIE" WILSON

Secretary of Defense Charles E. Wilson who issued the "roles and missions" memorandum in 1956 was the former chairman of General Motors. "Engine Charlie" Wilson is often remembered, courtesy of the media, as famously pronouncing at the confirmation hearings at the Senate that "what is good for General Motors is good for the nation."

In reality, however, Wilson's statement was somewhat different. During the hearings before the Committee on Armed Services on 15 January 1953, Senator Robert C. Hendrickson of New Jersey asked Wilson, "... if the situation did arise where you had to make a decision which was extremely adverse to the interests of your stock [in General Motors] and General Motors Corp. ... in the interests of the United States Government, could you make that decision?" Wilson replied: "Yes, sir; I could. I cannot conceive one because for years I thought what was good for our country was good for General Motors and vice versa. The difference did not exist. Our company [General Motors] is too big. It goes with the welfare of the country. Our contribution to the Nation is quite considerable."

transfer of a large part of the Army's research personnel, including JPL and von Braun's team at Redstone, to this new civilian organization would finally reduce the Army's role in long-range ballistic missiles and emerging space activities.

Ballistic vs Winged (Cruise) Missiles

The pace of the development of American rocketry began to gradually accelerate in 1950s, with the long-range delivery of atomic bombs being a major goal. Atomic weapons of the late 1940s remained relatively heavy, which raised a question of the optimum missile design for intercontinental surface-to-surface warfare. Two main concepts competed: a ballistic rocket and a winged, or cruise, missile. The ballistic rocket boosted a warhead into a high-altitude (partially outside the atmosphere) ballistic trajectory. In contrast, the cruise missile relied on the ambient air for propulsion and wings for lift and control. It was not clear at the time which approach would lead to a successful and effective weapon system.

Ballistic Missiles Win Competition

Simultaneously with pursuing ballistic missiles, the Air Force initiated development of intercontinental winged missiles, the supersonic missile Navaho (through contract to North American Aviation) and the subsonic missile *Snark* (contract to Northrop Aircraft). In addition three other guided cruise missile systems with shorter ranges were successfully developed and deployed in 1950s: Navy's *Regulus 1* (range 500 miles or 800 km; prime contractor Chance Vought Aircraft, Inc.) and Air Force's 700-mile (1100-km) range *Matador* and its successor *Mace*, both developed by the Glenn L. Martin Company. The rapidly advancing technology of ballistic rockets, demonstration of reliable atmospheric reentry, and significant reduction of the nuclear warhead mass and size ultimately led to cancellation of the Navaho in 1957 and termination of the Snark in 1961. Ballistic missiles had won the competition as a weapon of intercontinental warfare. The experience with the cruise missiles was not lost, however, and led to a highly successful modern system, the *Tomahawk*, with the range up to 1500 miles (2400 km). (The Tomahawk was developed by General Dynamics' Convair Division and McDonnell Douglas.)

The supersonic intercontinental cruise missile Navaho was terminated after spending almost $700 million. Although the Navaho was not successful as a

Fig. 12.20. Launch of subsonic intercontinental missile Snark on 3 March 1960. The Snark project had been initiated by the Northrop Aircraft Company in 1946, when the concepts of both winged and ballistic intercontinental missiles were explored. Snark was a failure as a weapons program, declared by President John F. Kennedy "obsolete and of marginal military value" in 1961. The program, however, advanced technology, provided experience, and prepared and trained the personnel of the Strategic Air Corps for the deployment of ICBMs. Photo courtesy of 45th Space Wing History Office Archives, Patrick Air Force Base, Florida.

Navaho's Techno-logical Heritage

weapon system, the program led to several major technological advances in propulsion, inertial guidance, materials, and fabrication processes. The other important program outcome was a transistor-based computer with etched circuit boards. In addition, a concept of parallel-mounted rocket boosters with aerodynamic separation had been demonstrated. Such a booster configuration would be favored later by the American *Space Shuttle* and Soviet *Buran* multiple-use vehicles.

Autonetics Division of North American Aviation

The Autonetics Division of the North American Aviation concentrated on advancing the emerging field of inertial guidance for the Navaho, including enabling components such as accelerometers and gyroscopes. The division would not only design the components and guidance systems, but it would also mass produce them for ballistic missiles. Autonetics aggressively recruited a large number of engineers, and it grew into a major production establishment that employed 36,000 people in Southern California at the peak of the ICBM program in early 1960s. (Autonetics would eventually become a part of Rockwell International.)

Seeds of Inertial Guidance

The XN6 inertial guidance system developed by Autonetics for the Navaho provided the basis for the future highly successful ship inertial navigation system (SINS). The world's first nuclear-powered submarine USS *Nautilus* demonstrated the reliability of inertial guidance on its historic trip to the North Pole in 1958. The emerging guidance technology would soon find a particularly important application on submarines armed with the fleet ballistic missiles Polaris.

ROCKET PROPELLANT RP-1

Kerosene substituted alcohol as a fuel of choice in combination with liquid oxygen in early 1950s. The specification limits on kerosene fuels used in aviation, such as the common JP-4, were very broad and clearly unacceptable for rocket propulsion. The combined efforts of the Air Force and the Rocketdyne Rocket Engine Advancement Program (REAP) led to tightening of the specifications and resulted in a standard RP-1 fuel, or Rocket Propellant - 1. In addition, the allowed fraction of sulfur and other minor components was limited to prevent effects harmful for regenerative cooling. The RP-1 is a hydrocarbon fuel with a nominal chemical formula $CH_{1.953}$, that is for each mole of carbon it contains 1.953 moles of hydrogen.

The power plants developed by North American Aviation for the Navaho's booster became a foundation to high-performance propulsion. Its original alcohol-based engine provided a stepping stone for the development of a new family of rocket engines based on kerosene and liquid oxygen. This propellant combination was selected by Convair in October 1951 for the new ballistic missile program that would eventually lead to the Atlas ICBM.

Rocketdyne Division of North American Aviation

North American Aviation concentrated the propulsion work in the new Rocketdyne Division organized in 1954 from its Aerophysics Laboratory. Rocketdyne, located in Canoga Park in the San Fernando Valley north of Los Angeles, California, employed 14,000 people by the late 1950s. Division president Samuel K. Hoffman and vice president of research and engineering Thomas Dixon led the development of several highly-successful rocket engines. Propulsion systems that powered Atlas (MA-1,-2,-3,-5), Thor (MB-1), Jupiter (S-3D), Delta (RS-27), and Saturn (F-1 and J-2), and the Space Shuttle Main Engine, were all built upon the advancements derived from the Navaho.

Fig. 12.21. Rocketdyne's Thomas Dixon (left) and Samuel Hoffman (right) led development of a new generation of liquid-propellant rocket engines in the Rocketdyne Division of North American Aviation in 1950s. Dixon's photo originally published in *Astronautics*, Vol. 2, No. 5, 1957, p. 20, and Hoffman's in the *American Rocket Society Roster*, 1959–1960. Copyright © 1957, 1959 by the American Institute of Aeronautics and Astronautics. Reprinted with permission.

Flow of Scientific Information

RPIA, SPIA, LPIA, CPIA, ICRPG, and JANNAF

Rocket propulsion in the United States was advanced by sometimes competing efforts of various government agencies, government research facilities, universities, and industrial contractors. Flow of the related scientific information among scientists and engineers was critical for the rapid progress of technology and preservation of the accumulated knowledge and expertise.

Recognizing importance of "a rocket intelligence agency" for rapid circulation of technical information, Navy's Bureau of Ordnance established the Rocket Propellant Intelligence Agency (RPIA) in 1946. RPIA's task was to abstract reports, organize filing and cross-indexing systems, and distribute information. RPIA, created in the Applied Physics Laboratory, was reorganized into the Solid Propellant Information Agency (SPIA) in May 1948. SPIA formed a number of specialized panels that addressed various technical issues including combustion, thermochemistry, propellant physical properties, and instabilities.

In 1958, the Navy organized, also in APL, the Liquid Propellant Information Agency (LPIA). The Interagency Chemical Rocket Propulsion Group (ICRPG) was formed in 1958 and was later reorganized as the Joint-Army-Navy-NASA-Air Force (JANNAF) Interagency Propulsion Committee. SPIA and LPIA were combined in 1962, forming the Chemical Propulsion Information Agency (CPIA). In addition to chemical rocket propulsion, CPIA covered electric and nuclear rocket propulsion, gun propulsion, and air-breathing propulsion. CPIA and JANNAF continue to play to this day an important role in advancing American research and development efforts in the area of propulsion.

Never-Ending Opposition

Encouraging successes of American rocketry in late 1940s and early 1950s and the growing expertise of the industry and research institutions failed to open a way for a full-scale development of long-range rockets, an indispensable step for a leap to space. The opposition to ballistic missiles seemed never ending, being fueled by a combination of complacency, poor judgement, and underestimation of the Soviet determination. "Logical" and incremental extensions of the capabilities of the existing weapon systems were often emphasized with the focus on perfection of jet aircraft, short-range missiles, and smaller atomic warheads.

The relatively heavy weight and small yield of early atomic bombs were among key obstacles for endorsing intercontinental ballistic missiles. The heavy weight of the bombs called for rockets of monstrous size and weight. The small yield of the existing bombs required an extraordinary accuracy of missile guidance to a target 5000 n miles (9000 km)

YIELD

The amount of explosive energy released by a nuclear weapon is usually described as *yield*. It is common to measure yield in terms of the quantity (mass) of common chemical explosive trinitrotoluene (TNT) that would produce the same amount of energy in explosion. Thus, the equivalent mass of TNT (and the weapon yield) is often expressed in units of kiloton (kt), or thousands of tons, and megaton (Mt), or millions of tons. One kt of TNT corresponds to the energy of 10^{12} calories or 4.186×10^{12} joules. The first atomic bombs detonated in 1945 were in the 10–20 kt range.

away. To make the guidance requirements realistic, the radius of the area effectively destroyed by a blast wave of an atomic explosion had to increase by a factor of 3 to 5. The radius of destruction is proportional to the cube root of the yield, or the amount of explosive energy released in an explosion. Consequently, the yield of nuclear warheads had to increase to 0.5–2.0 Mt to make the ICBM practical.

Low Yield and Heavy Weight

Fig. 12.22. A model of the kiloton-range atomic (fission) bomb Mk-5 in the National Atomic Museum, Albuquerque, New Mexico. The Mk-5 was the first attempt to build a smaller and lighter atomic bomb; it was deployed in 1952. Even this state-of-the-art compact and lightweight bomb was 11 ft (3.3 m) tall with diameter 43 3/4 in. (110 cm) and mass 3200 lb (1450 kg). Photo courtesy of Mike Gruntman.

Guidance, Maps, and Gravitational Field

The other, sometimes overlooked, difficulty was a lack of reliable maps of the Soviet Union. Locations of many places across the vast Soviet territory, particularly of those east of the Ural mountains, were known at that time with the accuracy of 10–15 miles. The Army-developed space mapping system *Argon* would successfully map the Soviet Union only in the early 1960s. In addition, precise targeting of warheads required an accurate knowledge of the Earth's gravitational field, allowing for nonuniform mass distribution. (The accurate knowledge of the gravitational field is required both for prediction of the warhead ballistic trajectories and for operation of inertial navigation systems.) Only analyses of satellite orbits in early 1960s would resolve the latter problem.

Hydrogen Bomb

The performance of nuclear weapons had to substantially improve to make a realistic ICBM possible. In 1948, the *Sandstone* tests at Eniwetok Atoll demonstrated the new design of atomic (fission) bombs with the superior efficiency of the utilization of expensive fissile material. Yields still remained unacceptably low and the bomb mass unacceptably high. Only in November 1952, a 0.5-Mt fission bomb had been demonstrated in the *King* shot at Eniwetok.

Fission-based atomic bombs were being overshadowed by the emerging much more promising and powerful weapon, the bombs on the basis of fusion, called thermonuclear or hydrogen (H-) bombs. The H-bomb required a well developed fission bomb to serve as a trigger and promised unlimited, in principle, explosive power. Although the physical concept of an H-bomb had been understood by the mid-1940s, the design of a practical weapon where a small fission device would ignite an arbitrarily large thermonuclear device remained elusive for several years.

"The Work of Many People"

Between January and May 1951, the workable H-bomb design had finally been found by physicist Edward Teller and his coworkers. It was, in Teller's words, "the work of many excellent people who had to give their best abilities for years

IN DEFENSE OF THE FREE WORLD

The first two atomic (fission) bombs dropped on Japan in 1945 were rather heavy, 8900 lb (4040 kg) and 10,800 lb (5000 kg). Only in the late 1940s the work began to improve the design of atomic bombs. The new bombs that entered service in the early 1950s remained heavy and had a relatively low explosive power, or yield (40–50 kiloton, or kt). Their significant improvement was in the efficiency of use of the scarce and expensive fissile material. Even the specially designed lightweight Mk-5 bomb had mass 3200 lb (1450 kg).

Delivering atomic bombs to their targets by intercontinental ballistic missiles did not look promising: a small yield of atomic weapons called for an extraordinary precision of missile guidance. After the intense and often bitter debates among the leading scientists and political and military leaders, President Truman directed, on 31 January 1950, the development of a much more powerful thermonuclear or hydrogen bomb (H-bomb), also called the superbomb. Among the scientists, Ernest Lawrence, Luis Alvarez, and Edward Teller passionately advocated the new powerful weapon as deterrence against Communist aggression.

Fig. 12.23. Physicist Edward Teller, 1908–2003, played the leading role in the development of American thermonuclear weapons. Teller relentlessly worked to strengthen defense and military preparedness of his adopted country. (He was naturalized in 1941.) Photo (mid-1950s) courtesy of Lawrence Livermore National Laboratory.

Only the availability of H-bombs with yield in a megaton range has significantly reduced the otherwise unrealistic requirements to guidance precision. At the same time the new much lighter warheads did not require rockets of monstrous mass and size. This breakthrough in nuclear weapons made the early ICBM practicable and led to the decision to proceed with the development of the Atlas.

Some scientists argued at that time for advancing the technology of fission bombs with higher yield (up to half a megaton) without developing the H-bomb, while trying to convince the Soviet Union — that is to convince the Communists on moral grounds — not to pursue the superbomb. It is clear that the postponement of the Truman's decision to achieve and demonstrate the H-bomb would have likely led to the delay in turning the struggling Atlas into a crash program of the top national priority and resulted in fielding the American ICBM a few years later than it happened otherwise. Such a delay would have produced a real and disastrous ICBM "missile gap" of several years vis-a-vis the aggressive Soviet Union that vigorously pursued its ICBM and nuclear weapon programs.

and who were all essential for the final outcome" (Teller 1955, 267). The road to practical thermonuclear weapons has thus been opened. As Teller later described, "the great military advantage of the hydrogen bomb turned out to be not the size of its explosion but the flexibility of its construction. The bomb material (deuterium) was cheap. That made it possible to optimize cost, yield, and weight" (Teller 2001, 346).

Cost, Yield, and Weight

The first American thermonuclear device, the predecessor of the hydrogen bomb, was detonated on 1 November 1952, in the *Mike* test at Eniwetok Atoll. The experimental device used liquid deuterium (a hydrogen isotope) and required extremely heavy cryogenic equipment. This crucial test produced yield of 10 megaton and verified the physics of the H-bomb.

"Wet" and "Dry" Designs

To weaponize the thermonuclear device, a new design was needed based on the fusion material in a noncryogenic, solid or "dry" (in contrast with the tested "wet" design) state at room temperature. The new "dry" bombs with solid lithium deuteride were successfully demonstrated during the *Castle* series of tests in March 1954. The first test of this series, the *Bravo*, yielded 15 megaton at Bikini Atoll on 1 March 1954, and convincingly proved that the Teller design would lead to a compact, lightweight, and efficient H-bomb.

Fig. 12.24. John von Neumann, 1903–1957, chaired several scientific advisory committees that significantly influenced development of the American nuclear weapons and long-range ballistic missiles. In addition to the extraordinary service to his adopted country (he became a U.S. citizen in 1937), von Neumann made fundamental contributions to mathematics (a special case of a Hilbert's problem), logic, quantum physics (operator theory), physics, design of atomic and thermonuclear weapons, game theory (the *minimax theorem*), and the emerging computer science and engineering. Photo courtesy of U.S. Air Force.

Von Neumann's Committee

A detonation of the thermonuclear device brought a critical technical question whether a practical powerful warhead can be developed under the weight constraints of a realistic intercontinental ballistic missile. The issue was so serious that the Air Force Science Advisory Board appointed a special committee in 1953 to evaluate nuclear weapons and means of their delivery. The Committee was headed by the brilliant scientist John von Neumann. In the words of Edward Teller, von Neumann was "one of the rare mathematicians who can descend to the level of a physicist" (Teller 1955, 270). This selected group included Hans Bethe, Norris Bradbury, David Griggs, George Kistiakowsky, L. Eugene Root, Herbert Scoville, and Edward Teller.

Von Neumann's committee concluded that powerful H-bombs could be achieved with manageable mass. In particular, the committee found that a 2-Mt bomb with the destruction area of 3–4 miles could weigh as little as 3000 lb. Such a large destruction area enabled intercontinental missiles with the dramatically reduced requirements to the accuracy of guidance, the latter being perhaps the most serious technical challenge. To further reduce the warhead weight, the committee then recommended reduction of the warhead yield down to 1 Mt. In September 1953, the Air Force Special Weapons Center independently confirmed the feasibility of 0.5-Mt fusion warheads with mass of 1500 lb (680 kg). Note that at that time, the state-of-the-art lightweight atomic bomb with the yield in a 40–60 kt range, the Mk-5, had mass 3200 lb (1450 kg).

Light-weight Thermo-nuclear Warhead Possible

A compact, lightweight, and powerful H-bomb could thus be delivered by a rocket with a reasonable size and weight. The yield in a megaton range allowed dramatic reduction of the otherwise extraordinary and unrealistic guidance accuracy. The advancement in the nuclear weapon technology has cleared doubt over the Air Force's program to develop an *intercontinental ballistic missile system*, or IBMS as it was called at the time. The term IBMS was soon changed to the *intercontinental ballistic missile*, or ICBM. Much heavier contemporary Soviet warheads — the reentry vehicle weighed 5400 kg, or 12,000 lb, on the first Soviet ICBM SS-6 — called for much larger and more-powerful rockets for their delivery. Consequently, the first Soviet ICBMs and first space launchers would be significantly larger than American.

Nuclear Warhead on ICBM

Soviet Ballistic Missiles

H-BOMB AND ICBM

Many people had been thinking about the ICBM and pushing it, but one thing that got it off top dead center was the advent of the thermonuclear weapon. I can recall attending an SAB meeting, an Air Force Science Advisory Board meeting, down at Patrick AFB [Air Force Base in Florida] in early 1953. At that particular meeting, both Dr. Teller and Dr. von Neumann made presentations on the thermonuclear shot that had been made in the Pacific earlier, and predicted verbally that it would be possible to build a thermonuclear warhead weighing not over 1500 lb with a yield of a megaton. Right then and there that changed the whole picture. An ICBM would be able to carry a low-weight, high-yield weapon — the thermonuclear warhead.

... [The von Neumann's Committee set up in the summer of 1953] confirmed the statements Teller and von Neumann had made at the Patrick meeting, and that really gave the ICBM an official status.

General Bernard A. Schriever
(Schriever 1972, 57)

MX-1953 and Atlas

In January 1951, the Air Force contracted Convair to build an intercontinental ballistic rocket, MX-1953, soon renamed *Atlas*. The experience gained in the MX-774 program helped Convair in winning the contract. The contract for the Atlas RP-1/LOX engines was issued to North American Aviation in February 1954. Them the Air Force awarded a propulsion contract for a backup engine system to Aerojet in January 1955. Before the ICBM became operational in 1959, the Atlas had gone through many ups in downs in priorities and funding and had experienced numerous changes in requirements and designs. It was a long and difficult road, with no guarantee of success.

Trevor Gardner

Special Assistant (for Research and Development) of the Secretary of the Air Force, later Assistant Secretary, Trevor Gardner had become a main champion of the American ICBM. A 37-year old, energetic and bright, Gardner knew how to operate in Washington and was dedicated to getting results. He felt it was his mission to help the nation to achieve the ICBM and thus save it from nuclear assault by the Communists.

Hard Choices

The early 1950s were a critical time for the American missile program. "By the end of 1953," wrote Gardner,

> the Air Force missile development program had produced prototype hardware and, in some cases, production hardware on which further progress could be made only by the expenditure of large additional sums. The situation was complicated by the fact that the Air Force jet aircraft program had arrived at an identical stage of progress at the very same time. Faced with hard decisions affecting the future of both programs, the Air Force was thereupon confronted by a new [Eisenhower] Administration and a new Congress both determined to effect major economies ... . At this critical juncture, the Air Force, not unnaturally, reacted to protect its newly-achieved manned jet combat units and, for the time being, re-programmed funds away from the debated missile area into the jet aircraft program. (Gardner 1958, 9)

Fig. 12.25. Assistant Secretary of the Air Force Trevor Gardner was the first civilian manager of the Air Force research and development activity. He skilfully navigated the ICBM program through the Pentagon bureaucracy. In February 1956, Gardner resigned his position in protest to reduction of Air Force's research and development programs leading, as he argued, to the "world's *second-best* Air Force." Photo courtesy of U.S. Air Force.

Gardner's Strategy

Gardner looked for a way to bring such attention to the sputtering Atlas program that the Department of Defense would have no choice but to make the program, irreversibly, its highest priority. He hoped that findings of a special committee, composed of scientific and technological luminaries, would do such a job.

"Teapot" Committee

Gardner's strategy worked. He began with forming a special committee, also chaired by John von Neumann, in October 1953. This *Strategic Missiles Evaluation Committee*, nicknamed the "Teapot Committee," specifically focused on nation's strategic missile programs. The committee included Hendrik W. Bode, Louis G. Dunn, Lawrence A. Hyland, George B. Kistiakowsky, Charles C. Lauritsen, Clark B. Millikan, Allen E. Puckett, Simon Ramo, Jerome B. Wiesner, and Dean Wooldridge. Colonel Bernard ("Benny") A. Schriever represented the Air

TREVOR GARDNER AND BUREAUCRACY

Even in 1954, the defense bureaucracy was so strong that it was difficult to speed a project through. It is far worse in the late 1980s, but the overlapping review system that grew quickly after WW II would have been enough then to force the ICBM project to move at a snail's pace had it not been for Trevor Gardner's disdain for lower-level military and civil-service personnel. They were determined to get on the ICBM management and Gardner was just as determined to deny them that opportunity. Again and again, the bureaucrats found that while they were still studying an action — perhaps to approve it, perhaps not, or more likely to demand more information and stall — it had already been implemented by Gardner. Were they then to insist on being involved, they would have had to seek a decision reversal, something they were loath to do because they might be accused of standing in the way of progress. They would give up.

Simon Ramo (1988, 109)

Force. The committee issued the report with the recommendations on 10 February 1954.

There was mounting, though inconclusive, evidence that the Soviet Union was ahead in development of the long-range ballistic missiles. Debriefing of German rocketeers returning from the USSR and reaching West Germany provided an important insight into the scope and focus of the Soviet missile program. In addition, it was recently determined that the Soviet Union successfully tested its first hydrogen bomb on 12 August 1953. The test was known in the United States as "Joe-4," after Stalin's given name *Iosif*, or Joseph (Joe) in its Anglicized form.

***Joe-4* Detected**

The secret Soviet test was promptly detected by the American air-collecting aircraft reconnaissance program. Systematic sampling of air along Soviet borders was established in late 1940s largely because of foresight and insistence of Commissioner Lewis Strauss of the Atomic Energy Commission (AEC). Later, Strauss chaired AEC. The Joe-4 test brought the feeling of urgency because, as the air samples revealed, radioactive debris indicated that the bomb was based on a dry thermonuclear design, that is, it was ready to be weaponized. (This Soviet test indeed used a solid nuclear fuel ready to be weaponized. The design of the detonated device, however, was not a truly hydrogen bomb but rather a so-called boosted nuclear weapon. The boosted nuclear bomb used thermonuclear reactions to obtain more efficient and powerful release of fission energy in contrast to a hydrogen bomb where a small fission charge could ideally ignite an arbitrary large fusion charge. The Soviet Union achieved the "real" H-bomb in a test explosion on 23 November 1955.)

Soviet H-Bomb

MISSILE COMMITTEE

The *Strategic Missiles Evaluation Committee* ("Teapot Committee") was established in October 1953 to evaluate and accelerate the Air Force's programs. The Committee was subsequently reorganized as the *Coordinating Committee on Guided Missiles* and continued its advisory role for various missile programs, including those of the Army and the Navy, reporting to the Secretary of Defense.

Soviet Missile Program

The Teapot Committee noted that "while existing [intelligence] evidence does not justify a conclusion that the Russians are ahead of us, it is also felt by the Committee that this possibility certainly cannot be ruled out." The Committee stated that "important aspects of the present long-range missile program [of the Air Force] consisting of three projects, Snark, Navaho, and Atlas, are believed to be unsatisfactory" (Neufeld 1990, 255).

"Long-Range Missile Program ... Unsatisfactory"

Fig. 12.26. Commander of Air Force's *Western Development Division* General Bernard A. Schriever, 1910–2005, supervised development of the first American ICBM Atlas, the highest research and development priority of the Air Force. Schriever also played an important role in the early military space program. In 1961, he assumed command of the newly established *Air Force Systems Command.* Photo (ca. 1956) courtesy of 45th Space Wing History Office Archives, Patrick Air Force Base, Florida.

System Engineering and Technical Direction

The Teapot Committee recommended a radical reorganization of the ballistic missile effort. The Atlas, with the projected range of 5500 n miles (10,200 km), would now rank a top priority of the Air Force and the actual development would begin. Setting up a special development-management agency for the entire Atlas program was deemed most urgent. The complexity of the ICBM was thus demanding emergence of a new technical area, *system engineering and technical direction*, that would become prominent in the future. The Committee stated that "the nature of the task for this new agency requires that over-all technical direction be in the hands of unusually competent group of scientists and engineers capable of making system analyses, supervising the research phase, and completely controlling the experimental and research phases of the program."

Atlas Could Achieve Operational Status in the Early 1960s

Two days before the Teapot Committee issued its report, another independent study confirmed the feasibility of the Atlas ICBM. The assessment by RAND's Bruno W. Augenstein stated that the Atlas could achieve the operational status in the early 1960s providing the stringent performance characteristics were somewhat relaxed and the program priority and funding increased. In an assuring development on 1 March 1954, the Bravo test demonstrated the feasibility of high-yield, compact, and low-weight nuclear warheads: the Atlas program was now possible within the state of the art.

The report of the Teapot Committee triggered a set of events that significantly accelerated the American ICBM program. First, the reorganization of the Air Force development effort followed. To manage the Atlas program, the special *Western Development Division* (WDD) was activated under command of General Schriever at 409 East Manchester Blvd., Inglewood, California, in July 1954.

Fig. 12.27. From humble beginnings to a *Fortune 500* company: Simon Ramo (left) and Dean Wooldridge outside the original location of their *Ramo-Wooldridge Corporation* in Westchester, California. Several years later the company would merge with *Thompson Products* forming *TRW, Inc*. Photo courtesy of TRW, Inc.

WDD, TRW, STL, and the Aerospace Corporation

Second, the role of the system engineering and technical direction was substantially expanded. The *Ramo-Wooldridge Corporation* (R-W), the predecessor of TRW, Inc., was founded by Simon Ramo and Dean Wooldridge to provide such services for the Air Force. (In contrast, the Army relied on its in-house expertise of the von Braun's group at the Redstone Arsenal for technical direction of the Army ballistic missile programs.) R-W's *Space Technology Laboratory* (STL) would thus become the main participant in the Atlas and other ICBM, IRBM, and space programs. The STL's role generated controversy, however, and, in several years, many functions in system engineering and technical direction would be taken over by the newly formed nonprofit *Aerospace Corporation*.

Development and deployment of such complex systems as the ICBM and future spacecraft required new management approaches. Concurrency, broadly understood as the parallel development and simultaneous completion of all of the tasks necessary for system development and deployment, became critically important. Bernard Schriever was among the pioneers of this emerging management concept when he advocated concurrency in a special Air Staff study in 1950.

Schriever's WDD grew, assuming later the responsibility for the Titan, Thor, and Minuteman missile systems as well as for the early military space programs, including the photoreconnaissance Corona satellite system. General Osmond J. Ritland became WDD's vice commander in April 1956. In June 1957, WDD was

TRW AND THE AEROSPACE CORPORATION

Hughes Aircraft Company had emerged as a guided missile powerhouse and major defense contractor in the early 1950s. Disagreements with Howard Hughes led to resignation of two leading specialists Simon Ramo and Dean Wooldridge, who formed in September 1953, with the financial help of *Thompson Products Company*, a new company, the *Ramo-Wooldridge Corporation* (R-W). R-W started with four employees, including the founders, and was located at first in a former barbershop on 92nd Street in Westchester near the Los Angeles airport. Thompson Products and R-W merged in 1958 to form *Thompson Ramo Wooldridge, Inc.,* the name officially shortened to *TRW, Inc.,* in 1965.

R-W became the main provider of system engineering and technical direction for the Air Force's Atlas ICBM. R-W's *Guided Missile Research Division (GMRD)* later expanded to provide system engineering for the Titan ICBM and Thor IRBM programs. After launch of *Sputnik*, GMRD was reorganized as a separate subsidiary corporation, the *Space Technology Laboratory (STL)*, with Simon Ramo as president, Louis Dunn as executive vice president and general manager, and James H. (Jimmy) Doolittle as chairman of board of directors.

The propriety of a for-profit company performing so exclusively services for the government was questioned by Congress. STL was in "an intimate and privileged position" for an Air Force contractor, being involved in evaluation of the bids from other companies. Internal barriers between STL and the parent company, TRW, did not prevent charges that such an arrangement gave TRW unfair competitive advantage. TRW was also not entirely happy because of the imposed limitations on the scope of hardware contracts the company could bid on. STL's rapid expansion, although being profit oriented, was perceived by many as inherently inappropriate for the technical direction of government programs.

Consequently the solution to the problem was found in the formation of a non-profit institution, *the Aerospace Corporation*, in June 1960 with Ivan Getting as the first Aerospace's president. Getting served in this position until 1977. The mission of Aerospace, according to the letter of contract, was "to aid the U.S. Air Force in applying the full resources of modern science and technology to the problem of achieving those continuing advances in ballistic missile and military space systems which are basic to national security."

By the end of 1960, the Aerospace Corporation had bought the recently finished STL's research and development center on El Segundo Boulevard and hired more than 1700 employees, one-third being scientists and engineers. Many specialists came to Aerospace from TRW's STL. Technical functions of Aerospace concentrated on general system engineering and technical direction (GSE/TD) of the Air Force's ballistic missile and space systems. Today, Aerospace operates as Federally Funded Research and Development Center (FFRDC) for the Department of Defense.

TRW has evolved into a *Fortune 500* company, a major defense contractor specializing in missile and space systems and defense electronics. The company continues to this day its original work on maintaining readiness of the nation's ICBMs. TRW had become the first industrial company to build an exploratory spacecraft, *Pioneer 1*, for NASA. TRW designed and built numerous military and civilian space systems, including *Defense Support Program* (DSP), *Tracking and Data Relay Satellite System*, deep space *Pioneer 10* and *11* spacecraft, and astrophysical space observatories *Compton* (γ-rays) and *Chandra* (X-rays). Northrop–Grumman acquired TRW in 2002, which became Northrop–Grumman Space Technology.

reorganized into the *Air Force Ballistic Missile Division* (AFBMD). Schriever built a unique military organization with a high level of education where "more than one third of the hand-picked officers held Ph.D.'s and Master's degrees."

Unique Military Organization

The Air Force assigned the highest priority to the Atlas program on 14 May 1954 and gave the full go-ahead in January 1955, with Convair's Astronautics Division in San Diego, California, as the prime contractor. Convair that had led the Atlas development was not entirely happy with the new arrangement that significantly downgraded its role in systems engineering and technical direction. Thus, the Teapot Committee led to a fundamental reorganization of the entire American strategic missile program. The Committee, in the words of General Schriever, "really pulled the cork and got the ICBM program underway" (Schriever 1972, 58).

Fig. 12.28. Simon Ramo, b.1913, began his research career at the General Electric Research Laboratories where he accumulated 25 patents before the age of 30. Ramo initially focused on the emerging microwave technology. His book (with John Whinnery) on electromagnetic fields and waves, first published in 1944, remains to this day a standard textbook in electronics. Ramo's work at Hughes Aircraft and later cofounding (with Dean Wooldridge) TRW significantly advanced electronics, missiles, and space technology. In 1950s he played a critical role in technical direction and system engineering in achieving the American ICBM. Photo (ca. 1986) courtesy of Simon

Top Air Force Priority

Another high-level government committee headed by James R. Killian, Jr. issued its findings on 14 February 1955. (Killian was President of MIT in 1948–1959 and served as the first special assistant to the President for science and technology when President Eisenhower established this office in 1957.) The *Killian Report* warned about America's vulnerability to a surprise Soviet attack and urged the elevation of the priority of the ICBM. Consequently, President Eisenhower approved on 13 September 1955 the recommendation of the National Security Council to designate the ICBM a national program of the highest priority. Incessant bureaucratic maneuvering and pressures, however, soon diluted this top priority status of the ICBM by assigning top priorities to other programs of the Air Force, Army, and Navy. The development of long-range ballistic missiles would eventually lead to major families of space launchers.

Killian Report

Technical Challenges

To achieve the Atlas, the performance of several key rocket subsystems had to be improved dramatically. The challenges included achieving low structural weight, high thrust and high-specific-impulse engines, efficient propellant utilization, precise engine cutoff, rocket staging, inertial guidance, and warhead atmospheric reentry with extraordinarily high velocities.

The advancement of Atlas structural design was based on Convair's experience with the MX-774. (Convair merged with the General Dynamics Corporation in 1954.) The original concept of pressurized tanks conceived by Charlie Bossart for the MX-774 was perfected and became one of the most prominent and recognizable features of the Atlas. The walls of the propellant tanks were so thin that even in storage they required pressurization by inert gas, usually nitrogen, with the 5–10 psi (0.3–0.7 atm) pressure to carry the structural burden and to prevent their collapse. The wall thickness varied from 0.01 in. (0.25 mm) to 0.05 in. (1.25 mm). Aluminum of the MX-774 was replaced by special extra full hard corrosion resistant stainless steel to withstand much higher thermal loads at the expected operational trajectories of the Atlas.

Steel Balloons

The 10-ft (3-m)-diam tanks were made by welding together 32-in. (81-cm)-wide bands. More than 100,000 spot welds and 400 ft of butt welding were needed to assemble a tank. Pressurized by nitrogen, the tanks earned the nickname of flying "steel balloons."

Fig. 12.29. Ivan A. Getting, 1912–2003, the founding president of The Aerospace Corporation. Beginning 1960, The Aerospace Corporation provided technical direction for the Air Force's ballistic missile and space programs. Getting who had a Ph.D. (1935) in astrophysics worked on applications of radar during the war. He was a founding member of the Air Force Science Advisory group headed by von Kármán. Photo (1961) courtesy of The Aerospace Corporation.

WD-40

Launches from Cape Canaveral called for additional protection of tank's stainless-steel skin against corrosion by humid and salty Florida air. A special coating introduced for tank protection became a basis for one of the most widely used and familiar products in the United States — WD-40 (*WD* stood for *water displacement*).

The propulsion systems had progressed by that time to the point where they were considered to be on track to the desired capabilities. The Rocketdyne Division projected that 150,000 lbf (667 kN) thrust engines based on a kerosene-type RP-1 fuel and liquid-oxygen oxidizer would soon be available.

Ignition in Weightlessness

The V-2 was a one-stage rocket with a limited range. A one-stage configuration made it extremely difficult to achieve high velocities. Rocket staging was thus needed to significantly increase the vehicle velocity. When the first stage consumed all of its propellant, it was jettisoned. The second rocket stage, with the warhead attached, continued its flight, and, after a short pause, the second stage would be ignited. Immediately after the separation of the first stage, the rocket would fly in weightlessness. Could one reliably ignite a liquid-propellant rocket in a state of

"EGGHEADS" IN DEFENSE AGAINST COMMUNISM

The quality of this space age military power and the rapidity with which we can achieve it depends fundamentally upon the kind of mind power we possess. Today, more than ever before, we need for national survival the disciplined and resourceful mind which makes possible intellectual performance of the highest order.

In my view it is a national disgrace that the term "egghead" as a synonym for intellectual excellence has become a derogatory expression. Let me tell you that it is the "eggheads" who are saving us — just as it was the "eggheads" who wrote the Constitution of the United States. It is the "eggheads" in the realm of science and technology, in industry, in statecraft, as well as in other fields, who form the first line of freedom's defense against communism.

General Bernard A. Schriever (1959, 37)

weightlessness? Where would the propellant exactly be in the propellant tank at that time? How to ensure that a gas bubble would not cover the propellant outlet thus preventing the second stage from proper ignition? Sloshing and vortexing of propellants in the tanks further complicated problems of attitude control and produced undesired loads on rocket structure.

One-and-One-Half-Stage Design

The uncertainty of the ignition of the second rocket stage under conditions of a free-fall environment (weightlessness) led to the selection of a conservative "one-and-one-half-stage" design for the Atlas. All Atlas's three engines started simultaneously at liftoff. Two side booster engines were jettisoned after two minutes of powered flight, while the third sustainer engine continued acceleration of the vehicle with the entire propellant tanks until the engine cut off. Interestingly, the concept of the first Soviet ICBM R-7 (SS-6) would be somewhat similar, with four side boosters.

Rocketdyne's Engines

As with any complex technological system, the performance characteristics of the Atlas power plant improved as the development progressed. The first development test vehicles, Atlas A, B, and C, used the MA-1 engine system consisting of the two XLR43-NA-3 booster engines and XLR43-NA-5 sustainer engine. The Atlas D, the first deployed ICBM, had the MA-2 engine system (LR-89-NA-3 boosters and LR105-NA-5 sustainer). The MA-3 system for the Atlas E and F ICBMs followed. The MA-5 system was a version of the MA-2 used for space launches. Thrust and specific impulse of the Atlas engines steadily progressed as shown in the following table:

Engine system	Sustainer (S) Booster (B)	Thrust, lbf (total)	Specific impulse, s
MA-1 (Atlas A,B,C)	S	54,000	210
MA-2 (Atlas D)	S	57,000	213
MA-3 (Atlas F,E)	S	57,000	214
MA-5 (launchers)	S	60,000	220
MA-1 (Atlas A,B,C)	B	300,000	245
MA-2 (Atlas D)	B	309,000	248
MA-3 (Atlas F,E)	B	330,000	250
MA-5 (launchers)	B	377,000	259

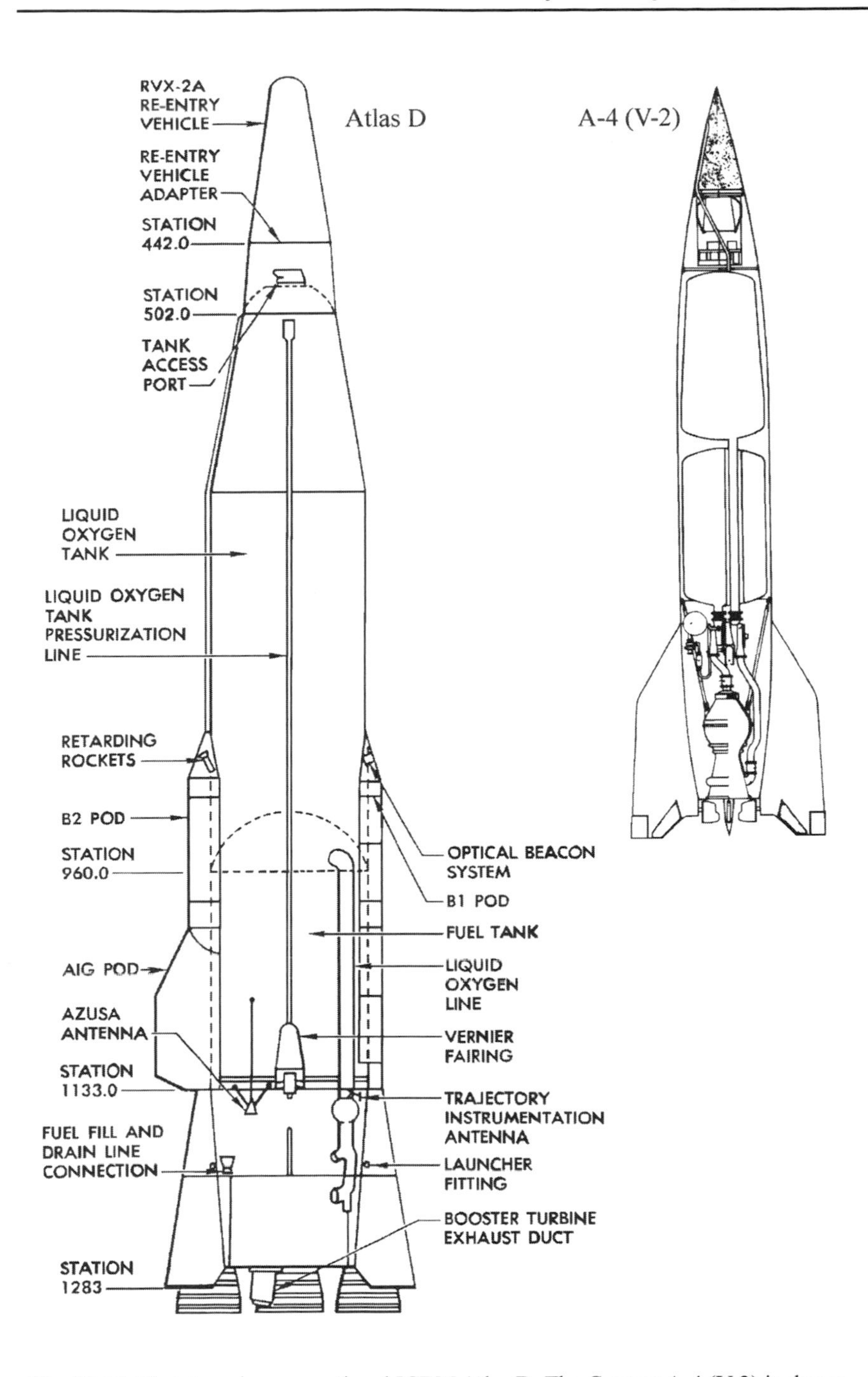

Fig. 12.30. First American operational ICBM Atlas D. The German A-4 (V-2) is shown in scale on the right. More efficient propulsion system, more propellant, and smaller structural weight allowed Atlas to achieve a 6800-m/s velocity at burnout (at 300-km altitude) compared to A-4's 1500 m/s (at 25-km altitude). The Atlas delivered a heavier warhead to a 10,000-km distance compared to A-4's 250 km. Based on drawings courtesy of U.S. Air Force and NASA.

In addition to Rocketdyne in Canoga Park, the production of engines for Atlas was organized at the plant at Neosho, Missouri, opened in 1957. The production was discontinued in 1968.

Propellant Utilization

With the extraordinary demanding requirements to rocket performance, the heretofore neglected problem of efficiency of propellant utilization rose to prominence. During engine operations, some propellant remained in tanks unused because of slight deviations of pumping speeds from nominal and was also trapped in the pipes, valves, and pumps. This remaining propellant became an inert, passive, ballast mass carried by the rocket. Propellants constituted most of the mass of a long-range ballistic missile (94–95% in case of the Atlas). If even just 1% of the propellant mass were not utilized, the rocket range would be substantially, by hundreds of miles, shortened. Alternatively, a corresponding significant reduction, up to 20%, of the warhead (payload) mass would be required.

Vernier Engines

The exact distance covered by an ICBM depended to a large degree on achieving precisely a certain velocity at engine cutoff. Large main rocket engines could not be stopped instantaneously, and the engine cutoff inherently involved uncertainties. Therefore, a new technique to control the missile range had emerged based on vernier engines. The vernier engines were small rocket thrusters working independently from the main engine. The main engine was stopped when the missile was close to the desired velocity but did not reach it yet. Small vernier engines continued to gently accelerate the missile for some time until the desired velocity was precisely achieved, and then these engines were cut off with a minimal uncertainty.

Radiocontrol and Inertial Guidance

Rocket guidance remained a most challenging area. The Atlas was initially designed to be controlled by a radio-inertial navigation system that relied on two ground radio stations. The radio signals transmitted by the rocket were received by two separated stations, and the rocket trajectory was thus calculated. The basics of this approach were developed for the MX-774 missile with the radio-inertial system called *Azusa*.

Inertial navigation that did not rely on active control from the ground offered serious advantages for missiles. In particular, inertial guidance was not, in contrast to radiocontrol, prone to radiointerference and jamming. Radiocontrol guidance was used on first test Atlas vehicles. Subsequently, this guidance system was modified to all-inertial, which was installed on the deployed ICBMs. Achieving all-inertial navigation required a breakthrough in science and technology.

Inertial Navigation

Several groups played a key role in advancing the highly promising inertial guidance in the United States after World War II. The Army supported the work conducted by von Braun's group of former Peenemünde engineers, first in Fort Bliss, Texas, and later in Huntsville, Alabama. The Air Force funded Charles Stark ("Doc") Draper and his *Instrumentation Laboratory* at MIT and two groups in industry, one at Northrop and the other at the Autonetics Division of North American Aviation. Hughes Aircraft was also involved, for a short time, in the initial work on guidance system components.

The Air Force first focused its development on inertial navigation for long-range bombers and cruise missiles and subsequently extended to ballistic missiles. (The Navy would later bring new challenges of precise ship navigation.) For example, the work at Northrop and at the Autonetics Division was originally driven by development of the Snark and Navaho cruise missiles, respectively. In

1954, Northrop practically terminated its work, while Autonetics aggressively pursued this new technology and ultimately emerged not only as a leader in navigation system development but also as a mass manufacturer of the systems for submarines and ICBM.

LOCAL VERTICAL IN THE OLD TESTAMENT

The Old Testament describes that the Lord used the "local vertical," as established by a plumbline, not only for erection of a wall but also for apparently navigational purposes:

"... the Lord stood upon a wall made by a plumbline, with a plumbline in his hand. ... Then said the Lord, behold, I will set a plumbline in the midst of my people Israel: I will not pass by them any more ... ."

Amos, 7:7–8

Local Vertical

Inertial navigation depends on the ability to separate forces acting on the navigational sensors caused by the gravitational field and caused by missile acceleration. The serious objection was initially raised as to whether such separation was physically possible because of the equivalence of the gravitational and inertial mass postulated by the general theory of relativity. A possible solution of establishing the true local vertical from a moving vehicle near the Earth's surface was found in the 1920s, and it required a pendulum system tuned to an undamped period of 84 minutes.

Performance Improvements of About 10,000 Times

The other serious problem was the grossly inadequate performance characteristics of the available sensors. Doc Draper wrote that

> a survey of the situation for specific force receivers (commonly called accelerometers) showed that indications of resultants of gravity and acceleration required instruments with performance improvements of about 10,000 times beyond what the behavior of commonly available devices. Few of the basic technologies required were available. The angular deviation receivers and specific force receivers did not exist, so they have to be developed. (Draper 1981, 455)

Atmospheric Reentry

The atmospheric reentry of warheads with hypersonic speeds also presented a major challenge. Several competing approaches to protect the warheads were studied, including heat sink, ablation, and transpiration cooling. H. Julian Allen of NACA's Ames Aeronautics Laboratory at Moffett Field, California, argued as early as in 1952 that the blunt nose cone shape was most suitable for ballistic missiles. Allen published together with A.J. Eggers, Jr. a classified research memorandum on the subject in 1953. It is unlikely that Charlie Bossart knew about this theoretical work when his Convair's team presented a missile design with a blunt nose cone to the Air Force in June 1953.

RECORD ALTITUDE BY X-17

During a test launch of an X-17 in April 1957, the rocket malfunctioned, and all three stages were fired on the upward part of the trajectory. As a result, the missile reached the record altitude exceeding 600 miles (960 km) and impacted 700 miles (1130 km) downrange.

X-17

In January 1955, a focused program to test various nose cone designs was initiated at Lockheed Aircraft Corp. A specially built three-stage missile, the X-17, was powered

INERTIAL GUIDANCE AND DRAPER LAB

Charles Stark Draper and his Instrumentation Laboratory at the Massachusetts Institute of Technology, Cambridge, Massachusetts, emerged in 1950s as a leader in the field of inertial guidance. The Laboratory draw on its prewar work on inertial navigation, particularly on studies by Walter Wrigley.

A breakthrough came in February 1953 when an Air Force B-29 bomber flew from Massachusetts to California guided by the *Space Inertial Reference Equipment* (SPIRE), developed by Draper. This 2250-n mile (4150-km) flight without information from the outside world spectacularly demonstrated the capabilities of the emerging technology. Inertial navigation offered serious advantages for missiles over commonly used radiocontrol systems, the latter being prone to radiointerference and jamming.

Fig. 12.31. Charles Stark Draper, 1901–1987, at his desk, ca. 1953. "Doc" Draper's Instrumentation Laboratory significantly advanced the science and technology of inertial navigation. Photo courtesy of The Charles Stark Draper Laboratory, Inc.

The Instrumentation Laboratory excelled in design and support of inertial guidance systems for strategic missiles of the Air Force (Thor, Titan, Peacekeeper) and the Navy (Polaris, Poseidon, Trident). It also developed highly successful guidance systems for the NASA's Apollo program. More than 2000 people were employed by the Laboratory in the late 1960s.

The Laboratory's heavy involvement in research and development in support of defense programs of the free world caught attention of the American political left in 1969. During the heat of the Cold War, the pressure and demonstrations of the leftists forced separation (the divestiture) of the Laboratory from MIT. In contrast with the technological breakthroughs, these years of turmoil did not boost the moral of scientists and engineers. After three years of planning and implementation, the *Charles Stark Draper Laboratory* (as it was named in January 1970) became an independent nonprofit corporation on 1 July 1973. Another attempt by the leftists to shut the Laboratory down by designating Cambridge a "Nuclear Free Zone" was decisively defeated by voters in 1983.

Today, the Charles Stark Draper Laboratory remains a leading research and development center in navigation and its diverse applications. The National Academy of Engineering administers the highly prestigious Draper Prize for innovative engineering achievements.

> ### *ARGUS*
>
> The modified X-17 rockets were used for a series of tests in August and September 1958 in which three small-yield nuclear charges were lifted to 300-mile (500-km) altitudes and detonated. The rockets were launched from the *USS Norton Sound* in South Atlantic. The Advanced Research Projects Agency (ARPA) conducted these tests, proposed by Nick Christofilos and called *Argus*, to study the physics of the Earth radiation belts and to determine whether an artificial belt could be created by nuclear explosions.

Mach Numbers 12–14

by Thiokol's solid-propellant motors. Normally, the X-17 with a test reentry vehicle was launched vertically, reached the altitude of 200 miles (320 km) using the first stage, then flipped over, and accelerated downward using the remaining two stages. The test nose cone reached velocities with the Mach numbers 12–14 at altitudes 7.5–9.5 miles (12–15 km), simulating the conditions of maximum heat load expected during ICBM warhead reentry into the atmosphere. For a realistic ICBM, the similar maximum heat load would occur at higher velocities (Mach number 25) but at higher altitudes characterized by lower atmosphere densities.

Reentry Problem

The reentry problem was approached in a methodical way. As General Schriever described it in Senate hearings in 1958, "we [the Air Force] roughed surfaces [of nose cones], we went through a very extensive program, the data we

Fig. 12.32. A technician at the Douglas Aircraft Company, California, inspects the polished copper nose cone of an Air Force's Thor IRBM, looking for flaws. The same type of the Mk-2 reentry vehicle with a 1.4 Mt W-49 warhead was installed on the first operational American ICBM Atlas. This heat-sink-type nose cone was developed by General Electric. The new generation of more efficient ablative cones would soon be introduced and replace the heat-sink reentry vehicles. Photo (1959) courtesy of National Atomic Museum, Albuquerque, New Mexico, and U.S. Air Force.

BLUNT NOSE CONE FOR ATMOSPHERE REENTRY

... In the design of long-range rocket missiles of the ballistic type, one of the most difficult phases of flight the designer must cope with is the re-entry into earth's atmosphere, wherein the aerodynamic heating associated with the high flight speeds of such missiles is intense. The air temperature in the boundary layer may reach values in the tens of thousands degrees Fahrenheit which, combined with the high surface shear, promotes very great convective heat transfer to the surface. Heat-absorbent material must therefore be provided to prevent destruction of the essential elements of the missile. It is a characteristic of long-range rockets that for every pound of material which is carried to "burn-out," many pounds of fuel are required in the booster to obtain the flight range. It is clear, therefore, that the amount of material added to protect the warhead from excessive aerodynamic heating must be minimized in order to keep the take-off weight to a practicable value. The importance of reducing the heat transferred to the missile to the least amount is thus evident ... .

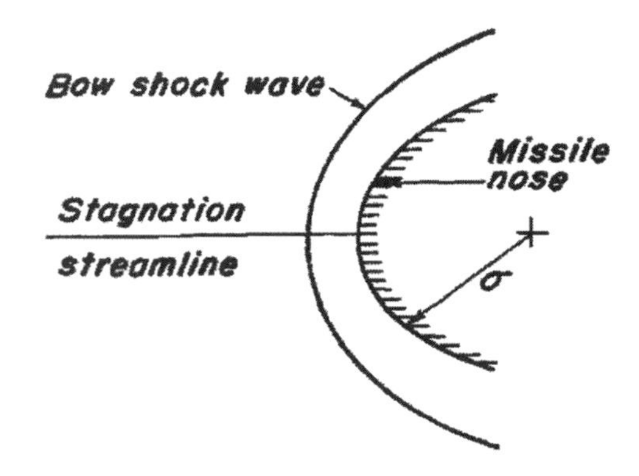

Fig. 12.33. Atmospheric reentry of a blunt body. Figure from NACA RM A53D28, 1953.

The elementary surface which is subject to the greatest heat transfer per unit area is, except in unusual cases, the tip of the missile nose which first meets the air. It seems unlikely that a pointed nose will be of practical interest for high-speed missiles since not only is the local heat-transfer rate exceedingly large in this case, but the capacity for heat retention is small. Thus a truly pointed nose would burn away. Body shapes of interest for high-speed missiles would more probably, then, be those with nose shapes having nearly hemispherical tips ... .

It is well known that for any truly blunt body, the bow shock wave is detached and there exists a stagnation point at the nose ... .

The region of highest local heat-transfer rate and, hence, probably greatest thermal stress was reasoned to be located at the forward tip of the missile in most cases. ... It was found that the magnitude of this stress was reduced by employing a shape having the largest permissible tip radius and over-all drag coefficient; that is to say, the blunt, high drag shape appears to have the advantage in this respect.

H. Julian Allen and A.J. Eggers, Jr.

"A Study of the Motion and Aerodynamic Heating of Missiles Entering the Earth's Atmosphere at High Supersonic Speeds"

Ames Aeronautical Laboratory, Moffett Field, California
NACA Research Memorandum RM A53D28, 25 August 1953
Declassified on 22 April 1957

got back as we carried on this program checked very, very closely with our theoretical calculations, with the data that we were getting out of our wind-tunnel testing, out of our shock-tube testing and so forth ... ."

Heat Sink and Ablative Technology

The first generation of nuclear-tipped IRBMs and ICBMs was equipped with the heat-sink nose cones. The technique required the cone's heat capacity to be sufficiently large to absorb heat during the atmospheric reentry. The approach was not very efficient, but it was based on a reliable and highly predictable technology. In parallel, a more complex ablative technology was being advanced. The ablative heat shields were made from a material that was being ablated, or vaporized, under the extreme temperatures of reentry. Because the atmospheric reentry was limited in time, the required thickness of the heat shield was also limited. Ablative heat shields were more efficient, but obviously more uncertain and risky, with the technology requiring thorough scientific study and experimental validation.

Fig. 12.34. Atlas rocket assembly at Convair's San Diego plant, ca. 1960. Photo courtesy of San Diego Aerospace Museum.

The first deployed Atlas missiles (Atlas D) carried the heat-sink Mk-2 reentry vehicles with 1.4 Mt W-49 warheads. The total mass of the reentry vehicle was about 3700 lb (1600 kg). The more efficient ablative technology soon replaced heat-sink shields. The introduction of the ablative reentry vehicle Mk-3 (Atlas E/F) reduced the total reentry vehicle mass to 2420 lb (1100 kg). Even with a 2.5 times more powerful W-38 warhead, the Atlas's ablative reentry vehicle Mk-4 had mass only 4050 lb (1840 kg).

Development of the Atlas ICBM and its deployment as a weapon system were a huge undertaking dwarfing the *Manhattan Project* of World War II. The complexity of the Atlas was in the required advancement in numerous areas of technology and operations. Vast ground facilities included various logistical and testing infrastructure and absorbed perhaps half of all of the program expenses.

Atlas Dwarfs Manhattan Project

At Convair alone, the work force involved in the Atlas rose from 300 in 1954 to 15,000 in 1959. Many thousands worked at various contractors. The rocket engines were fabricated by the Rocketdyne Division of North American Aviation; General Electric was responsible for the radio-inertial guidance system and for the reentry nose cone; AC Spark Plug built the entirely inertial guidance system; and Burroughs produced flight computers. The missile subsystems were designed and manufactured across the nation in such faraway places as Canoga Park, Syracuse, Philadelphia, and Detroit. It is no exaggeration to characterize the Atlas as a truly national effort.

From Sea to Shining Sea

Fig. 12.35. Atlas on a transport platform at the U.S. Space and Rocket Center in Huntsville, Alabama. Although considered as large at its time, the Atlas looks tiny compared with the first stage of the later Saturn V space launcher seen in the background (top left). Photo courtesy of Mike Gruntman.

Bumpy Road

The road to the ICBM was bumpy: the spending cuts enacted by the Eisenhower administration in 1956–1957 significantly affected the progress of missile programs. Assistant Secretary of the Air Force Trevor Gardner was so alarmed by the reduced role allocated to the research by the administration that he resigned in protest on 10 February 1956. The launch of the Soviet *Sputnik* in October 1957 helped, however, to reverse the trend and accelerated the sputtering missile programs. The first entirely successful test launch (first two launches were only partially successful) of the Atlas took place on 17 December 1957.

After a number of test launches, the Atlas completed its first full-range flight of 6325 miles (10,200 km) on 28 November 1958 and impacted in South Atlantic

30 miles (48 km) from the predicted point. In another Atlas first, on 18 December 1958, the 8700-lb (3950-kg) main stage of Atlas reached low-Earth orbit with the 168-lb (76-kg) payload named SCORE (*Signal Communications by Orbiting Relay Equipment*). The latter project was funded by the Advanced Research Projects Agency of the Department of Defense. The SCORE broadcast a Christmas message from President Eisenhower and stayed in orbit until 21 January 1959. In a communications experiment, the vehicle received several messages from the ground and transmitted them to various ground stations when commanded. The SCORE was the first Atlas's space launch, and it was the world's first active communications satellite. With time, the Atlas rocket system evolved into a highly successful family of space launchers.

SCORE

First Active Communications Satellite

Fig. 12.36. The first launch of operational Atlas 12D from Vandenberg Air Force Base on 9 September 1959. Photo courtesy of U.S. Air Force.

On 9 September 1959, a crew of the Strategic Air Command (SAC) successfully launched the first operational configuration of the Atlas 12D from Vandenberg Air Force Base to a target near Wake Island in the Pacific. Afterward, the SAC declared the Atlas ICBM operational, thus achieving this status one year earlier than the original projection by the von Neumann committee in 1954.

Atlas Declared Operational

Alternative Subsystems

Fielding the ICBM was considered so critical for national defense that the Air Force insisted on a highly conservative missile design and simultaneously contracted alternatives to

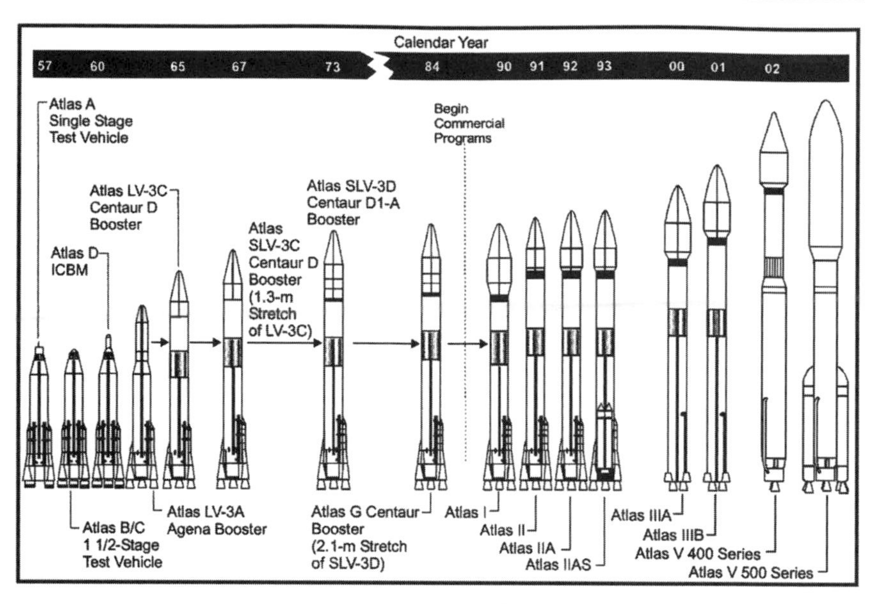

Fig. 12.37. Development of the first American ICBM, Atlas D, evolved into a family of space launchers. Figure courtesy of Lockheed Martin Corporation and International Launch Services.

all major rocket subsystems. For example the backup contract for the Atlas engine was let to Aerojet in January 1955. The alternative contractors had to be different from those involved in the Atlas. In addition, the Air Force decided to integrate these alternative subsystems into another, more sophisticated ICBM as a hedge to the entire Atlas program.

Titan

This new ICBM, called the Titan, relied on more advanced and consequently more risky subsystems. The Titan was conceived as a truly two-stage rocket system, with the second-stage ignition problem already under control. It had a monocoque frame and was designed for a launch from an underground hardened silos. General Schriever believed that the new rocket would achieve superior performance characteristics with the availability of better components and allowing design with a higher technical risk. The approach promised dramatic payoff.

At the Foothills of the Rocky Mountains

The Glenn L. Martin Company with the experience in the Viking and Matador programs was selected to build the Titan. (Two other companies, Lockheed and Douglas, also competed for the contract.) The manufacturing facilities were required to be located away from the coastal areas according to the industry dispersal policy of Eisenhower's administration. An inland site was selected at the picturesque foothills of the Rocky Mountains near Denver, Colorado. Bell Labs was responsible for the radio-inertial guidance system; American Bosch Arma built an alternative inertial guidance system; and Avco developed the reentry nose cone. TRW supported the program by system engineering and technical direction.

The Titan's kerosene-liquid-oxygen (LOX) engines, LR87 (first stage; specific impulse 249 s at sea level) and LR-89 (second stage; specific impulse 311 s in vacuum), were designed and fabricated by Aerojet. The first successful launch of the Titan occurred on 6 February 1959 at Cape Canaveral, followed by three more successful launches in February, April, and May. Then, in a setback, two launches ended in devastating explosions, and two more failed during the second-stage phases of ascent. Successful launches followed, and the deployment of the new ICBM, Titan I, began in 1962. A total of 54 Titan Is eventually entered service and were subsequently phased out in mid-1960s.

Storable Propellants

Titan Is were replaced by Titan IIs, introducing a major innovation. Aerojet converted the Titan I engines to noncryogenic propellants storable at room temperature. The rocket fuel was hydrazine (N_2H_4), or its derivatives such as unsymmetrical dimethylhydrazine (UDMH) and monomethylhydrazine (MMH). A 50-50% (by volume) combination of hydrazine and UDMH, known as Aerozine-50 (A-50), had been determined to be the best fuel by June 1959. Nitrogen tetroxide (N_2O_4), or NTO, served as oxidizer.

Rapid Launch

The new propellants allowed a significant increase of thrust and thrust-to-weight ratio of Titan II stages and reduced the number of active control components in the power plants. Propellant ignition was also simplified by a hypergolic combination. The most important advance of the new storable propellants was, however, in the rapid launch of the ICBM. Typically, it took from 15 to 30 minutes to prepare a Titan I for launch. Titan II reduced this time down to one minute.

Titan Family of Space Launchers

The new family of rockets based on storable propellants, that would include Titan II, Titan III, Titan 34, and Titan IV, had thus been born. All of these Titan rockets would serve as space launchers. Titan II was man-rated and performed

Fig. 12.38. Martin's rocket plant near Littleton, Colorado, mid-1960s. Photo courtesy of Lockheed Martin Corporation.

Fig. 12.39. Titan I launch failure on Pad 16 at Cape Canaveral on 12 December 1959: with the first-stage fuel tank rupturing, the second stage would fall back on the launch pad and explode. The road to the ICBM and to space was bumpy, with many violent and disastrous setbacks during development of powerful rockets. Photo courtesy of 45th Space Wing History Office Archives, Patrick Air Force Base, Florida.

flawlessly launching Gemini spacecraft with astronauts onboard. The role of Titan launch vehicles for manned space flight, however, ended with the cancellation of the military Manned Orbital Laboratory (MOL) program in 1969. Titan IV, launched for the first time on 14 June 1989, became the biggest U.S. expendable space launcher (after termination of the Saturn V in 1970s), specializing in deployment of heavy national security payloads. The last Titan IV launch is planned from Vandenberg Air Force Base, California, in 2005. It is anticipated that the

HYDROGEN-OXYGEN PROPULSION

The main launching systems had reached the capability of deploying satellites in orbit by late 1950s. Rocket dynamics instructs that the efficiency of launch could be significantly improved by higher specific impulse of rocket engines, especially on upper stages. A hydrogen-oxygen propellant achieves the most efficient practical chemical propulsion with specific impulse 450 s in vacuum.

Centaur and RL10

The Navy initiated the work on hydrogen-LOX propulsion for its High Altitude Test Vehicle in 1945–1946. Studies of various aspects of this challenging technology continued at Aerojet, JPL, Ohio State University, and Rocketdyne. NACA's Lewis Flight Propulsion Laboratory in Cleveland, Ohio, also joined the effort in late 1940s and steadily expanded its hydrogen work throughout 1950s under guidance of Laboratory's Associate Director Abe Silverstein. Cryogenic technology got a significant boost in 1950 by a decision to develop an H-bomb, which brought the support of the Atomic Energy Commission. Industry and government laboratories developed efficient technology for liquefying, storing, and transporting liquid hydrogen.

Fig. 12.40. Efficient and reliable hydrogen-oxygen RL10A remains the workhorse of the upper-stage propulsion. The engine allows multiple restarts in space and achieves specific impulse up to 450 s in vacuum. The regeneratively cooled nozzle, developed in late 1950s, is made of stainless-steel tubes brazed together with silver. The nozzle is further expanded by a skirt. Photo courtesy of Pratt & Whitney, A United Technologies Company; all rights reserved.

The focused effort on the H_2/LOX rocket engine began in 1958 at Pratt & Whitney Division of the United Aircraft Corporation. Earlier in 1956, the Air Force contracted Pratt & Whitney to develop a turbine engine, the 304, fueled by hydrogen for a CL-400 aircraft being built by Kelly Johnson at Lockheed's Skunk Works. The CL-400, conceived as the follow-on to the U-2 aircraft and known as the Suntan project, was cancelled in 1958. The developed technology of pumping liquid hydrogen at temperature 20 K proved, however, critically important for future space systems.

In August 1958, the Air Force authorized General Dynamics – Astronautics to build the Centaur, conceived as an upper stage for the Atlas. A brainchild of GD–Astronautics's Krafft A. Ehricke, the Centaur was based on two Pratt & Whitney's liquid-H_2/LOX RL10 engines, providing 15,000 lbf (66.7 kN) thrust each. The first full engine firing of the RL10 took place in July 1959. The engine achieved specific impulse at least 420 s. The tests were conducted at a new 10 square miles research facility established in a remote area near the Everglades in Florida. Finally, the first successful space launch of a rocket with the upper-stage RL10-based Centaur, performing flawlessly, took place on 27 November 1963. More than 350 of these remarkably reliable RL10s have since been flown.

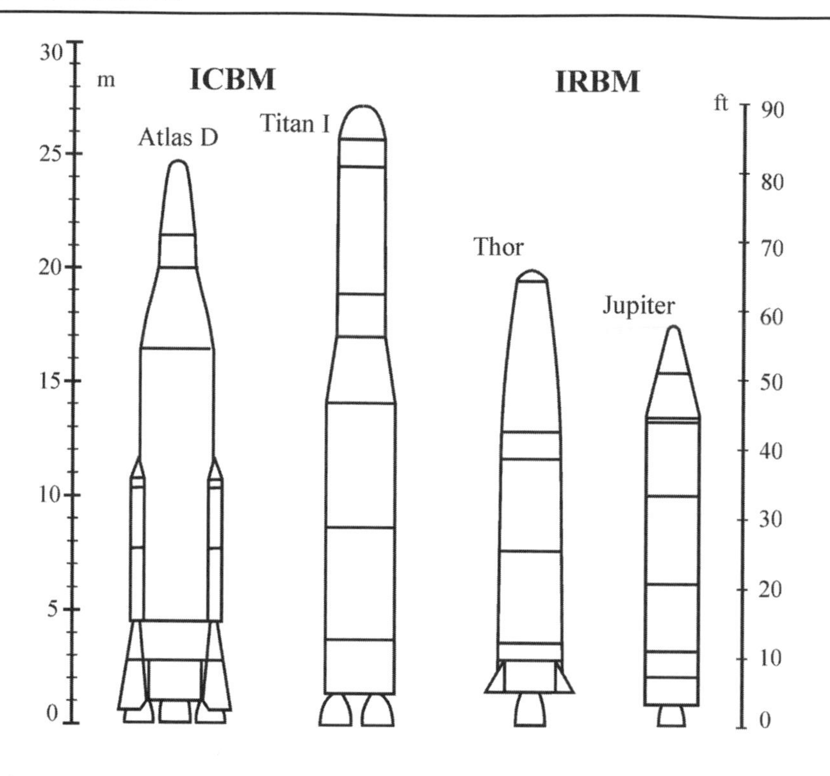

Fig. 12.41. First American ICBMs, Atlas D and Titan I, and IRBMs, Thor and Jupiter. The Air Force's Atlas, Titan, and Thor evolved into families of space launchers. Courtesy of Mike Gruntman.

new Atlas V and Delta IV rockets will provide the required heavy launch capabilities after that time.

Thor IRBM

In addition to Atlas and Titan, the Air Force assigned in December 1955 the highest priority to an intermediate-range ballistic missile, Thor, with a 1500-mile (2400-km) range. The Thor was rushed into development to provide — by deploying the missiles in Great Britain — a temporary counterweight to the possible deployment of Soviet ICBMs before the Atlas and Titan would be ready. The first Thor contract was let in December 1955 and in a record time — less than three years later — the first missile squadron was deployed in Great Britain.

The Thor program was initiated by the Air Force BMD study performed by Robert Truax and Adolph Thiel. The Vienna-born "Dolph" Thiel worked on the V-2 at Peenemünde and was a member of the von Braun's group at Redstone. (Adolph Thiel was not directly related to head of the V-2 engine development Walter Thiel who was killed in the British air raid on Peenemünde in 1943.) Subsequently he joined the R-W Company in 1955.

Two Rival IRBMs

Thus, two IRBMs, Thor and Army's Jupiter, were simultaneously under development in intense interservice rivalry. To complete the Thor program in a record time, the Air Force borrowed and adapted major missile subsystems from the Atlas. Another important requirement, the planned deployment overseas, limited the size of the Thor that had to be transported by the Douglas C-124 Globemaster.

CONTRACTORS OF THE FIRST AMERICAN LIQUID-PROPELLANT ICBMs (ATLAS AND TITAN) AND IRBMs (THOR AND JUPITER)

	ATLAS	TITAN	THOR	JUPITER
Prime Contractor (airframe)	CAD	Martin	Douglas	ABMA CC
Propulsion	NAAR	Aerojet	NAAR	NAAR
Guidance				
Radio-inertial	GE BC	BTL RRU	BTL	Ford
All-inertial	ABA	ACSP IL	ACSP	Ford
Reentry Vehicle				
Heat sink	GE,Avco			
Ablative	GE,Avco	GE,Avco	GE,Avco	GA,GE,Avco

ABA = American Bosch Arma Corporation, Garden City, New York
ABMA = Army Ballistic Missile Agency, Huntsville, Alabama
ACSP = AC Spark Plug Division, General Motors, Milwaukee, Wisconsin
Aerojet = Aerojet General, Sacramento, California
Avco = Avco, Inyokern, California
BC = Burroughs Corporation, Paoli, Pennsylvania
BTL = Bell Telephone Laboratories Allentown, Pennsylvania
CAD = Convair Astronautics Division, General Dynamics San Diego, California
CC = Chrysler Corporation, Ballistic Missile Division Detroit, Michigan
Douglas = Douglas Aircraft Company Manhattan Beach and Santa Monica, California
Ford = Ford Instrument, Long Island City, New York
GA = Cincinnati Test Laboratories, Goodyear Aircraft Corporation Cincinnati, Ohio
GE = General Electric Corporation, Pittsfield, Massachusetts
IL = Instrumentation Laboratory Massachusetts Institute of Technology Cambridge, Massachusetts
Martin = Martin Company, Denver, Colorado
NAAR = North American Aviation, Rocketdyne Division Canoga Park, California
RRU = Remington Rand UNIVAC, St. Paul, Minnesota

Fig. 12.42. The first Thor being prepared for launch at pad 17B at Cape Canaveral on 27 January 1957. Almost immediately after the liftoff, the oxygen valve failed, and the missile fell back and exploded, causing significant damage to the pad. Photo courtesy of 45th Space Wing History Office Archives, Patrick Air Force Base, Florida.

It was anticipated that one could actually fire a Thor missile in 15 minutes after the moment of the "go" signal.

Douglas Aircraft in Santa Monica

The Douglas Aircraft Company in Santa Monica, California, was selected as the prime contractor for the Thor with the liquid-propellant engines (RP-1 and liquid oxygen) supplied by Rocketdyne. Thor's propulsion system, the MB-3, was derived — by leaving only one engine — from a system developed for the Atlas. The reentry nose cone was built by General Electric, whereas the inertial guidance system was designed by MIT's Instrumentation Laboratory and built by AC Spark Plug.

The first Thor's launch was attempted on 27 January 1957. The rocket rose less than 2 ft before falling back and exploding on the pad. Three more failures followed before the launch on 20 September 1957 was declared satisfactory.

Thor-Able

With the addition of the second stage, the Thor evolved into a vehicle for testing reentry of nose cones and into a successful space launcher. Its importance for the space program dramatically increased after the launch of the Soviet Sputnik and the resulting requirement to accelerate development of American satellites. The first Thor's second stage was the Able, an upgraded second stage of the Navy's space launcher Vanguard. The Able was based on Aerojet's AJ10-40 engine with UDMH and nitric acid propellants. The Thor-Able combination was used for tests of nose cone reentry. A space launch with the modified Able was attempted on 11 October 1958 but failed. The lost Pioneer 1 spacecraft was equipped with scientific instrumentations and designed to reach the Moon. The spacecraft achieved, however, a record altitude of 70,700 miles (113,800 km) before falling back to Earth.

Record Altitude

Another Thor's second stage was the Agena powered by a rocket engine using inhibited red fuming nitric acid and kerosene propellants. The Agena was Lockheed's modification of the engine originally developed by the Bell Aircraft as a detachable thruster for the B-58 Hustler aircraft; the engine achieved specific impulse 268 s. The Thor-Agena combination was of an exceptional importance because one of its primary goals was deploying top-priority Corona photoreconnaissance satellites. The first Corona launch, announced as Discoverer I for cover purposes, failed on 28 February 1959. Numerous malfunctions during launches followed before achieving complete success, including payload recovery, in 1960.

Thor-Agena

In April 1959, NASA let the contract to Douglas Aircraft to provide launch vehicles for space missions. This was the first such contract to a private company by the recently created NASA. The new two-stage rocket, named Delta, used Thor as the first stage. The second stage was based on Aerojet's AJ10-118, using similarly to Able UDMH and nitric acid. This was the beginning of the highly successful family of space launchers, the "workhorse" of NASA programs. Today, this family that can be traced back directly to the Thor, includes Delta II in various modifications and the recently introduced Delta III. Delta IV, although carries the same name, is an entirely new vehicle with the different hydrogen-oxygen propulsion system, RS-68, developed by Rocketdyne.

Delta Family of Space Launchers

Merges and acquisition significantly changed the "namescape" of rocket industry. Titan's prime contractor, the Martin Company, merged with Marietta in 1961, forming Martin Marietta. Convair became Space System Division of General Dynamics in 1954, known as General Dynamics—Astronautics. Martin Marietta acquired General Dynamics' Space System Division in 1995 and then merged in the same year with Lockheed, forming The Lockheed Martin Corporation. Thus both, the Atlas and Titan families of space launchers ended up under the same corporate roof. Another important component of Lockheed Martin's rocket assets is the submarine-launched solid-propellant Tridents. Boeing added to its Minuteman missiles the Delta family of space launchers after acquiring McDonnell-Douglas in 1997.

Changing "Namescape"

While large liquid-propellant intercontinental missiles dominated the public view, solid-propellant rocketry also rapidly advanced in 1950s. Industrial companies that were engaged in solid rockets in early 1950s included Thiokol, Aerojet, Hercules Powder Company (Hercules also managed government-owned Allegany Ballistic Laboratory for the Navy), Atlantic Research Corporation, Grand Central Rocket Company (acquired by Lockheed in early 1960s), and Phillips Petroleum (Rocketdyne acquired rocket operations of Phillips — Astrodyne, Inc. — in 1959 and established Solid Rocket Division in McGregor, Texas, which was subsequently sold in 1978 to Hercules).

Solid-Propellant Ballistic Missiles

Solid-propellant ballistic missiles clearly had numerous advantages in mass production, deployment, and operations. For the Navy, solid rockets also offered a solution to a problem of an unacceptably slow, for ship launch, initial acceleration characteristic of liquid-propellant rockets. Arming submarines with long-range nuclear-tipped ballistic missiles promised awesome deterrent capabilities.

Navy Abandons Jupiter

At first, the Navy contemplated a concept of a liquid-propellant IRBM, the Jupiter. Actually, the Jupiter was initially a joint project with the Army Ordnance in a mutually beneficial arrangement. The Army got an important ally in its fight with the main missile rival, the Air Force, while the Navy succeeded in entering into the highly contested field of ballistic missiles. The Navy subsequently withdrew from the Jupiter collaboration and concentrated on the much more promising solid-propellant IRBM. The Air Force also embarked on exploring the technology of large-size solid-propellant rocket motors by initiating its own development, the *Air Force Large Solid Rocket Feasibility Program* (AFLRP), in December of 1955. These new programs would ultimately lead to the Navy's IRBM Fleet Ballistic Missile Polaris and Air Force's ICBM Minuteman.

Air Force Large Solid Rocket Feasibility Program (AFLRP)

Navy's rocket work dramatically accelerated with the appointment of Admiral Arleigh Burke as Chief of Naval Operations on 17 August 1955. The struggling Fleet Ballistic Missile (FBM) program was immediately elevated to the list of Navy's top priorities. To build a new revolutionary weapon system in the shortest possible time, a new organizational and management structure was required. Thus, the establishment of the *Special Projects Office* (SPO) followed three months later. (SPO continues to direct Navy's ballistic missile programs until this day.)

Special Projects Office (SPO)

Fig. 12.43. Rear Admiral William "Red" Raborn, Jr., 1905–1990, headed the Special Projects Office that developed in a record time the world first submarine-launched IRBM weapon system Polaris. Photo courtesy of U.S. Navy.

A naval aviator Rear Admiral William "Red" Raborn, Jr. received extraordinary powers as head of SPO. Raborn's "hunting license," issued by Chief of Naval Operations, opened many doors and helped to quickly resolve numerous bureaucratic problems. "Time was critical," recalled Raborn, "and money and other things were secondary. This didn't mean we wasted our money, but we did make quite a dent in the Navy pocketbook" (Raborn 1972, 63). Admiral Burke described in 1957 this unique situation "that

A Dent in the Navy Pocketbook

RABORN'S "HUNTING LICENSE"

2 December 1955

MEMORANDUM

1. It is quite evident that we must move fast on this fleet ballistic missile and that our present schedules for shipboard launching not good enough ... .

3. ... If [head of Special Projects Office] Rear Admiral Raborn runs into any difficulty with which I can help, I will want to know about it at once along with his recommended course of action for me to take. If more money is needed, we will get it. If he needs more people, these people will be ordered in. If there is anything that slows this project up beyond the capability of the Navy Department we will immediately take it to the highest level ... .

4. The Air Force has got a tremendous amount of enthusiasm which they demonstrate behind their [ballistic missile] project and we must have even more ... .

Arleigh Burke
Chief of Naval Operations

there is no restrictions on money and people. He [Raborn] has a blank check. He is the only man in the Navy that has a blank check" (*Hearings* 1958, 779).

Blank Check

To streamline development of exceptionally complex weapon systems, the SPO was organized outside of the traditional Navy's management structure. SPO performed a significant part of technical direction and system engineering in-house, relying on an abundance of technically educated naval officers. *Vitro Laboratories* provided support to SPO in system design, safety, and interface definitions.

Daunting Challenges

The challenges facing the Navy's IRBM were daunting. Three major technological advances were needed to achieve the Polaris: 1) lightweight reentry vehicle with a nuclear warhead with the desired yield; 2) propulsion system with specific impulse of 230 s; and 3) all-inertial guidance system six times lighter than the state-of-the-art system being built at that time for the Jupiter. In addition, problems of thrust termination and thrust vector control had to be solved, and the nozzle materials capable of withstanding

EDWARD TELLER AND POLARIS

Edward Teller was a consultant to Nobska [study on antisubmarine warfare conducted by the National Academy of Sciences at the Navy's request at Nobska Point in Massachusetts in summer 1956], and made a simple observation [regarding the Navy's FBM program] "Why use a 1958 warhead in a 1965 weapon system?" He went on to indicate that radically smaller and lighter warheads should be available on a compatible time scale to the submarine development. He could not spell these out in detail, but produced historical evidence of the trends in warhead dimensions ... . The germ of the whole idea — the thing which turned the FBM [Polaris] from a [unacceptably large and heavy] monster into a weapon ... — was in Teller's remark.

William F. Whitmore, Lockheed Missiles and Space Division (Whitmore 1961, 263)

Fig. 12.44. Dr. Edward Teller (far left, with the badge "escort required") and Vice Adm. Hyman G. Rickover (far right) at the controls of the nuclear-powered Fleet Ballistic Missile submarine USS *Sam Houston*, summer 1960. Dr. Teller vigorously pursued development of compact high-yield nuclear weapons, an enabling technology advancement toward a realistic submarine-launched ballistic missile. Admiral Rickover served as Director of Naval Reactors from 1949–1982, providing major impetus behind development of nuclear-powered surface ships and submarines, including Polaris submarines. Photo courtesy of General Dynamics' Electric Boat Division and Lawrence Livermore National Laboratory.

hot and highly corrosive exhaust gases of solid propellants had to be found. Similarly to the Air Force's ICBM, the concurrent development of FBM's subsystems was essential. In addition, the specifics of the submarine-launched ballistic missile (SLBM) Polaris brought a number of extraordinary technical challenges beyond those facing the missile itself.

Conference at Nobska

At the summer (1956) study sponsored by the National Academy of Sciences at Nobska, Massachusetts, Edward Teller predicted, based on the historic trend, the significant reduction of the weight of thermonuclear warheads. What was needed was, in essence, a new lightweight warhead integrating in one design the nuclear charge with the reentry cone to substantially reduce the warhead volume and weight. The argument whether such a warhead was feasible ensued between Teller and the present scientists from Los Alamos. Teller firmly stood by the statement that his recently established and still struggling Livermore National Laboratory in California could build a small nuclear warhead with the desired yield in five years.

At the urgent request of Admiral Burke, the Atomic Energy Commission confirmed Teller's predictions on 4 September 1956. (The detonation of an experimental nuclear device in July 1957 demonstrated the validity of the projections.) At the same time, a thorough study conducted by the Naval Ordnance Test Station also came in favor of the weapon system concept, which would later evolve into the Navy's first FBM. Nuclear-powered naval surface ships and submarines were becoming available at that time, championed by Admiral Hyman G. Rickover. Nuclear-powered submarines offered a promise of unprecedented autonomy and worldwide reach. All of these developments, combined with the

Converging Technology Advances

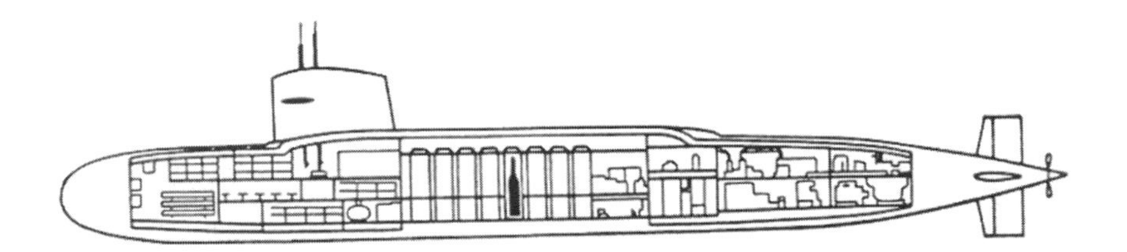

Fig. 12.45. Solid-propellant IRBM Polaris is being loaded into the USS *George Washington* (SSBN 598). The first Polaris (A1) was 28.5 ft (8.7 m) long, had mass 28,000 lb (12,700 kg), and achieved range 1200 miles (1900 km). The 598 class Polaris submarines were 380 ft (116 m) long with the submerged displacement of 6700 metric tons and crew of 10 officers and 100 men. The submarines were powered by water-cooled nuclear reactors and carried 16 Polaris missiles. Photo courtesy of U.S. Navy.

Fig. 12.46. German chemist Karl Klager, 1908–2002, worked in the research laboratories of *I.G. Farbenindustrie* during the war. The Navy brought Klager to California through the *Paperclip* program in 1948. At Aerojet, Klager developed the technology for mass industrial production of rocket fuel unsymmetrical dimethylhydrazine. Since 1954, he had become actively involved in solid-propellant motors. Photo courtesy of Aerojet-General Corporation.

independent assessment provided by Lockheed and Aerojet of the missile and of the propulsion system, respectively, gave confidence that a compact solid-propellant submarine-launched IRBM was possible. Secretary of Defense Wilson approved the fleet ballistic missile *Polaris* on 8 December 1956.

USS *George Washington*

The time needed to build a new type of submarine from scratch was unacceptably long. Therefore, the first nuclear-powered FBM submarines (class 598) were redesigned nuclear attack subs of the Skipjack class. The first Polaris submarine, the USS *George Washington*, was constructed from the USS *Scorpion* already laid at the General Dynamics Corporation's Electric Boat Division at Groton, Connecticut. The keel was cut in two, and an additional 130-ft (40-m)-long midship section was inserted for ballistic missiles. The USS *George Washington* was launched on 9 June 1959.

Fig. 12.47. Charles B. Henderson, b.1929, was the head of the group at Atlantic Research Corporation that proved in 1954 the effective utilization of aluminum as a high-performance solid-propellant ingredient. Henderson subsequently became the company's senior vice president for research and technology and stayed with Atlantic Research until his retirement in 1988. Photo (ca.1960) courtesy of Charles B. Henderson.

Aerojet's Solid Propellants

The future of solid-propellant ballistic missiles depended on the ability to fabricate large solid-propellant grains with high-performance characteristics. By 1953, Thiokol had succeeded to produce a record size, 31 in. (79 cm) in diameter, polysulfide-perchlorate grain with star-shape perforation. Then in late 1954, Aerojet initiated a crash program under the scientific lead of Karl Klager to develop a new composite propellant based on polyurethane. Aerojet, which worked on polyurethane since 1947, quickly adopted the recent discovery by the Atlantic Research Corporation that large amounts of the aluminum powder could be added to the grain, significantly boosting specific impulse. As a result, the superior specific impulse of 235 s had been achieved. Aerojet's team also showed that an addition of aluminum effectively suppressed

Aluminum Additive

SOLID-PROPELLANT TECHNOLOGY AND POLARIS

Advances in solid-propellant rocket technology greatly accelerated with the invention of composite propellants by von Kármán's group in 1942, followed by the introduction of Thiokol polysulfide binder. Specific impulse of solid motors improved to 210 s but was still insufficient for a compact submarine-launched fleet ballistic missile.

In late 1954, Aerojet initiated a crash program under Karl Klager to develop a more efficient composite propellant based on polyurethane. By the end of 1955, the new propellant combination had replaced the widely used Aeroplex (polyester as the fuel and binder and ammonium perchlorate as the oxidizer) in most of Aerojet's rocket units and demonstrated superiority to the Thiokol's polysulfide-based propellants. The 1950s witnessed the emergence of two main families of binders, polyurethane and polybutadiene. Hydroxy-terminated polybutadiene (HTPB) remains in use until this day.

It was known for some time that addition of a few percent of an aluminum powder to a rocket grain improved specific impulse. Chemical reactions of aluminum oxidation generated a large amount of heat, which translated into higher energy of the exhaust flow. In early 1955, the *Atlantic Research Corporation* (ARC) achieved a breakthrough in the effective utilization of aluminum as a high performance propellant ingredien. (Atlantic Research Corporation was formed in 1949 and excelled in small precise solid motors and gas generators for various rockets and spacecraft. Today, the company produces various propulsion systems.)

Under the contract to Navy's Bureau of Ordnance, a group headed by Charles B. Henderson demonstrated that, contrary to the accepted practice, aluminum could constitute 15–20% of the propellant grain, with the resulting significant gain in specific impulse. Proof testing involved static firings of 5-in. centrally perforated solid motors with experimental aluminum-rich propellants and measuring thrust and pressure as a function of time. Specific impulse values of 235 s were reliably achieved at a mean chamber pressure of 800 psi (54 atm). Henderson's group at Atlantic Research included chemist Erling Rosholdt and Evan Lawrence and was actively supported by company's President Arch Scurlock and Vice President Keith Rumbel.

The principal technical contact at the Bureau of Ordnance, Elliot Mitchell, briefed Aerojet's Klager on the breakthrough as soon as the improved specific impulse was demonstrated in static firings at Atlantic Research. Aerojet promptly adapted the innovation and soon also discovered that an addition of aluminum, especially in the form of a fine powder, effectively suppressed combustion instabilities. The most visible drawback of aluminum addition was increased exhaust smoke. Thiokol's polysulfide-based propellant could not adopt the use of aluminum additives because chemical reactions during storage led to emission of gases that could cause explosion.

The Navy contracted Aerojet to build propulsion systems for both stages of the first fleet ballistic missile Polaris. Aerojet achieved a specific impulse of 231 s in the first operational missile. (Modern solid motors perform at specific impulse 290–300 s in vacuum.)

In the next Polaris model, the *Hercules Powder Company* provided the second stage with the rotatable nozzle for thrust vector control and the higher-energy double base (nitrocellulose and nitroglycerin) propellant that also included aluminum powder and ammonium-perchlorate oxidizer. Double-based propellants could detonate and were thus ruled out at the initial stages of the Polaris program as too dangerous. As experience grew, double-based propellants gradually gained acceptance.

ALLEGANY BALLISTIC LABORATORY

Allegany Ballistic Laboratory (ABL) made important contributions to the technology of solid propellants. A section of the National Defense Research Committee headed by Clarence N. Hickman established ABL in the early 1944 to accommodate growing work in solid rockets. ABL was organized as a government-owned and contractor-operated research and development facility; it was located on the territory of the recently deactivated Pinto Branch of the Allegany Ordnance Plant in the eastern West Virginia. Ralph E. Gibson (from Carnegie Institution) and Alexander Kossiakoff (from Catholic University of America) directed the new establishment. After the war, Gibson and Kossiakoff joined the Bumblebee Program at the Applied Physics Laboratory and subsequently became APL directors in 1948–1969 and 1969–1980, respectively.

ABL excelled in the technology of high-energy double-based propellants. In particular, the Laboratory demonstrated addition of aluminum and ammonium perchlorate to the matrix of double-based propellant, thus significantly increasing propellant performance. Hercules Powder Company that managed ABL at that time for the Navy built highly efficient second stages for Polaris A2 and A3 missiles based on this new propellant combination.

combustion instabilities. The competing Thiokol's polysulfide-based propellants could not use aluminum because chemical reactions during storage led to emission of gas, which caused consequent explosions.

The Navy selected Lockheed Aircraft Corporation's Missile System Division in California as the prime contractor for the two-stage all-solid-propellant Polaris with the range 1200 miles (1900 km). The vehicle length was 28 ft (8.5 m), and its loaded mass was 28,000 lb (12,700 kg). Aerojet won the contract for the missile propulsion system.

Thrust Termination

The precise missile range, and ultimately its accuracy, was determined by the exact velocity of the rocket at the engine cutoff. In contrast to liquid-propellant engines, such as on the Atlas, one could not terminate thrust of a solid rocket by simply closing the valves supplying propellants to the combustion chamber. Combustion of the solid-propellant grain could not be stopped at will. A new approach to thrust termination had therefore been developed. Six pyrotechnically activated ports were installed at the front of the Polaris's

PROPELLANT DENSITY

Although liquid-propellant rocket engines exhibit a specific impulse superior to solid-propellant motors, there is another important missile characteristic — the propellant density. A denser propellant allows reduction of the volume of the propellant tanks and correspondingly of the missile size. The large propellant density was especially important for the Polaris because the missile had to be stored, transported, and launched from a submarine. Smaller missiles thus directly led to much smaller submarines and other related savings.

Solid propellants usually have relatively high density, or specific gravity, ~1.8 g/cm^3 (0.065 lb/in.^{3}). This density is, for example, a factor 25 higher than that of liquid hydrogen. In addition to specific impulse, another special term is thus introduced to characterize the properties of propellants, *density specific impulse*, equal to the product of the propellant average density and specific impulse.

Fig. 12.48. Crewmen work in the far end of the section housing ballistic missiles on one of the first Polaris submarines. The space is at high premium on submarines, making dense high-performing solid propellants critically important for submarine-launched missiles. Photo courtesy of Aerojet-General Corporation.

second stage. These ports were opened by a command at the desired moment, precipitously reducing pressure in the combustion chamber and effectively terminating thrust.

Thrust Vector Control

The new concept of thrust vector control had also to be proved. Lockheed's Willy Fiedler, a veteran of the German V-1 development, introduced the jetavator, a special surface inserted in the exit flow in a controllable way. The jetavator was a ring that could be rotated past the rim of the nozzle, partially blocking the exhaust flow and making it asymmetric. The rotatable nozzle was introduced a few years later on the second stage of the Polaris A2 to control the thrust vector. Then in December 1961, the United Technology Center demonstrated liquid-injection thrust vector control. This technique was based on injecting a gas or liquid asymmetrically into the exhaust flow of the rocket engine and producing asymmetry of the propellant flow and correspondingly of the thrust vector.

INGREDIENTS OF SUCCESS OF THE POLARIS SYSTEM

Admiral Raborn summarized the ingredients of the success of the Polaris at the hearings before the House's Committee on Science and Astronautics on 28 July 1959:

> Now, fortunately, about this time we were acquainted with the splendid work of the Atomic Energy Commission in getting a very large-yield atomic warhead in a small package. Concurrently with that, the Navy had an unusual success in its research efforts with the Atlantic Research Corp. to materially improve the specific thrust of the boost which we could get out of solid propellants. And the [nuclear-powered submarine USS] *Nautilus*, of course, was in existence, and it seemed a natural that we attempt to tie the three together in a very usable weapons system, and this was done [in the Polaris system].

Optimized Flight Path

Ballistic solid-propellant missiles could ascend in a flight path different from those typical for liquid-propellant rockets. The liquid rockets usually lifted off with a small acceleration and maintained a small angle of attack to avoid lateral structural loads in the dense ambient atmospheric air. In contrast, the solid rockets were structurally stronger and could employ higher acceleration. Aerojet's John Fuller showed that the higher initial acceleration and larger angles of attack allowed a significant increase in the missile range.

Submarine Launch

Demonstration of rocket launch from a submerged submarine was not simple. A hydrodynamically unstable missile had to be ejected with a sufficient velocity over the ocean surface, from a range of submarine operational depths. The first simplified rocket ejection test with compressed gas was successfully performed on a wharf at the Naval Shipyard in San Francisco: a redwood log imitated the rocket. An ejection of a full-size Polaris model from a submerged launch tube succeeded on 23 March 1958, at the underwater test range near San Clemente Island off the California coast. The Westinghouse Electric Corporation built the launch system for the Polaris.

Inertial Guidance

Beginning with the Polaris, Doc Draper's Instrumentation Laboratory designed guidance systems for Navy's fleet ballistic missiles. The Laboratory usually built and flight tested prototypes. Then, the industrial contractors built the guidance systems for operational missiles. For the Polaris, the Ordnance Department of General Electric was selected as a guidance system subcontractor.

Fig. 12.49. First-stage solid-propellant rocket motors for FBM Polaris A3 manufactured at Aerojet-General's plant in Sacramento, California. The 2000th unit is being packed (upper-left corner) for shipping to the Navy's Pacific Missile Facility in Bangor, Washington, where it would be integrated into an operational missile. Photo courtesy of Aerojet-General Corporation.

Launching Point Position

The precision with which a ballistic missile hits a target could not be better than the knowledge of the position of the launching point. For land-based missiles, the coordinates of launching sites were precisely known in advance. In contrast, a launch from a submerged submarine required precise knowledge of the continuously changing position of the submarine.

Ship's Inertial Navigational System (SINS)

Precise submarine navigation was among the most difficult challenges of the Polaris program. The Autonetics Division of the North American Aviation developed a specialized Ship's Inertial Navigational System based on a new type of gyroscope. The remarkable capabilities of the emerging navigation technology were demonstrated by the voyage of the USS *Nautilus* to the North Pole under the Arctic ice in 1958 and the USS *Triton* on the underwater cruise around the world in 1960. Many components of Autonetics' SINS were originally developed for and inherited from the Navaho intercontinental cruise missile program.

Absolute Fix and Space-Based Navigation

An inertial navigation system accumulated errors with time and required periodical fixes, or resets, of the submarine absolute coordinates. Initially, a Polaris submarine was required to conduct such resets every 8 h. Several specialized devices were carried by the submarines for such fix-taking, including a "Type 11" periscope (for ship heading), terrain-matching sonar (potentially betraying submarine presence), and Loran-C receiver with the trailing antenna. Signals from the Navy's Transit satellites, when they were deployed in 1960s were subsequently used for absolute fix. (The Transit program was remarkably successful and was officially retired in 1996.)

Fig.12.50. Richard B. Kershner, 1913–1982, first supervisor of the Space Department at the Applied Physics Laboratory. The Transit was the first spacecraft designed, built, and launched (1959, 1960) by APL. In Kershner's words, the Transit "was the original raison *d'être* for the [APL's Space] Department." Almost 60 APL spacecraft would follow over the next 45 years, demonstrating various technologies — such as (to name a few) yo-yo despin, magnetic torquing, and gravity-gradient stabilization — that are widely used today. Photo courtesy of Applied Physics Laboratory, Laurel, Maryland.

Polaris First Launch

Polaris rockets went through the extensive development program, with more than 25 test launches. Finally, the first Polaris was successfully launched on a full range by the submerged USS *George Washington* on 20 July 1960. The submarine fired the second Polaris three hours later. The Navy has thus firmly established its role in strategic deterrence.

On 6 May 1962, the USS *Ethan Allen* launched a Polaris missile from a submerged position in the Pacific. The missile hit "right in the pickle barrel" at the end of its 1000-plus miles trajectory at Christmas Island and detonated the nuclear warhead

TRANSIT AND NAVIGATION

The first Earth artificial satellite, the *Sputnik*, was launched in October 1957. The satellite transmitted continuous signal at a fixed frequency. Two specialists at the Applied Physics Laboratory, William Guier and George Weiffenbach, received Sputnik's transmissions Doppler shifted in frequency caused by the satellite motion. Encouraged by Frank McClure, Guier and Weiffenbach showed that the satellite orbit could be determined from Doppler observations.

Obviously, the process could be reversed: one should be able to obtain, from Doppler measurements, the position of an observer from the signals transmitted by a satellite in a known orbit. APL's McClure and Richard Kershner quickly designed the essentials of the satellite navigation. Soon, APL was busy, funded first by ARPA and then by Navy's SPO, developing the Navy Navigation Satellite System (NNSS) Transit. The system was especially attractive for submarines because it relied on measurements of frequency and time, both quantities independent of submarine motion.

Fig. 12.51. Weights and center of gravity test on Transit satellite, 28 Match 1960. Photo courtesy of 45th Space Wing History Office Archives, Patrick Air Force Base, Florida.

The first Transit 1A satellite, the first spacecraft built by APL, failed to deploy in the proper orbit in September 1959. The successful Transit 1B was launched on 13 April 1960. The uncertainty in the knowledge of the Earth's shape and mass distribution, and correspondingly of its gravitational field, translated into uncertainties of the evolution of satellite orbits. Consequently, first Transit satellites were used to perfect the knowledge of the Earth's gravitational field in order to achieve reliable predictions of satellite orbits. At the same time, NASA's Goddard Space Flight Center pursued similar approach with NASA satellites: thus a new field of research known as *dynamic geodesy* has been born.

The goal of precisely determining Earth's gravitational field had been achieved by 1964. Navigation at sea with the accuracy of better than 0.1 mile became possible, with the inherently unpredictable air drag responsible for the remaining uncertainty. (Note that the exact location of Australia was uncertain to several thousand meters in 1959.) The first space navigational system, the Transit, has thus become operational. The continuous Navy's interest, since the launch of the first Sputnik, in achieving perfect worldwide navigation would ultimately lead to the modern space-based global positioning system (GPS).

Fig. 12.52. Navy's Fleet Ballistic Missile Polaris A1 a few seconds after launch at the Air Force Missile Test Center at Cape Canaveral, Florida. Photo courtesy of Aerojet-General Corporation.

with an estimated yield of half a megaton. This was the only test of the American strategic missile with a real nuclear explosion. The Soviet Union conducted a similar test much earlier, in 1956, launching the land-based R-5M (SS-3) IRBM on a 1200-km (750-miles) distance with the live nuclear warhead. People's Republic of China would conduct a similar land-based ballistic missile nuclear test in 1966.

Nuclear Tests

With the development and deployment of the liquid-propellant Atlas and Titan ICBMs firmly on track, the Air Force focused on the solid-propellant ICBM. The development of this system, the *Minuteman*, got under way in 1958. The Boeing Company was selected as the prime contractor. The subcontractor team included Autonetics Division of North American Aviation, Inc. (guidance system); Avco Manufacturing Corporation (nose cone); Thiokol Chemical Corporation (propulsion system); and Hercules Powder Corporation (propellant).

Solid-Propellant ICBM Minuteman

The Minuteman ICBM was designed to be housed in dispersed hardened silos. The missile was smaller and cheaper than the liquid-propellant Atlas and Titan, required a smaller silo and simplified maintenance, demonstrated improved readiness, and dramatically reduced the number of people required for missile launch. The first flight test of the new missile took place at Cape Canaveral on 1 February 1961. The first squadron of Minuteman ICBM became operational in October 1962.

"Better" ICBM

By the early 1960s, the photoreconnaissance Corona satellites had accurately counted Soviet ICBMs. This intelligence breakthrough allowed the United States to time the deployment of its ballistic missile deterrence in response to the Soviet ICBMs. In particular, it became possible to avoid — without jeopardizing the national security — a very costly buildup and full-scale deployment of the existing and largely obsolescent liquid-propellant ICBMs, Atlas and Titan, and rely instead on the forthcoming much-more convenient solid-propellant Minuteman. This delay effectively ended the age of the American liquid-propellant ICBM almost as soon as it began. (A limited number of the liquid-propellant Titan IIs remained in service until 1987.) In contrast, the Soviet Union perfected the technology of handling storable noncryogenic propellants and maintains its families of liquid-propellant ICBMs until this day.

Deployment of Obsolescent ICBM Avoided

The advancement of solid-propellant technology also led to the first all-solid space launcher, the Scout. The missile was jointly developed by the Air Force and newly established NASA as a "poor man's rocket." The Scout was a four-stage space launcher with all four stages traceable to various military programs. The first test launch took place at Wallops Island on 1 July 1960. The Scout launch site at Wallops was used for launching small spacecraft in low-inclination orbits, whereas the Vandenberg Air Force Base on the West Coast served as a launch site for polar orbits. In total, 112 Scouts were launched, the last one in 1994. Nine missions were launched from the Italian San Marco platform off the Kenyan coast near equator. The Scouts were manufactured by Vought Astronautics (later became the LTV Aerospace and Defense Company).

First Solid-Propellant Space Launcher

13. ROAD TO SPUTNIK

Marxism emphasizes transformation of the society on the basis of the scientific understanding, revealed obviously to the true Marxists only, of the world. Consequently the communist government of the newly established Soviet Union elevated support for science and technology in the new socialist state. Rocketry was no exception.

First Soviet Rocket Lab

The first rocket laboratory was organized in Moscow by the Soviet military for Nikolai I. Tikhomirov (1859–1930) as early as 1921. Tikhomirov, an inventor and national authority on sugar production, first became interested in rockets in 1890s. His work focused on solid-propellant missiles, particularly using a new smokeless powder, cordite, as propellant. The first rockets, built by Tikhomirov, went into testing in 1925.

Gas Dynamics Laboratory (GDL)

The expanding activities of Tikhomirov were reorganized into a *Gas Dynamical Laboratory* (GDL) and moved to Leningrad. (Leningrad was known as St. Petersburg before World War I. The city's name was changed in the middle 1990s back to the original St. Petersburg after collapse of the communist regime in the Soviet Union.) GDL was formally established on 15 May 1929, and its work was supervised by the Military Scientific Research Committee of the Revolutionary Military Council of the USSR. The laboratory was located in downtown Leningrad, with the offices in the Admiralty building and the test station and mechanical workshop in the Peter-and-Paul Fortress. The laboratory's staff of almost 200 initially concentrated on solid-propellant missiles and powder boosters for heavy aircraft.

A graduate of the Leningrad State University Valentin P. Glushko (1908–1989) joined GDL in 1929. Glushko began his work with initiating studies in electric propulsion, the new technique that he had first investigated as a university student. In 1930, Glushko became an energetic leader of the development of liquid-propellant rocket engines. The first Glushko's liquid rocket engine, ORM (the Russian acronym for *Experimental Rocket Motor*), was tested in 1931. Valentin Glushko would become with time the leading developer, virtually a monopolist, of Soviet high-thrust liquid-propellant rocket engines.

Groups for Study of Jet Propulsion (GIRD)

In the 1920s the Communist Party established a *Society for Assistance to Aviation and Chemical Industry* (*Osoaviakhim)* in line with its policy to totally control all aspects of life and to militarize the country. (Osoaviakhim was later transformed into the *Volunteer Society for Assistance to Army, Aviation, and Navy — DOSAAF.*) Osoaviakhim organized a number of *Groups for Study of Jet Propulsion* (*Gruppa Izucheniya Reaktivnogo Dvizheniya*, or GIRD) in Moscow, Leningrad, and other cities in 1931. GIRD groups united rocket enthusiasts who built models, held exhibitions, and popularized rocketry. As a result, a new center, independent of GDL, of the Soviet rocket establishment has emerged as a part of the growing state-sponsored military effort.

Fig. 13.1. Soviet rocket pioneer Valentin P. Glushko in 1931. Glushko would become the leading designer of high-thrust liquid-propellant rocket engines. Photo courtesy of NPO Energomash, Khimki, Russia.

Moscow GIRD

Perhaps the most advanced GIRD group was in Moscow. A Soviet rocket pioneer, Fridrikh A. Tsander (1887–1933), was the first head of the Moscow GIRD, and he was later succeeded by Sergei P. Korolev (1906–1966). The Moscow GIRD consisted of four teams led by Tsander, Korolev, Mikhail K. Tikhonravov (1900–1974), and Yurii A. Pobedonostsev (1907–1973).

Tikhonravov worked on propulsion systems and designed the first Soviet hybrid rocket, *GIRD-09*, that used liquid propellant. Tsander concentrated on liquid-propellant engines, including the first entirely liquid-propellant rocket *GIRD-X* (*X* stands here for the Roman numeral 10).

JULES VERNE AND TSIOLKOVSKY

Valentin Glushko belonged to those boys whose imagination was ignited by Jules Verne's *From the Earth to the Moon.* The novels of Jules Verne, wrote Glushko, "shook me. The reading took away my breath, my heart pounded, and I was ecstatic and happy. It became clear that I have to devote all my life, without exception, to realization of these wonderful flights" (*Odnazhdy i Navsegda* 1998, 21).

The 15-year-old Glushko sent a letter to Konstantin E. Tsiolkovsky. He asked Tsiolkovsky for his publications that Glushko could not find in the libraries in his provincial home town Odessa in the southern Ukraine. The great Russian space visionary kindly replied and sent the publications. Half a year later, in 1924, Glushko wrote to Tsiolkovsky, "... regarding how much I am interested in interplanetary communications [travel] I can say you only that this is my ideal and the goal of my life, which I want to devote to this great purpose" (*Odnazhdy i Navsegda* 1998, 40).

Fig. 13.2. Sergei P. Korolev in the early 1930s. Korolev would become the leading Soviet rocket and satellite designer, achieving the first ICBM and placing the first artificial satellite and the first man into orbit. Korolev's last name is sometimes rendered from Cyrillic as *Korolyov*. Photo courtesy of Keldysh Research Center, Moscow.

Pobedonostsev studied powder rockets and jet engines and built a supersonic testing facility. Korolev was initially more interested in rocket-powered planes. He worked on a rocket plane, *RP-1*, that combined the glider designed by Boris Cheranovsky and the Tsander's rocket engine.

First Hybrid-Propellant Rocket

The first Soviet rocket that used liquid propellant was launched on 17 August 1933 at the Nakhabino test site near Moscow. The rocket, GIRD-09, was of a hybrid type. The oxidizer, liquid oxygen, was injected into the combustion chamber that contained 1.0–1.5 kg of the fuel, solidified gasoline. GIRD-09 was designed by Tikhonravov and built by Korolev's team. It was 2.4 m (7.9 ft) long with a mass of 19 kg (42 lb), including 5 kg (11 lb) of the propellant. The engine produced thrust between 245 and 320 N (55 and 72 lbf); the combustion chamber pressure ranged from 5–6 atm (73–88 psi). The missile reached a 400-m (1310 ft) altitude in an 18-s flight.

First Soviet Liquid-Propellant Rocket

The next rocket launch of the Moscow GIRD followed on 25 November 1933, at the same Nakhabino testing grounds. The GIRD-X was the first Soviet truly liquid-propellant rocket, and its engine used liquid oxygen and a mixture of alcohol with water. The rocket was designed by Fridrikh Tsander, who died eight months before the launch of the rocket at the age of 46. The GIRD-X was also, as the GIRD-09, built by Korolev's team.

The powerful Soviet military leaders realized early on, similar to their counterparts in Germany, the potential promise of rocketry. The Red Army moved to concentrate a large part of the missile

SOLID–LIQUID HYBRID PROPULSION

Hybrid propulsion systems combine, as the name suggests, the properties and, ideally, the advantages of liquid-propellant engines and solid-propellant motors. Usually, the rocket oxidizer is liquid, while the fuel is solid. Moscow GIRD initiated the work on hybrid propulsion in the early 1930s. In the United States, the Naval Ordnance Test Station conducted similar studies at about the same time. The work on hybrid propulsion systems continues to this day. No large hybrid rockets, including space launchers, have been demonstrated yet.

Fig. 13.3. Launch preparation at the artillery test grounds in Nakhabino some 20 miles (32 km) from Moscow. Researcher Nikolai I. Efremov fills the first Soviet hybrid-propellant rocket with liquid oxygen. Sergei P. Korolev, head of GIRD-Moscow, is on the left, and engineer Yurii A. Pobedonostsev is on the right. The rocket, GIRD-09, was launched on 17 August 1933, and reached an altitude of 400 m (1310 ft). GIRD-X (GIRD-10) would follow on 25 November 1933. Photo courtesy of Keldysh Research Center, Moscow.

PATRON SAINT OF RED ROCKETRY

Marshal Mikhail Tukhachevsky (1893–1937) is usually credited for initiating the large-scale rocket research in the USSR in 1930s and for supporting aviation, airborne operations, and other innovations and new technologies.

Tukhachevsky was a young tsarist officer who joined the Bolsheviks in 1918. This soldier, turned communist, led the Red Army offensive against Poland in 1920 and put down the Kronstadt and Tambov insurrections in 1921.

Tukhachevsky was a good communist whose brutality stood out even among the most ruthless Red commissars. He reportedly used poison gases for killing peasant rebels and prisoners in the death camps in the Tambov region in 1921. Tukhachevsky's devotion did not save him, however, from execution during a purge of the Communist ruling inner circle in 1937.

Jet Propulsion Scientific Research Institute (RNII)

research and development in a major center. On 21 September 1933, Deputy People's Commissar for Army and Navy Marshal Mikhail Tukhachevsky signed an order forming the *Jet Propulsion Scientific Research Institute* (RNII), merging GDL with the Moscow GIRD. The new institute, RNII, was located in the northern part of Moscow. Later RNII would be renamed *Scientific Research Institute N.1* (NII-1), then *Scientific Research Institute of Thermal Processes* (NIITP), and presently *M.V. Keldysh Research Center*.

RNII embarked on a large-scale research and development programs in solid- and liquid-propellant rockets. Among RNII's early achievements were solid-propellant unguided missiles that became famous as a barrage weapon during World War II, the *Katyusha*'s.

Fig. 13.4. GIRD engineers and technicians in 1933. First row, right to left: rocket and engine designer Fridrikh A. Tsander, glider designer Boris I. Cheranovsky, and Sergei P. Korolev. Photo courtesy of Keldysh Research Center, Moscow.

Fig. 13.5. Before launch of the liquid-propellant rocket GIRD-X at the testing grounds in Nakhabino, 25 November 1933. Standing left to right are Sergei P. Korolev, Nikolai I. Efremov, engine designer Leonid S. Dushkin, and engine designer Leonid S. Korneev. This rocket was originally designed by Tsander and built by Korolev's team. Tsander died from typhus earlier in March of that year.

GIRD-X was 2.2 m (7.2 ft) long with a mass of 29.5 kg (65 lb), including 8.3 kg (18.3 lb) of the propellants. Liquid-oxygen oxidizer and 78%-strong ethyl alcohol fuel were pressure fed. The rocket reached an altitude of about 240 ft (80 m) when the engine disintegrated. The combustion chamber pressure reached 10 atm (147 psi), and the specific impulse ranged from 162–175 s. Photo courtesy of Keldysh Research Center, Moscow.

Fig. 13.6. Tomb of a Soviet rocket pioneer Fridrikh A. Tsander in Kislovodsk, Russia. An artistic model of the GIRD-X rocket is at the top right corner of the stone. Tsander was fascinated with interplanetary flight since his childhood years in his native Riga (present Latvia). He graduated as engineer and went to work to Moscow and eventually became the head of the Moscow GIRD group. Tsander remained devoted to spaceflight through all his life. Even the names of his children were influenced by his passion: daughter Astra and son Merkuri. When on vacations at a resort at the North Caucasus, Tsander contracted typhus and died in 1933. The inscription on his tomb reads: "Pioneer of the Soviet Rocketry; Enthusiast of Interplanetary Flight; Fridrikh Arturovich Tsander; 1887-1933." Photo courtesy of Viktor Soloviev, Moscow, Russia.

(*Katyusha*, the nickname of the Soviet solid-propellant missile M-13, literally stood for an affectionate diminutive of the Russian girl's given name equivalent to *Katherine*.)

Katyusha

The M-13 projectiles, or Katyusha's, were 5.1 in. (132 mm) in diameter and 6 ft (1.8 m) long. The projectile mass was 92.5 lb (42 kg), including a 48.2-lb (21.9-kg) explosive warhead. The rocket range reached 3 miles (4.8 km). Katyusha missiles proved to be highly reliable and were used for many years after the war had ended. The missile also became a favorite heavy weapon of assorted Soviet-sponsored guerillas during the Cold War. Islamic and Palestinian militants had used the Katyusha's derivatives up to the late 1990s in attacks on Israel's towns and other civilian targets.

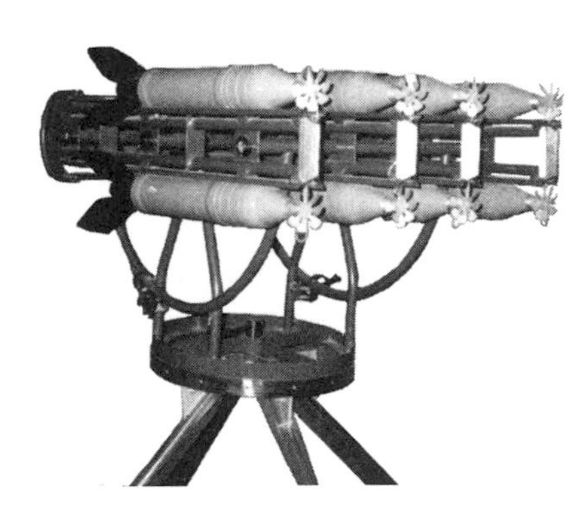

Fig. 13.7. Katyusha's little sister: smaller 3.3-in. (82-mm) rockets M-8 developed by the USSR during World War II. The mobile launcher, called *Stalin Organ*, could fire 30 to 48 such 17.5-lb (8-kg) rockets to a distance of up to 3 miles (4.8 km). A later missile model had mass 31 lb (14 kg) and a larger range. Photo courtesy of Mike Gruntman.

Rocketeers Shot and Imprisoned

During the purges in the late 1930s, many leading Soviet rocketeers were arrested and interrogated, with confessions often extracted by tortures, and then shot or imprisoned. RNII director Ivan T. Kleimenov and his deputy Georgii E. Langemak were shot in January 1938. Glushko was arrested in March 1938 and

"POST-SHOT" RECOGNITION

In June 1991, on the 50th anniversary of the first combat action of the M-13 Katyusha missiles, RNII's Ivan T. Kleimenov and Georgii E. Langemak, both executed in 1938, were posthumously given the highest nonmilitary state award of the USSR, the title of the Hero of Socialist Labor (that included the Medal of the Hero of Socialist Labor and the Order of Lenin), in recognition of their important contributions to the development of Soviet rocket weapons.

HIGHEST DECORATIONS

The *Order of Lenin* was the highest state decoration of the USSR. The titles of *the Hero of the Soviet Union* (for exceptional military exploits) and *the Hero of Socialist Labor* (for exceptional nonmilitary accomplishments) included special golden star medals and the Order of Lenin.

sentenced, after more than a year in prison, to eight years of hard labor. Korolev was arrested in June 1938 and, in three months, sentenced to 10 years of hard labor.

Sharashka

Those Soviet scientists and engineers who happened to be most lucky not to be tortured to death, executed, or sent to labor camps, ended in special prisons — *Sharashka*'s. Sharashka, an extraordinary invention of the dedicated and ruthless social engineers, combined a prison and research and design establishment. Thousands of imprisoned specialists worked in various Sharashka's, which at least provided some hope of survival. A very different fate awaited those millions sent straight to concentration labor camps, where malnutrition, excessive hard work, savage living conditions, starvation, and abuse by the guards took a tremendous toll. Many millions perished without a trace on this glorious march to the socialist paradise.

APPEALS WENT UNANSWERED

... The interrogators ... of [the People's Commissariat of Internal Affairs] NKVD by beatings and abuses forced me to write [and sign] the cooked-up intentionally false statements, which I declared to be false even before the trial ... . However, all my statements [about my false confessions extracted by the beatings that I had sent to the People's Commissar of Internal Affairs N.I. Yezhov and Prosecutor-General A.Ya. Vyshinsky] as well as my statement [about my false confessions] at the trial went unanswered and I was convicted to 10 years in prison, being absolutely innocent.

Sergei P. Korolev, in the Letter to the Chairman of the Highest Court of the USSR, 10 November 1938

... I was arrested on March 23, 1938, and was subjected to physical and moral pressure by the interrogating apparatus of [the People's Commissariat of Internal Affairs] NKVD; as a result of this violence I was forced to sign the protocol of my interrogation, with the nonsense, fictitious content.

Valentin P. Glushko, in the Letter to the Deputy People's Commissar of Internal Affairs, L.P. Beria, 7 December 1938.

After spending eight months in a transit prison in the town of Novocherkasks, Sergei Korolev landed in a labor camp in the harsh Kolyma region in the North-Eastern Siberia. Korolev was lucky. By a quirk of fate, he was plucked out from this land of no return and sent to a Sharashka directed by an airplane designer Andrei Tupolev. Tupolev, the designer of the famous family of the *Tu* airplanes, was a prisoner himself.

Rocketeers in Sharashka

In November 1942, Korolev was transferred to Sharashka in the town of Kazan' headed by another prisoner, his former RNII colleague Valentin Glushko. Both Glushko and Korolev were finally released in 1944.

Soviet Rocket Development Is Back

As the war was coming to an end, the Soviet Union had reestablished its rocket development activities. The successes of German scientists and engineers, particularly the development of the V-2 and jet aircraft, were a proof of the value and promise of the new propulsion technologies. The emerging atomic weapons made long-range missiles, even with a limited accuracy, especially important for the future warfare. The Soviet Union promptly began a massive effort to learn and evaluate the German rocket technology sending numerous specialists to the occupied Germany. The American interest in rocketry, with the first V-2 launched at the White Sands Proving Ground on 10 May 1946, also did not pass unnoticed. The American launch was open to the press, and the *Life* magazine did a spread on the shot.

Decree of 13 May 1946

The Soviet leaders acted swiftly. On 13 May 1946, the USSR Council of Ministers issued a special decree N. 1017-419 "Matters of the rocket weapons," signed by the Soviet dictator Joseph Stalin. This decree established the structure of the Soviet rocket and space establishment for many years to come. Five weeks later, another decree, N.12866-525, formed the top secret *Design Bureau 11* (KB-11) at the site *Arzamas-16* near the city of Gorkii. KB-11 headed by Yulii B. Khariton was tasked to develop Soviet nuclear weapons.

GERMAN AND SOVIET ROCKETEERS

Wernher von Braun was arrested and spent two weeks in a prison in an internal strife among the rival factions of the Nazi Germany. Two other Peenemünde rocketeers, Klaus Riedel and Helmut Gröttrup, were arrested at the same time and later also released. Von Braun's troubles, however, did not come even close to the sufferings of the Soviet rocketeers.

Many Soviet scientists and engineers, though absolutely loyal to the Soviet State, enthusiastic about the socialist paradise they were building, and devout Communist Party members themselves who had earlier, in the 1920s and early 1930s, cheerfully approved the extermination of millions of the "enemies of state" were now themselves arrested, tortured, murdered, imprisoned, and banished, after token trials or by executive order. The fate of these specialists was not any different from that of many millions of others, annihilated by that time by the enthusiastic socialist state (being aided, abetted, and admired, one should add, by numerous leftists in the West).

Many of those who survived the ordeal carried the fear through the rest of their lives, instilling slavish attitudes to the following generations of their children and grandchildren. Notably, some even preserved the cherished belief in the socialist ideas as the highest achievement of the human race.

COUNCIL OF MINISTERS OF THE USSR
DECREE N. 1017-419 (top secret, special file)
13 MAY 1946, MOSCOW, THE KREMLIN

Matters of the rocket weapons.

... the Council of Ministers of the USSR decrees:

... to establish the Special Committee on the rocket technology ... .

... all works ... performed in the rocket technology are controlled by the Special Committee on Rocket Technology. No departments, organizations, or persons can, without a special permission by the Council of Ministers, interfere in or inquire about the works on the rocket weapons.

... to establish the top priority goal to reconstruct, using domestically available materials, the rockets V-2 (long-range guided rocket) and Wasserfall (antiaircraft guided missile).

... to assign a top priority to the following tasks in the work on the rocket technology in Germany:

a) complete reconstruction of the technical documentation and samples of the long-range guided rocket V-2 and antiaircraft guided missiles Wasserfall, Rheintochter, Schmetterling;

b) reconstruction of the laboratories and test facilities with all equipment and instruments necessary for research and experimentation with the rockets V-2, Wasserfall, Rheintochter, Schmetterling and others;

c) training of the Soviet specialists who would master the design of V-2, antiaircraft guided missiles, and other rockets, methods of rocket testing, technology of manufacturing of the components and parts, and assembly of the rockets.

... to send to Germany the necessary number of specialists in various technical fields to learn the rocket weaponry, attaching the Soviet specialists to each German specialist to acquire the experience.

... to form in Germany a special artillery unit to learn and master the preparations and launching of the V-2 rockets.

... to transfer the Design Bureaus [established by the USSR in the occupied Germany] and German specialists from Germany to the USSR by the end of 1946.

... to allow ... the increased salaries for the German specialists brought to the work on rocket technology.

... to work out a plan to send a special Commission to the USA to place orders and purchase equipment and instruments for the laboratories of rocket scientific-research institutes, ... allocating 2 million dollars.

... to develop a plan for location and construction of the State Central Testing Range for rocket weapons.

... to provide plans on the increased salaries for the specially qualified specialists in the rocket technology.

... to arrange training of engineers and researchers in the rocket technology in the institutions of higher learning and universities ...

... to select ... 500 specialists, retrain them, and send them to work ... on the rocket technology.

Foundations of the Soviet Rocket and Space Establishment

FIRST ROCKET UNIT

On 15 August 1946, the Soviet Army formed a Special Brigade in the occupied Germany. The Brigade was tasked to master the servicing and launching of the German ballistic missile V-2. This was the first Soviet Army unit of the emerging Strategic Rocket Forces, which would be formally organized by the Decree of the USSR Council of Ministers on 17 December 1959.

Comprehensive Development Plan

The decree "Matters of the rocket weapons" provided a comprehensive plan marshalling enormous resources for the development of rocket weaponry. Various ministries were ordered to assign their numerous research, development, and manufacturing facilities to the new rocket effort. Typically for a totally controlled socialist society, the decree specified even the allocation of such minute items as food rations, transport, and living quarters. The plans to establish a dedicated rocket testing range and to form and train specialized military rocket units were drawn. The institutions of higher learning were instructed to prepare engineers and scientists for the emerging rocket technology.

Origins of Strategic Rocket Forces

The decree of 13 May 1946 required quick reconstruction, using the Soviet-designed and built parts and components, of advanced German weapons, particularly the long-range ballistic missile V-2 and the antiaircraft guided missile Wasserfall. A number of research institutions and design bureaus have since been specifically established for this task, mostly in Moscow and the nearby areas. Many aviation and artillery plants have been converted into the centers of rocket development.

Fig. 13.8. Valentin P. Glushko in 1970. Glushko's design bureau, OKB-456, later known as *NPO Energomash*, built powerful and reliable rocket engines that placed the first Soviet spacecraft, Sputnik, and the first cosmonauts in orbit. Photo courtesy of NPO Energomash, Khimki, Russia.

Kapustin Yar Missile Test Range

The Soviet Army formed the first specialized ballistic missile unit, a Special Brigade, in the occupied Germany with the task to master the launching of V-2s. A year later in 1947, the first large-scale missile test range, the *State Central* [Rocket] *Testing Range N.4,* was established near the settlement of Kapustin Yar on the eastern bank of the Volga river 70 miles (110 km) southeast from Stalingrad (Volgograd). Kapustin Yar would be the main center of Soviet missile testing for the next 10 years.

First V-2 Launched

The Special Brigade was transferred from Germany to Kapustin Yar in August 1947 and launched the first German A-4 (V-2) from the new range on 18 October 1947. Eleven A-4 rockets, assembled in Germany

HELPING HAND OF KGB AND GRU

Soviet rocket and space programs continuously benefited from the help of intelligence services. Actually, scientific-technical espionage in rocketry dates back to the days of the Russian Empire. The archives describe first successes in this area already in 1835 when the Russian Ambassador in Paris obtained "blueprints and description of a new type of incendiary rockets" (Primakov 1996, 147).

The recently published semiofficial history of the KGB's foreign intelligence (the publication was edited by Evgeny M. Primakov, who served consecutively as head of the Russian Foreign Intelligence, Foreign Minister, and Prime Minister in 1990s) describes the first achievements of rocket espionage in the United States:

> One of the first Soviet illegals on the American continent was an intelligence officer known only by his pseudonym, Charlie. His name [his file] has not been preserved in the archives of the [KGB] intelligence service. In 1938 he was recalled to the Soviet Union and repressed [i.e. executed] and his file was likely destroyed. ... [This] intelligence officer succeeded in the early 1930s in obtaining a report by an American scientist [R.H.] Goddard 'On the results of the work on development of a liquid-propellant rocket engine.' The document was presented to [Marshal Mikhail] Tukhachevsky who highly praised it. (Primakov 1997, 174)

While Germany kept its fast-growing ballistic missile development top secret, the KGB quickly learned about the program through its agent who served as a midranking official in the German Secrete State Police Gestapo. (This Soviet agent was uncovered and shot in 1942.) In November 1935, the agent reported development of solid-propellant rockets for delivery of chemical weapons. In addition, he identified Wernher von Braun and described that "in the forest, in a remote part of the [Kummersdorf] test range, permanent stands are erected for testing rockets working with the use of liquids [propellants]."

The report of the KGB agent was forwarded to the Soviet dictator Joseph Stalin and Marshal K.E. Voroshilov in December 1935 and then passed to Marshal M. Tukhachevsky in January 1936. The head of the Soviet military intelligence, who was briefed about German liquid-propellant rockets, asked the KGB for further information:

> Rockets and jet projectiles. a) Where does engineer Braun work? What does he work on? Are there possibilities to infiltrate his laboratory? b) Are there any possibilities to contact other workers in this area? (Primakov 1997, 344).

The rockets and space technology remained in the focus of the Soviet intelligence for the years to come. It was a sample of a solid rocket propellant and recruiting of a Thiokol's engineer in 1959 that launched a stellar career of Oleg D. Kalugin, who became the youngest general in the history of the KGB. Solid propellants are also prominently present in the recently published memoirs of Soviet intelligence officers.

Circumventing restrictions on export of military technologies established by the *Coordinating Committee for Multilateral Export Controls* (COCOM) was another key task of the KGB and especially of the military intelligence GRU. For example a large chamber for thermal-vacuum tests of spacecraft was obtained through a Japanese company that "ordered a chamber in Europe, shipped it to Japan, and there substituted it by a locally built [fake] chamber. The embargoed chamber was then delivered to the USSR, where Japanese specialists installed it" (Maksimov 1999, 108). Soviet intelligence officers even organized in Japan the development, based on the restricted Western technology, and manufacturing of photographic film with the desired characteristics for Soviet reconnaissance satellites.

GLUSHKO OR TSANDER?

Fridrikh Tsander performed first tests of a prototype of the liquid-propellant rocket engine in 1931, about five years after the launch of the first Goddard's liquid rocket. Glushko tested his ORM engine also in 1931. The question who built the first truly liquid-propellant rocket engine in the USSR, Glushko or Tsander, would remain a highly contested issue for many years.

and in the USSR from the original German components, were launched in October and November of that year. Five launches out of 11 were successful.

Khimki

Two areas in the immediate vicinity of Moscow, *Khimki* and *Podlipki*, have emerged as the major centers of the Soviet rocket and space establishment. A *Special Design Bureau N.456* (OKB-456) was organized on the outskirts of Moscow in the town of Khimki, on the main road to the Moscow's future international airport Sheremet'evo. The head of OKB-456, Valentin Glushko, made this design bureau the leader in development of Soviet high-thrust liquid-propellant rocket engines. Today the bureau is known as *NPO Energomash.*

Air and Missile Defenses at Khimki

Khimki is also a home of the *S.A. Lavochkin Science-Production Association* (*Lavochkin NPO*) that existed since 1937 as an aircraft design bureau. Originally called the *Special Design Bureau N.301* (OKB-301) and headed by Semyon A. Lavochkin (1900–1960), it was best known for the *La* family of Soviet fighter planes, such as La-5 and La-7, during World War II and first jet fighters in the late 1940s. It was the Lavochkin design bureau that built the aircraft that first reached the speed of sound in the USSR. After the war, OKB-301 also expanded into development of antiaircraft weapons and air defense systems, beginning with the reproduction of the German Wasserfall.

The surface-to-air missile (SAM) work was taken over in 1953 by a new independent establishment, the *Fakel Machine-Building Design Bureau,* also in Khimki, that grew out of Lavochkin's OKB-301. (The word *fakel* means *plume* in Russian.) Fakel, headed for many years by Pyotr D. Grushin, also pioneered Soviet rockets for antiballistic missile defense. The first rocket intercept of a ballistic missile warhead ("to hit a bullet with a bullet") was achieved in March 1961 at the Saryshagan test range by the Fakel-designed and built V-1000 rocket. Lavochkin NPO eventually took over development of interplanetary spacecraft from Korolev's design bureau in 1965.

U-2 OVER PODLIPKI AND KHIMKI

The second photoreconnaissance overflight of the Soviet Union by the U-2 aircraft took place on 5 July 1956. This was the only U-2 mission that flew over Moscow. Glushko's rocket engine plant in Khimki and Korolev's rocket development center in Podlipki were among the mission primary targets. Both locations were covered by clouds, and no photographs were taken.

Podlipki

The other area on the Moscow outskirts, Podlipki, has grown to a special prominence in Soviet rocket development. (The word *Podlipki* literally means *under the linden trees* in Russian.) Podlipki was about 15 miles (25 km) north-

northeast from the Moscow downtown. Within a few years, it became the area of the highest concentration of the Soviet rocket and space establishment.

NII-88, TsNIIMash, and TsUP

The Decree of 13 May 1946 established a new secret rocket research center, *Scientific Research Institute N.88*, or NII-88, about 2.5 km (1.5 miles) from the suburban railroad station Podlipki. Sergei Korolev was appointed the chief designer of long-range rockets at NII-88. Korolev's group was reorganized into a *Special Design Bureau N.1* (OKB-1) of NII-88 in 1950. NII-88 was subsequently renamed the *Central Scientific Research Institute of Machine Building*, or TsNIIMash. TsNIIMash became the leading Soviet research and certification institution in rocket and space technology. It is also a home of the Russian, former Soviet, spaceflight control center (*Tsentr Upravleniya Polyotami*, or TsUP).

SCALE OF THE KOROLEV'S ROCKET PLANT

Andrei D. Sakharov, one of the leading creators — "fathers" — of the Soviet thermonuclear weapons, wrote in his memoirs about a trip in 1953 "through a ballistic missile plant where I met Sergei Korolev, the chief designer, for the first time. We [in the nuclear weapons program] had always thought our own work was conducted on a grand scale, but this was something of a different order. I was struck by the level of technical culture: hundreds of highly skilled professionals coordinated their work on the fantastic objects they were producing, all in a quite matter-of-fact, efficient manner" (Sakharov 1990, 177).

Korolev's Design Bureau OKB-1

Korolev's design bureau became independent of NII-88 in 1956 when his OKB-1 was combined with the nearby *Plant N.88* to form the *Experimental*

Fig. 13.9. Entrance to the town of Korolev, former Kaliningrad, also known as Podlipki; November 1999. Podlipki remains the major center of the Soviet, and now Russian, rocket and space establishment. Photo courtesy of Mike Gruntman.

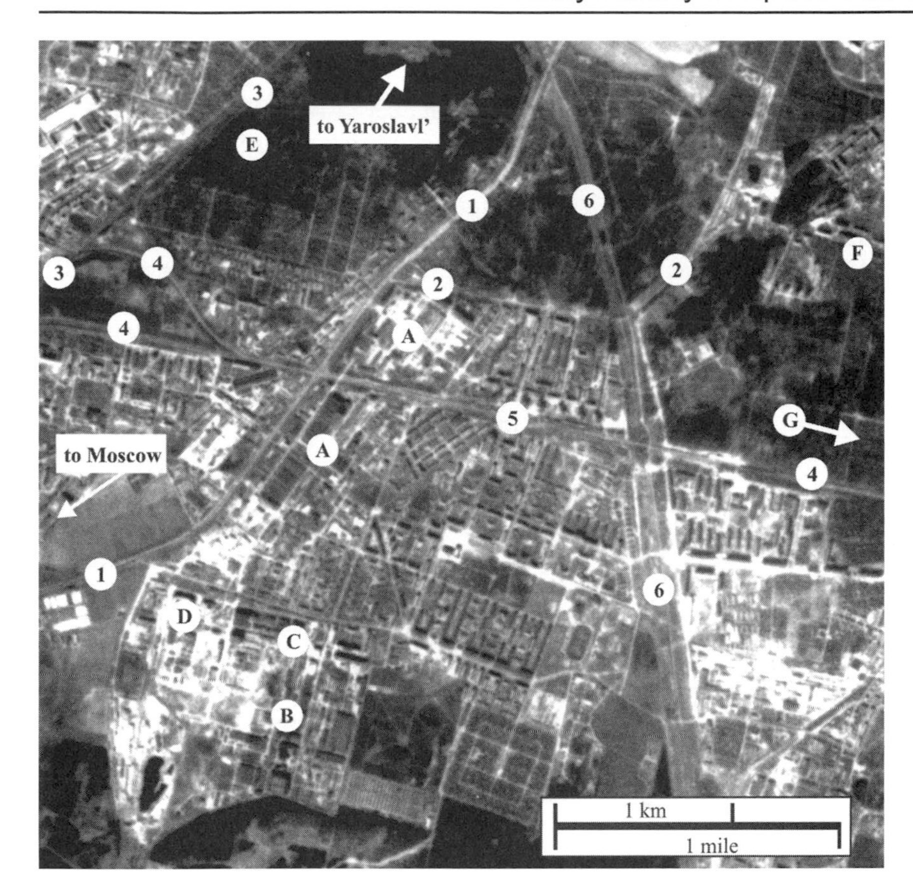

Fig. 13.10. *Corona* (Mission #1116) satellite photograph (22 April 1972) of the area with the highest concentration of the Soviet space research and development centers at *Podlipki* (after the name of the suburban train station), 25 km (15 miles) north-north-east from the Moscow downtown. Podlipki is also known as the town of *Kaliningrad*, recently renamed *Korolev*.

1) Highway Moscow-Yaroslavl'; 2) Bolshevo road; 3) railroad Moscow-Yaroslavl'; 4) railroad branch to Podlipki, Bolshevo, etc.; 5) suburban train station Podlipki (suburban trains reach the terminal station at the Moscow center in 30 minutes); 6) underground aqueduct with the restricted-access area above the ground. A) *Central Design Bureau of Experimental Machine Building* (TsKBEM), or the S.P. Korolev's Design Bureau, later known as *RKK Energia*; B) *Central Scientific Research Institute of Machine-Building* (TsNIIMash), with the *Space Flight Control Center* (TsUP) on its territory; C) *A.M. Isaev Design Bureau of Chemical Machine-Building* (KB Khimmash); D) *Scientific Research Institute of Measuring Techniques* (NIIT); E) *Moscow Technical Institute of Forestry* (MLTI, or *Lestekh*), originally an educational institution for the forest industry with a number of departments to train engineers, particularly in electronics and control, for Podlipki's space centers; F) *Scientific Research Institute N.4* (NII-4; in Bolshevo) of the Strategic Rocket Forces, Ministry of Defense; this institute would later split into NII-4 and the new co-located *NII Kosmos* of the Soviet Space Forces; G) direction toward the railroad stations Bolshevo (1 mile), Chkalovskaya (10 miles) with the *Yu.A. Gagarin Cosmonaut Training Center* at the nearby *Zvezdnyi Gorodok* (Star City), and Monino (15 miles), a home of the *Air Force Academy*.

Courtesy of Mike Gruntman.

Design Bureau N.1. After Korolev's death in 1966 the bureau was renamed *the Central Design Bureau of Experimental Machine Building*, or TsKBEM. Today, it is known as the *Rocket Space Corporation Energia* (RKK Energia).

Podlipki's Constellation of Rocket and Space Centers

Several other leading rocket and space research and development centers were set up in Podlipki, some spun off NII-88. The *A.M. Isaev Design Bureau of Chemical Machine Building* (KB Khimmash) is located next to TsNIIMash. KB Khimmash, named after its first director Aleksei M. Isaev, was initially a part of NII-88 and became an independent developer of liquid-propellant rocket engines in 1956. Another offspring of NII-88 is the *Scientific-Manufacturing Association of Measuring Techniques* (NPO IT). In 1966, NPO IT separated from NII-88 as the *Scientific Research Institute of Measuring Techniques* (NIIT). The main military research centers of the strategic rocket forces (NII-4) and of the space forces (NII Kosmos) are only one mile away from the Podlipki station.

Podlipki, Kaliningrad, and Korolev

A town grew up in the surrounding area to house almost 200,000 inhabitants, with the majority employed by rocket and space establishments of Podlipki's cluster. In the Soviet times, the town was called *Kaliningrad*, after one of the most trusted Stalin's henchmen, Mikhail I. Kalinin. It was renamed *Korolev* after

Fig. 13.11. Space flight control center (*Tsentr Upravleniya Polyotami*, or TsUP) in TsNIIMash, Podlipki (Kaliningrad, or Korolev), November 1999. A small model of the space station *Mir* is in the bottom-right. TsUP was formed on 3 October 1960 as the *Computational Center* of NII-88. The center was subsequently reorganized as the *Coordinating-Computational Center*, and finally as TsUP. It was this center that was routinely mentioned in the official announcements of the official Soviet TASS news agency about space launches: "Coordinating-Computational Center is processing the incoming [from the spacecraft] information." Deep-space missions were operated from the control center at Yevpatoria in the Crimea, a peninsula in the Black Sea. Now the Crimea is a part of the independent Ukraine. The Space Arm of the Strategic Rocket Forces operates the main military space control center in Golitsyno-2, or Krasnoznamensk, 15 miles west of Moscow. Golitsyno-2 is supported by a dozen command, control, and communication stations throughout Russia. Photo courtesy of Mike Gruntman.

SCIENTISTS AND ENGINEERS FOR PODLIPKI

Leading Soviet institutions of higher learning trained scientists and engineers for Podlipki's sprawling space complex.

The *Bauman Moscow Technical School* (Bauman MVTU) and *Moscow Aviation Institute* (MAI) had the branches in Podlipki. The nearby *Moscow Technical Institute of Forestry* (MLTI) established special departments for training specialists in several engineering areas, particularly in electronics and control, for the space centers. A number of senior students of the elite *Moscow Physical-Technical Institute* (MFTI, or *Fiztekh*) studied directly at TsKBEM and TsNIIMash for three years before earning their Master of Science degrees and continued as staff scientists and engineers after the graduation. Many other institutions and universities, including the *Moscow State University* (MGU), trained specialists for Podlipki.

collapse of the communist regime in the early 1990s. The railroad station for suburban trains is still called Podlipki.

First Ballistic Missile R-1

One of the top priority goals of the Decree of 13 May 1946, the reconstruction of the V-2 using the Soviet-designed and built parts, components, and materials, had been achieved by 1948. Korolev's team at NII-88 had designed and built such a rocket, the R-1. The rocket development required cooperation among a large number of the newly established Soviet research institutions and manufacturing plants. The rocket engine was built by Glushko's OKB-456; the inertial guidance system was produced by the *Scientific Research Institute N. 885* (NII-885) headed by Nikolai A. Pilyugin; the command and control equipment was developed by the *Scientific Research Institute N.10* (NII-10) led by Viktor I. Kuznetsov; and the launching site complex and fueling equipment were designed and built by the *State Design Bureau of Special Machine Building (GSKB Spetsmash)* headed by Vladimir P. Barmin. This early cooperation would remain the backbone of the Soviet rocket and space program for many years. In total, 13 research institutes and design bureaus and 35 factories and plants contributed to the development of the first Soviet long-range ballistic missile.

Core Team of the Rocket and Space Program

Fig. 13.12. Sergei P. Korolev in early 1960s. Korolev became the leader of the early Soviet rocket and space effort, while remaining unknown to the most of his countrymen. In the secrecy-obsessed totalitarian society, the Soviet people knew him only as an enigmatic "chief designer." Korolev's design bureau in Podlipki near Moscow was the major center of the early Soviet space effort and spun off several rocket and spacecraft research and development centers. Photo courtesy of S.P. Korolev Memorial House-Museum, Moscow.

Steep Learning Curve

The first R-1 was launched at Kapustin Yar on 17 September 1948. The rocket strayed away from its trajectory by 51 deg and crashed. Out of nine rockets, only one performed nominally and reached the target. This successful launch took place on 10 October 1948.

The learning curve was steep: 17 rockets out of 20 performed nominally in the next series of tests in the fall of 1949. One year later the R-1 was accepted by the Soviet Army for deployment under the military designation 8A11. The United States designated the new Soviet ballistic rocket as SS-1 or *Scunner*. (SS stands for surface-to-surface.)

The R-1 was just a beginning in the rapidly progressing development of increasingly powerful ballistic missiles. A new missile, a significantly improved R-1 with the 580-km (360-mile) range, would be the next. The Soviet Army deployed this new missile, the R-2, in 1952.

First Nuclear Missile

Korolev's rocket R-5M (8K51), or SS-3 (*Shyster*), became the first Soviet ballistic missile system capable of carrying nuclear warheads. The R-5M was based on the liquid-oxygen and alcohol propellants inherited from the V-2, and its engine developed specific impulse 215 s at sea level. The missile could reach a distance of 1200 km (750 miles) with a 1300-kg (2870-lb) warhead. On 2 February 1956, the R-5M was launched, for the first time in rocket history, with a live nuclear warhead that was detonated at the target. The same year, the rocket was rushed to deployment by the Soviet Army.

Ballistic missiles have become a major instrument of the Soviet foreign policy supporting expansion of communism worldwide. From now on Soviet nuclear warheads could reach most of the Western Europe. The Soviet Union was quick to use this new capability at the confrontation with Great Britain and France

ROCKET LAUNCH WITH LIVE ATOMIC BOMB WARHEAD

... [On 2 February 1956] the rocket R-5M carried through cosmos for the first time a warhead with an atomic charge. After flying the planned 1200 km, the warhead safely reached the ground in the area of the Aral Karakum [desert]. The impact triggered the fuse and a surface nuclear explosion ushered the beginning of the nuclear-missile era in the history of humankind.

We congratulated each other and consumed all champagne, which was vigilantly guarded in the special canteen ... . Later, when we returned from the [Kapustin Yar] missile range ... Korolev told us in a close meeting a great secret, "Do you know what I was told? This was an 80-kiloton explosion. This was four times greater than [the] Hiroshima [bomb explosion]."

[Head of the institute of radio guidance systems Mikhail S.] Ryazansky somberly joked, "Aren't you afraid that we will be tried one day as war criminals?"

Soon after the first successful launch of the R-5M with the live atomic warhead [S.P.] Korolev and [V.P.] Mishin were awarded the titles of the Heroes of Socialist Labor. Twenty more specialists of NII-88, including myself, received Orders of Lenin. The enthusiasm of all our institution was supported by the government decree to award the Order of Lenin to NII-88.

... It was the real holiday [celebrations] for us ... . The golden stars of the Heroes of Socialist Labor were also given to [V.P.] Glushko, [V.P.] Barmin, [M.S.] Ryazansky, [N.A.] Pilyugin, and [V.I.] Kuznetsov.

Boris E. Chertok (1994, 390)

BRITAIN WOULD CEASE TO EXIST

... If a war broke out ... Britain, which is often called in the West an unsinkable aircraft carrier, would cease to exist on the first day of the war.

Nikita S. Khrushchev, in speech at the General Assembly of the United Nations, 13 October 1960.

during the Suez Crisis in the second half of 1956. The world was entering a new era based on mutual assured destruction.

Korolev's "Empire"

Korolev initiated, led, influenced, and coordinated research and development in many Soviet academic and industrial institutions. His own design bureau, OKB-1, also continuously grew, establishing new programs in various areas of rocketry and space technology. OKB-1 developed new powerful rockets, including the first ICBM and the rockets that launched the Earth's first artificial satellite and the first man in space. Various spacecraft — scientific, communications, and photoreconnaissance — were also pioneered by Korolev's team.

With time, many of the activities, designs, technologies, and people of OKB-1 were transferred to new branches and other organizations. Korolev's "empire" thus grew, with the offsprings becoming independent centers of research, development, and production. This process was also a determined effort to disperse military important facilities throughout the country. With the advent of nuclear weapons, the high concentration of strategic technological, development, and production assets became especially vulnerable.

Fig. 13.13. Mikhail K. Yangel pioneered a family of ballistic missiles with storable, noncryogenic propellants. Yangel's OKB-586 in Dnepropetrovsk, Ukraine, also became a leading Soviet spacecraft development and production establishment. Photo courtesy of the Russian Academy of Sciences.

"Yuzhnoe" in Dnepropetrovsk

In 1951 the serial production of the R-1 was organized at Plant N.586 in Dnepropetrovsk, the Ukraine. The plant was the former *Dnepropetrovsk Automobile Factory* assigned to the rocket effort. Serial production of the new sophisticated weapon system required support by a strong engineering and design group. Consequently, a *Special Design Bureau N.586* (OKB-586) was created in April 1954, with Mikhail K. Yangel (1911–1971) as director.

Mikhail K. Yangel

Yangel worked for some time with Korolev and was director of NII-88 in 1952–1954. OKB-586 was later reorganized into the *Design Bureau "Yuzhnoe,"* and it is now known as *M.K. Yangel State Design Office "Yuzhnoe."* (The word *Yuzhnoe* stands for *southern* in Russian.)

Fig. 13.14. Commemorative plate (in the museum of the RKK Energia) presented to the Korolev's Design Bureau by the Yangel's Design Bureau "Yuzhnoe," Dnepropetrovsk (now in the Ukraine), showing two rocketeers dressed in the Ukrainian national garb. Photo courtesy of Mike Gruntman.

Missiles on Storable Propellants

Yangel's Design Bureau quickly embarked on development of a new type of liquid-propellant rockets. While Korolev excelled in powerful missiles based on liquid oxygen, Yangel concentrated on development of rockets with storable, noncryogenic propellants. Most common oxidizers of such rockets were nitric acid (and its derivatives RFNA and IRFNA) and nitrogen tetroxide, and most common fuels are kerosene and derivatives of hydrazine such as UDMH. Although these propellants were often highly reactive, corrosive, and poisonous, they could be stored at room temperature.

Yangel's R-12 (SS-4)

Yangel's first rocket R-12 (military designation 8K63), or SS-4 (*Sandal*), was launched for the first time in June 1957. The single-stage R-12 with mass 42,000 kg (92,600 lb) at launch had a range of 2000 km (1250 miles) and carried a 1600-kg (3500-lb) nuclear warhead. The liquid-propellant engine used nitric acid and kerosene and produced specific impulse 250 s at sea level. The R-12 also became the first Soviet ballistic missile with an entirely autonomous guidance system.

> **ROCKETS LIKE SAUSAGES**
>
> ... The Soviet Union is not afraid of war!
>
> ...We are not afraid of [the arms race]. We shall beat you [the Western World]! In our country [the USSR] rockets are produced on an assembly-line basis. Recently I visited a plant and saw rockets coming from it like sausages from an automatic machine. Rocket after rocket is coming off our factory lines.
>
> Nikita S. Khrushchev, in speech at the General Assembly of the United Nations, 11 October 1960

Emergence of Soviet Rocket and Space Establishment

Fig. 13.15. Sprawling Soviet rocket and space establishment created in 1950s and early 1960s.

a) *Central Specialized Design Bureau* (TsSKB) and plant *Progress*: photoreconnaissance and remote-sensing satellites, rockets (R-7, Vostok, Molniya, Soyuz); Kuibyshev (now Samara); b) *V.P. Makeyev Design Bureau* (KB Makeyev): SLBMs; Miass; c) *M. F. Reshetnev Scientific Production Association of Applied Mechanics* (NPO AM): communications and navigation satellites; Zheleznogorsk (Krasnoyarsk-26); d) *M.K. Yangel State Design Office "Yuzhnoe"* and plant: rockets (IRBMs, ICBMs, Kosmos, Tsiklon, Zenit) and spacecraft; Dnepropetrovsk, Ukraine; e) *Arsenal Design Bureau*: solid-propellant SLBMs; nuclear-powered radar satellites RORSATs (since 1969); Leningrad; f) *Production Association Polyot*: spacecraft, rocket engines, and rockets (ICBMs and Kosmos); Omsk; g) *Votkinsk Plant*: serial production of solid-propellant ICBMs; Votkinsk; h) *Fakel Experimental Design Bureau* (OKB Fakel), electric propulsion; Kaliningrad (another Kaliningrad in the western USSR, not Podlipki); i) *Pavlograd Mechanical Plant* (branch of "Yuzhnoe"): solid-propellant motors and ICBMs; Pavlograd, Ukraine; j) Kapustin Yar: first missile test range.

Moscow: 1) Podlipki; 2) *M.V. Khrunichev State Space Scientific Production Center* (GKNPTs): spacecraft and rockets (ICBMs and Proton); 3) *Salyut Design Bureau* (KB Salyut): heavy spacecraft and rockets (ICBMs and Proton); 4) *Central Scientific Research Institute Kometa* (TsNII): antisatellite systems, early warning satellites; 5) *Moscow Institute of Thermal Technology* (MITT, A.D. Nadiradze Design Bureau): solid-propellant IRBMs and ICBMs; 6) Khimki; 7) *Scientific Production Association for Machine Building* (NPO Mashinostroeniya), original design bureau and rocket plant of V.N. Chelomei: rockets (ICBMs and Proton); Reutovo; 8) *All-Russian Scientific-Research Institute of Electromechanics* (VNIIEM): meteorological satellites; 9) *M.V. Keldysh Research Center* (RNII, NIITP): research and development in rocket propulsion (including nuclear) and power.

Figure courtesy of Mike Gruntman.

In short, Yangel's IRBM met the requirements of the military. The SS-4 and its modifications became the core of the Soviet IRBM force and remained deployed until the late 1980s. The R-12 rockets were mass produced. They were the rockets that the Soviet leader Nikita S. Khrushchev referred to as coming from factory lines "as sausages."

Korolev's Competitors

The first Soviet IRBM and the first ICBM were achieved by Korolev's team. But it was Yangel's IRBMs that would constitute the overwhelming majority of the Soviet IRBM force in 1960s. Yangel has become a formidable competitor to Korolev. Yangel's design bureau in Dnepropetrovsk and the colocated plant became the main supplier of ballistic missiles for the Soviet Strategic Rocket Forces in 1950s and 1960s, mass-producing R-12 (SS-4), R-14 (SS-5), and R-36 (SS-18). With time, Yuzhnoe designed and built space launchers Kosmos, Tsiklon, and Zenit. In addition, Yangel's establishment grew into a major designer of Soviet satellites, developing probably the largest variety of spacecraft for military and civilian applications.

Fig. 13.16. Vladimir N. Chelomei became the main developer of the Soviet ICBMs in 1960s. His design and production establishment in Moscow substantially contributed to development of space stations and various military space systems. Photo courtesy of the Russian Academy of Sciences.

The deployed ICBMs would also be dominated not by Korolev's rockets, but by those designed and built by Yangel and by the other leading Soviet rocket and spacecraft designer, Vladimir N. Chelomei. Chelomei (1914–1984) entered the ballistic missile field in late 1950s. His original experience was in winged submarine-launched missiles. Chelomei was aggressive and innovative, and he enjoyed technological support of the aviation industry to which his design bureau administratively belonged. The newcomer to ballistic missiles, Chelomei gained some political advantage, which lasted until 1964, by employing the young engineer Sergei N. Khrushchev, the son of the Soviet leader Nikita S. Khrushchev.

Universal Rocket UR

Chelomei aggressively expanded into many areas of ballistic missiles, launchers, and spacecraft. He introduced a new family of long-range rockets UR, *universal'naya raketa*, or the *universal rocket*. The rockets were "universal" in the sense that they were designed to perform as both ballistic missiles and space launchers. Chelomei's silo-launched UR-100 (SS-11) carrying a single nuclear warhead to the distance of 12,000 km (7500 miles) would become the most

EARLY SOVIET BALLISTIC MISSILES: NOMENCLATURE

USSR Number		Western Number		Type	Designer
R-1	8A11	SS-1	Scunner	IRBM	Korolev
R-11	8A61	SS-1b	Scud-A	IRBM	Korolev
R-2	8Zh38	SS-2	Sibling	IRBM	Korolev
R-5	8A62	SS-3	Shyster	IRBM	Korolev
R-12	8K63	SS-4	Sandal	IRBM	Yangel
R-14	8K65	SS-5	Skean	IRBM	Yangel
R-7	8K71	SS-6	Sapwood	ICBM	Korolev
R-16	8K64	SS-7	Saddler	ICBM	Yangel
R-9A	8K75	SS-8	Sasin	ICBM	Korolev
R-36	8K67	SS-9	Scarp	ICBM	Yangel
UR-100	8K84	SS-11	Sego	ICBM	Chelomei

EARLY SOVIET BALLISTIC MISSILES: DEPLOYED

		Type	1960	1962	1965	1970
R-5	SS-3	IRBM	36	36	20	—
R-12	SS-4	IRBM	172	458	572	504
R-14	SS-5	IRBM	—	28	101	89
R-7	SS-6	ICBM	2	6	6	—
R-16	SS-7	ICBM	—	50	202	195
R-9A	SS-8	ICBM	—	—	26	26
R-36	SS-9	ICBM	—	—	72	270
UR-100	SS-11	ICBM	—	—	—	930

CRYOGENIC AND STORABLE PROPELLANTS

Noncryogenic propellants provided significantly improved readiness of ballistic missiles and facilitated their maintenance. No wonder that replacement of the first cryogenic liquid-propellant rockets was strongly favored by the military, in both the United States and the Soviet Union. For performance-driven space launchers, however, cryogenic propellants remained the top choice because of the superior energetic properties of liquid oxygen as oxidizer.

Similarly to the first Soviet ICBM R-7 (SS-6), the first American ICBMs, Atlas and Titan I, used liquid oxygen. American liquid-propellant ICBMs of the next generation, the Titan II, were based on storable propellants (nitrogen tetroxide and Aerozine-50, a 50%-50% mixture of hydrazine and unsymmetrical dimethylhydrazine). The United States rushed, in the late 1950s, the development of solid-propellant ballistic missiles, Polaris and Minuteman. Solid-propellant missiles would subsequently dominate American strategic rocket forces. The Soviet ballistic missile designers Yangel and Chelomei concentrated on perfection of the storable liquid-propellant rocket technology. Consequently, liquid-propellant ICBMs dominated the Soviet strategic rocket forces for many years. The Soviet Union had developed its own solid-propellant technology only much later, by the late 1960s.

numerous Soviet ICBM. In 1970, 930 ICBMs (of the total 1421) on the combat duty were UR-100s and their modifications. Chelomei's spectacular achievements would include guided missiles, space launchers (including UR-500 or *Proton*), space stations (*Almaz*), and various military space systems.

Solid-Propellant Ballistic Missiles

In contrast to the United States, development of solid-propellant intercontinental ballistic missiles did not become a high priority in the Soviet Union. Both Yangel and later Chelomei perfected rockets on liquid storable propellants. Only on 27 June 1959, a special government decree ordered Korolev's OKB-1 to develop a solid-propellant ballistic missile capable of delivering a 500-kg (1100-lb) warhead to a distance of 2500 km (1550 miles). A few days later the State Committee for Defense Technology assigned this task to OKB-1 and directed Korolev to absorb the nearby *Central Scientific Research Institute N.58* (TsNII-58). The main site of Korolev's OKB-1 and TsNII-58, located next to the railroad station Podlipki, were separated just by railroad tracks. A small bridge that could be seen on Fig. 13.10 would directly connect these two facilities 10 years later.

Takeover of Grabin's Design Bureau

The takeover of TsNII-58 was an important expansion for Korolev's OKB-1. The newly added institution was a design bureau of perhaps the most distinguished Soviet designer of wartime and postwar artillery systems, Vassilii G. Grabin. In 1950s, TsNII-58 was also actively involved in fabrication of automatic control systems for nuclear reactors and employed a number of specialists in instrumentation and control. Thus, OKB-1 absorbed Grabin's establishment with almost 5000 employees, including 1500 scientists and engineers and excellent fabrication facilities. The 59-year-old General Grabin was promptly retired to professorship of artillery systems in a technical institute in Moscow.

Solid-Propellant ICBM

The solid-propellant RT-1 was launched successfully for the first time on 18 March 1963. Simultaneously with the RT-1, OKB-1 initiated development of a much larger intercontinental solid-propellant missile RT-2. Several other research institutions and production facilities across the country supported the program. The RT-2 successfully lifted off for the first time on 26 February 1966 at Kapustin Yar. The rocket could deliver the 470-kg (1040-lb) warhead to a distance more than 10000 km (6200 miles); its propulsion system provided specific impulse 220 s. The Soviet Army accepted the RT-2 (SS-13 or *Savage*) for deployment on 18 December 1968.

Changes in the Kremlin

All of the three leading designers of Soviet ballistic missiles, space launchers, and spacecraft, Sergei P. Korolev, Mikhail K. Yangel, and Vladimir N. Chelomei, succeeded to a large degree because of the unwavering personal interest, enthusiasm, and support of the Soviet leader Nikita S. Khrushchev. Khrushchev was ousted from his office after a palace coup in October 1964. The new Soviet rulers led by Leonid I. Brezhnev remained committed

SOVIET ROCKET DESIGNERS AND THEIR PATRON

After his [Soviet leader Nikita S. Khrushchev] retirement, neither persevering Korolev, nor solid Yangel, nor amiable Chelomei ever telephoned him [Khrushchev] or congratulated [him] with [National] holidays [as was customary in the Soviet Union].

Sergei N. Khrushchev (2000, 616)

to rapid arming of the Soviet Union with the numerous and increasingly powerful ballistic missiles and to the expanding space effort, including the lunar program. The seemingly unstoppable quantitative growth of the Soviet ICBM force would be brought to a screeching halt only in 1980s by President Ronald Reagan who wisely shifted the strategic competition into a highly technological area of ballistic missile defense.

Spinning Off Technology

The transfer of the R-1 serial production to Dnepropetrovsk in 1951 would be the first major case of spinning off ballistic missile and spacecraft technologies developed by Korolev's OKB-1 and forming new independent research and development centers across the vast USSR. Shortly after Dnepropetrovsk's OKB-586 was created, another area of the Korolev's work was spun off and transferred to a new organization and remote place.

R-11 (SS-1b)

In the early 1950s, OKB-1 explored the possibilities of liquid-propellant rockets on noncryogenic propellants. So, the R-11 was designed on the basis of nitric acid and kerosene, and it was successfully launched on 21 May 1953. The rocket had the range of 270 km (170 miles) and was soon produced as a mobile missile complex.

Submarine-Launched Ballistic Missiles

The modified R-11, the R-11FM, was selected as the first Soviet submarine-launched ballistic missile. In 1955, development of the submarine-launched rockets was assigned to *Special Design Bureau N. 385* (SKB-385) near town of Zlatoust, in the South Ural region. The 30-year-old Viktor P. Makeyev, originally a lead designer of the Korolev's R-11FM rocket, became the head of the new center. (Makeyev also did a stint as a professional apparatchik in the Young Communist League, the Comsomol.)

Fig. 13.17. The 30-year-old Viktor P. Makeyev became in 1955 head of new design bureau to continue the work, initiated at Korolev's OKB-1, on the submarine-launched liquid-propellant ballistic missiles. Photo courtesy of State Rocket Center "Makeyev Design Bureau," Miass, Russia.

The new independent SKB-385 followed the familiar path. At first in 1949, the bureau focused on support of serial production of the R-1 missile. After the R-1 production and associated equipment were transferred to Dnepropetrovsk's OKB-586, the serial production of the R-11 was organized at the Zlatoust facility. Gradually becoming independent of its parent design bureau, Makeyev's SKB-385 concentrated on the submarine-launched IRBMs and later ICBMs. In 1959, SKB-385 was transferred to a new location, Miass, 30 km (20 miles) east.

SCUD

The Scud missiles that made so much publicity during the Gulf War in 1991 can be traced back to the Korolev's original R-11 rocket. Since the mid-1950s, Makeyev's SKB-385 worked on the submarine-launched R-11FM. In 1958, they expanded their work and initiated development of a new ballistic missile based, similarly to the R-11, on inhibited red-fuming nitric acid and kerosene as propellants.

Fig.13.18. Bottom — Scud missile in the Imperial War Museum in Duxford, England. The cuts in the missile body show the details of the missile design; a full-size model of a Patriot missile is seen on the right. The missile is 37 ft 4 in. (11.2 m) long, its diameter is 2 ft 11 in. (88 cm), and its mass (fully fueled) is 14,046 lb (6370 kg). Top left — the cut in the middle of the Scud shows propellant tanks: the propellant from the upper tank is delivered to the engine through the pipe along the axis of the lower tank. Top right — four aerodynamic surfaces in the engine exhaust flow for thrust vector control. The surfaces are moved hydraulically as shown by the arrow. Photo courtesy of Mike Gruntman.

The new missile, R-17, was only slightly longer than the R-11 but had much better performance characteristics, including a larger range up to 300 km (185 km) and improved reliability. The main innovation was an introduction of propellant pumps instead propellant expulsion to the combustion chamber by high-pressure gas. The missile plant in Votkins worked closely with SKB-385 on this project. The Soviet Army accepted the new ballistic missile R-17 (SS-1c or Scud-B) for deployment in March 1962.

The missile was also produced for export and the USSR provided this export version to its client states. Egypt was the first to fire Scuds, launching three missiles against Israel in 1973. Iraq and Iran widely exchanged Scuds during their 1980–1988 war. (Iran got its Scuds from North Korea.) Iraq actually developed an improved version — with the larger range — of the missile, the *Al-Husayn*, by enlarging propellant tanks. The largest number of Scuds, between 1000 and 2000, were launched by the communist Afghan government against Mujahideen guerilla fighters in 1989–1991.

On 16 September 1955, the first R-11FM rocket was launched in the presence of Korolev from a surfaced submarine in the White Sea. The rocket was accepted for deployment in 1959, thus beginning the submarine strategic rocket force of the Soviet Navy. On 10 September 1960, a ballistic missile was launched, for the first time, from a submarine submerged to a 30-m (100-ft) depth.

Underwater Rocket Launch

Korolev's OKB continued to grow and to establish its branches in other parts of the country. The first intercontinental ballistic missile R-7 was demonstrated in 1957–1958. A modified R-7 placed the Earth's first artificial satellite, Sputnik, into orbit.

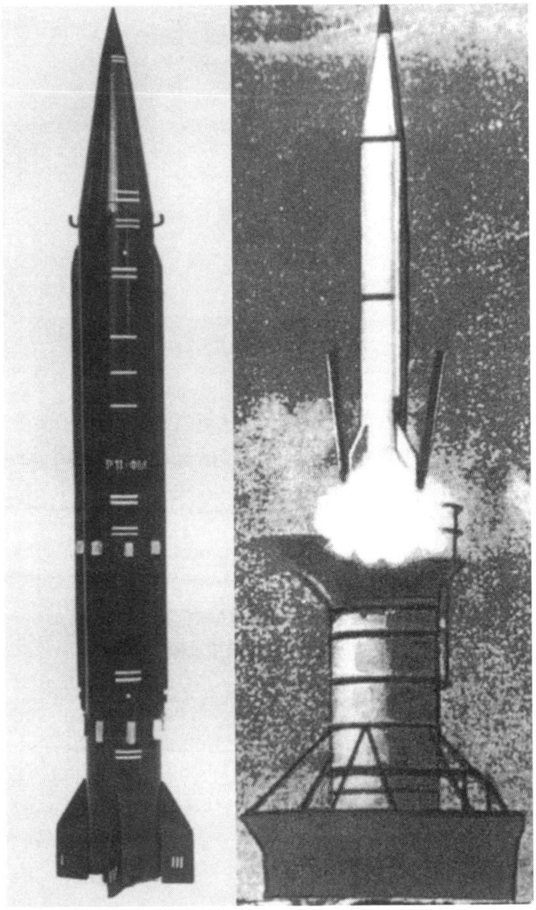

The Soviet Army accepted the R-7 ICBM (8K71, or SS-6, or *Sapwood*) for deployment on 20 January 1960. The serial production of the ICBM was established in the town of Kuibyshev. Kuibyshev, now renamed back to its prerevolutionary name Samara, is 500 miles (800 km) east-southeast of Moscow on the Volga river. The design group supporting rocket production in Kuibyshev was first organized as *Branch N.3* of OKB-1 in 1960, then reorganized as Branch of TsKBEM in 1967, and finally as the independent *Central Special Design Bureau* (TsSKB) in 1974. The main production facilities, the plant *Progress*, are colocated with TsSKB. Dmitrii I. Kozlov, the first head of Branch N.3, is still in charge of TsSKB.

Rocket and Space Center in Kuibyshev

Fig. 13.19. The first Soviet submarine-launched ballistic missile R-11FM was 10 m (33 ft) long and had mass 5440 kg (12,000 lb). The rocket carried 967-kg (2130-lb) warhead to a distance of 150 km (93 km) with probable errors ±1.5 km (range) and ±0.75 km (lateral). The liquid-propellant (nitric acid and kerosene) engine produced 218-s specific impulse at sea level. The rocket was designed at the Korolev's design bureau after the government authorized its development in January 1954. The missile production and further development were transferred to Viktor Makeyev's SKB-385 in Zlatoust in 1955. Photo courtesy of State Rocket Center "Makeyev Design Bureau," Miass, Russia.

The first Soviet optical image intelligence (IMINT)

Fig. 13.20. Dmitrii I. Kozlov, b.1919, headed the branch of Korolev's OKB-1 in Kuibyshev (Samara). Kozlov's TsSKB would become a leading development and production center of space launchers (Vostok, Soyuz, Molniya) and various spacecraft, particularly photoreconnaissance satellites. Photo courtesy of the Russian Academy of Sciences.

Photoreconnaissance Satellites

satellite, Zenit-2, was built at the Korolev's design bureau and successfully launched in 1962. The further development of reconnaissance satellites and their serial production were also transferred to the Kuibyshev branch. TsSKB built more than 900 spacecraft, mostly specializing in photoreconnaissance and remote sensing. Only NPO PM, another major spacecraft developer in Siberia, produced more satellites than TsSKB. The Kuibyshev plant also built rockets for more than 1600 space launches, probably more than any other company in the world. The launch vehicles included Vostok, Soyuz, Molniya, and their derivatives.

Fig. 13.21. Mikhail F. Reshetnev, 1924–1996, headed the Siberian branch of OKB-1. Reshetnev's NPO PM would become the leader in communications satellites. Photo courtesy of the Russian Academy of Sciences.

Missile and Space Center in Siberia

Another offspring of Korolev's design bureau was established in the middle of Siberia, 40 miles (65 km) northeast from town of Krasnoyarsk down the Yenisei river. This *Eastern Branch N.2* of the Korolev's OKB-1, was founded in 1959 in the town Krasnoyarsk-26. Kransnoyarsk-26 is a town with the controlled access, and it is today called Zheleznogorsk. The town is also a home of an important Soviet and Russian nuclear weapons facility for production and storage of plutonium.

Mikhail F. Reshetnev was appointed director of the Eastern Branch, or Branch N.2, and headed it from 1959–1996. The new establishment achieved "independence" in 1961 as *Specialized Design Bureau N.10* (OKB-10) and later renamed *Design Bureau of Applied Mechanics*. Today it is called *M.F. Reshetnev Science and Production Association of Applied Mechanics* (NPO PM).

Communications Satellites

The story of NPO PM also followed the familiar path. The design bureau was originally formed to provide design support for serial production of ballistic missiles on the associated *Krasnoyarsk Machine-Building Plant.* With time, NPO PM changed its specialization and concentrated exclusively on design of spacecraft and their fabrication. The Association focuses on communications, TV broadcasting, geodetic, and navigation satellites. NPO PM built more than 1000 spacecraft, more than any other establishment in the USSR.

Interplanetary Spacecraft

The last major spin-off of the work initiated by OKB-1 took place in 1965, one year before Korolev's death, when development and fabrication of all interplanetary spacecraft were transferred to OKB-301, or S.A. Lavochkin NPO, in Khimki.

14. GATEWAYS TO HEAVEN

Fig. 14.1. Erection of Atlas rocket at Pad 14 on Cape Canaveral, 1957. The complexity of the ground equipment had increased significantly since the days of the V-2. Photo courtesy of 45th Space Wing History Office Archives, Patrick Air Force Base, Florida.

To design and fabricate a rocket was only a part of the story. The rocket had to be assembled, tested, and launched, and its performance in flight had to be evaluated. Thorough testing was critical for development of better and more capable missiles, which continued to grow in size, power, and sophistication.

Complex Engineering Facilities

Complex and expensive infrastructure was needed for rocket testing and for eventual launch of satellites into orbit. Rocket-engine static, or captive, test firings, conducted at separate facilities, also involved complex experimental stands performing numerous engineering measurements and working under extreme conditions. The first real missile test site was established by William Congreve in the early 19th century at the artillery range of the Royal Laboratory in

Fig. 14.2. Dual-position Saturn I/IB test stand at the T-Stand at NASA's Marshall Space Flight Center, Huntsville, Alabama. Static firing of rocket engines plays an important role in development of new propulsion systems and requires complex and expensive engineering facilities. The test stands have to withstand violent mechanical, vibration, and heat loads. Photo courtesy of NASA.

Safety and Security

Woolwich, England. The German Peenemünde center provided an example of an early modern rocket test range. The space ports, or cosmodromes as the Russians call them, significantly grew in scope and technical complexity. Safety and security considerations usually called for building these sites in remote locations, which in turn required extensive development, construction, and engineering works under inhospitable conditions.

Facilities Span Thousands of Miles

Test-range engineering facilities usually included equipment for storage, handling, pumping, and preparation of propellants, often highly toxic. Cryogenic propellants required highly specialized machinery and handling procedures and were sometimes produced directly on the range. Vast and efficient networks of radio and optical systems for tracking missiles in flight, communications, and control could span thousands of miles.

The range infrastructure also included buildings for storage, assembly, fueling, and checking the rockets. With the advent of spacecraft, clean-room environment had to be added to store spacecraft before launch, perform spin and dynamic balance tests, payload checkout and to mate the spacecraft to the rockets.

Movable gantries were built for convenient access to erected rockets. In addition, the launchpad had to withstand multiple launches, with enormous

MISSILE TEST RANGE INFRASTRUCTURE

A rocket test range requires extensive infrastructure dispersed across large areas. Supporting radar and radio communication stations can be hundreds and thousands miles away. In addition, the range is usually equipped with extensive optical instrumentation.

Fig. 14.3. The *Bright Eyes* tracking telescope at White Sands Proving Ground achieved the first successful optical track over a 50 mile (80 km) distance in 1946. The M-45 gun mount was used as a telescope platform. This optical system was designed by the chief of the range optics branch, Clyde Tombaugh, the discoverer (in 1930) of the planet Pluto. Photo courtesy of White Sands Missile Range.

Fig. 14.4. Kinetheodolites at the Woomera missile test range in Australia. Photo courtesy of Defense Science and Technology Organization, Department of Defense, Australia.

Fig. 14.5. TLM-18 — typical of the 60-ft (18-m) telemetry antennas installed at Cape Canaveral, Grand Turk, and Ascension. These tracking antennas supported launches at the Eastern Test Range in 1950s and 1960s. The range facilities stretched from Florida to Puerto Rico to St. Lucia to Ascension Island. A number of picket ships supported rocket launches filling the gaps between the island stations. Photo courtesy of 45th Space Wing History Office Archives, Patrick Air Force Base, Florida.

mechanical and heat loads. Occasional and sometimes violent accidents always remained a possibility. Consequently, safe and convenient blockhouses had to be provided for launch operations.

A rocket body had to be protected from the rocket's own plume at liftoff to avoid damage by high-temperature exhaust gases. Exhaust gases can also cause large pressure variations around the rocket, leading to destructive lateral loads. It was desired to direct the exhaust flow away from the rocket and to somehow

Toxic Propellants

Enormous Mechanical and Heat Loads

Fig. 14.6. Fueling a booster of a surface-to-air BOMARC missile by red-fuming nitric acid at Cape Canaveral, September 1952. Testing and launching sites include facilities for storage and handling of propellants, sometimes highly toxic. Photo courtesy of 45th Space Wing History Office Archives, Patrick Air Force Base, Florida.

Fig. 14.7. Main rocket launch and test sites established in the United Stated and USSR by mid-1960s. The countries are shown in scale.

United States: a) Vandenberg Air Force Base, California; b) White Sands Missile Range, New Mexico; c) Cape Canaveral, Florida; d) Wallops Island, Virginia.

Soviet Union: e) Kapustin Yar, present Russian Federation, or Russia; f) Tyuratam (often called Baikonur), present Kazakhstan; g) Plesetsk, present Russia; h) Saryshagan, present Kazakhstan; i) an impact site at the Kamchatka Peninsula, present Russia, for tests of the first Soviet ICBMs launched from Tyuratam.

Spacecraft launch directions, and correspondingly orbit inclinations, are usually limited by safety considerations. Cape Canaveral and Tyuratam launch satellites into eastern directions resulting in low-inclination Earth orbits, while Vandenberg and Plesetsk launch satellites into polar orbits favored by many intelligence payloads. Consequently, the Soviet Plesetsk was for many years the busiest space port in the world placing Soviet reconnaissance satellites in orbit. Kapustin Yar and White Sands are used mostly for testing various missiles and sounding rocket shots, although a number of small satellites were put into orbit from Kapustin Yar in 1960s and 1970s. Saryshagan was the focus of the Soviet antiballistic missile defense development with the first rocket intercept by a rocket in 1961: a ballistic missile launched eastward from Kapustin Yar was detected, tracked, and shot down by a rocket from Saryshagan.

Figure courtesy of Mike Gruntman.

weaken it. Hence, specially built flame ducts accept and guide the plume away, and water sprays were sometimes injected into the hot gas entering the duct to moderate the effects of overpressure.

UNITED STATES

New Rocket Launch Sites Needed

The first major American test site, White Sands Proving Ground, was activated in 1945. WSPG provided excellent facilities for development of the first-generation relatively small missiles. Other important test sites were in California, the Navy's Guided Missile Test Center at Point Mugu south of Santa Barbara and the Naval Ordnance Test Station (NOTS) at China Lake. National Advisory Committee for Aeronautics launched its first rocket from a test station on Wallops Island off the coast of Virginia in 1945.

It became evident, however, that the future long-range ballistic and winged missiles, still on drawing boards, would require much larger and safer areas for testing. The rockets are very dangerous. They sometimes explode, and the first stages of multistage rockets come back to the ground downrange (that is, along the flight path of ascending rockets).

WALLOPS ISLAND

NACA's Langley Aeronautical Laboratory (Hampton, Virginia) established a missile research *Auxiliary Flight Research Station* on Wallops Island in 1945. More than 14,000 various rockets were subsequently launched from Wallops.

The first space launch took place from Wallops on 16 February 1961. Wallops-launched Scout rockets placed into orbit twenty satellites. The Scout was the first U.S. launch vehicle with exclusively solid-propellant propulsion. The last Scout lifted off Wallops in December 1985. Today, the *Wallops Flight Facility* is a part of NASA's Goddard Space Flight Center and responsible for suborbital research programs that include sounding rockets and high-altitude balloons.

Cape Canaveral

In one accident in 1947, a V-2 missile fired from White Sands deviated from its preset path and passed over El Paso, Texas, and the Mexican town Ciudad Juarez across the Rio Grande. Fortunately the missile hit a barren hill and nobody was hurt. New bigger and safer testing sites were needed to support development of future ICBMs and to eventually launch satellites to space.

CAPE CANAVERAL AND BALTIMORE GUN CLUB

Jules Verne's *Baltimore Gun Club* fired a manned projectile to the Moon in his novel *From the Earth to the Moon*. Jules Verne's launch site was 15 miles inland from Tampa, Florida, only 100 miles from Cape Canaveral. A different kind of spaceship would take men from Cape Canaveral to the Moon 104 years later.

The choice for the new rocket range fell on Cape Canaveral on the Florida's Atlantic coast. The cape had been known to Spanish seamen as Cabo de Cañaveral (which meant the *cape of reed* or *cane*) since the 16th century. This area was settled much earlier than many other parts of Florida, and the cape housed a lighthouse since 1843.

By the end of World War II, Cape Canaveral had remained a

SELECTION OF CAPE CANAVERAL

The site selection for the interservice joint proving ground began in 1946. Brigadier General William L. Richardson, Chief of the Joint Long Range Proving Ground Committee of the Joint Research and Development Board submitted a report "on 20 June 1947 ... recommending that a National Guided Missile Range be established as soon as possible. First choice for the launching site was in the vicinity of El Centro, California, firing southeast, and second choice was firing southeast from a launching site [at Cape Canaveral] near Banana River, Florida. The Joint Research and Development Board approved the report in July 1947 and delegated the implementation to the War Department, who in turn directed the Air Force to handle the matter ... " (Status Report, Long Range Proving Ground Committee, 2 January 1948).

The first choice site near El Centro was 100 miles (160 km) east from San Diego, California, just across the border from the Mexican town Mexicali. Because the launch from El Centro involved the overflight of the foreign territory, the Mexican government had to be approached.

"In December 1947, the Acting Secretary of State, Mr. Lovett, addressed a communication to the U.S. Ambassador to Mexico, briefly outlining the project and requesting the Ambassador to suggest the best method of approaching the Mexicans on the question of the El Centro – Gulf of California Range. On 9 January 1948, the Ambassador replied that, in his opinion, the best approach would be direct to the President of Mexico, Senor Miguel Aleman. The Ambassador pointed out that the President of Mexico would have to approve the proposition in the long run and it would be best to avoid entangling the project in the lower echelons of the Mexican government before reaching the final authority. The Ambassador requested and received permission from the State Department to approach President Aleman privately and informally on the project. The results of this interview indicated that it would be impossible to secure favorable agreement with Mexico. The Secretary of Defense was notified to this effect by the State Department on 27 January 1948" (Status Report, Long Range Proving Ground Committee, 2 February 1949).

The second best site at Cape Canaveral had to be discussed with the British military authorities because of the vicinity of the Bahama Islands. "On 28 January 1948, the proposal was submitted to the Office of Deputy Chief of the RAF Mission to the United States. The initial reaction was favorable."

Public Law Signed by President Truman

AN ACT

To authorize the establishment of a joint long-range proving ground for guided missiles, and for other purposes.

Be it enacted by the Senate and House of Representatives of the United States of America in Congress assembled, That the Secretary of the Air Force is hereby authorized to establish a joint long-range proving ground for guided missiles and other weapons by the construction, installation, or equipment of temporary or permanent public works, including buildings, facilities, appurtenances, and utilities, within or without the continental limits of the United States, for scientific study, testing, and training purposes by the Departments of the Army, Navy, and Air Force ...

There is hereby authorized to be appropriated, ... not to exceed $75,000,000 to carry out the purposes ... of this Act ... .

Approved 11 May 1949.

poorly developed area of sand and scrub palmetto. The isolated and relatively uninhabited place and the deactivated *Banana River Naval Air Station* 20 miles (32 km) south favored this Florida site. In addition, weather allowed launches all year around, with the occasional interruptions brought by hurricanes and thunderstorms. A deep-water port at the southern edge of the cape permitted convenient transportation of large-size rocket components and made possible development of submarine-launched missiles. On 11 May 1949, President Truman signed *Public Law 60*, authorizing the Secretary of the Air Force to establish the *Joint Long Range Proving Ground* at Cape Canaveral.

Allowable Launch Azimuth

Cape Canaveral offers an important advantage of safe launches over the ocean within a wide range of directions, the so-called allowable launch azimuths. The launch of a spacecraft has to be conducted along a precisely defined trajectory to achieve the desired orbit inclinations. (Orbit inclination is the angle between the satellite orbital plane and the plane of the Earth's equator.)

Launch directions, or azimuths, from 35 to 120 deg (counted clockwise from the north direction) are possible from Cape Canaveral. This azimuth range includes the most energetically favorable direction due east, which takes the full advantage of the Earth's rotation at a given location. If a rocket is launched outside this allowable azimuth range, its falling parts could hit Newfoundland and the Eastern Seaboard (azimuth <35 deg) or Cuba (>120 deg).

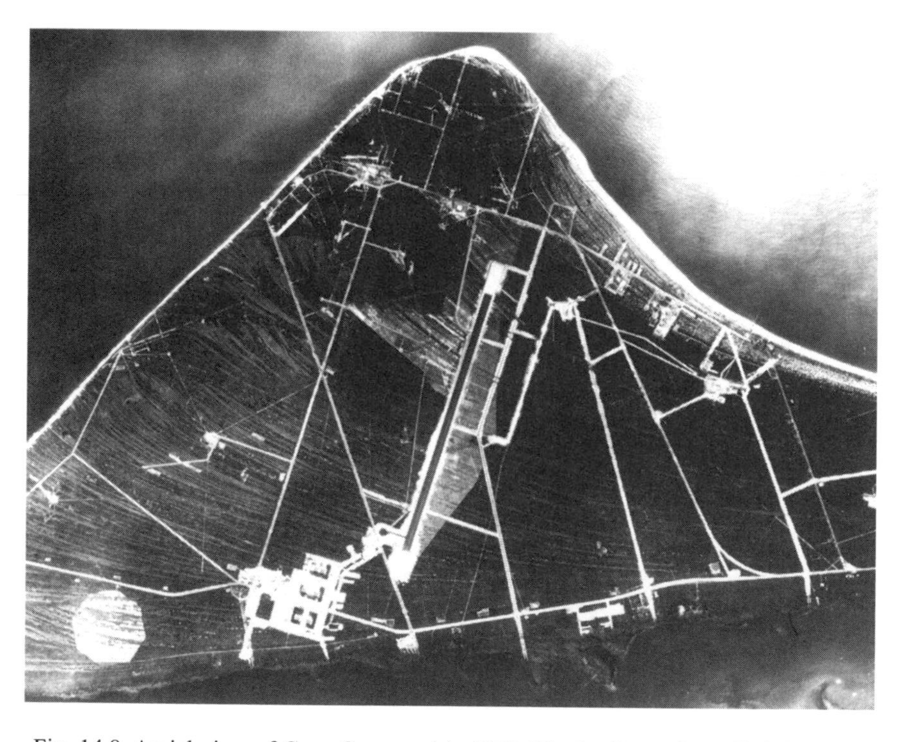

Fig. 14.8. Aerial view of Cape Canaveral in 1955. The landing strip, called the "skid strip," for the X-10 vehicle developed as a part of the *Navaho* surface-to-surface guided missile program is in the center. The strip was expanded with time to accommodate landing of heavy Air Force's cargo carriers. Photo courtesy of 45th Space Wing History Office Archives, Patrick Air Force Base, Florida.

On 1 October 1949, the new joint proving ground was activated at Cape Canaveral under the supervision of the Air Force. The construction of launching pads, rocket-processing facilities, and instrumentation sites began next year. The site would become known as the *Air Force Eastern Test Range* (AFETR) and later as the *Eastern Space and Missile Center*. The Air Force also took over the nearby Banana River Naval Air Station and reorganized it into the *Patrick Air Force Base*.

Air Force Eastern Test Range (AFETR)

> **PATRICK AFB**
>
> Patrick Air Force Base was named in honor of Major General Mason M. Patrick, the chief of the Air Service of the American Expeditionary Force in World War I and the postwar head of the Army Air Service.

The share of the tests of the civilian space vehicles and civilian launches was gradually increasing at Cape Canaveral, especially with the initiation of the Apollo program and its gigantic Saturn V launcher. The National Aeronautics and Space Administration was established in 1958. The growing civilian space activities at the Cape led to the government acquisition of the northern part of the nearby Merritt Island. In addition a large submerged area in the Mosquito Lagoon was obtained from State of Florida, dredged, and filled in.

New NASA's Kennedy Space Center at Cape Canaveral

A new NASA test and launch facility, called the *Launch Operations Center*, was formed on the newly acquired land in July 1962. The new field center was organized on the basis of the Cape Canaveral's *Launch Operations Directorate* of NASA's Marshall Space Flight Center. Dr. Kurt H. Debus, one of the Peenemünders who had come to Fort Bliss with the von Braun team, became the first director of the new NASA facility. The center was renamed the *John F. Kennedy Space Center* in December 1963.

First Rocket Launch at Cape Canaveral

The first launch at the newly established proving ground at Cape Canaveral took place on 24 July 1950. The Bumper 8 missile was launched on that day, and Bumper 7 followed five days later. During the first launch attempt, Bumper 7 collected moisture within the missile and was moved back to a hangar to dry out and be rechecked. Thus, Bumper 8 was launched the first.

Fig. 14.9. Kurt H. Debus was the head of Army's *Missile Firing Laboratory* responsible for launch operations in 1950s. Debus was appointed first Director of NASA's Kennedy Space Center. Photo (ca. 1958) courtesy of 45th Space Wing History Office Archives, Patrick Air Force Base, Florida.

The launching and control facilities at this first launch were primitive: a painter's scaffold was used as a gantry, and the control blockhouse was arranged in a sandbagged tarpaper bathhouse and in an old Sherman tank. An alligator slipped into one of the bunkers before the launch bringing additional excitement to the countdown. The launch and monitoring crews would move into special fully instrumented and hardened blockhouses as the rockets grow in size and complexity.

Bumpers 8 and 7 were the last launches culminating this Army missile program. (The Bumpers were initially tested at White Sands.) The next ballistic missile to be fired at Cape Canaveral would be the Army's Redstone rocket: this launch took place three years later on 20 August 1953. The facility did not remain idle, however, concentrating in the meantime on antiaircraft and cruise missile tests. The first to follow was the *Lark*, an antiaircraft missile designed by the Fairchild Aircraft Company for the Navy. The Lark was fired in October and November of 1950.

Fig. 14.10. The first launch at Cape Canaveral. The *Bumper 8* rocket lifts off on 24 July 1950. Photo courtesy of 45th Space Wing History Office Archives, Patrick Air Force Base, Florida.

Lark, Matador, Snark, Navaho, and others

Testing of various rocket systems was gradually being transferred to Cape Canaveral from other places. Among the first was the Air Force's surface-to-surface *Matador*, a "pilotless bomber" missile, that was under development by the Glenn L. Martin Company. After 15 tests elsewhere, the Matador program was moved to Cape Canaveral, with the first launch on 20 June 1951.

Other intense test programs included intercontinental cruise missiles, subsonic Snark and supersonic Navaho. Ninety-seven Snarks were launched downrange from 1952–1960. A joke was around at that time about the "Snark-infested

GEMINI / TITAN II
FIRST FLOOR BLOCKHOUSE
COMPLEX 19

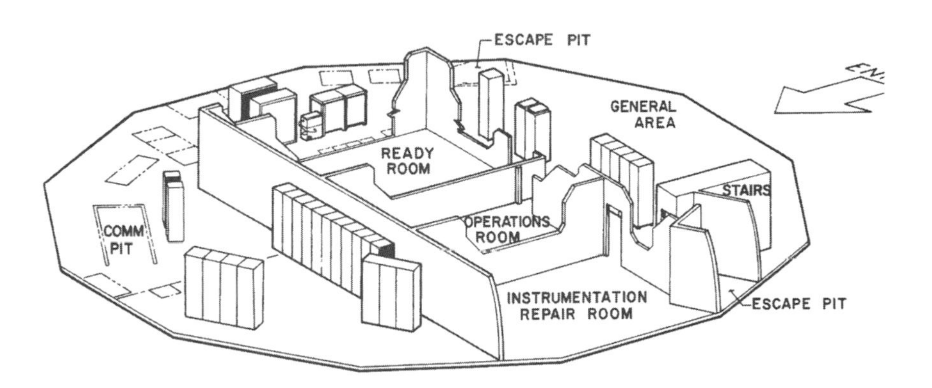

Fig. 14.11. Complex 19 blockhouse at Cape Canaveral after remodeling, August 1963, shows how dramatically the test-range infrastructure has improved since the first launch of Bumper 8 in 1950. The first manned Gemini mission (with astronauts Virgil I. Grissom and John M. Young) was launched from this complex in 1965. Photo courtesy of 45th Space Wing History Office Archives, Patrick Air Force Base, Florida.

waters" of the Atlantic surrounding the Cape, thus testifying that not all flights were entirely successful.

Missile Firing Laboratory

Cape Canaveral was the primary range for testing Army's ballistic rockets. Thirty-seven Redstones were fired there during the five years of the rocket development. In addition the Army's short-range missile *Pershing* was tested at the Cape. Von Braun's associate Kurt H. Debus was in charge of the *Missile Firing Laboratory* responsible for ABMA's launch operations.

Space Port

With the advent of the space age, Cape Canaveral became the primary launching site for the first satellites. It was Cape Canaveral from where the Army launched a modified Jupiter C, the Juno 1, that placed the first American satellite Explorer I into orbit. Cape Canaveral is the nation's major space port for launches of military, scientific, and commercial satellites. All American manned launches, including the lunar Apollo missions, were performed from the Cape. Cape Canaveral is the home of the space shuttle fleet. Weather permitting, the space shuttle lands at the Cape.

IRBMs and ICBMs

Towards the end of 1950s, testing and launches of the Air Force's IRBM Thor and ICBMs Atlas, Titan, and Minuteman came to dominate Cape Canaveral. In addition, the Navy's fleet ballistic missiles, Polaris, Poseidon, and Trident, were regularly launched at the Cape during their development. The solid-propellant Minuteman required

Fig. 14.12. Two-stage liquid-propellant Titan I rises from Pad 16 at Cape Canaveral on 28 April 1960. The rocket reached the planned impact area near the Ascension Island in South Atlantic. Photo courtesy of 45th Space Wing History Office Archives, Patrick Air Force Base, Florida.

development of the technology of launch from silos, with the first silo launch taking place on 30 August 1961.

Another rocket test center had been emerging at this time, focused on ICBM launches, including those from silos, and on personnel training. This new site was located in Southern California at Point Arguello near Lompoc in western Santa Barbara County. This area had been used for cattle and horse grazing during the 19th century. The Union Oil Company purchased the ranch in 1904 and drilled for oil in 1920s. The U.S. Army bought the land from Union Oil in 1941 and established Camp Cooke for armor and infantry training during World War II and the Korean War.

Fig. 14.13. Complex 17 (Pads A and B in foreground) on Cape Canaveral, January 1961. Photo courtesy of 45th Space Wing History Office Archives, Patrick Air Force Base, Florida.

Vandenberg Air Force Base

The Air Force took over Camp Cooke in November 1956 to build a West Coast base for ballistic missile launches and for personnel training. The base was transferred from the Air Research and Development Command to the Strategic Air Command on 1 January 1958, and reorganized into the Vandenberg Air Force Base, named after General Hoyt S. Vandenberg, on 4 October 1958.

Launch into Polar Orbits

The Air Force's Thor was the first missile launched from Vandenberg on 16 December 1958. This successful launch was performed as a part of combat training by a "blue suit" crew (that is a crew consisting of the Air Force personnel without participation of rocket developers and contractors). Today, Vandenberg AFB tests and evaluates ballistic missile systems and launches satellites, mostly into polar orbits favored by remote sensing and intelligence payloads and many scientific missions.

Fig. 14.14. Modern launching pad at the Vandenberg Air Force Base, California: movable rocket service tower (right); concrete rocket exhaust flame duct (bottom). The insert (top left) shows another view of the flame duct. Photo courtesy of Mike Gruntman.

SOVIET UNION

In May 1946, the USSR Council of Ministers issued a decree making the development of the Soviet rocket weapons a top national priority. Among the urgent tasks specified by the decree was establishing of the *State Central Test Range*. A group of Army officers headed by General Vassilii I. Voznyuk surveyed seven promising areas. Finally Moscow selected the site near the settlement *Kapustin Yar* on the eastern bank of the Volga river 70 miles (110 km) southeast from Stalingrad (Volgograd). The first Soviet major missile test range, the *State Central* [Rocket] *Testing Range N.4* (GTsP N.4), was subsequently activated at Kapustin Yar, with General Voznyuk as its first commander.

Missile Test Range at Kapustin Yar

First Army construction units moved in in July 1947 to build the missile range, supporting facilities, barracks, and the settlement. *Kap-Yar*, as the site was often called by those in the know, was at the edge of the uninhabited semidesert with cold winters and hot summers. The Soviet Army formed its first brigade specializing in long-range ballistic missiles in the occupied Germany in 1946. After familiarizing itself with the V-2, the brigade was transferred to Kap-Yar in August 1947. The first V-2 was launched from this new Soviet rocket range in October of the same year.

First V-2 Launch at Kap-Yar

Kapustin Yar was the primary testing ground for Soviet missiles in 1940s and early 1950s. Not only did it remain an important development facility after the establishment of other Soviet test sites at Tyuratam, Saryshagan, and Plesetsk, but Kap-Yar also became a space port, the cosmodrome, for launching small satellites.

FLIGHT OVER KAPUSTIN YAR

German specialists began to return from the Soviet Union in the early 1950s. Those who succeeded in reaching West Germany were debriefed by the American and British authorities. The Germans pointed at a major missile development effort at Kapustin Yar. The Kap-Yar range was considered so important that, reportedly, the British RAF attempted in 1953 a bold high-altitude reconnaissance overflight of the Soviet territory to specifically examine this site.

The specially equipped two-engine Canberra aircraft performed this risky mission. (The modified Canberra was built also under license in the United States by the Glenn L. Martin Co. as the B-57.) Apparently, the RAF plane was either detected by the improved Soviet radar or the mission was betrayed by the Soviet agents inside the British intelligence. The Soviet fighter aircraft intercepted the Canberra and nearly shot it down.

It was claimed that the Canberra took off from Germany, photographed Kap-Yar, and, though badly damaged, succeeded in reaching Iran and landing there. It is still not revealed whether the mission had achieved its goal.

The first space launch from Kap-Yar took place on 16 March 1962 when the satellite *Cosmos-1* was placed into orbit. The launch was performed from a silo-based complex *Mayak-2*. Kap-Yar was the site from where Indian (1975 and 1979) and French (1977) spacecraft were launched by Soviet rockets. Sounding rockets *Vertikal'*, a derivative of the IRBM R-5 (SS-3), were shot many times up from Kap-Yar. After a 12-year hiatus in 1980s and 1990s, Kap Yar resumed space launches in May 1999.

New Missile Test Range Needed

The Kap-Yar facilities were adequate for flight tests of the early Soviet missiles with the range not exceeding 1600 km (1000 miles). Then new programs were initiated to build intercontinental weapon systems. The government decree of 13 February 1953 authorized the new 8000-km (5000-mile) winged, or cruise, missiles. These missiles *Burya* (*Tempest* in Russian) and *Buran* (*Blizzard*) were under development by the design bureaus of S.A. Lavochkin and V.M. Myasishchev, respectively. Many features of the Burya and Buran were not unlike those of the American Navaho. Then another decree authorized on 20 May 1954, the development of the first ICBM R-7. Testing of the emerging intercontinental weapons called for a much larger range.

Three Promising Sites for the New Test Range

General Voznyuk was again assigned a task of heading a survey group selecting areas for the new missile range. Three sites looked most promising. The first site was in the wooded sparsely populated area 1600 km (1000 miles) almost exactly east from Moscow in the European part of the USSR. The high geographical latitude of this site, 57° N, would have made the location unfavorable for the future space launches.

The second promising site was at the northern part of the western shore of the Caspian Sea. This location was excellent: well-developed transport, acceptable geographical latitude, and relatively favorable climate. In addition, the launches would go over the desert and semidesert areas and over the Caspian Sea.

Korolev's R-7 ICBM required, however, three posts of radio control, or RUPs (*Punkt Radioupravleniya*). Two RUPs, controlling the flight direction, had to be

RUPs for R-7

positioned symmetrically from the launching site, 150–250 km (90–150 miles) left and right from the flight trajectory. The third RUP at the distance of 300–500 km (180–300 miles) opposite to the flight direction performed Doppler measurements of the rocket velocity in order to generate the engine cutoff command, which determined the distance to the warhead impact site. The launch site infrastructure thus "grew the [two lateral symmetric RUP] 'radio-whiskers' and the [RUP] 'radio-tail'" (Chertok 1994, 406) when drawn on the map.

All RUPs needed unobtsructed lines of sight to the ascending rocket immediately after the liftoff, precluding RUP locations in mountainous areas. The requirements imposed by this radio control system had consequently excluded the area on the Caspian Sea shore from consideration for a new rocket test range. Ironically, the advent of autonomous inertial guidance made radio control obsolete in only a few years.

130TH DIRECTORATE

The Soviet Army activated a new *130th Directorate of Engineering Works* (initially called the *130th Directorate of Special Works*) in 1950 for the construction of one of the sites at the Kap-Yar missile range. The directorate was transferred in 1951 to Tashkent, the center of the *Turkestan Military District.* (Tashkent, an empire outpost, was the center of the Russian penetration to the Central Asia since the 19th century. Today, this city with a four-million population is the capital of a new state, Uzbekistan.)

The 130th Directorate supervised construction of numerous military installations dispersed throughout enormous territory of the military district in the Soviet Central Asia. In 1954, the directorate was assigned to and built the Tyuratam (Baikonur) cosmodrome.

Selection of Tyuratam

On 12 February 1955, the USSR Council of Ministers authorized establishing a new rocket test range at the third considered location near the train stop *Tyuratam* on the Turkestan railroad line east of the Aral Sea. The new range was called the *Scientific-Research Test Range N.5* (NIIP-5). The Army's *130th Directorate*

DECREE OF THE COUNCIL OF MINISTERS OF THE USSR

On 12 February 1955, the Chairman of the Council of Ministers of the USSR Nikolai A. Bulganin signed the decree (N.292-181) "On the new testing range for the Ministry of Defense." The Decree approved a proposal

> ... to establish in 1955-1958 the scientific-research and testing range of the USSR Ministry of Defense for flight tests of [the first Soviet ICBM] R-7, [long-range winged missiles] "Burya" and "Buran" with the location of
>
> the main part of the range in the Kzyl-Orda and Karaganda administrative regions of the Kazakh SSR [Soviet Socialist Republic] between [the towns] Novo-Kazalinsk and Dzhusaly;
>
> the region of the warhead impact in the Kamchatka region of the RSFSR [Russian Soviet Federal Socialist Republic] near Cape Ozernoi;
>
> ... and the region of the impact of the first stages of the R-7 [rocket] on the territory of the Akmolinsk administrative region of the Kazakh SSR near Lake Tengiz ... .

Fig. 14.15. Major General Georgii M. Shubnikov, 1903–1965, Commander of the 130th Directorate of Engineering Works. Photo courtesy of Mike Gruntman.

of Engineering Works (Upravlenie Inzhenernykh Rabot, or UIR) was assigned the task to build the new range from scratch. Several brigades of construction troops and various specialized engineering units were attached to the directorate.

130th Directorate

Thc commander of the 130th Directorate, Major General Georgii M. Shubnikov, was responsible for all aspects of life and work of the military units subordinated to his command, including construction, political indoctrination, personnel, logistics, readiness, and medical care. The engineering and construction work was supervised by Directorate's Chief Engineer Engineer-Colonel Alexander Yu. Gruntman. Both Shubnikov and Gruntman were decorated World War II veterans and highly experienced specialists. Gruntman already worked on the construction of the Kap-Yar missile test site as Chief Engineer of the newly formed 130th Directorate in 1950–1951.

Harsh Climate

Thousands of officers and men toiled on construction at Tyuratam, or TT (pronounced *Teh-Teh*), as the site was often informally called, in exceptionally difficult conditions. And those conditions were inhospitable indeed. The surrounding semi-desert had a harsh continental climate with lengthy hot and dry summers and short winters with some snow. Annual precipitation was about 5 in. (127 mm), and average temperatures varied from 10°F (-12°C) in January to 79°F (29°C) in July. Temperature could climb up higher than 104°F (40°C), with omnipresent and all-penetrating dust, in the summers and drop down to -40°F (-40°C), with strong winds, in the winter. Each year dust storms raged for several days usually in April, blocking sunlight and turning day into night.

In short, the site selected for the new Soviet range was no Florida. (To

Fig. 14.16. Chief Engineer of the 130th Directorate of Engineering Works Engineer-Colonel Alexander Yu. Gruntman, 1912–1975, supervised all of the construction work at the Tyuratam cosmodrome, supporting facilities, military installations, settlement, and infrastructure. Photo (ca. 1962) courtesy of Mike Gruntman.

TYURATAM PIONEERS

In the middle of March, 1955, ... we [the group of lead officers] disembarked from the train [at the Tyuratam stop]. There was only a lonely figure of the assistant station-master on the platform. Nobody else. From the platform, one could see the whole settlement consisting of a small station building, two small brick houses, school, dormitory of the rotating train crews, water tower, and several dry-mud houses of the station workers. There were three trees near the station building and no other vegetation at all. A monotonous snow-covered desert stretched beyond the railroad tracks.

... [This was] a place deep in the south of our country. It was the middle of March and we expected to feel the spring's warm. We were met, however, by the penetrating wind and 5°F (-15°C) cold. The wind burned the open face, the winter overcoat did not protect against the wind, and cold reached the bones. The local climate was harsh, befitting for the surroundings.

We were lucky to persuade the house keeper of the railroad crews' dormitory to let us [lodge] in a small room: narrow metal beds, benches, rickety table, but a very good stove with plenty of coal. This room was a real asset for settling in the new region ....

The next morning we drove to the area of the planned construction of the launching site and the assembly-testing building [with a 4000 square meter main hall]. There were no roads. During construction of the Turkestan railroad [leading to Tashkent] in the beginning of the 20th century, there was a sand pit at the place of the future launching site. Sand was hauled over a narrow-gauge railroad spur to the main line for the construction of the railroad bed. When the construction was over and the sand source exhausted, the narrow-gauge track was dismantled but its low earth foundation remained and was well preserved during the last fifty years. This landmark was also a perfect, though only, navigational aid [in the otherwise monotonous desert]. ... Finally we reached the goal of our trip [in 30 km (19 miles)]. We did not see anything special there. Almost the same landscape, and here was a [surveyor] pole marking the center of the launch complex.

It was hard to imagine and to believe that in only one year and a half many thousands of soldiers and officers would build here far away from the railroad, in the region with no living quarters, no water, and no elementary roads, the facilities from where the cosmos would be stormed. There was no time for well planned [gradual] deployment of the people and initiation of the construction. Everything had to be started from scratch and simultaneously, to build the main launching complex, electric power supplies, water supply installations, roads, railroads, supporting facilities and living quarters, storage buildings, parking and maintenance facilities for numerous construction mechanisms and automobile transport, factories and [production] sites for prefabricated construction components, etc....

Soldiers ... not involved in surveying were sent to store ice in two open concrete cavities at the pumping facilities of the railroad station ... This ice allowed storage of food supplies for officers and men [during the following hot summer months].

Almost every week, military construction units scored a big or small victory: opening the first [general] store and the first officers canteen, beginning the operation of the first bath-and-laundry train, baking the first bread in the first bakery, providing elementary conditions for the medical personnel, removing the first appendix by surgeons in the new hospital, though still in crude barracks. And so with everything, the first, the first, the first. This was the joy, the reward for our hard work.

The summer of 1955 was over and many things had been accomplished.

Engineer-Colonel Alexander Yu. Gruntman (*Tyuratam*, unpublished, 1970)

be fair, the Banana River station was no paradise either: this hard-luck Florida post was known among American naval aviators in 1940s as a "bug capital" of the world.) The considerations of safety and security tended to place rocket test sites in relatively uninhabited areas. And there were very good reasons why those areas were not favored by the natives. The French test sites at Hammaguir and in Guiana and the Australian site at Woomera were also not

Fig. 14.17. Tyuratam with a harsh continental climate and cold and windy winters was definitely no Florida. The author of this book plays in the snow in Tyuratam in February 1959. Photo courtesy of Mike Gruntman.

No Florida

Fig. 14.18. Unforgiving land met the first settlers in Tyuratam. The photo on the left (the author of this book with his mother Raya, ca. 1958) shows barren soil without a trace of vegetation with a row of prefabricated wooden-plank houses recently erected for the families of servicemen. The all-penetrating and omnipresent dust was everywhere. Water, tireless effort, and care have turned the desert into an oasis. The photo on the right shows the same spot, as on the left, four years later. Not only the trees and brush appeared but the author of this book (shown with his brother Sergei on the photo) has grown as well and become a second-grade pupil in the primary school. (The photo was taken on 1 September 1962, the first day of classes). Photo courtesy of Mike Gruntman.

unlike Tyuratam in this respect.

The region selected for what would become the first Soviet space launch site, the cosmodrome, did not have any infrastructure except the main railroad line, Orenburg-Tashkent, connecting the European part of the Soviet Union with its Central Asian parts. Tyuratam was a God-forgotten hamlet with the passenger trains stopping several times a day just for two minutes. The Syr-Dar'ya river brought some life to the otherwise monotonous semidesert. Its muddy brownish waters carried a lot of sand and the flow was exceptionally strong at that time, making swimming dangerous. Sheatfish, pike, perch, carp, and chub were plentiful in the river, and saigas, hares, foxes, ducks, and wild geese were abundant in the surrounding areas. In the spring, an explosion of wild tulips turned the desert, briefly, into a bright carpet of red, yellow, and violet flowers.

God-Forgotten Place

DRY GARRISON

The [Tyuratam] garrison was under an absolute dry law ... [banning] all alcoholic beverages including beer, but it [the dry law] was inefficient because everything at the testing range from [electrical] contacts [at telemetry stations] to rocket [propellant] tanks were cleaned by [ethyl] alcohol [that was routinely drunk]. The surrounding [local inhabitants] Kazakhs [also] offered their services [in procuring liquor], so a phrase "Arak bar?" (*Do you have vodka?*) in Kazakh language became common.

Vladimir V. Poroshkov,
(Gerchik 1998, 106)

M.G.: Pure ethyl alcohol was a common drink and highly valued "barter" product in the socialist paradise of the old Soviet Union, where one half-liter bottle of vodka cost about one-day average salary. In the workplace and in private lives, one could routinely procure services of a plumber or obtain plywood or some specialized electronic components in short supply by "paying" with alcohol. Heads of units in R&D institutions received alcohol for maintenance of vacuum, optical, or electronic equipment and controlled its use. Only part of the precious liquid was used for "official" purposes though. Alcohol was guarded in safe boxes among most valuable assets and used as a "hard currency" for barter and/or for parties. Alcohol was usually mixed with water (40–50% concentration, similar to vodka) but some liked it straight, 200-degree proof.

"Stadium"

The construction of the main pad that would launch the first ICBM, the first satellite, and the first cosmonaut began in September 1955. This prodigious structure required excavating a pit, nicknamed "the stadium," with the volume of almost 1.4 million m^3 (50 million ft^3), and pouring 25,000 m^3 (900,000 ft^3) of concrete. The first several meters of soil were easily removed by bulldozers. Then, layers of clays and compressed sand were uncovered beneath. With temperatures occasionally falling down to -40°F (-40°C) during the winter, explosives had to be used to fragment the soil. The pit was finally ready by March of the next year, and the soldiers began to pour concrete.

One day Sergei P. Korolev visited the site and went to the "stadium" to take a look at the sight from the bottom of the pit. Chief Engineer Gruntman recalled that Korolev told to him as they climbed back to the surface, "You know, Colonel, from here we will launch not simply rockets. From here, people will conquer space on our rockets and fly to the Moon" (A.Yu. Gruntman, *Tyuratam*, unpublished, 1970). Military ballistic missile units had been activated at Tyuratam by that time and were already busily preparing for the arrival and testing of the first intercontinental rockets. The former director of the military missile research institute NII-4 General Aleksei I. Nesterenko became the first commanding general of the Tyuratam test range.

"From Here the People Will Conquer Space ... and Fly to the Moon"

Fig.14.19. Aerial view of the launchpad for the first ICBM and the gigantic flame pit at Tyuratam in 1957. The first Earth artificial satellite and the first cosmonaut were launched from this site. Photo courtesy of Videocosmos.

Simultaneously with constructing the main launching site, the army units built the test-range infrastructure, including the thermal and power-generating plant, liquid-oxygen plant, water-supply pipelines, airstrip, automobile roads, railroads, storage areas and depots, rocket assembly buildings, army barracks, and living quarters for families of servicemen and civilian personnel. Several remote radio control and telemetry stations were also deployed, some of them constructed more than 2500 km (1500 miles) from Tyuratam.

Growing Infra-structure

At the peak of the construction, 20,000 officers and soldiers worked at Tyuratam. In addition, to engineering and construction brigades, a large number of specialized army engineering units — automobile, railroad, mechanical, communications, and aerodrome battalions — were deployed. Many officers, with the Tyuratam experience under the belts, were later assigned to the construction of other Soviet rocket centers at Saryshagan (antiaircraft and missile defense) and Plesetsk, the latter becoming the busiest world space port for many years.

As the construction progressed, army engineering units were joined by the growing number of civilian specialists who assembled the arriving ground equipment. Vladimir Barmin and his *State Design Bureau of Special Machine Building* designed and built the launch complex for the R-7 ICBM that would be tested at Tyuratam. For the first time the rocket of such size and sophistication had to be transported, erected, fueled, and prepared for launch. Later Barmin's design bureau developed mobile and stationary launch complexes, including silo-based, for R-12 (SS-4) and R-14 (SS-5) IRBMs and for UR-100 (SS-11) ICBMs, and initiated (in 1960s) the design of the permanent lunar base.

Launch Complex

The first launch of the new truly intercontinental ballistic rocket, Korolev's R-7 (SS-6 Sapwood), was attempted on 15 May 1957. The test launch was unsuccessful, and the second and third launches in July also failed. Finally on the fourth launch on 21 August the rocket reached its target area on the Kamchatka Peninsula. The official Soviet news agency TASS declared to the world on 27 August 1957 that the USSR had the ICBM, a rocket capable of delivering a nuclear warhead practically anywhere in the world.

First ICBM Launch

The Soviet Union kept the location of its new testing site at Tyuratam in total secrecy. One of the Soviet military test engineers, Vladimir V. Poroshkov, recalled that

> to preserve secrecy [of the rocket test range] the trains on the [passing nearby] main railroad line Moscow-Tashkent were stopped [on May 15, 1957, during the first launch of the R-7]. The effect [of this action] was exactly the opposite. Passengers of the trains stopped at [the adjacent] stations Dzhusaly and Kazalinsk [about 90 km (55 miles) to the east and west of Tyuratam, respectively] had nothing to do, went out to the [station] platforms, and perfectly observed the rocket launch, although it is difficult to say how they understood what they saw (in those days, the print media wrote nothing about rockets and even the military were not informed). (Gerchik 1998, 101)

The location of the Tyuratam missile center did not remain unknown for too long, however. The American signal intelligence pointed at the increased activities at the newly established rocket test range somewhere east of the Aral Sea in Kazakhstan. The repeated launch attempts of the new ICBM at a secret site, combined with the tough talk of the Soviet leaders, led to the Presidential approval of a new series of American high-altitude overflights of the Soviet territory by the U-2 reconnaissance aircraft. It was obvious that the test site must be located near a major railroad line somewhere east of the Aral Sea.

Secret Location

On 5 August 1957, the U-2 pilot followed the main Turkestan railroad line

> **BAIKONUR**
>
> The first artificial satellite, Sputnik, was launched on 4 October 1957, in about a month and a half after the successful launch of the ICBM. The TASS announcement did not specify the location where the ICBM or Sputnik launches took place.
>
> Later, when the first cosmonaut, Yurii A. Gagarin, was launched on 12 April 1961, the Soviet Union registered this record flight with the *International Aviation Federation*. The Federation is responsible for registering international flight records, which requires specifying the launch site. In a clumsy attempt to preserve the secrecy of the location — that was precisely known to the United States since August 1957 — the Soviet officials deliberately placed the launching site, the cosmodrome, 200 miles (320 km) away from Tyuratam. This remote place, a small coal-mining town Baikonur, gave the official Soviet name of the cosmodrome.

Fig. 14.20. Vladimir P. Barmin, 1909–1993, designed launch complexes for major Soviet rocket systems, including the cosmodromes, mobile launchers, and first silos. Barmin, a specialist in refrigerators, got involved with the rocket launchers during World War II when he organized mass manufacturing of launching stands for Katyusha missiles. Barmin's tombstone in Moscow clearly shows his contribution to the Soviet rocketry. Photo courtesy of Mike Gruntman.

(Orenburg-Tashkent) when he noticed a spur. The spur track led north to a sprawling launch complex 15 miles (25 km) away. The secret Tyuratam test range has thus been found. The first photograph of the launching site was obtained from a large distance and at an oblique angle.

Tyuratam Test Site "Found"

The next two U-2 overflights on August 20 and 21 concentrated on the area further east from Tyuratam, including the nuclear testing ground near Semipalatinsk (now Kazakhstan). Had these flights covered the Tyuratam rocket site, they could have "caught" the Soviet ICBM being transported to or erected on the launching pad in preparation for its first successful launch on August 21.

U-2s Over Kazakhstan

In the meantimc the U-2 mission on August 21 scored a different "hit" by flying over the Semipalatinsk grounds just four hours before a half-megaton nuclear test explosion. In addition, the pilot flew for the first time over another rapidly growing missile development center on the western end of Lake Balkhash. This site, Saryshagan, would emerge in a few years as the main test center of the Soviet development efforts in antiballistic missile defense systems.

Missile Range "Named"

The next U-2 flew on August 28 directly over the Tyuratam launching site, only a week after the first successful launch of the Soviet ICBM. The most reliable maps (available in the United States) of the area were German military maps captured at the end of World War II. The location of the new rocket center was identified on the map near the *Bf.*, or *Bahnhof* (*station*), Tyuratam. The chief information officer of the CIA's *National Photographic Interpretation Center* (NPIC) Dino A. Brugioni named the launching site *Tyuratam*.

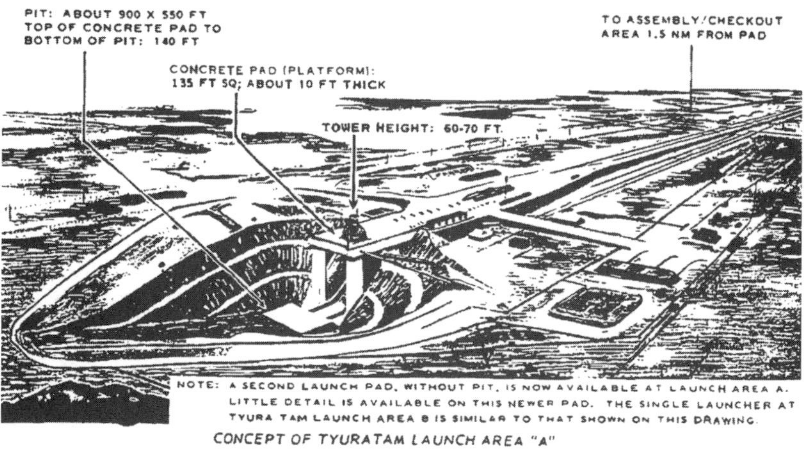

Fig. 14.21. Top: one of the first U-2 photographs of the Tyuratam launching pad in 1950s. The first ICBM, the first satellite, and the first cosmonaut were launched form this site. Bottom: sketch of this launching site in 1961 based on U-2 and Corona satellite photographs. Figures courtesy of Central Intelligence Agency.

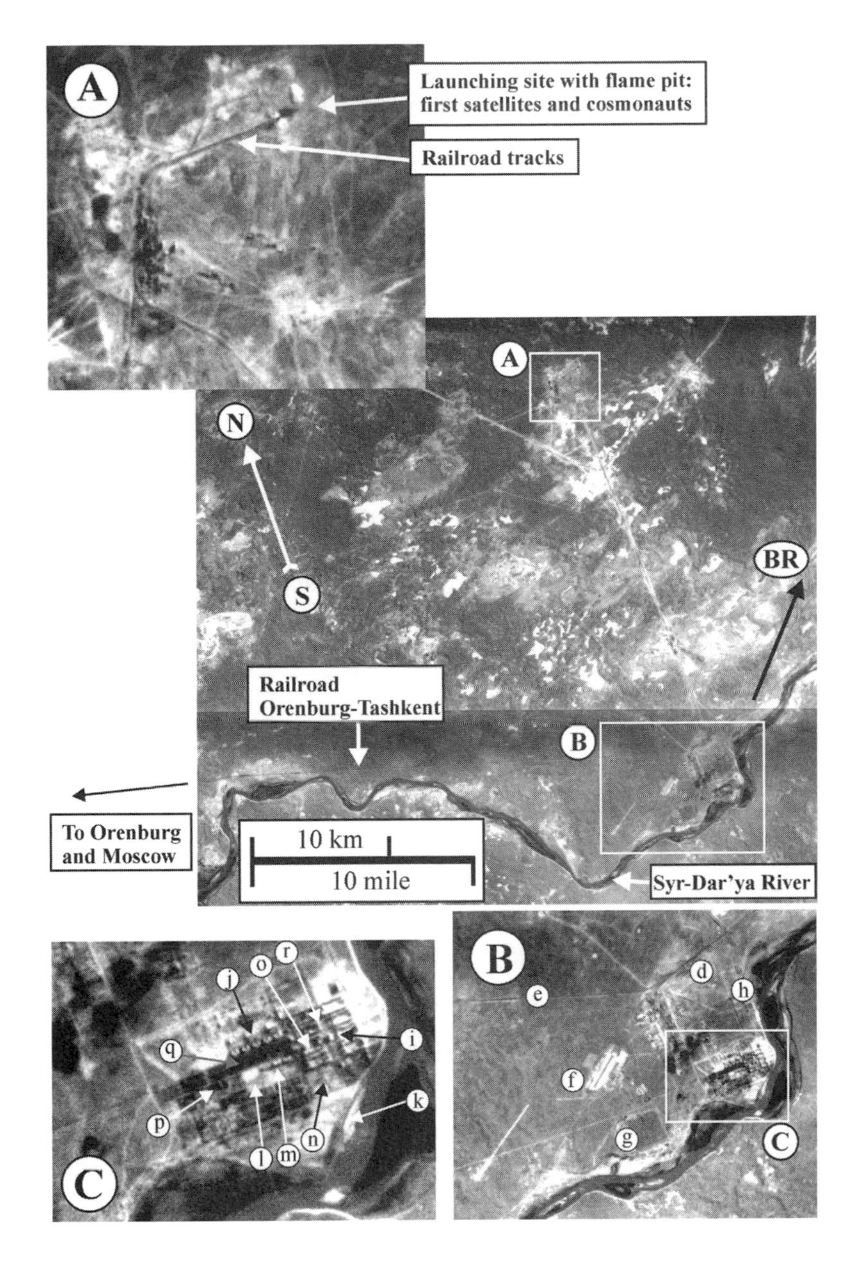

Sprawling Tyuratam Complex in 1962

Fig. 14.22. Composite satellite image (center) of the Tyuratam launch complex, the cosmodrome, obtained by Corona (Mission 9035) on 30 May 1962. A) First space launching pad; B) and C) settlement, later called *Leninsk*; arrow "BR" shows the direction toward Baikonur 320 km (200 mile) away; d) Tyuratam railroad station; e) railroad Orenburg-Tashkent; f) temporary airstrip (airfield *Krainii* under construction 3 km to the west); g) agricultural farm organized by the military to supplement rations; h) water pumping station; i) garrison cultural center; j) military barracks; k) river beach; l) stadium; m) hospital; n) *Soldatskii* (soldiers') *Park*; o) school; p) headquarters of the Construction Engineering Directorate and the summer movie theater; q) park; r) headquarters of the commander of the test range and rocket units. Courtesy of Mike Gruntman.

Brugioni followed the intelligence community practice of naming installations after the nearby towns.

After the overflights of 1957, the U-2 had revisited Tyuratam two years later on 9 July 1959, when two more launching pads were spotted under construction. Francis Gary Powers was the last who flew the U-2 over Tyuratam, provoking an international accident, when the Soviet surface-to-air missile shot his plane down on 1 May 1960.

Launch Telemetry Intercepted

Powers's fateful flight ended deep overflights of the USSR by the U-2. Signal intelligence thus became particularly important for gaining insight into the activities at this main Soviet rocket test center. Usually, when major tests were prepared at Tyuratam, American electronic intelligence aircraft cruised along the Soviet-Iranian border. On 9 June 1959, the CIA's U-2 and Air Force's RB-57D Canberra intercepted, for the first time, 80 s of the telemetry stream from the first stage of the Soviet ICBM launched from Tyuratam.

Tyuratam From Space

The hiatus in photoreconnaissance of Tyuratam did not last long. The Corona program was finally approaching the operational phase. The first time Tyuratam was photographed, this time from space, in December 1960 by *Discoverer XVIII*. Since then, the test site was regularly observed by reconnaissance spacecraft. Tyuratam veterans also recalled that efforts in camouflage against space reconnaissance started at the test range some time in 1960.

24 October 1960

Tragedy Struck

A major accident occurred at Tyuratam on 24 October 1960. The second stage of the fully fueled R-16 ICBM (SS-7) was accidentally activated during the preparation for a test launch. One-hundred tons of propellant ignited, and gas bottles of the propellant tank pressurization system exploded. Seventy-four military and civilian personnel died in the fire, and 49 were burned and injured, many of the latter died later in hospitals. In total, 92 men died as a result of the accident. Among the dead were Commander-in-Chief of the Soviet Strategic Rocket Forces Marshal Mitrofan I. Nedelin and deputies of the rocket designer M.K. Yangel. This accident led to a thorough review of safety procedures. On the same day of the year in 1963, a fire broke out during fueling of the R-9A ICBM (SS-8) in a silo. All eight technician working in the silo perished.

Fig.14.23. All Soviet manned space missions originated from Tyuratam, with cosmonauts spending some time there in preparation for their flights. First woman-cosmonaut Valentina Tereshkova (left) on 13 June 1963, three days before her space flight. Photo courtesy of Mike Gruntman.

With the disintegration of the Soviet Union, the Tyuratam cosmodrome became a part of the territory of the new independent state, Kazakhstan. The Russian Federation now leases 6000 km^2 (2300 sq. mile) enclosing the cosmodrome from Kazakhstan. The Soviet Strategic Rocket Forces and the Russian Space Agency manage the facility.

Tyuratam in Foreign Country

> **FIRST ICBMs TARGETED**
>
> ... [The rocket launch base for the first R-7 ICBMs] was established ... near the railroad station Plesetsk. Equipping the first two launch sites of the first phase of the base development was completed in 1959. The first test launch was performed on 30 July, and the [two ICBM] rockets were put on combat duty on December 15, 1959. Two more launch sites were added in 1961. All of them served until the middle of 1964 when the rocket base was reorganized into a cosmodrome [for space launches] ... .
>
> The [first ICBM] rockets were targeted one at New York City and the other at Washington, D.C. [Two other cities] Chicago, Ill., and Los Angeles, Calif., were considered as the targets for two other rockets. I did not see the maps with the targets, but there was a lot of talk in the [Soviet] rocket community about the first cities-hostages in the United States. My father [Nikita S. Khrushchev] also mentioned them.
>
> Sergei N. Khrushchev (2000, 255)

The other most important Soviet launch site is located in Russia's north. This site, Plesetsk, was activated first as a base for Korolev's R-7 ICBMs on combat duty. The shortest trajectory to reach the "main adversary," the United States, was across the Arctic ocean. Therefore, the government decree of 11 January 1957 authorized a new supersecret rocket base in the north, called *Object Angara*, for operational ICBMs. The new site was several miles off the Moscow-Archangel railroad line, 500 miles (800 km) north from the Soviet capital.

Object Angara

Plesetsk

On 15 December 1959, the first Soviet rocket unit armed with the R-7 ICBMs (SS-6) was declared operational at Plesetsk. Two days later, the USSR Council of Ministers formed a new branch of the Soviet armed forces, the Strategic Rocket Forces.

First ICBMs

Beginning with two ICBMs in combat readiness in the early 1960, Plesetsk added two more operational ICBMs by July 1961. The first ICBMs could carry 3-Mt nuclear warheads to a distance of 5000 miles (8000 miles) and targeted the most important centers in the United States. It took 12 h to prepare the missile for launch; the rocket could stay fully fueled (kerosene and liquid oxygen) for no more than 8 h. During the Cuban crisis, the ICBMs were put on heightened alert.

World's Busiest Space Port

Because of its high geographical latitude (62.8° N) Plesetsk is at a disadvantage for spacecraft launches into low-inclination orbits, particularly for launches into geostationary orbit. However, Plesetsk allows safe space launches into northern directions placing spacecraft into polar orbits. Such orbits are favored by intelligence payloads.

The first satellite, Cosmos-112, was launched from Plesetsk on 17 March 1966. With time Plesetsk cosmodrome has become the busiest spaceport of the world. During its history, more than 1500 launchers have placed 1900 spacecraft into orbit. In addition, more than 500 ICBM tests were conducted in Plesetsk.

15. THE BREAKTHROUGH

SPUTNIK

The Soviet breakthrough into space can be directly traced back to the government decision on 4 December 1950, authorizing a feasibility study of intercontinental missiles "with the range 5000–10000 km [3100–6200 miles] and warhead mass 1–10 tonne" (Semenov 1996, 73). Deputy Prime Minister Vyacheslav A. Malyshev narrowed the target technical specifications in October 1953, requiring delivery of a nuclear charge with a mass 3000 kg (6600 lb) and a total warhead mass of 5500 kg (12,200 lb).

Fig.15.1. Air Force Colonel Mikhail K. Tikhonravov (photo ca. 1951) began his rocket career in GIRD in 1930s. After World War II, Tikhonravov significantly contributed to the development of the concepts and designs of the first ICBM R-7, first artificial satellite *Sputnik*, first manned spaceship *Vostok*, and first Soviet reconnaissance satellite *Zenit-2*. Photo courtesy of Videocosmos.

Origins of Soviet ICBM

The design bureau of Sergei Korolev advocated this initiative to develop an ICBM capable of carrying a nuclear warhead to any area in the world and especially to the territory of the main adversary. (The Soviet vernacular reserved the term "the main adversary" for the United States.) Leading Soviet specialists and their vast design bureaus, research institutions, and manufacturing facilities would contribute to the program concentrating on propulsion, guidance, navigation, control, communications, telemetry, and other key rocket systems. The government decree of 20 May 1954 made the ICBM program a top national priority.

The new intercontinental ballistic missile, designated R-7 (SS-6 Sapwood), had one central stage and four side sections attached to the central core. Each section was equipped with its own engine. A veteran Soviet rocketeer Mikhail K. Tikhonravov, who was one of the GIRD's team leaders in 1930s, described a concept of several mechanically joined rockets in a "packet" or "package" (*paket* in Russian) in July 1948. (Tikhonravov advanced Tsiolkovsky's thoughts on this issue dating back to 1934.) In a technical report three years later, Korolev proposed a package of rockets of different sizes, in contrast to the identical rockets of the original concept. Starting in the late 1940s, Tikhonravov has worked in a military missile research institute, NII-4, where his group laid theoretical foundations for the practical work that would follow on the first space launcher and the first artificial satellite. Tikhonravov joined Korolev's OKB-1 in 1956. There, he significantly contributed to the design of the first satellite, first manned spaceship, and first Soviet space reconnaissance system. It was reportedly Tikhonravov who argued for using the word *kosmonavt*, or *cosmonaut*, in Russian instead of the American *astronaut* (*astronavt* in Russian). Tikhonravov retired in 1960.

"Rocket Packet"

Mikhail K. Tikhonravov

Fig.15.2. The first ICBM R-7 at the Tyuratam missile test range in May–June 1957. Two tall service towers that are shown detached from the rocket body allowed access to all parts of the rocket on the launching pad. Four inclined side masts held the rocket at the "waist," at a point somewhat higher than the rocket's center of gravity. Photo courtesy of Videocosmos.

All five rocket engines of the R-7 were started simultaneously. The central sustainer section was designed to burn for 283 s while the firing time of the four side sections was 115 s. All of the side sections separated at the same time after their cutoff. The R-7 was customarily called a two-stage rocket; it was similar, however, to the "one-and-one-half-stage" configuration of the American Atlas.

R-7 ICBM

Valentin P. Glushko's OKB-456 designed and built liquid-oxygen-kerosene engines for the R-7. The engines produced 80–90 tonnes of thrust each and achieved specific impulse 250 s at sea level. The fully fueled rocket weighed 280 tonnes at launch, with the dry weight 27 tonnes. The warhead section, the "payload," accounted for about 5000 kg. The projected rocket range was 8000 km (5000 miles) with the impact accuracy ±10 km (±6 miles). Following the idea of Korolev's deputy Vassilii P. Mishin, the R-7 was suspended on the launching pad from the four side masts that held the rocket body at a point somewhat higher than its center of gravity. This clever suspension method decreased loads acting at the base of the fully fueled rocket standing on a pad and consequently allowed reduction of the structural weight of the rocket body.

Rocket Suspension

Fig.15.3. Sergei P. Korolev was the main driving force behind the first ICBM, first artificial satellite, first manned spaceflight, and many other first Soviet satellite systems. Photo (early 1960s) courtesy of S.P. Korolev Memorial House-Museum, Moscow.

R-7 and Atlas

The R-7 was significantly heavier and taller than the Atlas ICBM. When the Atlas program was restarted as the MX-1593 in January 1951, it called for delivery of a 8000-lb (3600-kg) warhead by a rocket 160 ft (49 m) tall with a launch mass 670,000 lb (304,000 kg) and powered by one central (sustainer) and four or six side (booster) engines. In the spring of 1953, the Atlas had "become," on paper, smaller as a result of the projected reduction of the warhead mass and improved engine performance, the latter because of a more efficient kerosene-oxygen-propellant combination. The then-envisioned rocket had height 110 ft (33.5 m), mass 440,000 lb (200,000 kg), one central sustainer, and four jettisonable side engines. This variant of the American missile was strikingly similar to the configuration and the height of the R-7, though the Atlas was somewhat lighter. As the nuclear weapons technology advanced further, the mass of the American warheads became smaller, and their higher yields led to relaxation of the missile accuracy requirement, the circular error probable (CEP), to 1 mile (1.6 km) from the original 1500 ft (460 m). These improvements in warheads allowed reduction of the number of Atlas's side engines to two and consequently made the American ICBM significantly smaller and lighter than the Soviet R-7.

DECISIVE 20 YEARS IN ROCKET DEVELOPMENT

Comparative Performance Characteristics of the First Operational Soviet (R-7) and American (Atlas D) ICBMs and the German A-4 (V-2)

	R-7	Atlas-D	A-4
Nominal range			
Nominal range, km	8000	10,000	250
Nominal range, nautical miles	4300	5500	135
Length			
Length, m	34.2	24.7	14.0
Length, ft	112	81	46
Total (fueled) rocket mass			
Total mass, kg	280,000	119,000	12,700
Total mass, lb	617,300	263,000	28,000
Propellant mass			
Mass, kg	253,000	112,000	9700
Mass, lb	558,000	248,000	21,400
Mass fraction	0.904	0.943	0.764
Reentry vehicle (payload) mass			
Mass, kg	5400	1700*	1000
Mass, lb	12,000	3700*	2200
Mass fraction	0.0193	0.014*	0.079
Structural mass (total mass minus propellant and payload)			
Mass, kg	21,500	5100	2000
Mass, lb	47,000	11,300	4400
Mass fraction	0.0768	0.043	0.157
Thrust			
At sea level, kN	3960	1640	245
At sea level, lbf	890,000	368,000	55,000
Thrust-to-weight ratio	1.44	1.40	1.96
Specific impulse			
At sea level, s	250	252**	210

* Mk II reentry vehicle with W-49 warhead
** booster engines

Korolev rushed the development of the R-7. The first nonflight model of the missile arrived to the Tyuratam test site for training of the ground crews in December 1956. The first flight model followed on 3 March 1957. Everything was now ready for the ICBM test launch on a full range. The target impact area was instrumented 6314 km (3924 miles) away at the Kamchatka Peninsula. Fifteen tracking and observation stations with radars and optical equipment spanned a huge distance across the Soviet Central Asia and Siberia to monitor the flight.

R-7 Development Rushed

Fig.15.4. The R-7 ICBM being readied for launch at Tyuratam in May–June 1957. The first ICBM, the first artificial satellite, and the first cosmonaut were launched from this site. This installation is operational until this day and is used for the Soyuz space launchers. The launch platform stands out as a cliff over the gigantic flame pit beneath. Photo courtesy of Videocosmos.

The first launch of the R-7 was attempted on 15 May 1957. The service crews did not have experience in handling such a large and complex rocket and prelaunch ground testing lasted almost ten days. Tests of the electrical systems alone took more than 110 hours. The ambitious first test flight was nominal during the first 70 s. The later report to the Central Committee of the Communist Party of the USSR stated that

First Attempt

> beginning with the 97th second [of flight], large angular deviations of the rocket [orientation] appeared because of the loss of control due to fire in the tail section of one of the side engines which had begun from the moment of the launch. Because of this development, the engines were automatically shut down ... at the 103th second of flight. (Derevyashkin and Baichurin 2000, 68)

BURNED WARHEAD AND DESIGNER SOLIDARITY

Head of a design bureau Grigorii V. Kisun'ko led in 1950s and early 1960s development of the Soviet antiballistic missile defense system that demonstrated the first intercept of a ballistic warhead in 1961. Kisun'ko writes in his memoirs (Kisun'ko 1996) that he was the person who pinpointed the cause of the warhead disintegration in the Korolev's first successful full-range flight of the R-7 on 21 August 1957.

According to Kisun'ko, defense officials ordered him to Tyuratam in late August 1957 to provide expert help to Korolev (who was at the site after the R-7 launch) in finding what happened to the warhead. No "warhead traces or debris" were found at the impact area in Kamchatka, the telemetry from the simulated warhead was lost simultaneously on all radio frequencies, and Korolev was confident that the quality of the heat shield "was beyond any doubt."

Kisun'ko began with examining the telemetry from the warhead motion sensors recorded after the separation. He discovered that the separated warhead turned alternatively to one side and then to the other. Kisun'ko then concluded that "the velocity of separation of the warhead from the rocket [second stage] body was very small ... [and] because of its spin it [the warhead] hit the [rocket] body by the [warhead's] one side, changed the direction of the spin to the opposite and hit the [rocket] body by the other side ... swinging like a pendulum ... . The impacts ... destroyed the heat shield and, unprotected, it [the warhead] reentered the atmosphere and ultimately burned there" (Kisun'ko 1996, 372).

Kisun'ko describes that he met with Korolev the next day and told him, without witnesses, about his conclusions. "Did you tell it to anybody?" asked Korolev. "You are the first to whom I told and now I have forgotten [lost from my memory] all my findings," answered Kisun'ko and added, "[it is] the designer solidarity" (Kisun'ko 1996, 373).

Kisun'ko recalled that some time later at a banquet celebrating a new successful test flight of the R-7, "Korolev took ... [Kisun'ko] aside to the [Korolev's] table, poured cognac in two tumblers, and offered a toast, 'To the designer solidarity!'" (Kisun'ko 1996, 376).

Difficult Launches

The rocket fell 400 km (250 miles) downrange. The cause of the fire was traced down to a leak in a kerosene fuel pipe. The evaluation of the rocket failure involved examination of the telemetry recorded on oscilloscope photographic films 20 km long.

The second R-7 test launch was scheduled for 11 June 1957, and was tried, unsuccessfully, three times. Frozen liquid-oxygen valves aborted the first two attempts. Then, the incorrectly mounted valve for nitrogen purge of liquid-oxygen pipes led to failure. The rocket was safely removed from the launching pad and returned to the assembly building.

ICBM Achieved

The third launch was attempted on 12 July. The liftoff was again successful, but the rocket was destroyed after 33 s because of uncontrollable spin caused by the failure of control electronics.

Finally, the fourth launch on 21 August 1957 succeeded, and the R-7 reached its target area at the Kamchatka Peninsula. The simulated warhead did not impact the ground, however, and disintegrated in the atmosphere, although the warhead heat shields were believed to be highly reliable. (Soviet scientist Vsevolod S.

TASS ANNOUNCEMENT

A launch of a super-long-range, intercontinental, multistage ballistic rocket was recently performed.

The rocket test was successful and it has confirmed the correctness of the calculations and of the selected design. The rocket flew at a very high, not-achieved-before, altitude. After covering in a short time a huge distance, the rocket had impacted in the planned area.

The obtained results show that there is a possibility of launching rockets to any area in the world. The solution of the problem of building intercontinental ballistic rockets would allow reaching remote regions without strategic aviation which is presently considered vulnerable to the modern antiaircraft defense ... .

Recently, a number of [test] explosions of atomic and thermonuclear (hydrogen) weapons were conducted in the Soviet Union. The explosions were performed at a high altitude in order to assure safety of the population. The tests have been successful ... .

TASS Announcement, 27 August 1957

Warhead Problems

Avduevsky directed development of heat shields protecting warheads during atmospheric reentry.) The failure was attributed to a likely collision of the warhead with the rocket second stage after the warhead separated with a too small velocity.

The next successful full-range rocket flight followed on 7 September 1957, but the warhead again disintegrated. The problem was corrected during the next several months by modifying the warhead design and improving the separation system, which included an increase in the delay between the main engine cutoff and warhead separation from 6 to 10 s. The first completely successful ICBM flight was finally accomplished on 29 March 1958.

SOVIET SATELLITE

Alexander Nesmeyanov, president of the Soviet Academy of Sciences, reports in a recent issue of [the official newspaper of the Communist Party] *Pravda* that Soviet scientists "have created the rockets and all the instruments and equipment necessary to solve the problems of the artificial earth satellite." Nesmeyanov said also that the first satellite launched by the Soviet Union would follow an orbit not more than a few hundred kilometers above the earth's surface (a kilometer is roughly five-eights of a mile). The *Pravda* article gave no date for the launching of the satellite; however, preparations have aimed at a target date sometime within the International Geophysical Year.

Science, 21 June 1957

Six days after the first full-range flight of the R-7 on 21 August 1957, the official Soviet news agency, TASS, declared to the world that the USSR had demonstrated the ICBM. Simultaneously, the successful recent nuclear tests were also announced. The Soviet Union had thus realized the nuclear-tipped ICBM. The first unit armed with the R-7 ICBMs, called SS-6 in the West, was declared operational in December 1959. At the same time, a new branch of the Soviet armed forces, the Strategic Rocket Forces, was activated. On 20

R-7 (SS-6) Deployed

January 1960, the Soviet Army formally accepted the R-7 for deployment and combat duty.

Artificial Satellite

While concentrating on the development of the R-7 ICBM, Korolev continued to advocate launching an artificial Earth-orbiting satellite. Tikhonravov, whose group worked on the satellite since 1948 and who later would become a scientific advisor to the satellite design program, aided Korolev with the "scientific ammunition" for lobbying on behalf of this idea. In 1954 Korolev forwarded Tikhonravov's report "About Artificial Earth Satellite" to the government requesting the permission to establish a special research department in his design bureau focused on the satellite problem.

Plans Announced

The Soviet military were not particularly thrilled by the satellite and understandably worried that it might distract Korolev's resources from achieving the ICBM. In July and August of 1955, both the United States and the Soviet Union announced their plans to launch scientific satellites as part of the International Geophysical Year (IGY) scheduled for an 18-month period from July 1957 to December 1958. After these announcements, Soviet officials periodically confirmed the plans to launch a satellite. The Soviet government decree of 30 January 1956 put the Korolev's program in high gear by directing him to design, build, and launch an artificial satellite using a modified R-7 rocket.

INTERNATIONAL GEOPHYSICAL YEAR (IGY)

The International Geophysical Year (IGY) was conducted from July 1957 to December 1958 under the aegis of the International Council of Scientific Unions (ICSU). A group of American scientists led by Lloyd V. Berkner began to work on the concept of the IGY in 1950. In 1952, ICSU approved the idea and appointed a committee to make preparations for a broad synoptic study of various geophysical phenomena. The IGY followed the pattern of the first (1882–1883) and second (1932–1933) International Polar Years.

More than 70 countries participated in the IGY, which included studies of cosmic rays, auroras, geomagnetism, ionosphere, and upper atmosphere, and research projects in meteorology, seismology, and oceanography. The timing of the IGY coincided with maximum of the 11-year cycle of solar activity and thus presented a special interest for study of space environment. The IGY was marked by launch of the first Earth-orbiting scientific satellites by the Soviet Union and the United States.

Object D

The Korolev's satellite, called *Object D*, was designed to have a total mass 1000–1400 kg (2200–3090 lb), including 200–300 kg (440–660 lb) of scientific payload, and be powered by body-mounted solar cells. The initial launch plans were soon in danger, however, because the R-7's engines demonstrated specific impulse only 304 s in vacuum instead of the projected 309 s. In addition, the development of the scientific payload quickly fell behind the schedule. The new launch date was fixed for the spring 1958.

Pushing All the Right Buttons

Knowing about the announced American program to launch a satellite during the International Geophysical Year, Korolev's OKB-1 proposed to quickly launch — in order to beat the American competition — a much simpler and much lighter satellite. On 5 January 1957, Korolev wrote to the USSR Council of

WE ARE ASKING FOR PERMISSION

We are asking for permission to prepare and launch two [R-7] rockets modified to the variant [of launchers] of artificial earth satellites during the period of April-June 1957 before the official beginning of the International Geophysical Year conducted from July 1957 to December 1958.

At this time, the [R-7 ICBM] rockets are being tested on the ground and, according to the test program, the first rocket will have been prepared for launch by March 1957.

... By introducing certain modifications, the rocket can be made in a variant [allowing launch] of an artificial earth satellite with a payload weight of 25 kg for [scientific] instruments.

It is thus possible to launch to an orbit of an artificial earth satellite with an altitude 225–500 km the central [sustainer] stage of the rocket with weight 7700 kg and separating spherical container of the satellite with a diameter about 450 mm and weight 40–50 kg.

A special shortwave transmitting [radio] station with power sources for 7–10 days of operation can be installed among the satellite's instruments.

Two [R-7] rockets modified to this [satellite] variant could be ready in April-June 1957 and launched immediately after the first successful launches of the intercontinental [ballistic] missile.

The launch of the [earth satellite] rockets will allow simultaneous flight verification of a number of [technical] questions that were scheduled for the [ICBM] flight test program (launch; functioning of the side and central power plants; functioning of the flight control system; separation; etc.)

... At the same time, a very intensive preparation has been under way in the United States of America for launching an artificial earth satellite. The most well-known project is called "Vanguard" and it is based on a three-stage launcher with the rocket "Redstone" as the first stage in one of the variants. The satellites are a spherical container with a diameter 50 cm and weight about 10 kg.

In September 1956, the USA attempted to launch a three-stage rocket with a satellite at the Patrick [Air Force] Base, Florida, keeping the launch in secret.

The Americans failed to launch the satellite, and the third stage of the rocket, probably with a sphere-like container, had flown over a distance of about 3,000 miles, or about 4800 km, which was then publicized in the press as an outstanding national record. They emphasized that the American rockets flew higher and farther than all rockets in the world, including Soviet rockets.

From separate reports from the press, the USA is preparing new attempts of launching an artificial earth satellite in the nearest months, desiring to be the first [country to achieve the satellite] at any price ... .

S.P. Korolev, 5 January 1957, Memo to the USSR Council of Ministers

---

M.G.: The attempted launch in Florida referred to by Korolev was a test flight of the Jupiter C conducted by the U.S. Army on 20 September 1956. The missile was actually a four stage configuration with the last fourth stage inactive and carried as ballast. Had the Army been permitted to use the "live" stage, as was advocated by ABMA's Medaris and von Braun, the vehicle could have obtained sufficient velocity to place the world first artificial satellite in orbit at that time.

Ministers asking for the permission to launch such a satellite. To obtain a positive decision on his request, Korolev was pushing all the "hot buttons" in the Kremlin, emphasizing the progress of the rival American program. The Soviet government acted promptly and approved the OKB-1's proposal on 15 February 1957. The new satellite was called *Object PS*, (*Prosteishii Sputnik*), or the *simplest satellite*. (In Russian, *sputnik* literally means a *fellow traveller* or a *travel companion*.) The development of the originally planned Object D also continued, and it would ultimately be launched on 15 May 1958, as Sputnik 3, with mass 1327 kg (2924 lb) and operated for 692 days.

Simplest Satellite PS

The new PS-1 was built as a hermetically sealed sphere with a diameter 58 cm (22.8 in.) and pressurized by dry nitrogen at 1.3 atm (19 psi). Two pairs of antennas were 2.4 m (7.9 ft) and 2.9 m (9.5 ft) long. The radio-system transmitter had 1 W of power and sent signals with the duration 0.4 s alternatively at 7.5-m and 15-m wavelengths (approximately 40 and 20 MHz). Three silver-zinc batteries provided power for the satellite and were expected to last for two weeks.

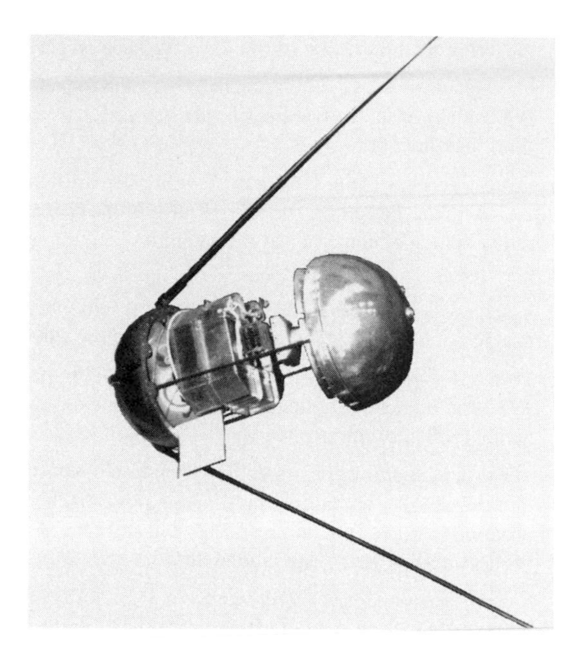

Fig.15.5. First artificial satellite Sputnik 1 in the museum of the Energia Corporation. The spacecraft consists of two hemispheres made of aluminum alloy and attached together by 36 bolts. The sphere diameter is 58 cm (22.8 in.) and the satellite mass is 83.6 kg (184.3 lb). One can see two horizontal bars supporting the expanded spacecraft and the hanging technical specifications sheet. Photo courtesy of Mike Gruntman.

The top-level breakdown of the satellite total mass of 83.6 kg (184.3 lb) was as follows: structure — 13.9 kg (30.6 lb), antennas — 8.4 kg (18.5 lb), and payload — 58.4 kg (128.7 lb). The spacecraft power unit, with mass 51.0 kg (112.4 lb), accounted for 87% of the payload mass.

4 October 1957

Launched from Tyuratam on 4 October 1957, the SP-1 has reached the orbit together with the sustainer stage of the rocket. The side sections of the modified R-7 separated from the sustainer on the 116th second of the flight. The main engine of the sustainer, or the second stage, was cut off at an altitude 228.6 km (142.1 miles). The satellite separated from the rocket 20 s later on the 315th second after launch. In addition to 2.73-m/s (9-ft/s) separation velocity, the rocket body was slowed down a little by venting gas remaining in the oxidizer tanks through valves opened in the forward direction.

SPUTNIK LAUNCH

... As a result of big intensive work by scientific-research institutes and design bureaus, the first-in-the-world artificial satellite has been created. On October 4, 1957, the first satellite was successfully launched in the USSR .... The successful launch of the first man-made earth satellite makes a most great contribution to the treasure-house of world science and culture .... Artificial earth satellites will pave the way to interplanetary travel, and our contemporaries are apparently destined to witness how the freed and conscientious labor of the people of the new socialist society makes the most daring dreams of mankind a reality.

TASS Announcement, in *Pravda*, 5 October 1957

Sputnik in Orbit

The Sputnik launch direction followed the trajectory of the ICBM test flights toward Klyuchi at the Kamchatka Peninsula, resulting in an orbit with inclination 65 deg. It was the fifth launch of the R-7 and the first space launch. The first artificial satellite of the Earth, SP-1 or Sputnik 1, had thus been born, and the Russian word "sputnik" entered many languages.

Sputnik orbital parameters:	Planned	Achieved
Perigee altitude, km (mile)	228 (142)	228 (142)
Apogee altitude, km (mile)	1450 (901)	947 (589)
Period, minute	101.5	96.2

The year 1957 was the year of solar maximum of the 11-year solar cycle. Solar activity reaches maximum during this phase of the cycle, and the enhanced

UNDER HIS REAL NAME

A commemorative meeting was held on 17 September 1957 in downtown Moscow to celebrate the one-hundredth anniversary of Konstantin E. Tsiolkovsky. Sergei P. Korolev addressed a screened audience of party functionaries, scientists, and engineers. Korolev stated in his speech that "in the near future first test launches of artificial earth satellites will be conducted in the USSR and USA for scientific purposes."

The sanitized version of Korolev's presentation was published on the same day in the official Communist Party newspaper *Pravda*, 17 September 1957. One-fourth of this issue was devoted to Tsiolkovsky's anniversary. Korolev did not mention artificial satellites in the article but noted that "many high-altitude rocket flights will be conducted for research purposes according to the program of the [USSR] Academy of Sciences during the current International Geophysical Year." The article focused on the achievements and legacy of Konstantin Tsiolkovsky. Korolev also complained that "a large part of foreign scientists, who widely use the ideas and works of Tsiolkovsky, purposefully do not mention the name of the author."

It is interesting that this Korolev article was signed by his own name, *S. Korolev, Corresponding Member of the USSR Academy of Sciences*. Being shrouded by the veil of secrecy, his later publications (numerous after the launch of Sputnik) would be signed by a cover name *Prof. K. Serge'ev*.

solar output in the X-ray and extreme ultraviolet spectral ranges heats the upper atmosphere and ionosphere to the highest temperatures during the 11-year period. As a result, the atmosphere expands outwards, increasing aerodynamic drag on satellites in low-Earth orbit and consequently reducing their lifetime.

Solar Maximum

The initial perigee of the Sputnik orbit was rather low at an altitude 228 km (142 miles), and the atmospheric drag was correspondingly high. To make things worse, solar maximum in 1957 was characterized by an unusually high solar activity, much higher than typically observed during solar maxima. As a result, the satellite stayed in orbit only until 4 January 1958, making 1440 revolutions. The rocket sustainer stage made 882 revolutions and reentered the atmosphere on 2 December 1957.

Fig. 15.6. Chief designers of space systems on 4 October 1957, in Tyuratam, after the launch of the first artificial satellite of the Earth, Sputnik. Left to right: Aleksei F. Bogomolov (b.1913), space telemetry systems; Mikhail S. Ryazansky (1909–1987), radio guidance systems; Nikolai A. Pilyugin (1908–1982), autonomous guidance systems; Sergei P. Korolev (1906–1966), rockets and spacecraft; Valentin P. Glushko (1908–1989), rocket engines; Vladimir P. Barmin (1909–1993), ground launch systems; and Viktor I. Kuznetsov (1913–1991), inertial guidance and gyroscopic and control systems. Photo courtesy of NPO Energomash, Khimki, Russia.

Two New "Stars"

Two new artificial stars thus appeared in the sky. The substantially larger sustainer was seen as a 100 times brighter object in the night sky, and it was much easier to observe it by the naked eye than the barely visible Sputnik. Apparent visual magnitudes of the R-7 sustainer stage and Sputnik were $m = +1$ and $+6$, respectively. (The scale of stellar magnitudes assigns smaller values m to brighter stars. The average unaided eye would see stars with apparent visual magnitudes $m = +5$ and brighter, i.e., $m < +5$, under typical conditions.)

The frequencies of Sputnik's transmitter, 20 and 40 MHz, came as a surprise to scientists and engineers because these frequencies were different from the

Unexpected Radio Frequencies

108 MHz (wavelength 2.77 m) agreed upon by the IGY's committees. Consequently, the *Minitrack* tracking stations being prepared and deployed for the American satellite program were not able, initially, to track Sputnik. The engineers and technicians improvised and did their best to quickly design, build, and deploy new antennas and to adjust electronic equipment. By mid-October, several American stations already tracked Sputnik, and the whole Minitrack network was ready when the second Soviet satellite, Sputnik 2, was launched in early November.

Crowning Achievement

The launch of Sputnik was a crowning achievement of the powerful Soviet missile and space establishment. Numerous design bureaus, research institutions, and industrial plants actively contributed to the program. Many top Soviet specialists and managers who led this effort had met and established their professional links during the days when they spent time in the after-war Germany as a selected group of trusted scientists and engineers learning about Germany's scientific and technological achievements.

Fig.15.7. The Energia–Buran vehicle combination engraved on the tombstone of Valentin P. Glushko at a cemetery in Moscow.

After Korolev's death in January 1966, his deputy Vassilii P. Mishin succeeded him as head of the design bureau TsKBEM. In 1974 Mishin was relieved from his duties, and the development of the gigantic N-1 space launcher for the manned lunar expedition was terminated.

In a major reorganization of the Soviet program, Glushko was appointed head of a new enterprise that combined both the original Glushko's establishment in Khimki (present NPO Energomash) and Korolev's design bureau and plant in Podlipki (present RKK Energia). With the enormous resources concentrated in his hands, Glushko led development of the world's most powerful rocket *Energia* and the space shuttle *Buran*. (The Buran plane was built by NPO Molniya under Gleb E. Lozino-Lozinsky, b.1909.) This was a "symmetric response" of the Soviet leadership to the American Space Shuttle. The Energia successfully carried Buran, without a crew, into orbit in November 1988. This achievement culminated an outstanding career of Glushko in space technology. After Glushko's death in 1989, NPO Energomash and RKK Energia split and became again independent corporations.

Photo courtesy of Mike Gruntman.

The relations within the Soviet rocket establishment were not simple and serene, however. The bitter rivalry between Sergei P. Korolev and Valentin P. Glushko continued unabated. The eternal question remained obviously unresolved: What is more important, a "horse," that is, a rocket engine (Glushko), or a "cart," that is, integration of a rocket system (Korolev)? Another Soviet missile prince, Mikhail K. Yangel had already established his base in Dnepropetrovsk in the Ukraine and flexed the muscles. Yangel was emerging as the main challenger to Korolev in ballistic missiles, focusing on storable noncryogenic propellants and, as a result, favored by the military brass. In addition, Vladimir N. Chelomei would soon challenge both Korolev and Yangel with his "universal rocket," UR, resulting in a family of ICBMs and leading eventually to the today's powerful Proton space launcher.

Rivalry in Rocket and Space Establishment

Fig. 15.8. Monuments to Sergei P. Korolev (left) and Mstislav V. Keldysh (right) in Moscow. In a secrecy-obsessed totalitarian society, they were known to most Soviet people as the enigmatic *Glavnyi Konstruktor* (Chief Designer) and *Glavnyi Teoretik* (Chief Theoretician), respectively, with their true identities revealed only after their deaths. Keldysh had a distinguished scientific career and served as president of the USSR Academy of Sciences in 1961–1975. Photo courtesy of Mike Gruntman.

Veil of Secrecy

All leading Soviet rocketeers remained covered by the veil of secrecy of the totalitarian state. The anonymous Korolev was known to the public only as the legendary *Chief Designer*. Another key participant, Mstislav V. Keldysh, who coordinated research and development in space science and technology and contributed to most major decisions shaping the Soviet program was known as the *Chief Theoretician of Cosmonautics*.

The launch of Sputnik began the glorious space history of the R-7, or *semyorka*, an affectionate Russian word for *seven* or *little seven*. With the additional third stage, it sent a spacecraft to the Moon. The modified and improved R-7 launched the first cosmonaut. The rocket became a workhorse of the Soviet space program, being further advanced and produced in series by the rocket plant in Kuibyshev (Samara). Today's Russian *Soyuz* space launcher is a direct descendant of the R-7.

Glorious History of the R-7 *Semyorka*

The Soviet launch of the first artificial satellite had a tremendous effect on the world. It was not only a feat of science and technology but also a potent weapon in a sharp ideological confrontation of the Cold War. In the words of the Director of Central Intelligence Allen W. Dulles, "Khrushchev had moved all his propaganda guns into place" (National Security Council meeting on 10 October 1957).

SOVIET LEADER DELIGHTED

Sergei N. Khrushchev recalls that his father, the Soviet leader Nikita S. Khrushchev, learned about the launch of Sputnik during the meeting with the Ukrainian leadership in Kiev. Nikita S. Khrushchev said to the assembled Communist Party chieftains and military brass:

> ... I can tell you a very pleasant and important news. Korolev has called me. He is the designer of our rockets. Please keep in mind that you should not mention his name; this is a state secret. So, Korolev reported that the [first] artificial satellite of the Earth had been [successfully] launched this evening.
>
> ... The Americans told all the world that they were going to launch an [artificial] satellite. Its size was just like an orange. We kept silent, and now our satellite, not tiny but [with the weight of] 80 kg, spins around the planet. (Khrushchev 2000, 224)

A Load-star Speaking for Socialism

The Soviet mouthpiece in the West, *The New Times*, 10 October 1957, promptly stated that the Sputnik launch "proved anew, the enormous advantage of the socialist system" and pointed out that "the first artificial moon in the world was made and launched in the first socialist state" and "circling in cosmic space, it [Sputnik] is a lodestar speaking for socialism." Then, *The New Times*, 2 January 1958, did not miss the opportunity to remind "that the Soviet Sputniks have proved to the world that the USSR has the intercontinental missile, that 'absolute weapon,' while America has only intermediated-range missiles." Korolev himself wrote in the official Communist Party newspaper *Pravda*, 10 December 1957, that "two bright stars of peace [Sputnik 1 and 2] launched by a mighty hand of the Soviet people ... testify about the greatest achievements of the socialist state, Soviet science, technology, and culture."

Foreign Policy Reactions "Pretty Somber"

The launch of the Soviet Sputnik had become a major political event with serious consequences to the U.S. relations with foreign countries. At the meeting of the National Security Council on 10 October 1957, Undersecretary of State Christian Herter "described the first foreign policy reactions as 'pretty somber.' The United States will have to do a great deal to counteract them and, particularly,

American Reaction to Sputnik

INTERNATIONAL BASKETBALL

News of the successful launching by the USSR of the first man-made satellite has been met in most responsible quarters with an attitude of scientific good sportsmanship. Certainly no one who espouses the cause of space flight can deny the magnitude of the feat nor hesitate in offering congratulations for what can only be regarded as an achievement of which mankind as a whole may well be proud.

While we in the United States should not panic at the thought of Russian success, neither should we be complacent about the implications of the feat. The first score by the Russians in the international technological basketball game clearly gives them an early lead in the contest for mastery of space.

While the Russians may be scientific good fellows, their politics are another matter. They must not be allowed to win this game – a game with far-reaching political, social and economic consequences.

The early lead gained by Russia thus far in the game, and its stated intention of widening this lead by rapid expansion of such probing to include even more spectacular endeavors in the field of space flight, should certainly prod us into a thorough examination of our own limited and short-term program.

The partial measures, hit-or-miss planning and confused organization that have marked our own work in this field to date are not likely to permit us to eliminate or even cut into the lead Russia has built.

Unless we as a nation recognize that astronautics will be to the last half of the present century what aeronautics was to the first half, and unless our space flight activities are organized so as to adequately exploit its potential over the long haul, the gap between our present position and that of the Soviet Union, still relatively small, will widen, rather than close – with results that could ultimately be disastrous.

Editorial, *Astronautics*, American Rocket Society, November 1957

GREATER CHALLENGE THAN PEARL HARBOR

... Our country [U.S.] is disturbed over tremendous military and scientific achievements of Russia. Our people have believed that in the field of scientific weapons and in technology and science, that we were well ahead of Russia. With the launching of Sputniks 1 and 2, and with the information at hand of Russia's strength, our supremacy and even our equality has been challenged ... .

It is not necessary to hold these [Senate] hearings to determine that we have lost an important battle in technology. That has been demonstrated by the [Sputnik 1 and Sputnik 2] satellites that are whistling above our heads. But to me, and I think to every other American, a lost battle is not a defeat. It is, instead, a challenge, a call for America to respond with the best that is within them ... .

In some respects, I think [that launch of Sputniks] is an even greater challenge [than Pearl Harbor]. In my opinion, we do not have as much time as we had after Pearl Harbor.

Senator Lyndon B. Johnson, Texas
Hearings of the Preparedness Investigating Subcommittee of the Committee on Armed Services on 25 November 1957

to confirm the existence of our own real military and political strength." Herter added that

> the reaction of our [U.S.] allies had been pretty firm and good, though even the best of them require assurance that we have not been surpassed scientifically and militarily by the USSR. The neutralist countries are chiefly engaged in patting themselves on the back and insisting that the Soviet feat proves the value and the wisdom of the neutralism which these countries have adopted.

Poor State of Science Education

The Soviet Sputnik had shocked the general public in the United States and awakened it from complacency. As a result of the Sputnik, a poor state of science education in American schools was brought into focus of the public attention, and the much-needed alarm was sounded. Irwin Hersey, the editor of *Astronautics*, a publication of the American Rocket Society, described in the November 1957 issue that "the reactions of leading American scientists and officials, registered within the first few days after the Soviet feat, ranged all the way from amazement and admiration to bitter charges of administrative bungling" (Hersey 1957, 24).

Preventive and Remedial Actions

Soul-searching followed while many politicians and media types used the opportunity, not surprisingly, to pin the blame on their political opponents. This was the time when President Eisenhower tried to guide the nation, in the words of his Special Assistant for Science and Technology James R. Killian, Jr., "into the space age and ... [seek] presidential control of military and space technology against all of the rampant forces, good and bad, of the military-industrial-congressional complex as they impinged on his presidency" (Killian 1977, xviii). James Doolittle thoughtfully observed on the occasion that an

> American trait is an unwillingness to accept the unpleasant until we are hurt. When we are hurt enough we react soundly, but frequently, we fail to take preventive action and have to take remedial action ... . We incline to operate like a pendulum ... . We seldom stop in the middle. This has been reflected in our national policies. (*Hearings* 1958, 117)

Space Pearl Harbor

The Soviet success has actually helped the American missile program by preventing the planned further reductions of military funding by the Eisenhower administration. (The reader may recall that Assistant Secretary of the Air Force Trevor Gardner resigned in February 1956 in protest of reduction of research and development programs.) As Vannevar Bush put it at the after-Sputnik hearings in the Senate, "The Sputnik was one of the finest things that Russia ever did for us. I am very glad they fired the thing off. It has waked this country up. There has not been an awakening like this, I was going to say since the days of prohibition, but let me say since the days of Pearl Harbor." Such comparisons with the Pearl Harbor were not uncommon.

Chose to Remain Uninformed

The Soviet launch did not come, however, as a surprise to people in the know. Many American media personalities and politicians conveniently chose to remain uninformed. The Soviet Union announced its intention to launch a satellite as early as in 1955. In the fall of 1956, the Soviet National Committee for the International Geophysical Year confirmed that the preparations for launching satellites were under way. A leading American scientific magazine, *Science*, quoted the president of the USSR Academy of Sciences in July 1957 that the Soviet "scientists have created" the required rockets and instruments for the satellite. In

SOVIET AND AMERICAN EDUCATION AND SCIENCE

... The Russians are catching up with us [the United States] and have surpassed us in certain fields. Their rate of advance is considerably faster than the rate of our advance. I think that this is perhaps more clear in science than in any other field. They have stressed science very greatly, and they have built up their scientific manpower in a most admirable, most remarkable manner ... .

The philosophy that the Russians profess puts a very great weight on scientific and technical accomplishments. ... To build up a scientific capability must come first. Technology is the second step, and also necessarily the slower step because more material means are involved.

This has led in Russia to the situation where I think the scientist is in quite unique position. Most people in Russia starve. If you do not want to starve in Russia, then there are two ways to get an agreeable life. One is to get a lot of responsibility, for instance, become a more or less prominent member of their one [Communist] party.

The other is to become a scientist. If you get a lot of responsibility, then as far as I understand, you also lead a comfortable life, but by no means a secure one. If, on the other hand, you become a scientist, you have as good a life and as secure a life as you possibly can have in Russia. Now that is only part of the story. It is not only true that you are comfortable. It is also true that the Russian scientist is greatly honored. Scientific books in Russia have a sale which outstrips the sale of scientific books in the rest of the world by something like a factor of 10. It is an amazing phenomenon. Their scientists are honored, and to some extent even understood, by the general public ... .

The word "scientist" just has a connotation [of respect] in Russia which is not present in our country. A Russian boy thinks about becoming a scientist like our young girls dream about becoming a movie star. It is just the best thing you can do; the best, the most active minds are attracted to this field, and they are given both the means to live decently and all the honors they can dream of.

Now, under these conditions, of course, and with practically no limit on expenditures, it is not a surprise that Russia has gone ahead ... . The Russians have shown in many examples that they are quite willing to support ... impractical science, which is the breeding ground of the practical science which in turn is the foundation for the real technological advances. They took a broad program and they were not saving any effort, and that is the way they have beaten us ... .

I am sure that ... [the missions to the planets and the moon] will have a great practical advantage, and I am sure a great peaceful practical advantage and I am sure it will have also a great military advantage ... . I cannot tell you in black and white what will be advantages ... . But the fact is that if you make such a very big step like going to the moon, it will have both amusing and amazing and practical and military consequences.

The spirit in our schools is such that a kid who is interested in science is ridiculed by his fellow students, and he will rapidly lose that interest ... .

In Russia, the scientist is driven forward by the whip of hunger, by the fear of misery, and by the reward of honor. We [in the United States] are living in a world which is successful in eliminating, progressively, misery. We cannot and we should not drive our scientists ahead with any sort of whip ... . I also do not want to say that any group of people, and the scientists in particular, should be put on a pedestal as they are put on a pedestal in Russia. That, to my mind, is distasteful.

Edward Teller, Hearings of the Preparedness Investigating Subcommittee of the Committee on Armed Services, U.S. Senate, 25 November 1957

CURIOUS INDIFFERENCE

... The near-hysterical reactions to Sputniks I and II ... [and] the surprise that greeted Sputnik ... did not result from a failure of American intelligence or a lack of knowledge in the scientific or diplomatic community about what the Soviets were up to. The CIA had detected preparations for the Soviet satellite, and the Soviets had freely boasted about their plans. But there was a curious indifference and lack of technological sensitivity among the American public and political leaders, and even within the scientific community, toward these clear indications of what was to happen. More significantly, few anticipated what the reaction would be to what was to happen.

I have come to ascribe the American panic in part to the vapors of the Cold War and to the excessive security that enveloped our ballistic missile technology. People were woefully ignorant of how much qualitatively advanced and forehanded rocket technology had been under development by the [U.S.] Department of Defense under the direction of far-seeing military officers in the army and air force.

James R. Killian, Jr.
(Killian 1977, xvii)

the same month, Soviet scientific officials wrote in *Nauka i Zhizn'*, a Soviet magazine similar to *Scientific American*, that "artificial satellites will be launched [during the IGY] in the Soviet Union and United States."

Sputnik Anticipated

The U.S. government was also fully aware of the Soviet advances. As early as November 1956, the American intelligence community "estimated that the Soviets would be capable of launching an earth satellite any time after November 1957." It was expected that there would be launches by both the United States and the Soviet Union during the International Geophysical Year, but the exact timing remained unknown. Director of Central Intelligence Allen W. Dulles wrote to Deputy Secretary of Defense Donald A. Quarles on 5 July 1957 that "information concerning the timing of the launching of the Soviet's first earth-orbiting satellite is sketchy, and our people here [at the CIA] do not believe that the evidence is sufficient as yet for a probability statement on when the Soviets may launch their first satellite." Dulles also added that "the Russians like to be dramatic and could well choose the birthday of Tsiolkovsky [on 17 September 1957] to accomplish such an operation, especially since this is one hundredth anniversary of his birth."

Impact Under-estimated

The American space effort was not geared at beating the Soviet Union in what would become the space race. The Eisenhower administration rather focused, quietly, low key, without attracting attention on establishing the principle of freedom of space by a clearly nonmilitary program. Freedom of space was considered to be of critical importance for future space reconnaissance missions. This policy has obviously underestimated the impact of the satellite launch on the nation and on the world.

Lack of Priority

Although the technological capability of deploying a satellite by the American Jupiter C rocket existed for some time, the interservice rivalry, Air Force's focus on the ICBM, and administration's policy and decision to clearly separate the military and civilian space efforts opened an opportunity for Korolev to achieve a historic victory in the competition. Simon Ramo noted that "the spectacular achievement of putting Sputnik in orbit in October 1957 involved little [Soviet] advance over the available technology in the U.S. at the time. They [USSR] launched Sputnik when they did because they chose priorities different from ours" (Ramo 1988, 80). Andrew Haley also pointed at the same problem, "The Soviet Union beat the United States with the first earth satellite, not because this country failed to spend enough money, or because Americans lacked any of the technological knowhow, but because the United States chose to be beaten" (Haley 1958, 155).

Chose to Be Beaten

The things would become worse for the United States before they turn to the better in the space race. After the success of Sputnik 1, Sergei Korolev pressed on. The second satellite, Sputnik 2, was successfully launched on 3 November 1957. This launch was rushed by a personal request of the Soviet leader Nikita S. Khrushchev who wanted the satellite to be in orbit by the 40th anniversary of the Great October Socialist Revolution on 7 November. And Korolev delivered. The new Sputnik 2 had mass 508.3 kg (1121 lb), which emphasized again the alarming capabilities of the Soviet launcher as an ICBM. The future American satellite, the Vanguard, was originally planned to be tiny in comparison, 9–11 kg (20–25 lb), and had to be further reduced to 1.5 kg (3.25 lb). In addition, Sputnik 2 carried a live space passenger, a dog *Laika*. (The name *Laika* could be translated as *Barker*.)

Korolev Presses Forward

Then Object D was launched as Sputnik 3 on 15 May 1958 after the first attempted launch had failed to reach the orbit on 27 April. Sputnik 3 weighed 1327 kg (2924 lb) and demonstrated the first Soviet solar panels. The Soviet state clearly showed, as it had done earlier with the nuclear weapons, that by concentrating its resources in the selected area it could meet any single American challenge and produce spectacular results. However, as Killian put it, "they [USSR] have not, in the field of technology, proved capable of meeting *all* challenges that the American economy can offer" (Killian 1977, 5).

Object D Launched

The American public demanded a response and three entirely different space launchers (Juno, Vanguard, and Atlas), developed by three different organizations, would place satellites in orbit by the end of 1958. In addition two more American launchers, Thor and Scout, would become operational in 1960. Only at that time the Soviet Union would try to launch a satellite by a rocket (Yangel's R-12) built by an other-than-Korolev's establishment. All of this time since the first historical launch of Sputnik, it was Sergei Korolev who led and sustained the Soviet space effort. And then he achieved another historic first by putting the first man in space.

American Rockets Close the Gap

The R-7 capabilities were further significantly enhanced by increasing thrust of the engines and by adding the third stage, called *Block E*, powered by a new kerosene-liquid-oxygen engine built jointly by Korolev's OKB-1 and a design bureau, OKB-154, headed by Semyon A. Kosberg (1903–1965) in Voronezh, 500 km (300 miles) south of Moscow. Block E was designed to be "hot-started" in space to deal with the problem of engine ignition in weightlessness as yet unsolved. Therefore, Block E was ignited before the complete cutoff of the engine of the sustainer stage and before its separation. Block E was later replaced by the improved *Block I* in the space launcher that would become known as the *Soyuz*.

Block E and Block I

The R-7 with the Block E third stage allowed sending a spacecraft to fly by the Moon on 2 January 1959. The continuing improvements of the R-7 and the reliable third stage made possible deployment of a spacecraft with mass 5000 kg (11,000 lb) in low-Earth orbit, which thus enabled launching man in space. The improved rocket launcher and the spaceship to carry a cosmonaut were both named *Vostok* (*east* in Russian). Actually, a Soviet government decree of 22 May 1959 directed Korolev to develop an experimental spaceship (Vostok-1) to provide the foundation for further advancement toward a reconnaissance satellite Vostok-2 for image and signal intelligence and a spaceship Vostok-3 for manned spaceflight.

Manned Spaceflight Enabled

The space race between the United States and the Soviet Union to launch the first man to space had been on by that time. Korolev methodically pressed forward with testing the launcher and the spaceship for the Vostok program. The simplified spacecraft Vostok (Vostok-1KP) without life support systems was successfully launched on 15 May 1960. The next success came when a spacecraft with two dogs, Belka and Strelka, successfully reached orbit on 19 August 1960 and reentered the atmosphere the next day and safely landed. The Vostok-2 part of the program would eventually lead to the Soviet photoreconnaissance satellite Zenit-2.

Vostok

The rush to orbit man culminated on 12 April 1961, when the space launcher *Vostok* (8K72) put into orbit the spaceship *Vostok* (Vostok-3KA) with the first cosmonaut, an Air Force pilot Yurii A. Gagarin. The spaceship weighed 4725 kg (10,400 lb) that constituted 1.65% of the 287,000-kg (633,000-lb) rocket weight

First Man in Space

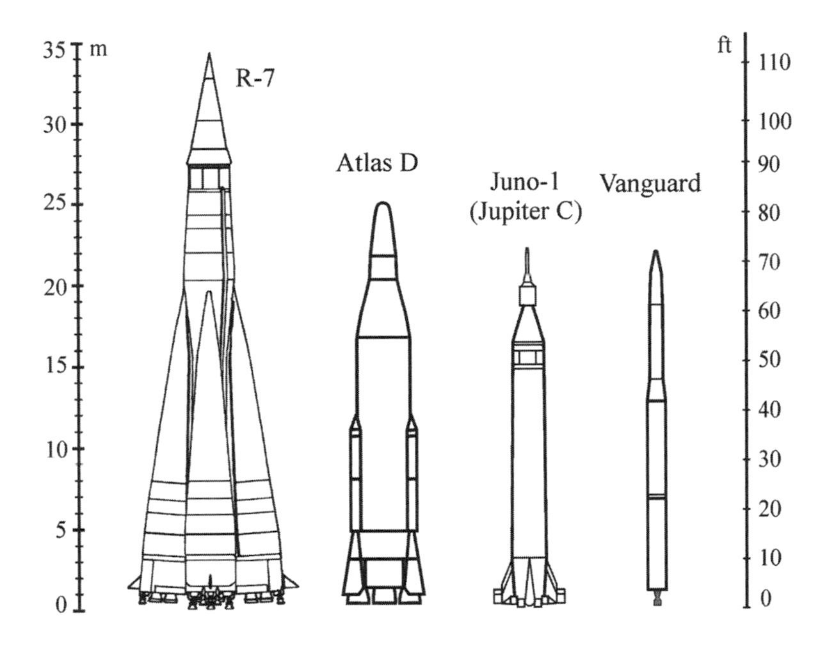

Fig.15.9. The first Soviet ICBM R-7 was significantly larger and heavier than the first American ICBM Atlas. The modified R-7 deployed the first artificial Earth satellite Sputnik and later launched the first cosmonaut Yurii Gagarin. The first American satellite Explorer I was put into orbit by the Juno-1, a variant of the Jupiter C modified for satellite launch. By the end of 1958, all three shown American rockets, Juno-1, Vanguard, and modified Atlas, launched satellites into Earth orbit. Figure courtesy of Mike Gruntman.

Fig. 15.10. Vostok rocket that launched the first man into space on display in Moscow. One can see the third stage clearly separated from the rocket sustainer ("second") stage with four booster (side) sections. Photo courtesy of Mike Gruntman.

Fig.15.11. First cosmonaut Yuri A. Gagarin in Tyuratam on 12 June 1963, more than two years after his historical space flight. Gagarin is in the uniform of an Air Force lieutenant-colonel. Major Andrian Nikolaev, the third Soviet cosmonaut, is on the left. Photo courtesy of Mike Gruntman.

at launch. The orbit was low with perigee at a 187-km (116-mile) altitude and an apogee altitude 327 km (203 miles). After one orbit and 108 min in space, the first cosmonaut safely landed in Russia.

Gagarin's flight was a clear political victory for the communists. Congratulating Soviet space and rocket establishment with the outstanding achievement, the Kremlin did not miss an opportunity for launching an ideological offensive:

Tireless Care of the Communist Party

> ... Our free, talented, and hard-working people, inspired to conscientious historical creativity in [the] October [Revolution] 1917 by the Party of Communists headed by the great leader and teacher of all working peoples of the world Vladimir Il'yich Lenin, show today to all world the greatest advantages of the new socialist system in all aspects of life of the society.
>
> Space flight of man is a result of the successful implementation of the grandiose program of building Communism and of tireless care of the Communist Party, [its] Leninist Central Committee, and the Soviet Government about continuous development of science, technology, culture, and well-being of the Soviet people.
>
> ... Long live the glorious Communist Party of the Soviet Union, the great inspirer and organizer of all victories of the Soviet people! Long live Communism!

EXPLORER AND VANGUARD

The road to the American satellite was bumpy, to say the least. The early enthusiasm about Earth-orbiting satellites in the immediate after-the-war years had gradually dissipated as a result of lack of government support. Not all activities, fortunately, stopped: dedicated space enthusiasts continued to publish scientific articles advocating small research satellites, and RAND was evaluating the utility of spacecraft for overhead reconnaissance.

Bumpy Road

The same exigencies of the Cold War that had called for the ICBM pointed to satellite reconnaissance as a top national security priority. The studies performed by RAND with close participation of industrial contractors culminated in a report, issued in March 1954, on the Project Feed Back describing a military satellite equipped with a television camera. In addition, technology was rapidly advancing in many areas important for spacecraft, such as invention of the transistor in 1947 and practical (silicon-based with reasonable efficiency) solar cells in 1953, enabling significantly more efficient and capable satellites.

Converging Develop- ments

Several converging developments in the early 1950s would lead to the American satellite. In 1952, the *International Council of Scientific Unions* approved the concept of the International Geophysical Year. Subsequently, the National Academy of Sciences formed the United States National Committee for the IGY. The Committee was chaired by Joseph Kaplan who headed the University of California's *Institute of Geophysics*, later known as *Institute of Geophysics and Planetary Physics* (IGPP), since its foundation in 1944.

IGY

An American scientist, Lloyd V. Berkner, was a key figure in formulating and advancing the concept of the IGY. Subsequently he served as vice president of a special international committee arranging and coordinating the IGY activities. At its meeting in Rome, Italy, in October 1954, the committee accepted a proposal by

PLAN FOR IGY

The plan for a third International Polar year, later broadened in scope and renamed the International Geophysical Year 1957-1958, originated on April 5, 1950, at a small dinner party of geophysicists at my home [in] ... Silver Spring, Maryland. The basic concept was put forward by Lloyd V. Berkner. He and Sydney Chapman were principally responsible for developing and enlarging the concept to a persuasive level of detail and potential implementation, with the help of suggestions by others present: Ernest H. Vestine, J. Wallace Joyce, S. Fred Singer, my wife Abigail, and myself ... .

Following the dinner, as we were all sipping brandy in the living room, Berkner turned to Chapman and said, "Sydney, don't you think that it is about time for another international polar year?" Chapman immediately embraced the suggestion, remarking that he had been thinking along the same lines himself. The conversation was then directed to the scope of the enterprise and to practical considerations of how to contact leading individuals in a wide range of international organizations in order to enlist their support. The year 1957-1958, the 25th anniversary of the second polar year and one of anticipated maximum solar activity, was selected. By the close of the evening Chapman, Berkner, and Joyce had agreed on the strategy of proceeding.

James A. Van Allen, 1982
(Van Allen 1984, 49)

Fig. 15.12. Redstone (left) and Jupiter C (right) missiles at the U.S. Space and Rocket Center in Huntsville, Alabama. The Jupiter C consisted of the elongated Redstone as the first stage and added upper stages of clusters of scaled-down solid-propellant Sergeant rockets. The Jupiter C was developed for the Army program to test the nose cone reentry and demonstrated that it was capable of launching an Earth satellite in 1956. The Army proposed to use this launcher for Project Orbiter in 1956. Photo courtesy of Mike Gruntman.

Small Satellites for Scientific Purposes

American scientists (Berkner, Kaplan, Fred Singer, Homer E. Newell, Jr., James Van Allen, and several others) to recommend "that the thought be given to the launching of small satellite vehicles, to their scientific instrumentation, and to the new problems associated with the satellite experiments ..." (Green and Lomask 1971, 23). The National Academy of Sciences actively advocated and lobbied through various parts of the Eisenhower administration the idea of preparing and launching American scientific satellites as part of the IGY.

Project *Orbiter*

In September 1954, von Braun's group in Huntsville produced a report entitled "The Minimum Satellite Vehicle Based upon Components Available from Missile Development of the Army Ordnance Corps." The report argued that a 5-lb (2.2-kg) Earth-circling satellite could be placed in orbit using existing Army missile hardware by adding clusters of solid-propellant Loki rockets to the modified Redstone rocket serving as the first stage. In a couple of months, the Army and the Navy joined the resources of the ABMA's organization in Huntsville, Jet Propulsion Laboratory, Office of Naval Research, and several industrial contractors in what was become known as *Project Orbiter*. The Navy took the responsibility for the payload and tracking facilities, while the Army's tasks included modifying the Redstone and developing the Loki rocket clusters.

Sergeant

Army's JPL conducted a feasibility study in support of Project Orbiter and suggested the substitution of the scaled-down Sergeant solid-propellant rockets, under development at JPL at that time, for the Loki rockets. This improvement should

Fig. 15.13. Donald A. Quarles, 1894–1959, (right) being sworn in as Secretary of the Air Force on 15 August 1955. (Quarles would become Deputy Secretary of Defense on 1 May 1957.) Secretary of the Army Wilbur M. Brucker (left) administers the oath of office; Secretary of Defense Charles E. "Engine Charlie" Wilson looks on. Wilson, Brucker, and Quarles played important roles in American ballistic missile and space programs. Photo courtesy of National Archives and Records Administration.

have allowed increase of the Orbiter's satellite payload to 18 lb (8 kg). The promising project would come, however, to a screeching halt in August 1955.

NRL Proposals

In early March 1955, the Naval Research Laboratory submitted two proposals to the Department of Defense, one by Milton W. Rosen advocating use of the modified Viking rocket for launch of a satellite and the other by John T. Mengel and Roger L. Easton for a "minimum trackable satellite (Minitrack)." Before submitting their Viking proposal, assistant head of the NRL's Rocket Sonde Branch John W. Townsend and Rosen had an opportunity to see the earlier Army proposal for Project Orbiter.

Thus by the middle of 1955, there had been satellite programs under consideration by the National Academy of Sciences, by the Army, and by the Navy. In addition, the Air Force continued studies of large and complex reconnaissance satellite systems. Launching a satellite was also supported by the American Rocket Society that in November 1954 approached the director of the National Science Foundation, Alan T. Waterman, suggesting a study of the utility of an Earth-circling vehicle.

Killian Report

On 14 February 1955, the *Technological Capabilities Panel* (of the Scientific Advisory Committee of the Office of Defense Mobilization) headed by James R. Killian, Jr., issued a report "Meeting the Threat of Surprise Attack." This was the same *Killian Report* that played an important role in accelerating development of the ICBM. The report recommended an immediate initiation of a program aimed at launching an American scientific satellite with the goal of asserting the principle of freedom of space.

By May 1955 the National Security Council endorsed and President Eisenhower approved a decision to launch artificial satellites as a U.S. contribution to the International Geophysical Year. The new policy specifically required that satellites should not distract resources from the Air Force's top-priority ICBM. Testing the principle of freedom of space by a scientific satellite was considered critically important for future satellite reconnaissance, and the new program had to demonstrate the right of overflight by spacecraft. Satellite reconnaissance became a top national security priority especially in the light of the rejection of the Eisenhower's *open skies* proposal by the Soviet leader Khrushchev at the Geneva summit of the Big Four powers in July 1955.

Eisen-hower's Policy

CHANGING TIMES: PEENEMÜNDE AND 1950s

Scientists from [German] universities and research institutes were called in [to help the V-2 development program during World War II] only if there was a scientific problem to solve which we [working at Peenemünde] could not solve ourselves. When they were called upon, it was to solve a problem, not to discuss it or philosophize about it.

It's a far cry from today [in 1958 in the U.S.], when a simple question asked of a scientist brings a 500-page report on the subject, containing a good deal of highly interesting theory and speculation, and, possibly, the answer, which often is contained in a footnote on page 357.

Walter R. Dornberger
Technical Assistant to the president of Bell Aircraft Corporation
(Dornberger 1958, 58)

American plans to launch scientific satellites were announced on 29 July 1955, when President's Press Secretary James C. Haggerty made the following statement: "On behalf of the President, I am now announcing that the President has approved plans for this country for going ahead with the launching of small earth-circling satellites as part of the United States participation in the International Geophysical Year." The statement avoided any link to the national security rationale and designated the civilian National Science Foundation to direct the program, with "logistic and technical support" of the Department of Defense. American scientific satellites were thus clearly decoupled from any military applications.

President's Announcement

Two days later, the Sixth Congress of the International Astronautical Federation opened in Copenhagen, Denmark. This was the first IAF Congress attended by the official Soviet delegation as observers. Soviet representatives promptly responded to the American announcement. The official Soviet newspaper *Pravda*, 5 August 1955, described the statement made by academician Leonid I. Sedov, who said that

Soviet Response

> recently, a lot of attention is devoted in the USSR to research problems related to interplanetary communications [travel], in the first place to the problems of creation of an 'artificial earth satellite.' Engineers, designers, and researchers dealing with and interested in rocket technology already know very well the reality of technical projects of artificial satellites.

Sedov added that in his opinion "it will be possible to launch an 'artificial earth satellite' during the next couple years" and that "the realization of the Soviet project [to launch an artificial satellite] could be expected in the relatively near future."

American government deliberately played down the role of satellites in the future and in funding priorities. The military particularly had been instructed to avoid any mentioning of military space applications in public in order not to trigger debate over freedom of space. Many years later Bernard Schriever was still fuming about this policy and its consequences. "In 1957," recalled Schriever in 1972,

"... Never to Use the Word Space ..."

> I made a speech at a joint symposium in San Diego about how the missile program was really creating the foundation for space. The day after I made the speech I got a wire signed by Secretary [of Defense] Wilson telling me never to use the word space again in any of my speeches. In October [1957], Sputnik came along, and for the next 18 months or so after, I was going back and forth to Washington at least four times a month testifying before committees or meeting in the Pentagon as to why we couldn't move faster in the missile program. (Schriever 1972, 60)

Although it might have been a wise policy for assuring the acceptance of future space reconnaissance missions, the Eisenhower administration clearly underestimated the prestige, national pride, and psychological factors of a satellite launch.

Stewart Committee

Eisenhower's approval of launching scientific satellites had thus set the American policy, and the Department of Defense was charged with its implementation. Deputy Secretary of Defense Donald A. Quarles appointed a special panel, the *Ad Hoc Advisory Group on Special Capabilities*, under JPL's Homer J. Stewart, to consider satellite proposals from the Army, based on Project Orbiter, and from the NRL, based on advancement of its Viking sounding rocket. (The Air Force also offered the Atlas B as a launcher should other satellite proposals be found unacceptable to meet the IGY goals. The Committee shelved this suggestion because of the danger of interference with the top-priority ICBM program.) In a contentious five-to-two vote in the early August, the Stewart Committee selected the NRL proposal, which would become known as the Vanguard program.

Selection of Vanguard

Project Orbiter advocated by ABMA's John Medaris and Wernher von Braun had thus been terminated. As Medaris described it later in Senate hearings, "the decision was made that the national satellite effort would be the Vanguard effort, and no funds were available for any further work [by the Army Ordnance]" (Hearings 1958, 1699). When Medaris was asked whether it was "correct that at one point when expressed orders were sent down to the Army not to launch a satellite, auditors ... checked on your agency [ABMA] to make certain that the orders were obeyed," he replied that "there were some people who came down [to Huntsville] to take a look and be sure that I was not fudging" (*Hearings* 1958, 557).

Project Orbiter Terminated

The NRL was assigned the role of technical direction of the Vanguard program, with the Glenn L. Martin Company that built the Viking sounding rockets responsible for the new space launch vehicle. The NRL selected John P. Hagen, a superintendent of the NRL's Astronomy and Astrophysics Division, to head the Vanguard. Hagen was assisted by a number of leading NRL specialists: J. Paul Walsh (deputy director), Milton W. Rosen (technical director), Homer E. Newell (science programs coordinator), Thomas Jenkins (budget), Leopold Winkler (mechanical

NRL Team

design), James M. Bridger (launch vehicle and senior NRL representative at the Martin Co.), Kurt R. Stehling (power plant), Frank H. Ferguson (control and guidance), Daniel G. Mazur (telemetry), John T. Mengel (tracking and guidance), Joseph W. Siry (orbital mechanics), James Flemming (data processing), and Robert Mackey (electronic instrumentation). At the peak of the program, the entire NRL group included 180 people.

Martin Team

The Martin team responsible for providing the Vanguard launch vehicle was led by N. Elliot Felt, Jr. (operations manager), Donald J. Markarian (project engineer), Robert Schlechter (head of field tests), Leonard Arnowitz (flight-path control), Mel Ruth (manufacturing manager), Guy Cohen (quality control), Joseph E. Burghardt, Russell Walters, and Sears Williams. The relations between the NRL and Martin were initially characterized by sharp differences on how the program should be structured and how much supervision the NRL team should exercise. Financially, the space launch vehicle was not especially attractive to Martin, particularly after it won a major Air Force contract in September 1955 to build the Titan ICBM. The company began promptly relocating many its best men to a new facility in Colorado.

New Launch Vehicle

The new Vanguard was unique among the early Soviet and American space launchers that deployed satellites in the late 1950s. It was the only rocket that did not emerge from the military missile programs and was specifically designed as a

Fig. 15.14. Director of Project Vanguard Dr. John P. Hagen (center) with the staff members of Project Vanguard. Left-to-right: Joseph W. Siry (Theory and Analysis Branch), Daniel G. Mazur (Vanguard Operations Group — VOG — at Patrick Air Force Base, Florida), James M. Bridger (Vehicle Branch), Winfred E. Berg (Navy Program Officer), Hagen, J. Paul Walsh (Deputy Project Director), Milton W. Rosen (Technical Director), John T. Mengel (Tracking and Guidance Branch), and Homer E. Newell, Jr. (Science Program Coordinator). Photo courtesy of NASA.

Fig. 15.15. Project engineer Donald J. Markarian (left) and operations manager N. Elliot Felt, Jr. (right) led development of the Vanguard launch vehicle at the Glenn L. Martin Company. Originally published in "The Martin Team" by D. Cox, *Astronautics*, Vol. 2, No. 3, p. 33, 1957. Copyright © 1957 by the American Institute of Aeronautics and Astronautics, Inc. Reprinted with permission.

space launcher. The Vanguard launch vehicle consisted of three stages with the total fully-fueled mass 22,600 lb (10,250 kg).

Early in the program, Martin's engineers decided to switch to a new liquid-propellant engine for the first stage, the General Electric's X-405, instead of the engines built by Reaction Motors, Inc., for the Viking. The X-405 was a regeneratively cooled turbopump-fed gimballed engine that used kerosene and liquid oxygen to produce 27,000 lbf (120 kN) thrust. Kerosene was used for regenerative cooling and decomposition of hydrogen peroxide powered the turbopump. The contract for the second stage was awarded to Aerojet. The Aerojet engine AJ10-118 provided 7500 lbf (33.4 kN) thrust and used pressure-fed unsymmetrical dimethylhydrazine as fuel and white fuming nitric acid as oxidizer. The second stage also carried the Vanguard's guidance and control system manufactured by the Aeronautical and Ordnance Division of the Minneapolis-Honeywell Regulator Company.

Power Plants

Thiokol found the requirements to a solid-propellant motor of the third stage unrealistic, and the Grand Central Rocket Company was subsequently selected to build the motor with 2800 lbf (12.5 kN) thrust. In addition, the Allegany Ballistic Laboratory developed a backup solid motor that flew in the last launched Vanguard vehicle in September 1959. The third stage was mounted on a turntable that was spun up by small solid-propellant thrusters produced by the Atlantic Research Corporation. The Allegany's solid motor, known as the Altair X-248, and

Aerojet's AJ10-118 would become the basis of the third and second stages, respectively, of the new Delta space launcher in the early 1960s.

ABMA and von Braun Excluded

The Army's ABMA was practically excluded from the Vanguard development, and Wernher von Braun was not even invited as a consultant. Other Army components, however, such as the Signal Corps and Corps of Engineers significantly contributed to the program by building the network of satellite tracking stations and even operating some of them. The Air Force provided services and facilities at its missile test range at Cape Canaveral supporting the preparation and launches of the Vanguards.

Comprehensive Program

As John Hagen described it, the goals of the project Vanguard were "first, to put a satellite in an orbit within the time of the IGY; second, to so instrument it and so arrange here on the earth that it could be observed and proven to be in an orbit and third, to so instrument it that useful scientific work could be done in the satellite" (*Hearings* 1958, 143). The Vanguard was a large, comprehensive program that established many procedures, infrastructure, and the framework for space exploration of the future. For example, a new type of contracting between the government and the industry was worked out. In the scientific payloads, the criteria for selection of satellite experiments were formulated; a special panel solicited proposals for scientific investigations and selected the experiments for first Vanguard satellites.

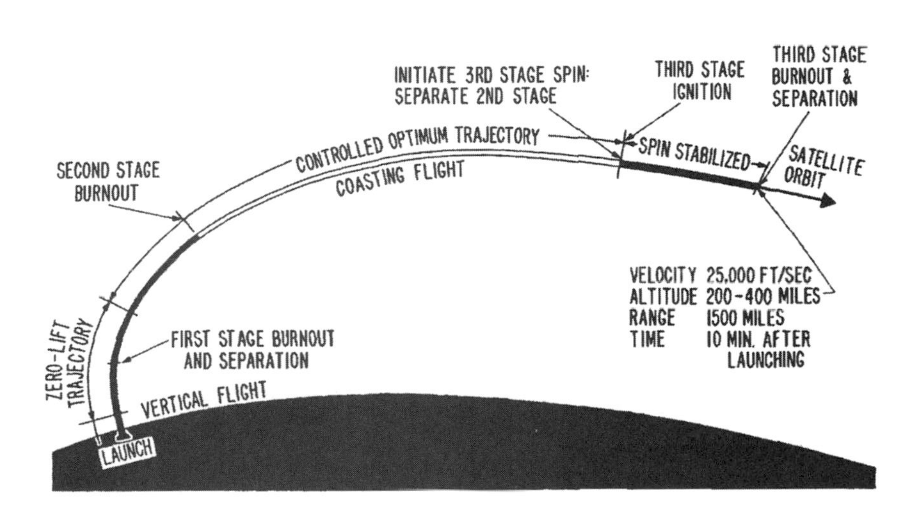

Fig. 15.16. Launch sequence of the three-stage Vanguard rocket. Figure from *Hearings* (1958, 174).

Minitrack System

The Vanguard program developed a worldwide network of Minitrack ground tracking and communication stations. Finding a satellite in the sky and tracking it to establish its orbit was a challenging problem. Orbit parameters were expected to evolve with time in a way that was difficult to precisely predict because of the poor knowledge of the gravitational field of the Earth. Precise measurements of the orbit evolution were particularly desired because they opened a way to better determine the gravitational field and the time-varying altitude profile of the atmospheric density, the latter responsible for satellite drag.

Satellite tracking relied on the onboard transmitters with the output 5 and 10 mW at the 108-MHz frequency, agreed upon by the committees of the International Geophysical Year. The propagation of radio waves at this frequency was expected to be minimally affected by the conditions in the ionosphere, which was important for precise tracking. (Conversely, this frequency was poorly suited for probing of the ionospheric properties, in contrast with Sputnik's radio frequencies.) The Minitrack ground stations were designed to determine the direction toward the satellite using the interferometric effect of signals received by separate antennas. To increase the accuracy, antenna arrays were used. (The experience with the reception of Sputnik signals transmitted at fixed frequencies would lead the Applied Physics Laboratory to tracking based on measurement of Doppler shifts rather than directions; this alternative concept would be implemented in the Navy Navigation Satellite System Transit built in 1960s.)

Worldwide Network

Each Minitrack station covered a flat 23-acre area with the surface gradient not exceeding 1 deg. A typical station usually tracked a satellite with an accuracy up to a few minutes of arc. The line of stations, about 800 miles (1300 km) apart, spread along the 75th meridian from Maryland down south all of the way to Chile

Fig. 15.17. Minitrack station near Quito, Equador. Photo from *Spacecraft Tracking* (Corliss 1969, 5).

providing a high probability for communicating with a satellite in low Earth orbit, when the satellite crossed this Minitrack station "fence." In addition to stations at Blossom Point, Maryland, and Savannah, Georgia, the Minitrack stations were established in Havana, Cuba; Quito, Equador; Lima, Peru; and Antofagasta and Santiago in Chile. The Air Force missile test range at Cape Canaveral contributed to space vehicle tracking with its own system that included remote stations on Grand Bahama Island, Antigua, and Grand Turk. Additional Minitrack stations were built in San Diego, California, near Johannesburg, South Africa, and at Woomera, Australia. This worldwide network of Minitrack stations laid the foundation for the later *Spaceflight Tracking and Data Network* (STDN) operated by NASA in support of its missions until 1980s.

Predecessor of STDN

Optical Tracking System

In addition to Minitrack radio tracking, a large international network of sites for optical satellite tracking had been organized by the Smithsonian Astrophysical Observatory (SAO) and directed by Fred L. Whipple, J. Allen Hynek, and Karl C. Henize. The specially designed Baker–Nunn telescope-tracking cameras (named after James G. Baker and Joseph Nunn) were built for this purpose capable of observing a 15-in. (38-cm) sphere at a 1000-mile (1600-km) distance. A Baker–Nunn camera with a 20-in. (51-cm) aperture and 30-deg field of view could photograph a satellite and the surrounding stars. Because star positions were known with high precision, the direction to the satellite could then be established with the accuracy of up to 2 arcsec, much better than by the Minitrack. Accuracy of time measurement and recording was critical for precise orbit determination by optical observations. A special clock designed by Robert Davis of Harvard University and built by Ernst Norrman Laboratories of Williams Bay, Wisconsin, provided a ±0.001-s accuracy.

Precise Time

On the down side, optical tracking required laborious and time-consuming processing of the recorded images, so that the two approaches — radio and optical — were truly complementary. Among the 12 tracking stations established worldwide, only an optical station at Woomera in Australia shared its site with a Minitrack station. Other optical stations were located at various locations in the United States (Hawaii, Florida, and New Mexico) and in Peru, Argentina, Spain, South Africa, Japan, India, Iran, and Netherlands West Indies.

Computers for Satellite Tracking

Two highly specialized computing facilities based on the state-of-the-art IBM 704 computers were organized at Cape Canaveral and in Washington, D.C. The radar data were electronically fed into the computer at the Cape during the launch. The computer then transferred the processed data to the central facility in Washington, where preliminary predictions of spacecraft orbits were calculated and forwarded to Minitrack stations for tracking. This entire system would be baptized during the launch of the first American satellite, ironically not the Vanguard but the Army's Explorer I.

Scientific Instruments

More than 30 scientific experiments were proposed to the Vanguard program. The selected four scientific instrumentation packages focused on study of the solar ultraviolet radiation, cosmic rays, Earth's magnetic field, and radiometry in support of meteorological research. The requirements for the instrumentation were much more stringent then those used for sounding rocket experiments and placed a premium on miniaturization, power efficiency, and reliability. Particularly, the electronics had to be transistorized to the maximum degree possible. In 1957, the Van Allen's cosmic-ray instrumentation package would be transferred to the Explorer I and replaced by a meteorological experiment of the Army's Signal Engineering Laboratory. The originally planned Vanguard satellites were designed as 20–25-lb (9–11-kg) spheres with a 20-in. (50-cm) diam convenient for optical tracking.

Development of the Vanguard launch vehicle, an entirely new rocket system, progressed in a methodical fashion. The first test vehicle (TV), the TV-0, was successfully launched on 8 December 1956, from Cape Canaveral to an altitude 126.5 miles (203.5 km). The TV-0 was a familiar one-stage Viking, and its launch allowed the NRL-Martin team to establish the project infrastructure at Cape Canaveral and familiarize with the requirements and operations of the Air Force missile test range. In addition, the flight demonstrated reception of Minitrack signals for telemetry and tracking purposes.

Success of TV-0 and TV-1

The next test vehicle, TV-1, consisted of the last available old Viking rocket as the first stage and a prototype of the new third-stage motor built by the Grand Central Rocket Company. The solid-propellant motor served as the second stage of the TV-1. The launch fully succeeded on 1 May 1957. The original Vanguard plan called for the first attempt to launch a satellite only on the test vehicle TV-4. In July 1957, however, the NRL, encouraged by the success of the TV-1, decided that beginning with the TV-3 all Vanguard test launches should carry a satellite that could be deployed in orbit as a bonus.

Baby Satellite

The primary objective of TV test launches was to validate the vehicle design and evaluate its performance; therefore, these vehicles carried additional instrumentation. Consequently, the NRL designed a new smaller and lighter "baby satellite" that weighed only 3.25 lb (1.5 kg) and was 6.4 in. (162 mm) in diameter. The actual flight weights of the first launched Vanguard satellite (Vanguard I) and the separation mechanism would be 3.211 lb (1457 g) and 0.830 lb (376 g), respectively. The decision to design and build the new smaller satellite was made already in January 1957. To meet the tight schedule, the satellites had to be fabricated in the NRL's machine shops, which were not equipped to machine the originally planned magnesium. Consequently, the satellite spheres were made of aluminum.

Transmitters and Solar Cells

The Vanguard baby satellite carried a transistorized transmitter at the frequency of 108.00 MHz with the power output of 10 mW. The transmitter was powered by a set of seven Mallory mercury-cell batteries in a hermetically sealed case. The set was designed to last for two weeks. The second transmitter worked at the frequency 108.03 MHz and had the output power 5 mW. The power for this transmitter was supplied by solar cells provided by the Army Signal Research and Development Laboratory at Fort Monmouth, New Jersey. Six clusters of the solar cells that would become the first ever used in space were distributed over the satellite sphere. Six 12-in. (30.5-cm) antennas were pointed at mutually perpendicular directions. Four antennas were connected to the battery-powered transmitter, and the remaining two antennas were connected to the solar-powered transmitter. Two thermistors to measure internal and surface temperatures were also carried onboard. The heaviest parts of the satellite were the solar cells (0.557 lb or 253 g) and the batteries (0.670 lb or 304 g).

The next launch of the Vanguard test vehicle was rapidly approaching. The TV-2 consisted of the new first stage with the X-405 engine and dummy upper stages. As the preparation for launch progressed at Cape Canaveral, the Soviet

Union placed Sputnik into orbit. Suddenly, the American space program, and the Vanguard in particular, had been catapulted into the focus of the nation's attention. The TV-2 successfully went up on 23 October 1957, to an altitude 109 miles (175 km) and distance 335 miles (540 km) downrange. But this was not enough — the American public now demanded the Earth-orbiting satellite.

Vanguard in the Focus of Attention

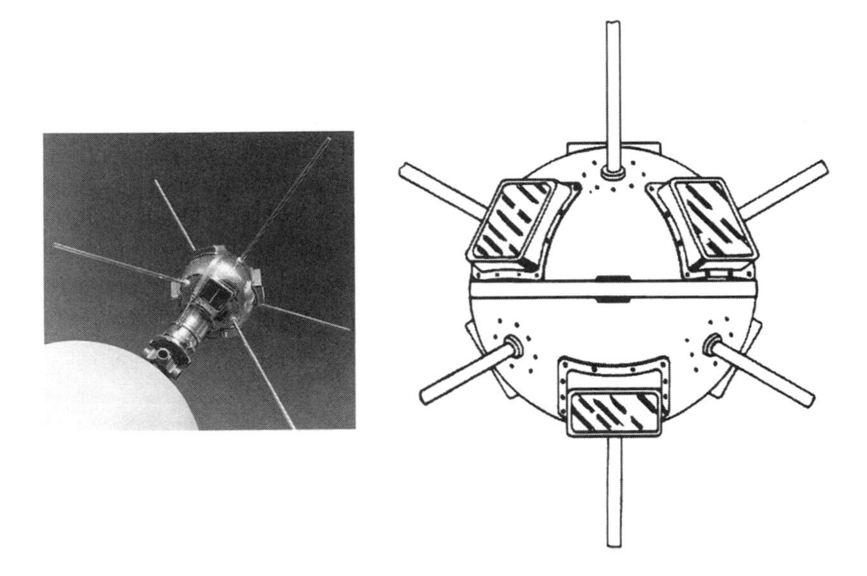

Fig. 15.18. Baby satellite (Vanguard I) attached to the third stage through a separation mechanism based on the strap, pull-pin and girth-ring method (left). The satellite was separated from the carrier in orbit. The satellite consisted of two aluminum hemispheres joined by a reinforcing ring (right). Six mutually perpendicular antennas were 12 in. (30.5 cm) long. Each antenna was made of an aluminum-alloy tube with 5/16 in. (7.94 mm) outer diameter and walls 0.02 in. (0.51 mm) thick. Six clusters of solar cells were distributed over the surface of the sphere and provided power for seven years. Photo (left) courtesy of NASA; figure (right) from *Vanguard I Satellite Structure and Separation Mechanism* (Shea and Baumann 1961, 8).

Jupiter C

While the Vanguard development was steadily advancing, the Army was waiting for its chance. The Army Ordnance was well prepared for satellite launch. When ABMA was ruled out of the satellite effort in 1955, its activities on the Project Orbiter were redirected to testing reentry vehicles. This reentry test vehicle (RTV) program required a multistage missile to provide sufficient velocities for validation of the technology for the nose cones being developed for the Jupiter IRBM. A special new missile Jupiter C (C stood for a *composite* vehicle) was authorized in September 1955.

The Jupiter C consisted of an elongated Redstone as the first stage and two upper stages of clustered scaled-down Sergeant rockets, similar to the original Project Orbiter concept. The Rocketdyne Division of North American Aviation Company introduced a new more efficient fuel for the Redstone engine, hydyne

(60% UDMH and 40% diethylenetriamine). As von Braun described it, hydyne "yields from 10 to 15 per cent more specific impulse than does alcohol and can be used in an engine designed for alcohol and liquid oxygen without major modification" (von Braun 1959, 127).

Hydyne

Eleven modified solid-propellant Sergeant rockets formed the second-stage annular ring of the Jupiter C. The third stage, a cluster of three Sergeant rockets, filled the inner space of the ring. The fourth stage was a single modified Sergeant motor. Two battery-powered electrical motors spun up the upper stages before launch in order to reduce dispersion of the thrust vector, which was especially important in the case of clustered motors with the inevitable differences in thrust and alignment of individual motor units. In addition, the spinning upper stage provided gyroscopic stability. Spinning of solid-propellant motors remains a standard technique still in use today to minimize thrust dispersion and to avoid complications of active thrust vector control.

Upper Stages

Fig. 15.19. Juno 1, a modified Jupiter C rocket with an elongated Redstone as the first stage ready for launch of the first U.S. satellite Explorer I on 31 January 1958. A similar concept, Project Orbiter, was proposed by the Army in 1956. One can see oxygen vapor exiting through the vent valve. Before launch the valve would be closed and the liquid-oxygen tank pressurized. The upper stages are in the cylindrical structure on the rocket top, which would be spun up before launch. Photo courtesy of NASA.

A high point of the Army's RTV program was a Cape Canaveral launch on 20 September 1956 of the multistage Jupiter C vehicle for a separation test. The follow-on launches would carry test nose cones. On this launch, however, the payload consisted of 20 lb (9 kg) of instrumentation attached to an inactive fourth stage. The missile reached an altitude 682 miles (1097 km) and impacted the area 3335 miles (5366 km) downrange in the Atlantic Ocean. If the ABMA were allowed to use a Sergeant missile instead of the inactive fourth stage, as it would later do launching the Explorer I, a satellite could have been deployed in orbit on that day. The ABMA had thus clearly demonstrated the capabilities of placing small satellites in orbit by September 1956, but was not permitted a space launch. This Army achievement was what Senator

20 September 1956

Russian Comments on the American Satellite Project

The undersigned [Ernst Stuhlinger, a close collaborator of Wernher von Braun in Peenemünde and Huntsville], while attending the Eighth Congress of the International Astronautical Federation in Barcelona, Spain, had the opportunity to meet members of the Russian delegation to the Congress. The head of the delegation, Professor Leonid Sedov, made some comments [in conversations on 7 and 8 October 1957] regarding the American Satellite Project ... . The conversation between Prof. Sedov and the undersigned were held in German. Notes of the talks were written immediately afterwards. While they may not be entirely accurate word by word, the opinions and thoughts expressed by Prof. L. Sedov are reflected very closely by these notes ... .

Leonid Sedov (LS): We could not understand why you people [Americans] picked such a strange design for your Satellite carrier [Vanguard]. It was complicated, difficult to develop, and very marginal. The development time which you allotted to the project appeared much too short. Why did you try to build something entirely new, instead of taking your excellent engines from your military projects, such as the REDSTONE or the IRBM? You would have saved so much time, not speaking of troubles and money. This design would have given you also a very good growth potential, whereas the VANGUARD will always be limited to 20 lbs. One wants to have more weight in a Satellite, and a design based on one of your big engines would have given you that. After all, we are only at the beginning of a new and very great development. Why did you not choose this very natural, straightforward approach? Why did Dr. von Braun select this other design instead?

Ernest Stuhlinger (ES): Dr. von Braun? He did not decide this. He is not a member of the VANGUARD Committee; in fact he is even not a consultant or an adviser for the American VANGUARD Satellite.

LS: Is that true? Was he really not connected with the VANGUARD Project? Well, if that is so, I think that we can understand a few things better. When we first learned about the VANGUARD Project and the design of the vehicles, we were really stupefied. But why was Dr. von Braun not a member of the VANGUARD Project?

ES: He would have loved to be, as you can imagine, and his team would have loved too. But he was busy with other assignments, while the VANGUARD Project was given to the Navy.

LS: In our Country [USSR], we gave the Satellite Project highest priority, because we considered it to be of utmost importance, not only for scientific reasons, but from the political angle. We felt that it was really a national project of the first order. We started our Satellite project less than two years ago, and we concentrated our best forces on it. We took the engines we had from other programs, which we knew thoroughly, and we combined them with other well-proven components. We worked very hard until we finally succeeded. In particular, we avoided any novel designs as far as possible, but rather made the approach as logical and straightforward as could be. Why did you not do the same? It would have been the natural choice, and you were in an excellent position with all your missiles. After all, Dr. von Braun has done such an outstanding development job with his early V-2, and he has been in business ever since.

Ernst Stuhlinger, ABMA, 29 October 1957
(*Army Ordnance Satellite Program* 1958, Appendix 10)

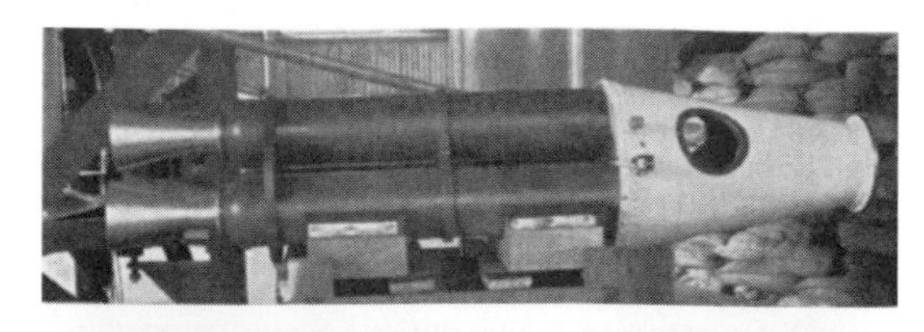

Fig. 15.20. Second (bottom) and third (top) stages of Jupiter C. The second stage was powered by an annular ring of 11 modified Sergeant solid-propellant motors. The third stage was a cluster of three identical motors fitted inside the ring of second-stage motors. Originally published in "Explorer Rocket Research Program" by G. Robillard, *ARS Journal*, Vol. 29, No. 7, p. 493, 1959 (Figs. 2 and 3). Copyright © 1959 by the American Institute of Aeronautics and Astronautics, Inc. Reprinted with permission.

Missed the Boat Back in 1956

Estes Kefauver of Tennessee meant in January 1958 when he said during senate hearings that "with the ABMA program ... we [the United States] missed the boat back in 1956" (*Hearings* 1958, 1717).

After the successful 20 September 1956 shot, the Army repeatedly tried to get a permission to launch a satellite. As General Medaris described it,

> we [the Army] had on hand a backup missile for that one still in the original [four-stage] satellite configuration, and at varying times during this period [after September 1956] we suggested informally and verbally that if they [Department of Defense] wanted a satellite we could use that backup missile as a satellite ... . In various languages our fingers were slapped and we were told to mind our own business, that Vanguard was going to take care of the satellite problem. (*Hearings* 1958, 1700)

Fingers Slapped

All attempts by the Army's ABMA to get permission to launch a satellite were in vain so far. In the meantime, the Vanguard team was preparing for the first flight test of the fully assembled launch vehicle with a baby satellite. This TV-3 test was the first launch of the complete Vanguard vehicle with all three live stages and with the complete guidance and control system. It was also the first time that the second stage was to fly. Obviously, the probability of success for such an untried new and complex vehicle was low. But in the wake of the Sputnik shock, the media brought the nation's attention to Cape Canaveral and to the launch, fueling the expectations. Words of caution had been offered by some, however, prior to launch. *The New York Times*, 6 December 1957, for example, on p. 33 quoted the warning by Martin's Vice President George S. Trimble, Jr. that "failure usually results at least three times out of seven" with testing of the type of the Vanguard launch.

TV-3 Explodes

On 6 December 1957, the TV-3 rocket raised a few feet over the launchpad when the first-stage engine lost thrust. The rocket then settled back and exploded on the pad in a fireball. A low pressure in the fuel tank allowed some of the high-temperature gases to enter the fuel system through the injector head from the

Fig. 15.21. An attempt to launch the Vanguard test vehicle TV-3 ends in failure on 6 December 1957 at Cape Canaveral. The engine of the first stage loses thrust less than 1 s after liftoff, and the vehicle explodes on the launchpad. Photo courtesy of NASA.

combustion chamber, causing fire and destruction of the injector and a consequent complete loss of thrust.

Newspaper headings next day were gloomy, and the public felt humiliated. *The New York Times* proclaimed on the front page that "Failure to Launch Test Satellite Assailed as Blow to U.S. Prestige." The lead article in *The Los Angeles Times* was titled "Vanguard Fiasco Deals Heavy Blow to American World Prestige." A wire service reported that "news from Cape Canaveral, Fla., about the rocket failure hit the State Department with a kind of sickening thud. Officials had been watching for two or three days a buildup of ridicule in Europe of the [technical problems during the preparation to the] satellite test launching" (*Los Angeles Times*, p. 2, 7 December 1957). Perhaps most clearly the sentiments manifested at the New York Stock Exchange. "The Martin Company, prime contractor for the Vanguard missile," reported *The New York Times* next day

Blow to Prestige

BLACK YEAR FOR IMPERIALISM

... The United States sustained a severe moral and political defeat when the two Soviet Sputniks were sent into space and when it failed to launch its own satellite, a 1.5-kilogram aluminum sphere [Vanguard satellite on 6 December 1957].

The New Times [Soviet magazine distributed in the West], No. 1, p. 7, 2 January 1958

New York Stock Exchange Reacts

> had the most violent reactions of any stock ... . Following the report of the failure of the satellite firing, a flood of sell-orders caused governors of the exchange to suspend trading in Martin stock at approximately 11:50 A.M. ... Trading was resumed at 1:23 P.M. ... It was the most actively traded issue on the Big Board. With the exception of Lockheed Aircraft ... other aircraft and missile stocks also showed losses. (*New York Times*, p. 8, 7 December 1957)

Army Leaders at Redstone

The launch of the first Soviet satellite on 4 October 1957 found the newly appointed Secretary of Defense Neil H. McElroy and Secretary of the Army Wilbur M. Brucker visiting the ABMA in Huntsville, Alabama. John Medaris and Wernher von Braun took the opportunity, dramatized by Sputnik in orbit, to plead again for the permission to proceed with the launch of a satellite. Three space-capable vehicles based on the Jupiter C were left over from the RTV program, and they were stored in Huntsville.

Medaris Charges Ahead

To save time and anticipating the go-ahead from Washington, General Medaris boldly ordered to start the preparation of the launcher. A few weeks elapsed with large sums of money being spent on an unauthorized program without any word from the Department of Defense. Medaris would have likely ended up court-martialed had the decision been negative or significantly delayed or had the satellite launch failed. In the meantime the second Soviet satellite was put in orbit on 3 November. Finally, the Secretary of Defense directed the Army, on 8 November 1957 to launch two satellites. The revived Army satellite program was soon expanded to a series of Explorer satellites.

The first Army launch vehicle was designated the Juno 1, and ABMA and JPL embarked on a crash program to prepare the launcher and to build the satellite. For tracking and telemetry, the satellite relied on both the Navy's Minitrack and on the Microlock system designed and built by the Army's JPL. The Microlock was a combination of a simple, low-weight, low-power onboard transmitter and a complementary ground-receiving station; the system provided an angular tracking accuracy up to 4 arcmin. Microlock stations were located in California, Florida, Singapore, Nigeria, and other places.

Microlock

One of the scientific instrumentation packages prepared for the Vanguard was modified to take a ride on the Juno 1. JPL engineers helped to quickly adapt the instrumentation to the new satellite and telemetry. The payload of the Army satellite that would become known as Explorer I included a cosmic ray instrument, two types of micrometeorite detectors, and temperature sensors. The cosmic ray instrument was provided by James van Allen of the State University of Iowa. Van Allen was a most experienced veteran of sounding rocket programs and one of the leading and active advocates of space research in the United States. His cosmic ray instrument was based on a halogen-quenched Geiger–Mueller counter with a scaling electronic circuit.

Discovery of Radiation Belts

The story of the Van Allen's experiment is well known: the instrument measured very high count rates at altitudes greater than 1100 km, and the Geiger tube was saturated, or "jammed," by the unexpectedly high fluxes of energetic particles. The measurements on Explorer I, though initially inconclusive, would ultimately lead, with the follow-on experiments, to a discovery and characterization

Fig. 15.22. Members of the Army team with a model of Explorer I. Left to right: Eberhard Rees, Willie Mrazek, John B. Medaris, Walter Haeussermann, Wernher von Braun, Ernst Stuhlinger. 1957. Photo courtesy of U.S. Army.

RADIATION BELTS

... In regard to cosmic radiation [radiation belts measured by Van Allen], it is interesting to note that, although the [first] Russian satellites [Sputniks] also carried instruments for measuring this effect, the preliminary announcements of scientific results made by the Russian scientists did not indicate that they had discovered this high intensity region. Their preliminary information was confined to altitudes below 600 miles ... . Also, the Sputniks carried continuous transmitters which sent the information to Earth steadily to any receiver which the satellites happened to be passing over. The cosmic-ray measurements were being transmitted on the continuous transmitter in the same way that information is telemetered on [American] Explorer I and IV, but the cosmic-ray information was not put on the tape recorders as was done with [the Van Allen's experiment on] Explorer III. Thus the Russians could observe cosmic-ray data only when the satellite was passing near its lowest point. When the satellite was passing through the high altitude region, and presumably detecting the intense cosmic-ray field [that was observed in the Van Allen's experiments], the satellite was not over Russia. Instead, when the satellite was passing through this interesting region, it was over some part of the world near or below the equator.*

It is true that many radio receiving stations over these portions of the world were recording the signals sent by the Russian Sputniks. However, these signals were meaningless without telemetering code, and this code was not published by the Russians. Furthermore, in spite of the request by this country [U.S.] and others, the Russians would not enter into any arrangements to exchange preliminary satellite data. Thus, although the Sputnik was observing this surprising high intensity cosmic-ray field [a region of the radiation belt], the Soviet government, by its own decision, refused to accept this data or make it possible for anyone else to interpret it. As a result, the Soviet scientists were the first to launch an artificial Earth satellite, they were the last to learn about this vital discovery of the high intensity cosmic-radiation field. Instead, it remained to the American scientists to discover and announce this completely new geophysical phenomenon.

G. Robillard, California Institute of Technology
(Robillard 1959, 496)

* M.G.: Launch of Soviet satellites in the direction of Kamchatka culminated by the final rocket burn at perigee of the final Earth-circling orbit somewhere over Siberia, with the orbit apogee located in the southern hemisphere.

of the Earth radiation belts, commonly referred to today as the Van Allen radiation belts.

Micrometeorite Sensors

The abundance of micrometeorites in the orbit environment was not well known, and characterizing these potentially dangerous particles was critically important for future satellite design and applications. Explorer I carried two types of micrometeorite detectors developed by the Air Force Cambridge Research Center. Maurice Dubin built a piezoelectric sensor, an acoustic microphone, that detected impacts upon the cylindrical satellite body by large size and large momentum microparticles. W. M. Alexander and J. L. Bohn contributed greatly to the design, construction, and calibration of this instrument.

The second instrument, a set of grid abrasion detectors, was built by Edward R. Manring with the help of F. Dearborn and Arthur Corman. The sensors were

Fig. 15.23. Director of the Jet Propulsion Laboratory William H. Pickering (1910–2004) holds a prototype of the Army satellite Explorer I, December 1957. Originally published in *Astronautics*, Vol. 3, No. 1, p. 75, 1958. Copyright © 1958 by the American Institute of Aeronautics and Astronautics, Inc. Reprinted with permission.

thin cards with 17-μ enameled nickel-alloy wires wound around. It was anticipated that a wire would be severed and the event consequently recorded, if a micrometeorite 10 μ or larger struck the gauge, fracturing the wire grid and destroying electrical continuity.

153 Events

One or two grid abrasion card sensors were broken during the launch, but one event was apparently recorded during the flight. Thus, it was possible to establish an upper limit on the micrometeorite flux. Dubin's instrument detected 153 events, allowing one to probe the day-to-day variation of the particle environment.

Passive Thermal Control

Another onboard experiment focused on satellite technology, validating passive thermal control. One internal and three external temperature sensors measured temperature in various places of the spacecraft. By manipulating absorbing (in the 0.3–1.5 μ wavelength range) and emissive (in the 5–20 μ wavelength range) characteristics of surfaces, it was possible to achieve the desired temperature of the spacecraft periodically exposed to the direct solar illumination during one part of its orbit and being in the Earth's shadow during the other.

The shell of the instrument compartment of the satellite was made of 0.025-in. (0.64-mm) thick stainless steel. The surface was sandblasted, and eight white stripes (aluminum oxide) were deposited on the satellite body to achieve the desired thermal performance. This passive thermal control successfully maintained temperature in the instrument compartment, as measured by sensors, between 43 and 104°F (6 and 40°C) during the flight.

Spacecraft Spin

The cylindrically shaped spacecraft was spin-stabilized with the spin rate 750 rpm, providing gyroscopic stability. The spin of the upper stages commenced 13 min prior to launch. At takeoff, the nominal spin rate was 550 rpm, and it gradually increased to 750 rpm in 115 s. This very high spin rate of the satellite raised concerns about the loads on and performance of the miniaturized data tape recorder developed for the original Vanguard instrumentation package. Therefore, the recorder was removed. As a result, only the measurement data obtained by the spacecraft flying over receiving stations were recorded on the ground. Good signals would be received for up to 10 min on a single pass with the spacecraft in apogee. The payload was mounted directly on the vehicle fourth stage, a scaled-down Sergeant solid-propellant motor, and they would stay together in orbit.

Real-time Data Transmission

Fig. 15.24. Explorer I satellite with the fourth-stage scaled-down Sergeant rocket, January 1958. Four wires of the turnstile antenna point sideways. These wires would efficiently dissipate energy of the spin-stabilized satellite, facilitating change of the direction of the spin axis with respect to the satellite body in orbit. Eight white strips on the spacecraft surface serve for passive thermal control. Photo courtesy of NASA.

Explorer Transmitters

Explorer I carried two fully transistorized continuously operating transmitters with the 10- and 60-mW outputs. The high-power 60-mW transmitter supported a turnstile antenna consisting of four wires perpendicular to the payload section of the satellite. The 108.03-MHz circularly polarized radiation was amplitude-modulated for data transmission to the Minitrack stations. The transmitter was designed to operate at least two weeks.

The lower-power transmitter provided a 10-mW power output from a dipole antenna at the frequency 108.00 MHz . The antenna was formed by electrically spliting two halves of the payload. The linearly polarized radiation was phase-modulated for data transmission to the Microlock receivers. It was anticipated that the low-power transmitter would operate for two months.

TV-3BU Launch Scrubbed

The Air Force test range could support only one launch at a time, and its services had to be shared by the Navy's and Army's launch crews. The day for the Army launch was set for 29 January 1958. In the meantime, the Vanguard team worked frantically throughout January attempting to launch the TV-3 backup missile, the TV-3BU. Bad weather and malfunctioning of the rocket and ground equipment frustrated the Navy effort and scrubbed all launch attempts. On 26 January the Vanguard was removed from the launching pad — its next launch attempt was scheduled for 3 February.

The Army team had now a very brief launch window until 31 January. Rain and strong high-altitude jet streams delayed launch for a couple days. Finally on

Fig. 15.25. Juno 1 on a launching pad on 31 January 1958. The first stage was powered by Rocketdyne's A-7 liquid-propellant engine with thrust 83,000 lbf (369 kN) for 155 s of burning time; specific impulse — 235 s; oxidizer — liquid oxygen; fuel — hydyne (60% UDMH and 40% diethylenetriamine); the turbopump was driven by hydrogen peroxide. Stages 2, 3, and 4 consisted of 11, 3, and 1 scaled-down Sergeant solid-propellant rockets, respectively. Each Sergeant produced thrust 16,500 lbf (73.4 kN) for 6.5 s of burning time; specific impulse — 235 s. Fully fueled launcher weight was 64,000 lb (29030 kg) with the propellant mass fraction 83.97%. Mass of the payload with the burned-out fourth stage (deployed in orbit) was 31.5 lb (14.28 kg), or 0.05% of the initial mass. Photo courtesy of U.S. Army.

Fig. 15.26. A model of Explorer I displayed by jubilant (left to right) William H. Pickering (Jet Propulsion Laboratory), James A. Van Allen (State University of Iowa), and Wernher von Braun (Army Ballistic Missile Agency). Photo courtesy of NASA.

the last day of the window, 31 January 1958, and 84 days after the go-ahead from Washington, the Juno 1 lifted off eastward from Cape Canaveral. The first stage burned out on the 150th second. After 240 s of coasting, the upper stages were fired by the command sent from the ground control center. The satellite was supposed to reach the orbit in 7 min after the liftoff. The ground-tracking station at Antigua reported clear reception of the signals from both Explorer transmitters. Quick estimates showed that the satellite reached the orbit with a 106-min period.

Juno 1 Lifts Off

More than 1 h of anxious waiting followed for a confirmation from a receiving station in California. The satellite did not appear, however, at the predicted moment of time. Then 8 min later, the tracking station in California reported receiving signals from a spacecraft approaching from the west. It turned out that its orbit had a 114.7-min period. Wernher von Braun later described that "those 8 minutes difference, during which we waited in vain for the signals to be picked up by our receiver station in California ... were the longest 8 minutes of my life!" (von Braun 1959, 140). The first American artificial satellite of the Earth was in orbit.

Longest 8 Minutes

Explorer I in Orbit

The achieved Explorer I orbit had perigee 360 km (224 miles), apogee 2551 km (1585 miles), orbital period 114.7 min, and 33.3-deg inclination. The payload (without the fourth-stage motor) had mass 18.13 lb (8.22 kg). The satellite was 6 in. in diameter and 36 in. long between the nose cone and the motor. The nickel-cadmium batteries powered the transmitters and the payload. The high-power

EXPLORER I SPIN AXIS

The Explorer I spacecraft was of a roughly cylindrical shape with diameter $d = 6.5$ in. (16.5 cm) and length $L = 80.75$ in. (205 cm). After separation the spacecraft initially rotated at a very high angular rate of 750 rpm about the axis of symmetry A-A as shown in figure. After some time in orbit, the observation of temporal variations of received radio signals suggested that the spacecraft was spinning with a period of 7 s, or 8.5 rpm. What happened?

In orbit, angular momentum of spacecraft changes very slowly subject to usually exceptionally weak environmental torques (gravity gradient, magnetic, drag, solar radiation pressure). One can often assume that the spacecraft angular momentum direction and magnitude are nearly conserved. Angular momentum of Explorer I was thus $H_0 = I_{AA}\,\omega_0$, where I_{AA} was the moment of inertia about the axis A-A and ω_0 was the angular rate (750 rpm initially after launch), with the spin axis being the axis with the lowest moment of inertia of a cylindrical body.

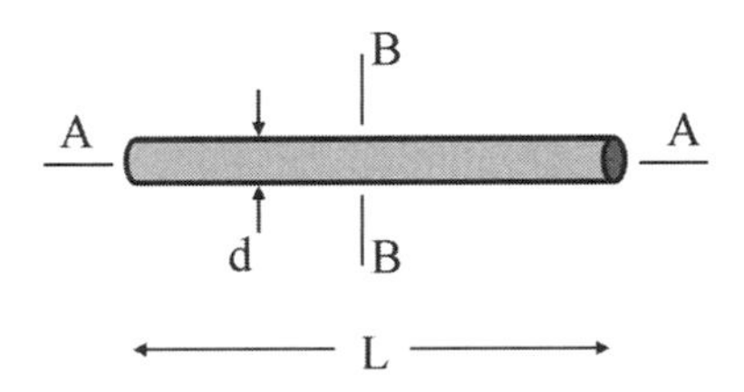

Fig. 15.27. Simple model of Explorer I: a solid cylinder of uniform density with diameter $d = 6.5$ in. (16.5 cm) and length $L = 80.75$ in. (205 cm).

In contrast to angular momentum, spacecraft energy could be dissipated (through thermal emissions) and the spacecraft tended to settle in a state with minimum energy. Energy E_0 of a spacecraft rotating about one of the principal axes of inertia (as A-A or B-B in figure) with moment of inertia I would be equal to

$$E_0 = \frac{H_0^2}{I}$$

For a given (and conserved) H_0, minimum energy would therefore be achieved when a spacecraft rotates about the axis with the maximum moment of inertia I. In case of a cylindrical Explorer I, the axis B-B would be such an axis if one makes a realistic assumption of uniform density of the spacecraft body. As Explorer's energy dissipated with time and the angular momentum vector was preserved in inertial space, the spacecraft spin axis had to change its orientation (with respect to the spacecraft body) from the axis A-A to the axis B-B. Because angular momentum was conserved, the spin rate of the spacecraft about the B-B axis ω_1 should be equal to

$$\omega_1 = \frac{I_{AA}}{I_{BB}}\omega_0$$

For a cylinder with uniform density,

$$\frac{I_{BB}}{I_{AA}} = \frac{1}{2} + \frac{2L^2}{3d^2} = 103.4$$

which gives $\omega_1 = 7.25$ rpm, close to the observed 8.5-rpm spin rate of Explorer I.

> **THIRTEEN MONTHS AND TWO SPUTNIKS LATER**
>
> Thirteen months before the first Sputnik shocked and startled the free world, the men of my command [ABMA] knew they had the ability to launch a satellite. We had built the hardware; we had seen it perform — its upper stages loaded with sand (by directive) instead of the few extra pounds of propellant that would have put its payload into orbit. ... For months we begged and pleaded for the chance to put up the first earth satellite, flying the American flag. Could we obtain that permission? Could we get a decision? Only thirteen months and two Sputniks later!
>
> John B. Medaris (Medaris 1960, viii)

(60-mW) transmitter functioned, as expected, for two weeks until 12 February. Five days later it unexpectedly resumed operation and provided useful signals for several more days. The low-power transmitter operated continuously until mid-April, with the acceptable signal-to-noise ratio until the end of March. Explorer I would stay in orbit for 12 years and reenter the atmosphere in 1970.

Dancing in the Streets of Huntsville

The news of the successful launch of the first American satellite started celebrations on the streets of Huntsville: happy people drove to the Army base and danced on the streets. President Eisenhower announced America's first satellite to the nation by issuing a statement on 1 February 1958 that "the United States has successfully placed a scientific earth satellite in orbit around the earth. The satellite was orbited by a modified Jupiter C rocket." (Medaris bitterly noted later that "neither then — nor later — was there any reference to the Army, to ABMA, or to any of us [on the Army team] as individuals" in the president's announcement (Medaris 1960, 225). President and Mrs. Eisenhower gave a formal "white-tie" dinner on 4 February 1958 on the occasion of the satellite launch. In addition to the top military brass and top scientific officials, the dinner was attended by von Braun, Pickering, Van Allen, and Hagen. ABMA's Medaris was not invited.) In Washington, jubilant Wernher von Braun, William Pickering, and James Van Allen displayed the model of the satellite to the gathered journalists at the National Academy of Sciences, the moment recorded in the memorable photograph (Fig. 15.26). It was a long way for the American satellite, and finally it was in orbit and the main participants were proud of their achievement.

Vanguard and Explorer Fail

Now, it was the Navy's turn again at Cape Canaveral. The new attempt to launch the TV-3BU failed on 5 February 1958. After 57 s of nominal flight, the missile disintegrated as a result of a malfunctioning control system. The spurious electric signals overwhelmed the control system, and the vehicle broke up after exceeding an angle of attack of 45 deg. The launch of the Army's second Explorer followed on March 5, but the vehicle's fourth stage failed, and the satellite was lost.

Vanguard I in Orbit

Finally, the test vehicle TV-4 successfully put in orbit the Vanguard I baby satellite and the 53-lb (24-kg) third-stage motor case on 17 March 1958. The battery-powered transmitter worked for two weeks, and the sun-powered transmitter provided signals for seven years. In addition, the spacecraft was tracked by the worldwide network of optical telescopes. The nearly spherical shape of the satellite made it a perfect tool to probe atmospheric densities and their variations

Fig. 15.28. This perfect launch from Cape Canaveral on 17 March 1958 deployed the Vanguard I satellite in orbit and demonstrated the new space launch vehicle. After the failures in the next four launches, the rocket would deploy the initially designed 23.7-lb (10.8-kg) Vanguard satellite in February 1959.

Vanguard launch vehicle	Stage 1	Stage 2	Stage 3
Thrust, sea level, lbf	28,000	—	—
Thrust, sea level, kN	124.5	—	—
Thrust, altitude, lbf	30,600	7343	3100
Thrust, altitude, kN	136	32.7	13.8
Specific impulse, sea level, s	252	—	—
Specific impulse, altitude, s	275	268	252
Burning time, s	143.6	119.7	38

The Vanguard program demonstrated a record fast development of an entirely new space launcher, with only 30 months from the vehicle authorization in August 1955 to the first successful launch in March 1958. Photo courtesy of Naval Research Laboratory.

Fig. 15.29. NRL personnel on the top of the gantry crane with the Vanguard I satellite at Cape Canaveral in early 1958. Left-to-right: Roger L. Easton, unidentified, Robert C. Baumann, Joseph Schwartz. The satellite — the world's oldest spacecraft — still remains in orbit after its launch on 17 March 1958. Photo courtesy of Naval Research Laboratory.

in the low-Earth orbit environment from the evolution of the spacecraft orbital parameters.

Vanguard I orbital parameters	Initial	In 2003
Perigee altitude, km (mile)	660 (410)	655 (407)
Apogee altitude, km (mile)	3958 (2460)	3841 (2387)
Eccentricity	0.190	0.185
Period, minute	134.3	132.9

The Oldest Man-Made Object in Orbit

The Vanguard I satellite, the oldest man-made object in space, is still in orbit. The orbit apogee changes with time most notably caused by drag by the tenuous Earth's upper atmosphere and ionosphere. The orbit inclination, in contrast, changes very little. Because drag is higher at lower altitudes (near perigee), its effect is especially pronounced in lowering apogee and decreasing orbit eccentricity, making the orbit more circular. As the satellite orbit lowers, the air drag effect is increasing. It is expected that Vanguard will reenter the atmosphere in a couple hundred years.

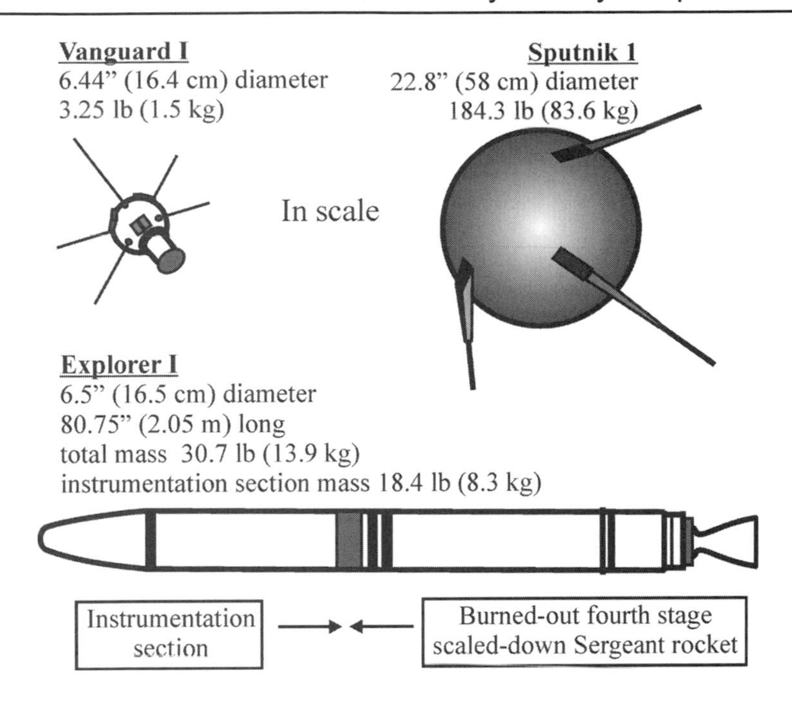

Fig. 15.30. Comparative sizes and masses of the first three Earth satellites, Sputnik 1, Explorer I, and Vanguard I. Figure courtesy of Mike Gruntman.

BIRTH OF NASA

Launches of neither first Soviet Sputniks nor American Explorer and Vanguard drew any diplomatic protests about space vehicles flying over sovereign territories. The principle of the freedom of space for satellites has thus been accepted in practice. The American satellite reconnaissance program, the primary space objective of President Eisenhower, had the road cleared for implementation.

Freedom of Space Accepted

Several special interest groups pushed for a more aggressive space agenda in the United States. One such group included government and industrial scientists and managers with the vision of strong American military presence in space in order to fight and win the battles of the Cold War. Another "pressure" group consisted of university and government scientists interested in space exploration with the experience in high-altitude sounding-rocket research, coordinated for years by the Upper Atmosphere Rocket Research Panel. The members of this group had a growing appetite for more sophisticated and expensive space experiments. Yet another lobbying group comprised spaceflight enthusiasts in government, universities, and professional societies, who passionately advocated manned spaceflight, space stations, colonization of the moon, travel to planets, and other wonderful futuristic ideas leading, ultimately, to interstellar travel.

Emerging Space Lobby

Space advocates had limited success in advancing their agenda so far although Eisenhower unambiguously supported measured progress in space technology. The Republican president stayed focused on a clearly defined goal of meeting national security objectives in space while containing expenses. The

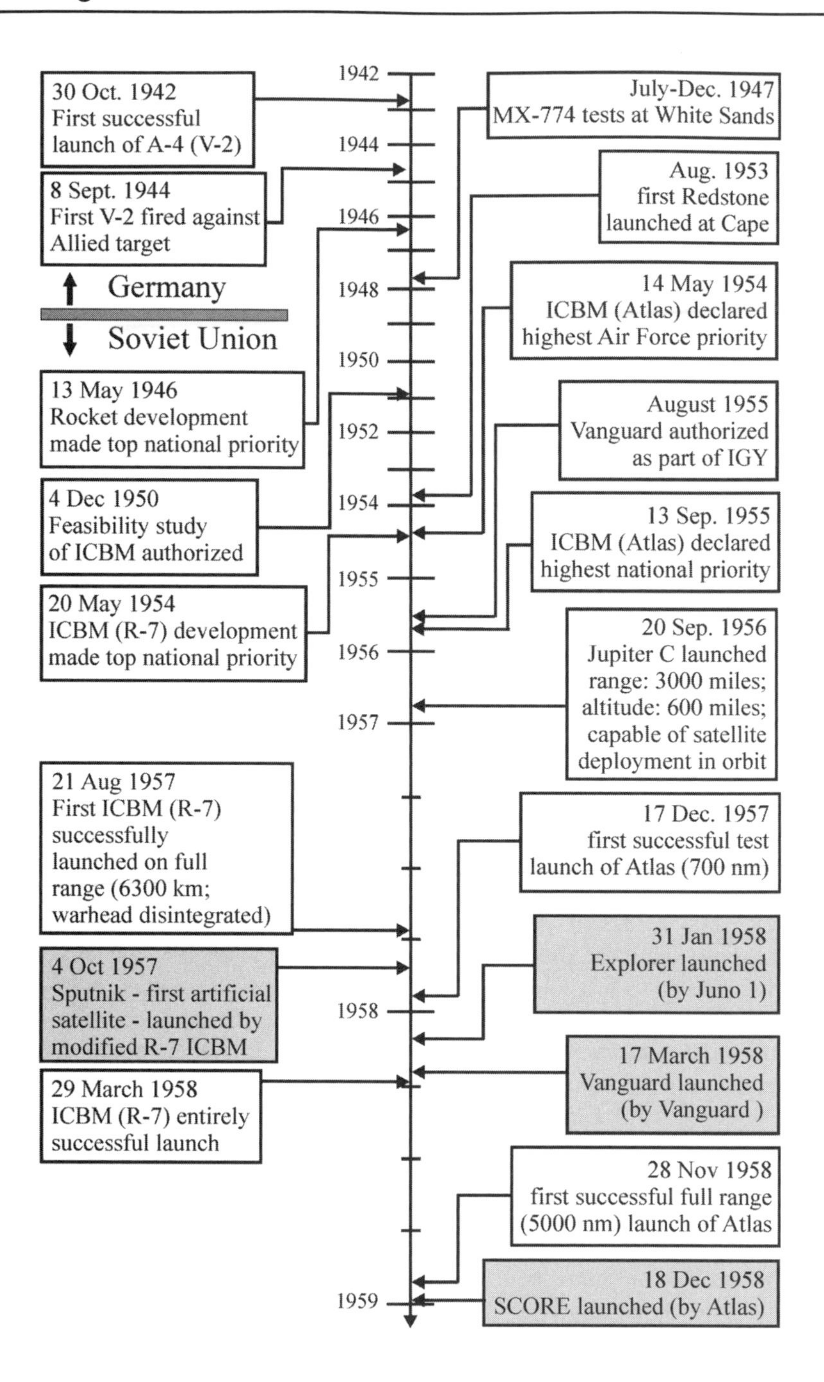

Fig. 15.31. Timeline of major developments on the road to the ICBM and first satellites (shaded boxes): left — Germany and the Soviet Union; right — the United States. The USSR (Korolev's design bureau) was the first to achieve the ICBM and to launch a satellite. By the end of 1958, the United States had placed spacecraft into orbit by three entirely different launchers. It would take more than three years for the USSR to launch a satellite by a rocket designed by an other-than-Korolev's establishment. Figure courtesy of Mike Gruntman.

launch of Sputnik and the following public shock and perception of the technological gap opened a unique opportunity for the proponents of space. Now the space enthusiasts got significant help from the Democrats in Congress who saw an opportunity — politics as usual — to assault the Republican administration, often branding it as incompetent, and promote their own policies.

National Space Effort

Reorganization of the national science effort, especially in the space area, was unavoidable to address real problems and to satisfy the public demand for action. One of the first steps made by the president was creating a position of a presidential science advisor on 9 November 1957, and appointing MIT's President James R. Killian, Jr. to this position.

Science Advisor

Much of the American space research and development was concentrated within the military establishment that was reorganized. The new Advanced Research Projects Agency was formally established in February 1958 in the Department of Defense and took over the funding and direction of the growing national space effort. Then the Army consolidated its programs by forming the Army Ordnance Missile Command on 31 March 1958. The AOMC included ABMA, Redstone Arsenal, Jet Propulsion Laboratory, and White Sands Proving Ground. John Medaris became the Commanding General of the AOMC, responsible for all Army Ordnance programs in guided missiles and ballistic rockets and in space field.

ARPA and AOMC

Two main questions of the ensuing national debate were whether the space effort should be directed by the military or by civilians and whether a new dedicated government organization was needed or the program could be managed by the existing government entities. Politicians and leaders of the scientific, military, and industrial establishments vigorously argued about the way to organize the space enterprise. Interestingly, little discussion focused on the mission of the new effort.

National Debate

On 4 February 1958, President Eisenhower asked his science advisor, James Killian, to convene the President's Scientific Advisory Committee (PSAC) to consider the organization of the national space program. The PSAC came up with a plan for a new agency to direct all nonmilitary space activities, based on the reorganized and expanded National Advisory Committee for Aeronautics and putting under one roof civilian programs in aeronautics and space. The interest groups also actively pressed for creation of such a new agency, and the PSAC supported the idea. In addition, the Senate and the Congress took action forming committees to work on the necessary legislation. Therefore, the president had to endorse, reluctantly, a new government space agency — though not as powerful as many space enthusiasts would have wanted it.

PSAC's Advice

The warning by President Eisenhower in his farewell address to the American people on 17 January 1961, against "the acquisition of unwarranted influence, whether sought or unsought, by the military-industrial complex" is well remembered. What is conveniently forgotten by many, however, is that in the same address he also gravely regarded "the prospect of domination of the nation's scholars by Federal employment, project allocations, and the power of money" and called to be alert to "the equal and opposite danger that public policy could itself become the captive of a scientific-technological elite." Eisenhower's experience with the establishment of the space agency likely contributed to this latter warning.

Scientific-Technological Elite

> **NATIONAL AERONAUTICS AND SPACE ACT OF 1958**
>
> ... The Congress declares that the general welfare and security of the United States require that adequate provision be made for aeronautical and space activities. The Congress further declares that such activities shall be the responsibility of, and shall be directed by, a civilian agency exercising control over aeronautical and space activities sponsored by the United States, except that activities peculiar to or primarily associated with the development of weapons systems, military operations, or the defense of the United States ... shall be the responsibility of, and shall be directed by, the Department of Defense; and that determination as to which such agency has responsibility for and direction of any such activity shall be made by the President ... .
>
> There is hereby established the National Aeronautics and Space Administration ... .

National Aeronautics and Space Act

The president formally proposed the new National Aeronautics and Space Administration in his message to Congress on 2 April 1958. Formation of the new civilian agency would require transfer of large parts of military research and development organizations, particularly those of the Army, under the civilian control. Not everybody agreed with division of space into military and civilian domains. The outspoken General Medaris argued against the transfer, "I cannot in conscience endorse an independent agency. I believe ... that there is no need for erecting a separate agency with operating characteristics outside the Defense Department for doing this job" (*Hearings* 1958, 1710).

T. Keith Glennan

President Eisenhower signed the National Aeronautics and Space Act into law on 29 July 1958, and NASA was formed on 1 October of the same year. The president chose T. Keith Glennan to head the new space agency. A fiscal conservative Glennan tried to contain the growth of the new organization and prevent it from becoming another large bureaucracy with entrenched interests and a voracious appetite for public funds. The space enthusiasts were not completely satisfied with the new arrangement. Andrew Haley, the former president of Aerojet and president of the International Astronautical Federation at that time, accurately sensed the reluctance of the administration when he commented that "importantly, America established a national civilian space flight agency (NASA), with broad powers and a generous budget. Although this was done by the Government, more because of the embarrassing pressure of public opinion than because of any firm conviction that it was necessary, it is still a significant move in the right direction" (Haley 1958, 263).

Fig. 15.32. T. Keith Glennan, 1905–1995, became the first NASA administrator in 1958. Photo (ca. 1960) courtesy of NASA.

OMNIPRESENT KGB

The first NASA administrator T. Keith Glennan caught attention of the Soviet spies and collaborators in the United States in the early 1950s when President Truman appointed him to the Atomic Energy Commission.

An FBI's double agent and well-known Hollywood producer and musical director Boris Morros described that his KGB controller "instructed ... [Morros] to investigate the new AEC commissioner's [Glennan's] political leanings and the state of personal finances." Morros was tasked, as a Soviet spy, to "also find out, of course, whether there were any shady incidents in his [Glennan's] past. Was he a heavy drinker, for example? Did he have a mistress? Who was she? What about his secretary? Was Keith honest? Could he be bribed?" (Morros 1959, 143)

Emerging NASA

When NASA was activated, it absorbed 8000 scientists, engineers, and technicians of the 43-year-old National Advisory Committee for Aeronautics. The operating NACA's facilities included Wallops Station (now Wallops Flight Facility), Wallops Island, Virginia; Langley Memorial Aeronautical Laboratory (now Langley Research Center), Hampton, Virginia; Lewis Flight Propulsion Center (now Glenn Research Center), Cleveland, Ohio; Ames Aeronautical Laboratory (now Ames Research Center), Moffett Field, California; and High Speed Flight Station (now Hugh L. Dryden Flight Research Facility), Edwards Air Force Base, California.

NACA Centers

Transfer of JPL

Administrator Glennan promptly requested transfer of the ABMA's technical personnel and JPL to NASA. This was another major blow to the Army, which rivalled in severity of consequences the Secretary Wilson's 200-mile missile range limitation in 1956. The Army's space capabilities and ambitions that were propelled into the public focus by the successful launch of Explorer I and follow-on satellites were threatened again. As a first step, JPL was transferred from the Army jurisdiction to NASA on 3 December 1958.

Marshall Space Flight Center

In the meantime the von Braun group at Redstone studied, with the support of ARPA, the possibility of clustering several engines in order to determine whether it would be possible to produce thrust of one million pounds (2.4 MN). The Army refused to give up this prominent technical group for the new space agency, and a protracted bureaucratic battle ensued. In August 1959, the Defense Department decided that such large rocket engines would not be needed for military objectives and finally signed an agreement with NASA in December 1959. On 1 July 1960, von Braun and his organization of 4000 employees were transferred from the Army to NASA. Von Braun was appointed director of the newly formed *Marshall Space Flight Center* (MSFC) in Huntsville, Alabama.

Beltsville Space Center

Another NASA field center emerged in the vicinity of Washington, D.C., when NASA absorbed 150 personnel from the Project Vanguard, creating the NASA-Vanguard Division. This group soon combined with more NRL space engineers and scientists and with the groups from other institutions, forming the *Beltsville Space Center*, now *Goddard Space Flight Center*, in Greenbelt, Maryland. Beginning with 216 employees at the end of 1958, the new center had grown to almost 3500 by 31 December 1963.

Building on the satellite programs inherited from the Navy and the Army, NASA initiated a number of programs focused on study of space environment

Fig. 15.33. A 100-ft (30.5-m)-diam passive communication satellite Echo I during the inflation test in 1959. Designed by the NASA Langley Research Center and built by General Mills of Minneapolis, Minnesota, Echo I was one of NASA's first programs in application satellites. In addition, a simple aerodynamic form of the satellite allowed one to explore the vertical density profile of the atmosphere and ionosphere at high altitudes from the satellite tracking data. The satellite mass was only 150 lb (68 kg), and it was successfully launched on 12 August 1960 and inflated in space. Photo courtesy of NASA.

Science and Applications

and space exploration. The new science satellites measured properties of the upper atmosphere, ionosphere, magnetosphere, and solar wind; characterized cosmic rays, astrophysical X-ray sources, and solar emissions; and sent first spacecraft to the Moon and to the nearby planets, Venus (*Mariner 2* in 1962) and Mars (*Mariner 4* in 1965).

In addition, the development of technologies for promising satellite applications was undertaken. One such area was communications, where NASA engineers explored a concept of passive communications with radio transmissions reflected from a satellite. The special satellites were launched, with the largest Echo I being 100 ft (30.5 m) in diameter. The Air Force's SCORE program pioneered earlier an active communications technology with the signal repeater. Communications via spacecraft looked so promising and ripe for practical implementation that a special *Communications Satellite Corporation*, jointly owned by the telecommunications companies and the public, was established in 1962.

Communication Satellites

Although President Eisenhower and the first NASA administrator Glennan restricted the scope and size of the new government space organization, the achievements in space were numerous by the end of the outgoing Republican administration. In his annual report to the Congress on aeronautics and space, the space "report card," the president listed several major achievements in developing satellite systems. One prominent success story in space was not highlighted, however, and the public knew nothing about it. The breakthrough in vital national security applications of satellites had been achieved by that time with the Corona spacecraft conducting highly successful photoreconnaissance missions. This largely "invisible" and unknown to the public space program would continue to play a significant role in development of space technologies in the United States for many years.

Space Report Card

In addition to pursuing scientific and application satellites, NASA started a manned spaceflight program. In 1958, a group of engineers from the NACA's center in Langley under Robert R. Gilruth proposed a detailed plan for putting man in orbit. This plan initiated activities that would evolve into the Mercury program. The development of the manned spaceship began, and a group of the first seven astronauts was selected for training. The Gilruth's group of more than 650 people, organized at first as part of GSFC, moved in 1962 into the newly formed *Manned Spacecraft Center*, later known as *Johnson Space Center*, in Houston, Texas.

Manned Spacecraft Center

The new Kennedy administration was not especially enthusiastic about the emerging space program. In his first annual message to Congress on 30 January 1961, President Kennedy mentioned space only in the context of a possible area

Kennedy Administration Lukewarm

Fig. 15.34. The original seven Mercury astronauts were selected in 1959. The selection criteria were 1) less than 40 years old; 2) less than 5 ft 11 in. (180 cm) tall; 3) excellent physical condition; 4) bachelor's degree in engineering or equivalent; 5) test-pilot school graduate; 6) minimum of 1500 h of flying time; 7) qualified jet pilot. Left to right: Virgil "Gus" Grissom, Alan Shepard, Scott Carpenter, Walter Schirra, Donald "Deke" Slayton, John Glenn, and Gordon Cooper. Photo courtesy of NASA.

"of cooperation with the Soviet Union and other nations." He was reluctant to support the proposal of the new NASA administrator James E. Webb for significant increase in space funding with the focus on landing man on the Moon before the end of the decade.

Pressure on Mercury Program

Fig. 15.35. Alan B. Shepard in the Freedom-7 Mercury spacecraft before launch on 5 May 1961. The Mercury Redstone rocket would send Shepard on a 15-min suborbital flight, reaching the altitude of 115 miles. This was the first U.S. manned space flight, 23 days after the Gagarin's flight. Photo courtesy of NASA.

The new Soviet space feat of placing the first man, Yurii Gagarin, in orbit on 12 April 1961 resulted again in the public perception that the Soviet Union was winning the Cold War. The Mercury program was under pressure to fly an American in orbit. The program steadily progressed overcoming numerous technical difficulties. The first flight in space by Alan B. Shepard lasted 15 min on a suborbital trajectory (he splashed 300 miles

SPACE "REPORT CARD" FOR 1960

I am transmitting herewith [to the Congress] the third annual report on the Nation's activities in the fields of aeronautics and space.

As this report testifies, 1960 witnessed a vast expansion of man's knowledge of the earth's atmosphere and of the limitless regions of space beyond. The Vanguard, Explorer, and Pioneer spacecraft have added substantially to our knowledge of the earth's environment and of the sun-earth relationship. Experiments with [passive and active communications satellites] Projects Echo and Courier, [Television Infra-Red Orbiting weather Satellites] TIROS I and II, and [navigational satellites] TRANSIT I and II have shown promise of spacecraft applications in the fields of communications, meteorology, and navigation. Among the outstanding accomplishments in technology were a series of successful recoveries from orbit of capsules of the DISCOVERER satellites [the cover name for the photoreconnaissance Corona satellites] and the increasing degree of reliability in stabilizing these satellites in the required orbit.

Significant advances were made in the manned space flight program and in the preparation of a small fleet of powerful launch vehicles to carry out a wide variety of space missions.

Dwight D. Eisenhower, 18 January 1961

... I BELIEVE WE SHOULD GO TO THE MOON ...

... If we are to win the battle that is now going on around the world between freedom and tyranny, the dramatic achievements in space which occurred in recent weeks should have made clear to us all, as did the Sputnik in 1957, the impact of this adventure on the minds of men everywhere, who are attempting to make a determination of which road they should take. ... Now it is time to take longer strides — time for a great new American enterprise — time for this nation to take a clearly leading role in space achievement, which in many ways may hold the key to our future on earth.

... Recognizing the head start obtained by the Soviets with their large rocket engines, which gives them many months of lead-time, and recognizing the likelihood that they will exploit this lead for some time to come in still more impressive successes, we nevertheless are required to make new efforts on our own. For while we cannot guarantee that we shall one day be first, we can guarantee that any failure to make this effort will make us last ... this is not merely a [space] race. Space is open to us now ... . We go into race because whatever mankind must undertake, free men must fully share.

I therefore ask the Congress ... to provide the [additional] funds which are needed to meet the following national goals:

First, I believe that this nation should commit itself to achieving the goal, before this decade is out, of landing a man on the moon and returning him safely to the earth. No single space project in this period will be more impressive to mankind, or more important for the long-range exploration of space; and none will be so difficult or expensive to accomplish. We propose to accelerate the development of the appropriate lunar space craft. We propose to develop alternate liquid and solid fuel boosters, much larger than any now being developed, until certain which is superior ... . In a very real sense it will not be one man going to the moon ... in a very real sense it will be an entire nation. For all of us must work to put him there.

Secondly, an additional 23 million dollars ... will accelerate development of the Rover nuclear rocket. This gives promise of some day providing a means for even more exciting and ambitious exploration of space, perhaps beyond the moon, perhaps to the very end of the solar system itself.

Third, an additional 50 million dollars will make the most of our present leadership, by accelerating the use of space satellites for world-wide communications.

Fourth, an additional 75 million dollars ... will help give us at the earliest possible time a satellite system for world-wide weather observation ... .

It is a most important decision that we make as a nation. But all of you have lived through the last four years and have seen the significance of space and the adventures in space, and no one can predict with certainty what the ultimate meaning will be of mastery of space.

I believe we should go to the moon ... .

This decision [of going to the moon] demands a major national commitment of scientific and technical manpower, materiel and facilities, and the possibility of their diversion from other important activities where they are already thinly spread. It means a degree of dedication, organization and discipline which have not always characterized our research and development efforts ... .

John F. Kennedy, 25 May 1961
Special Message to the Congress on Urgent National Needs

downrange) on 6 May 1961. The second American suborbital flight of "Gus" Grissom followed on 21 July 1961. Finally, John Glenn made three orbits circling the Earth in the Mercury spacecraft, built by the industrial prime contractor McDonnell Aircraft Corporation. The flight took place on 20 February 1962, half a year after the second Soviet cosmonaut German Titov spent more than one full day in space orbiting the Earth.

Fig. 15.36. President John F. Kennedy with Wernher von Braun, 19 May 1963. Courtesy of U.S. Army.

Another Blow to Prestige

Gagarin's flight on 12 April 1961 was viewed by many as another serious blow to the prestige of the United States. This new Soviet first was followed several days later by the disastrous Bay of Pigs attempt to overthrow the communist regime in Cuba. The president, a pragmatic cold warrior, had to act quickly.

Kennedy Challenges the Nation

On 25 May 1961, President Kennedy addressed the Congress on the urgent national needs, describing specific objectives of the American space program and asking for additional funding. He clearly presented the rivalry in space as part of the global competition between the free world and forces of communism, though nonmilitary in its nature but still vitally important. Kennedy emphasized landing man on the moon and advancing several application satellite programs. Typically for the information-controlled propaganda-filled totalitarian society, in its report about President Kennedy's speech the official Soviet Communist Party newspaper *Pravda* mentioned only that the president requested additional funding of one-and-a-half billion dollars for military purposes "as well as for expanding activities in space."

Kennedy's decision to go to the moon led to the Apollo program. This initiative together with the accelerated work on other rocket and satellite projects resulted in a major expansion of NASA and shaped the agency's focus for many years to come.

16. OPENING THE SKIES

Military Space

Ballistic missile programs driven by the possibility of carrying nuclear warheads along with civilian initiatives such as the Vanguard are widely recognized as major contributors to the breakthrough to space. The military space effort of the late 1950s had, in contrast, remained largely unknown until early 1990s when the photoreconnaissance Corona satellites were declassified. Corona and other early national security programs dominated the space effort in 1950s and played a vital role in paving a road to conquest of space and determining to a substantial degree the way the civilian space effort was conducted. Many critically important space technologies were pioneered by the national security space effort and essentially contributed to the civilian space programs.

Growing Tensions

The post World War II world was not peaceful. In the words of President Dwight D. Eisenhower, "every agreement the Soviets entered into at [the] Teheran [Conference] in 1943 and [the] Yalta [Conference] in 1945, was ruthlessly broken, save for those palpably to their advantage. The same holds true for the Potsdam Conference of 1945" (Eisenhower 1963, 504). A confrontation between the free world and totalitarian communist states became a dominating feature of the political and military situations in Europe and Far East.

Intelligence Failure Unacceptable

Development of powerful long-range ballistic missiles and nuclear weapons rapidly accelerated in the 1950s threatening devastating consequences should the Cold War turn into a full-scale military conflict. New technologies allowed no time for preparation for hostilities and mobilization and made an intelligence failure such as Pearl Harbor absolutely unacceptable. Therefore, monitoring military developments and posture of the adversary and accurate knowledge of its offensive potential and deployment of forces became a key to avoiding a fatal miscalculation and hence reducing the risk of war.

Big Four Summit

After the death in 1953 of Communist dictator Joseph Stalin, new leadership had emerged behind the Iron Curtain: Nikolai A. Bulganin, Nikita S. Khrushchev, and Georgii M. Malenkov. The ensuing power struggle led to elevation of Khrushchev to a position of the supreme Soviet leader. Soon, the new rulers of the USSR met Western leaders at a summit of the Big Four powers — the United States,

Fig. 16.1. President Dwight D. Eisenhower (left) proposed an open skies policy and initiated a vigorous program in overflight reconnaissance. Secretary of State John Foster Dulles stands next to the president. Photo (8 May 1957) courtesy of National Archives and Records Administration.

Soviet Union, Great Britain, and France. This important meeting was held in Geneva, Switzerland, on 18–23 July 1955.

Reducing Dangers of Surprise Attack

The United States finally agreed to the conference in Geneva when the signing of the *Austrian State Treaty* establishing a neutral Austria without foreign occupying troops brought a glimmer of hope that the new Soviet leaders "might be genuinely seeking mutually acceptable answers" to dangerous confrontation. The main concern of the American president was to avoid "frightful surprises, whether by sudden attack or by secret violations of agreed restrictions." Any disarmament, or arms control agreement must be verifiable, argued the Eisenhower administration, in contrast to devout communist fellow travellers and many other Western socialists and pacifists who were eager to accept Soviet pronouncements about their sincere and peaceful intentions and deeds without verification. The wise Russian saying, *doveryai, no proveryai*, that is *trust but verify,* would be accepted by the Soviet Union much later as the basis for agreements in arms control. This change in attitude happened only after the Kremlin faced an unmistakable resolve of the United States in 1980s.

In preparation for the summit of the Big Four powers, President Eisenhower endorsed a suggestion to propose a bold new doctrine allowing reciprocal aerial reconnaissance overflights of the American and Soviet territories. This concept, called the "open skies" plan, was enthusiastically advocated by President's Special Assistant for Cold War Strategy Nelson A. Rockefeller. At the same time Rockefeller strongly argued for launch of a scientific satellite, advancing the idea that would be realized in the Vanguard program.

Fig. 16.2. Nikita S. Khrushchev emerged as the leader of the Soviet Union after the death of Joseph Stalin. Khrushchev's "Nyet!" at the Geneva Conference sealed the fate of the Eisenhower's open skies plan in 1955. Khrushchev was instrumental in shifting the emphasis of the Soviet military power to new emerging strategic weapons such as long-range rockets and missiles, hydrogen bombs, and submarines. He vigorously promoted development of the ICBMs and championed a Soviet thrust into space. Photo (18 September 1959) courtesy of Franklin D. Roosevelt Library, Hyde Park, New York.

Concept to "Open the Skies"

On 21 July 1955, Eisenhower proposed the open skies concept during the summit meeting in Geneva. The president characterized the concept "both patently practical and fair" and emphasized an objective of reducing dangers of a surprise attack, which could be achieved by reciprocal overflights of the Soviet Union and the United States by reconnaissance aircraft.

Soviet *Nyet!*

The American proposal received a flat *Nyet! — No!* — from the Soviet leader. Eisenhower later recalled that "he [Nikita S. Khrushchev] said the idea was nothing more than a bold espionage plot against the USSR, and to this line of argument he stubbornly adhered" (Eisenhower 1963, 521). Soviet Premier Nikolai I. Bulganin formally responded to the American president on 19 September 1955, as noted in the *Department of State Bulletin*, p. 645, 24 October 1955, that "the problem of aerial photography [and mutual overflights proposed by you at the Geneva Conference] is not a question which, under present conditions, would lead to effective progress toward insuring security of states and successful accomplishments of disarmament." Ironically, a tacit "common-law" agreement would tolerate routine overhead photography from satellites in only five short years.

Peacetime Overhead Reconnaissance

The Soviet refusal and continuing Cold War confrontation accompanied by belligerent Kremlin rhetoric left no choice to the Eisenhower administration but to initiate a program of overflights of the Soviet Union. Peacetime overhead reconnaissance of the denied areas had thus become a national policy, and its

OPEN SKIES PROPOSAL

... In recent years the scientists have discovered methods of making weapons many, many times more destructive of opposing armed forces — but also of homes and industries and lives — than ever known or imagined before. These same scientific discoveries have made much more complex the problems of limitation and control and reduction of armament.

... I should address myself for a moment principally to the delegates from the Soviet Union, because our two great countries admittedly possess new and terrible weapons in quantities which do give rise in other parts of the world, or reciprocally, to the fears and dangers of surprise attack.

I propose, therefore, that we take a practical step, that we begin an arrangement, very quickly, as between ourselves — immediately. These steps would include:

... to provide within our countries facilities for aerial photography to the other country — we to provide you the facilities within our country, ample facilities for aerial reconnaissance, where you can make all the pictures you choose and take them to your own country to study; you to provide exactly the same facilities for us and we to make these examinations — and by this step to convince the world that we are providing as between ourselves against the possibility of great surprise attack, thus lessening danger and relaxing tensions.

Dwight D. Eisenhower, Big Four Conference, Geneva, 21 July 1955
(from the *Department of State Bulletin*, pp. 173, 174, 1 August 1955)

Objectives of Overflight Program

implementation would eventually make possible verifiable arms control and limitation agreements between the USSR and the United States.

The three major intelligence objectives of American aerial reconnaissance in the mid-1950s were the Soviet long-range bombers, guided missiles, and the nuclear weapons program. Lack of accurate information about the closed information-controlled socialist state led to the perceived *bomber gap* and later to the *missile gap*, feared large-scale deployment of Soviet strategic bombers and dangerous superiority in deployed ICBMs, respectively. These expensive misconceptions were corrected only by the successful overhead reconnaissance programs initiated and implemented by President Eisenhower.

Richard S. Leghorn

The importance of peacetime overhead reconnaissance in the postwar world was recognized by a number of military officers, scientists, and industrial leaders. An American veteran of World War II aerial reconnaissance, Richard S. Leghorn, became a leading proponent of strategic overflights. The new intercontinental weapons were changing the very nature of armed conflict. As early as 1946 when he was invited to speak at the dedication of the Boston University Optical Research Laboratory (BUORL), Leghorn argued for importance of defensive reconnaissance overflight of the territory of potential adversaries, if needed without their permission. (BUORL was later renamed Boston University Physical Research Laboratory — BUPRL — and it played an important role in early space reconnaissance.)

BUORL and BUPRL

In 1951–1952, the Air Force conducted a special study that focused, in part, on national reconnaissance needs. This important assessment, known as the *Beacon Hill study*, was chaired by an Eastman Kodak physicist Carl F.P. Overhage

"WE WILL ALLOW NOBODY TO SPY IN OUR HOME"

Sergei N. Khrushchev, the son of the Soviet leader Nikita S. Khrushchev, described (Khrushchev 2000, 140–142) the Soviet rejection of the open skies concept during the summit of the Big Four powers in Geneva in July 1955:

> In May 1955 both sides [i.e. the USSR and U.S] agreed that no agreement related to arms reduction was possible without effective monitoring of its implementation. ... [The negotiations] stumbled in this area for many years. Just a thought of allowing foreigners to access secret installations [in the USSR] looked to our [Soviet] generals as a monstrosity ... .
>
> Even the father [Nikita S. Khrushchev] did not allow himself to go that far. He understood that verification was necessary. But how could one realize it without revealing military secrets?
>
> ... The plan of the "open sky" was included [by the U.S. delegation] as one of the main proposals in a package [of initiatives] that the president [Eisenhower] took with him to Geneva [summit meeting] in July 1955. It was decided [by the Americans] to present it for negotiations not during the first day [of the summit]; it was desirable at first to get to know each other better .... The relations were developing favorably, and it was agreed not to take political disagreements beyond the doors of the meeting room. Eisenhower suggested to visit a bar after plenary sessions of the meeting in order, as he put it, to relieve hard feelings ... .
>
> My father [Nikita S. Khrushchev] decided to take advantage of the informal situation in the bar. During the cocktails, he found the president ... .
>
> [Khrushchev said to Eisenhower,] "... Your proposal [of the open skies plan] is a legalized espionage. It is important not what guided you but why do you want to fool us? Such a proposal is highly advantageous to you, since the American air forces will obtain all the necessary information, while we will get nothing."
>
> Eisenhower tried to convince my father, but he [Khrushchev] remained firm: "We will allow nobody to spy in our home." ... Being assured that President understood the seriousness of his statement, my father did not argue when Eisenhower asked him at the end "not to throw the idea out the window."
>
> From Geneva the father brought [home] a colorful ... brochure, presented by [President] Eisenhower promoting the "open skies concept." He gave me the brochure to take a look, accompanying it with his words of admiration of the achievements of modern technology. The photographs were impressive indeed. On an [aerial] photograph from a 10-km (33,000-ft) altitude one could see at first a general view of a town; on the next photo one could distinguish houses, and then cars. All was concluded by [a photo of] a blurred body of a man siting in a chair in his backyard with a newspaper in hands.
>
> The capabilities of the [American] photo technology convinced my father even more that the American planes should not be allowed into our air space. [The Soviet government] procrastinated [with the reply] until the end of the year, and on November 25, 1955, Chairman of the Council of Ministers [Nikolai A.] Bulganin [formally] rejected the plan for the open skies in his message to President Eisenhower as unacceptable and violating the sovereign rights of the Soviet Union.

and included, in addition to Leghorn, James Baker (astronomer, prominent optical designer, and codesigner of the future Baker–Nunn cameras for the Vanguard program) of the Harvard University Observatory, Edwin Land (founder of Polaroid), Richard Perkin (cofounder of Perkin–Elmer), physicist Louis Ridenour, Nobel prize winner physicist Edward Purcell of Harvard University, and Allen Donovan of Cornell Aeronautical Laboratories. Baker, Land, and Purcell would make especially important contributions to satellite reconnaissance throughout 1950s. The Beacon Hill report strongly endorsed aerial reconnaissance in peacetime and urged development of high-altitude aircraft that would operate, if required, in the denied Soviet airspace.

Beacon Hill Study

Fig. 16.3. Leading postwar advocate of high-altitude reconnaissance Richard S. Leghorn. During World War II, Leghorn served in aerial reconnaissance units in Europe. After the war Leghorn returned to a civilian career at Eastman Kodak. He was recalled to active duty with the outbreak of the Korean War and was again actively involved in advancing nation's reconnaissance capabilities. Leghorn played an important role in establishing the U-2 and Corona programs. In 1956, he headed the Scientific Engineering Institute in Cambridge, Massachusetts. With the financial backing of Laurance Rockefeller, in 1957 Leghorn founded Itek Corporation, which built the first satellite photoreconnaissance cameras. After resigning from Itek in May 1962, Leghorn continued his participation in the Pugwash movement and worked in executive positions in telecommunications, cable television, and research and development companies. Photo courtesy of Central Intelligence Agency.

Balloon Overflight Program

The overflights by balloons, Project GENETRIX, would come first however. The balloon program, also know as WS-119L, started in November 1955 and lasted for half a year. After release, the balloons reached altitudes between 50,000 and 100,000 ft (15–30 km), above the contemporary fighter plane capabilities, and drifted across the Soviet Union and communist China. It was planned to recover the balloons over friendly territory.

Cover for GENETRIX

The GENETRIX balloon development was coordinated with other balloon research programs of the U.S. Air Force studying airflows over the continental United States. As operational ceiling of aircraft was continuously increasing, one needed to know the conditions at high altitudes. Studies of air currents provided a reasonable cover for balloon development for reconnaissance and could plausibly explain carrying photographic cameras allegedly for recording cloud patterns. High-altitude balloons required production of quality thin polyethylene film without weak spots. Mastery of this critical technology did not come easily because industry saw little incentive

"AT ALL COSTS TO KEEP THE USSR A CLOSED SOCIETY"

Despite all the futility of doing so [arguing with Khrushchev about the advantages of the open skies plan], I continued the discussion. I urged upon him [Khrushchev] the value of an effective plan in which both sides could trust. In view of his assertions that the USSR had nothing but peaceful intentions while the NATO powers were planning aggressive war, it was greatly to his advantage, I said, to use every legitimate opportunity to keep his government informed of all NATO moves in return for giving NATO powers the same important privileges in the USSR. I told him also that we in the United States would accept the Soviet plan of fixed posts if they could accept ours of aerial inspection. His protests were, of course, spurious. Khrushchev's own purpose was evident — *at all costs to keep the USSR a closed society*. He would permit no effective penetration of Soviet national territory or discovery of its military secrets, no matter what reciprocal opportunities were offered to him. Of course, he was aware that without agreement of any kind there was already available to the Soviet government a vast volume of information about us which was constantly being accumulated at little or no cost from United States newspapers, road maps, aerial photographs, magazines, journals, and government reports — some of it of types that could not be obtained even from aerial reconnaissance.

Dwight D. Eisenhower
(Eisenhower 1963, 521, 522)

to tackle the problem, with the big money being in the wraps for groceries and not in esoteric military and reconnaissance applications.

Balloon Campaign

The first balloons were launched for flights over Soviet and Chinese territories in January 1956. By the end of February, a total of 516 balloons had been sent in the sky. The program produced a storm of international protests and adverse publicity. Winds often carried balloons away from the target areas, and many balloons were shot down or landed in wrong places. Only 34 balloons, or 7% of the total number, succeeded in obtaining useful photographs covering more than one million square miles. President Eisenhower pragmatically summarized that "the balloons gave more legitimate grounds for irritation than could be matched by the good obtained from them" (Pedlow and Welzenbach 1998, 86).

Kelly Johnson's CL-282

The balloon program had been terminated, and a new, more powerful means of photoreconnaissance was coming. Clarence L. "Kelly" Johnson's team at the Lockheed's *Skunk Works* in Burbank, California, was completing crash development of a revolutionary high-altitude aircraft CL-282 that would become known as the U-2. The Central Intelligence Agency (CIA) had the primary responsibility for the U-2 program with vital contributions by the Air Force. This was a most unusual arrangement for a major national security program, and President Eisenhower thus established a seed of a civilian organization, reporting to secretary of defense, to direct overhead strategic reconnaissance. This new management structure eventually evolved into a civilian-directed office in the Department of the Air Force and was subsequently reorganized into the National Reconnaissance Office (NRO). CIA's Richard M. Bissell, Jr. was put in charge of the U-2 program (called initially Project AQUATONE) since its inception, and he would later direct the follow-on A-12 (SR-71) aircraft (Project OXCART) and Corona satellite reconnaissance programs.

Seeds of the NRO

The U-2 was a high-altitude unarmed aircraft carrying primarily optical equipment for photoreconnaissance. The cameras used high-acuity lenses based on the technology pioneered by James G. Baker and Richard S. Perkin in the mid-1940s. The *Hycon Manufacturing Company* of Pasadena, California, established by Trevor Gardner, built the cameras for the CIA with Perkin–Elmer and Baker's *Spica, Inc.*, as subcontractors.

Bomber Gap Re-evaluated

The aircraft became operational in a record time, and a new overhead reconnaissance program was baptized by deep-penetration flights over Soviet territory on 4 and 5 July 1956. These first U-2 missions immediately proved their enormous and unique value by revealing that the Soviet Union did not have a large number of long-range Myasishchev-4 "Bison" heavy bombers and that the feared bomber gap did not really exist.

U-2 Shot Down

Sporadic reconnaissance flights, each requiring prior presidential approval, continued in spite of vigorous Soviet protests until that fateful day of 1 May 1960, when the Soviet air defenses finally succeeded in shooting a U-2 aircraft down deep inside the USSR, with the pilot Francis Gary Powers captured alive. In the exemplary information-controlled socialist state, Soviet people had been kept in the dark until that time about American overflights and embarrassing inability of Soviet air defenses to stop intruders. Shooting down of the U-2 and the follow-on publicity and show trial of Powers had stopped aircraft overflights. By that time, however, an alternative means, a secret crash reconnaissance satellite program Corona was close to producing first results.

> **INTRUSION IN SOVIET AIRSPACE**
>
> If United States planes continue to intrude into our [Soviet] airspace and we are impelled to shoot them down, the peoples will be awakened at some tragic hour by a thermonuclear war.
>
> Nikita S. Khrushchev
> in speech at the General Assembly of the United Nations, 13 October 1960

RAND Satellite Studies

The origins of this satellite program can be traced back to 1946 when RAND initiated first studies of the utility of satellites at the request of the Air Force. In the early 1950s, RAND revisited satellite reconnaissance under direction of James Lipp and issued a series of reports. With the help of major industrial contractors such as North American Aviation and the Radio Corporation of America (RCA), RAND evaluated the feasibility of satellite guidance, attitude determination and control, television imaging (based on RCA's orthicon) with data relay, and nuclear power sources. The latter study of power sources would eventually lead to development of radioisotope thermoelectric generators, or RTGs, flown for the first time on a Transit 4A navigational satellite in June 1961. RAND reports issued in 1953 and 1954 confirmed the feasibility of military satellites.

WS-117L

The U.S. Air Force issued system requirements for the Advanced Reconnaissance System (ARS) on 27 November 1954 and formulated general operational requirements on 16 March 1955 followed by a one-year design study by three competing contractors, Lockheed Aircraft, Glenn L. Martin, and the Radio Corporation of America. The ARS was designated the weapon system 117L, or

WS-117L, and it was managed by the program office under Navy Captain Robert C. Truax in General Schriever's Western Development Division.

A question of international legality of Earth-circling space vehicles remained a major concern of the Eisenhower administration. The Soviet Union vigorously condemned reconnaissance flights *along* Soviet borders at that time and attacked and even shot down some American aircraft with loss of life. After

Fig. 16.4. Richard M. Bissell, Jr. (second from the left) with President John F. Kennedy and CIA directors Allen Dulles (left) and John McCone (right), in March 1962. Bissell directed the U-2, A-12 (SR-71), and Corona reconnaissance programs. "In vital areas of scientific intelligence collection," stated President Kennedy, "his [Bissell's] achievements have been unique" (Bissell 1996, 204). Bissell became CIA's Director for Plans in 1959, while preserving control of overhead reconnaissance. He had to take the blame for a failure of the Bay of Pigs operation in 1961 and resigned from the CIA in February 1962. Photo courtesy of Central Intelligence Agency.

Freedom of Space

considering the principle of freedom of space, the National Security Council concluded on 28 March 1955, that "the present possibility of launching a small artificial satellite into an orbit about the earth presents an early opportunity to establish a precedent for distinguishing between 'national air' and 'international space,' a distinction which could be to our advantage at some future date when we might employ larger satellites for intelligence purposes." This decision was subsequently translated into a policy announced by the president in July of the same year and led to establishing the Vanguard program to launch small nonmilitary scientific satellites.

Lockheed Selected

One year later, in June 1956, the Air Force selected Lockheed's Missile Systems Division in Sunnyvale, California, to build reconnaissance satellites. In addition to the specified Air Force requirements, Lockheed proposed a new payload, an infrared telescope with sensors allowing detection of hot exhaust gases produced by jet engines of heavy bombers and by plumes of ballistic missiles during their powered ascent after launch. The infrared satellite-based detection system was advocated by Lockheed's engineer Joseph J. Knopow who would later play an important role in development of advanced communication satellites.

Rocket Launch Detection in Infrared

Detection of rocket-engine hot plumes in the infrared spectral band was first discussed in 1948 by NRL's J.A. Curcio and J.A. Sanderson and considered in detail in 1955 by RAND's William Kellogg and Sidney Passman. Kellogg and Passman outlined the unavoidable difficulties of missile launch detection by "picket" aircraft patrolling near Soviet borders and suggested satellite-based sensors. Following Lockheed's proposal, the infrared sensing concept would first evolve into *Missile Detection and Alarm System* (MIDAS) satellites of the early 1960s and be subsequently implemented as the *Defense Support Program*, a constellation of satellites in geostationary orbit. DSP satellites provided early warning of ballistic missile launches for more than 30 years; they are still in operation today and will be replaced in the future by the new *Space Based Infrared System* (SBIRS).

The envisioned WS-117L was a large, comprehensive, and progressively sophisticated system with photographic, electronic, and infrared capabilities being implemented in three operational series of satellites. The program was under

MISSILE LAUNCH DETECTION IN INFRARED

The total radiation from uniform bright areas of two acid-aniline [i.e., nitric acid and aniline] flames of 400-pound-thrust motors were measured with a total-radiation pyrometer consisting of a radiation thermocouple sensitive to all wavelengths between 0.2 μ and 15 μ and located at the focus of a concave mirror ... .

These measurements of total radiation intensity and others made by the Aerojet Engineering Corporation have been used in making estimates of night detection ranges of the rocket motors with a lead sulfide cell in the focus of a 12-inch-diameter mirror ... .

The following conditions are chosen. Let the optical system be 30 cm in diameter and 30 cm in focal length. At useful ranges, the optical system will form an image of the flame smaller than the sensitive area of the lead sulfide cell. It is assumed that the radiation characteristics of the rocket in flight are those which have been measured on test stands [by NRL and Aerojet] ... .

... Under favorable nighttime conditions in the absence of haze, it was estimated that a 12-inch-diameter optical system with lead sulfide cell detector would exhibit signal-to-noise ratios of 50 against a 200-pound-thrust motor at horizontal range 8.8 nautical miles at 30,000 feet altitude, or 50 against a 400-pound-thrust motor at range 13 miles at 10,000 feet, or 200 against a 1000-pound-thrust motor at range 25 miles at 40,000 feet, or 9 against the same motor at range 91 miles at 20,000 feet.

J.A. Curcio and J.A. Sanderson (1948, iv, 2, 20)

"MILITARIZATION" AND "CIVILIANIZATION"

... Some writers have asserted that there is a trend toward "militarization" of space after its development for civil purposes. This does nothing less than stand history on its head.

... We find a preponderance of "military" space missions that were transformed into civilian applications ... .

RAND research of the 1940s and 1950s played a role in transferring concepts initially explored for the U.S. Air Force, and implemented in related programs of all three military services and the Office of the Secretary of Defense, into broad-ranging civil applications. If one had to choose between the "militarization" and the "civilianization" of space to capture broad trends in U.S. space policy, the latter would more closely approximate reality.

Merton E. Davies and William R. Harris (1988, 108–110)

Tight Spending Limits

exceptionally tight spending limits: although a decision to proceed with the WS-117L full-scale development had been made in April 1956, it was initially funded at the level of one-tenth of the required budget. The launch of Sputnik on 4 October 1957, in spite of all the accompanying uproar, did not immediately bring significant increase in funding, partially because of initial bureaucratic uncertainty and complications introduced by creation of ARPA.

Sputnik "Comes to Rescue"

Ironically, the Soviet space successes had finally resolved the lingering issue of the space overflight rights that so concerned the Eisenhower administration. Deputy Secretary of Defense Donald A. Quarles stated at the meeting of the National Security Council on 10 October 1957, that "another our objective in the earth satellite program was to establish the principle of the freedom of outer space — that is, the international rather than national character of outer space. In this respect the Soviets have now proved very helpful [by launching Sputnik six days ago]. Their earth satellite has overflown practically every nation on earth, and there have thus far been no protests." To further reduce a possibility of hostile international and especially Soviet reaction to satellite overflights, the first ARPA director Roy Johnson instructed the Air Force in October 1958 to emphasize the defensive, nonaggressive nature of its space programs by ceasing use of the WS, or weapon system, designations of military satellites.

Project *Corona*

In response to Soviet ICBM and Sputnik successes, the WS-117L was reorganized in the early 1958. One of its original components that offered an early, essentially interim reconnaissance capability was established as a separate covert program, designated *Project Corona*. Similarly to the highly successful U-2, Corona was managed jointly. This time, however, in addition to the CIA and the Air Force, ARPA joined the management of the program. CIA's Bissell directed Corona with Vice Commander of the Western Development Division Brigadier General Osmond J. Ritland representing the Air Force. ARPA assumed a significant role in funding of the project. By the end of 1959, the military satellite system WS-117L had split into three separate programs: missile launch warning infrared-sensing platform MIDAS and two photoreconnaissance systems, Corona (with the cover name *Discoverer*) and heavier and larger *Satellite and Missile*

Corona, MIDAS, and SAMOS

Observation System (SAMOS). (The first MIDAS and SAMOS satellites would be placed in orbit on 24 May 1960 and 31 January 1961, respectively.)

Agena Upper Stage

Deployment of reconnaissance satellites depended on the availability of reliable space boosters. Two of the Air Force's ballistic missiles, Thor and Atlas, had a promise of eventually evolving into space launchers. The Thor-based two-stage space booster that was initially planned for an interim satellite system of the WS-117L was selected for Corona. (The more powerful Atlas-based launcher was planned for the heavier SAMOS.) A new upper stage, the Lockheed's Agena, was selected to serve as the second stage for orbiting Corona satellites. (A considered alternative was the Aerojet-built second stage of the Vanguard launch vehicle.) After reaching orbit, Agena did not separate from the Corona payload and provided it with attitude control, power, and thermal control during the mission.

Fig. 16.5. Major General Osmond J. "Ozzie" Ritland, 1909–1991, with a model of an Agena upper stage, 1960. (The Lockheed-made Agena served as the second stage with the first-stage Thor for launches of Corona satellites.) Vice Commander (since 1956) of the Western Development Division Gen. Ritland concentrated on the early military effort in space. He worked as deputy to Richard Bissell in both the U-2 and Corona photoreconnaissance programs. As Commander of the *Ballistic Missile Division*, he oversaw the early Corona, MIDAS, and SAMOS programs. In 1961, Ritland assumed command of the newly created *Space Systems Division* of the *Air Force Systems Command*. Photo courtesy of U.S. Air Force.

Film Recovery and Talk-Back Systems

In 1956 and 1957, RAND revisited the earlier concept of physical film recovery from reconnaissance spacecraft. Such a concept obviously did not allow delivery of images in near real time in contrast to a "talk-back" system either with direct television imaging (with or without temporary storage of data) or with the photographic film being processed onboard and subsequently scanned and transmitted to the ground through a television channel. Physical recovery of photographic film, however, provided a significantly larger amount of information with much higher image resolution than an alternative television-based system.

AGENA

In October 1956 Lockheed Aircraft initiated development of the Agena for the Air Force as an upper stage for the Atlas space launcher. In 1958, providing Agena as the second stage for a Thor-based launcher became the top priority.

The Lockheed Agena was based on the engine developed by the Bell Aircraft for takeoff assist of the B-58 Hustler bomber. The first Bell's engines (model 8001) used, as originally designed, jet propellant JP-4 as fuel and inhibited red-fuming nitric acid as oxidizer. In 1958, UDMH substituted JP-4. The engine provided thrust 15,500 lbf (68.9 kN) for 120 s.

The first Agena A was followed by a modification Agena B and finally by standardized Agena D. The latter model used the modified engine (model 8096) with HDA (44% of nitrogen tetroxide and 56% nitric acid) as oxidizer and USO (UDMH with addition of 1% silicon oil) as fuel. Evolving in time, engine specific impulse improved from 265 s of the original model to 300 s and its firing duration increased (with the corresponding increase in carried fuel) from 60 s to 240 s. In addition, multiple restart capability became available.

In the Corona program, the Agena was more than an upper stage: it did not separate in orbit and remained attached to the payload section, providing power, attitude determination and control, and thermal control during the mission. Agena was in fact the first American mass-produced satellite bus carrying various payloads.

Agena became one of the most successful upper stages with the total of 361 vehicles launched by the time it retired in 1987. The launch rate peaked at 41 in 1966. It has flown with Thor, Atlas, Delta, and Titan. In addition to numerous military payloads, the Thor-Agena launched a number of early civilian satellites such as a passive communication satellite *Echo II*, weather satellite *Nimbus 1*, and the first Canadian satellite *Alouette 1*.

General Electric's Capsule

The practicality of the film-recovery concept was also supported by the recently demonstrated warhead reentry into the atmosphere under much more demanding conditions than expected for the Corona. In addition, implementation of the concept did not have to wait for television and videotape data storage technologies to be perfected. It was thus decided in the late 1957 to base Corona on physical recovery of the exposed film, and General Electric was contracted to design and build recovery vehicles. In retrospect, the decision to rely on film recovery from orbit proved to be fortunate because electronic readout technology of the desired quality had not become available in 1960s.

Discoverer Cover

The Corona project was conducted under tight security and with elaborate cover. Parts of the vast satellite program were publicly announced as an Air Force scientific program Discoverer. The stated Discoverer objectives included studying environmental conditions in space and conducting biomedical experiments. This cover program allowed procurement of launch vehicles and construction of ground control facilities through open funding.

Itek Wins over Fairchild

The initial Corona design called for a spin-stabilized satellite with an optical camera built by *Fairchild Camera and Instrument Company* (FCIC). The projected ground resolution was between 40 and 60 ft (12 and 18 m). In April 1958, however, the Fairchild's camera was replaced by an alternative optical system proposed by a recently formed company, *Itek Corporation*. Itek took over scientists and engineers of Boston University Physical Research Laboratory, and its camera design was based on the BUPRL's *Hyac* optical system that offered a promise of better resolution on the ground (20 ft or 6 m) and a potential for further improvements in the future.

Absorption of BUPRL made Itek overnight a major competitor for the space reconnaissance camera. BUPRL was a national technological asset and a leader in the field of aerial reconnaissance. The laboratory origin went back to James Baker who established a large group in Harvard University Observatory making high-quality lenses for aerial photography during World War II. After the war, Harvard University refused to house the group's activities. To preserve the unique expertise of critical importance for national security, a "father" of American aerial reconnaissance then Colonel (later Brigadier General) George W. Goddard found another home for Harvard's optical specialists in the nearby Boston University. A young scientist Duncan MacDonald was made director of the new BUORL.

From Harvard to Boston University

Fig. 16.6. Upper-stage Agena A, ca. 1960. The Agena was 5 ft (1.5 m) in diameter and 15.5 ft (4.7 m) long with fully fueled mass 8400 lb (3800 kg). The later modifications improved performance characteristics, carried more propellants, and allowed multiple restarts in space. Photo courtesy of National Reconnaissance Office.

Main Contractors

Lockheed thus contracted Itek to provide the optical system, and Fairchild remained in the project as a subcontractor to Itek manufacturing the cameras. Douglas Aircraft became another important Lockheed subcontractor, providing Thor missiles for the Corona program; General Electric was responsible for the satellite recovery vehicle and Eastman Kodak for the photographic film and its processing.

The Itek optical system required a more complex three-axis-stabilized satellite. The panoramic camera had a scanning telescope in a nadir-pointed (Earth-pointed) pod. The telescope rotated over a 70-degree range in the direction perpendicular to the satellite flight path, producing an image on a stationary film of a ground strip approximately 10 by 120 miles (16 by 190 km). This so-called C camera would be retroactively called KH-1, where KH stood for "keyhole," the system designator of the overhead reconnaissance product security system. The Itek camera reliably met the 25-ft (8-m) ground-resolution goal, and the resolution was subsequently improved to 6 ft (1.8 m) through better optics, photographic film, and spacecraft attitude control.

Panoramic Camera

The Thor–Agena launches took place at the Vandenberg Air Force Base on the Pacific Coast placing spacecraft in the desired near-polar orbits. Such high-inclination orbits passed over the Soviet Union and were convenient for film recovery over Pacific. Gyroscopes and infrared horizon sensors of the upper-stage Agena that remained attached to the payload in orbit were used for attitude determination and cold-gas thrusters for attitude control. Corona achieved a 0.2-deg attitude accuracy about all three axis; the knowledge of attitude was further improved by using images from horizon and star cameras during image processing.

Attitude Control

Recovery of film capsules was a complex operation and required establishing a special organization in the Air Force. Many elements of the technique were

EASTMAN KODAK'S FILM FOR CORONA

In the early 1950s, common film for aerial photography was high speed and relatively low resolution, the characteristics compatible with the vibrating airplane platforms. In space photoreconnaissance, spacecraft vibrations were much smaller and distances to the ground were much farther, favoring high-resolution fine-grained films similar in this respect to the existing microfilm emulsions. In addition, the films had to be high contrast and most sensitive in red to penetrate haze.

Eastman Kodak had developed the acetate-based films meeting the WS-117L requirements. The WS-117L optical reconnaissance system operated in a controlled environment of the satellite under atmospheric pressure because the film had to be developed and electronically readout onboard. The Corona was essentially different. It was designed to operate without pressurization in vacuum of space to conserve weight. The acetate-based photographic film quickly cracked and jammed the camera under vacuum conditions.

Meeting the challenge, Eastman Kodak developed a new film support based on DuPont's Mylar ®. This new polyester-based film support was called *Estar*. Estar was 4 mil (0.1 mm) thick and perfectly adapted for operation in vacuum. One side of the film was coated with a photographic emulsion optimized for space photoreconnaissance. The other side was coated by a pelloid to prevent film from curling and sticking to each other and provided the desired friction characteristics.

For first Corona flights Eastman Kodak produced film of a 6000-ft (1800-m) continuous length without splices. In another technical feat, the film thickness was maintained within one-half of one micron. With time, the Estar film thickness was reduced to 2.5 mil (0.063 mm), and the later Corona spacecraft usually carried rolls of film of 16,000 ft (4900 m) in length each, spliced roughly every 6000 ft.

OPTICS FOR AERIAL RECONNAISSANCE AND BUORL

... If the importance of aerial reconnaissance [after World War II] was seen by some in War department, it was not seen by others outside of it, at least in the same context as it had been viewed during the fighting. I found this out with a vengeance when I went over to visit Jim Baker at Harvard [University Observatory].

His progress had been remarkable. With his staff of one-hundred scientists and technicians, he had nearly completed the first 100-inch ultra-precision lens ever to be used in an aerial camera. Baker's machine shop had made the barrel to accommodate the lens, and it was equipped with temperature and pressure compensation features. It was exciting for me to see this wonderfully efficient team operating so smoothly and effectively, and after a thorough review of their work, I paid a call on [Director of Harvard Observatory] Dr. Harlow Shapley to discuss the continuance of the operation.

The doctor heard me out and then said, "I am sorry, Colonel, but I can't go along with you. I want Jim Baker to get back to teaching as soon as he can. As long as there was a war on we were happy to do our part, and in emergency we will be happy to help you again, but for now I'm afraid we'll have to close the shop."

... Following the horrendous news, I held a meeting with Jim and his principal assistants in an effort to consider contingency plans, for I was fiercely determined that even without Jim or Harvard, the work must go on. At the meeting I became most impressed with Dr. Duncan MacDonald, a young scientist with a broad background in teaching astronomy and physics. Dunc, short, gregarious and pipe smoking, was also an outstanding administrator, I learned, and since there was no more chance of getting Jim Baker away from Dr. Shapley than there was of turning President Truman into a Republican, I decided that Dunc would be the ideal man to take over the management of the project.

In time through the good offices of Dr. David L. Marsh, president of Boston University, and Dr. Chester M. Alter, dean of the Graduate School, we were successful in making the move from Harvard to Boston University and in doing so, we were able to keep most of the top people with the operation. ... However, I realized that a special vote of thanks was due to Drs. Marsh and Alter in agreeing to take on the Harvard optical facility. They did so mainly because they saw the need and were willing to assist the government in its fulfillment.

The formal opening [in 1946] of the Boston University optical research laboratory [BUORL] was attended by members of the photographic industry, faculty members and staff officers of the Air Force – including Lieutenant General Curtis E. LeMay, Chief of the Strategic Air Command, Major General Alden R. Crawford, Chief of Research and Development at Wright Field, and Major General Laurence C. Craigie, Chief of Air Force Matériel ... . The presence of LeMay, Crawford and Craigie was without question most opportune for my cause. They got the picture. Afterward I never experienced any trouble in getting their financial support.

... One of Duncan MacDonald's first assignments involved the research and development of the largest and most advanced aerial camera ever flown in airplane ... . At thirty thousand feet it could take large detailed pictures one hundred miles distant or more. Its main purpose was for long-distance oblique photography ... . Dunc's camera was the forerunner of others whose need and usage became increasingly more necessary as the cold war set in.

Brigadier General George W. Goddard (1969, 349–351)

perfected earlier in the GENETRIX balloon reconnaissance program. When the film was exposed onboard the Corona satellite, the satellite recovery vehicle (SRV) had to be placed on the precise trajectory leading to landing in an impact area of the 200 by 60 miles (320 to 96 km) recovery zone. The tracking ground station on Alaska sent a command to a southbound orbiting spacecraft to begin descent. The SRV was separated from the Agena, oriented in the desired direction, spun up, the solid-propellant retrorocket motor then fired to reduce velocity by 1300 ft/s (400 m/s), and the SRV despun. During reentry, the SRV was protected by a heat shield that was jettisoned at about 60,000 ft (18 km) and the parachute was opened afterwards. Specially trained Air Force crews were flying several C-119 cargo planes (operating from Hawaii) in the recovery zone. Homing on the capsule, the aircraft had to snare the parachute in midair and reel the capsule into the plane.

Satellite Recovery Vehicle

Fig. 16.7. Air Force's C-119 catching a Corona capsule in the air. Photo courtesy of National Reconnaissance Office.

Difficult Development

The development of the top-priority Corona was difficult and marked by many setbacks. Numerous problems plagued the Thor booster, Agena upper stage, and the payload. The first Corona launch announced as Discoverer I failed to reach orbit on 28 February 1959 as a result of malfunction of the Agena. In 1959 only, eight Corona launches failed for a number of reasons: Agena did not achieve expected performance characteristics; cameras failed; retrorocket malfunctioned; SRV did not separate; or descending capsule was not recovered.

ITEK CORPORATION

Richard S. Leghorn founded Itek Corporation in October 1957 several days after the launch of Sputnik, with Laurance Rockefeller as the largest shareholder. The initial company's mission was information — and document-processing equipment — therefore the company's name, Itek, based on the words "information" and "technology." Leghorn originally planned to concentrate on processing graphic information (photographs, maps, drawings, etc.) with the emerging computer systems. A specific goal that was not publicly discussed was to support processing of the enormous amount of graphical information expected from the SAMOS space photoreconnaissance system. The business strategy included licensing Kodak's Minicard system built around miniaturized photographs attached to computer punch cards.

The original Itek plan was quickly modified as the new opportunities appeared. Funding cuts of 1957 put Boston University Physical Research Laboratory in danger of closing. In addition, the relations between Boston University and the research-and-development-oriented BUPRL were far from harmonious. The laboratory focused on development of optical reconnaissance systems and was a true national scientific and technological asset. Itek Corporation had taken over the personnel, equipment, and contracts of the laboratory by 1 January 1958. (Trevor Gardner's Pasadena-based *Hycon* that made photoreconnaissance cameras for the U-2 also made an attempt to acquire BUPRL.)

With the new capabilities, the Itek business strategy began to shift. Using BUPRL's expertise in reconnaissance systems, Itek bid on cameras for Corona. The proposal was based on the new camera design Hyac (standing for "high acuity") conceived by Walter J. Levison, with the heritage in BUPRL's optical payloads developed for balloon applications. In late March 1958, CIA's Bissell selected the Itek design as backup to the camera proposed earlier by Fairchild. These two cameras required two very different spacecraft, three-axis stabilized and spinning, respectively. A few days later, Bissell reconsidered his decision and made the Itek camera the only system to be developed, with Fairchild assigned manufacturing of these optical systems using the Itek-built lenses.

Fairchild was obviously unhappy with such a turn of events and tried to regain its position by proposing a new camera design in 1961. Itek however won over Fairchild again, this time with the new more complex and technically risky Petzval camera design (known later as C'''), and promptly took over profitable camera manufacturing from Fairchild. (Edwin Land played a crucial role in explaining the advantages of the new Itek design to the decision makers, including to President Eisenhower.) Thus, Itek designed and built the following generations of Corona lenses and cameras (C''' or KH-3, M or KH-4, J-1 or KH-4A, J-3 or KH-4B) until program termination in 1972.

Itek's growth was phenomenal: from about a half-dozen employees in the end of 1957 to more than 500 one year later; its stock soared. Leghorn led Itek on an acquisition spree, which nearly brought the company down. Consequently, Frank Lindsay replaced Leghorn as company president in May 1962.

Serious disagreement between Itek and CIA, aggravated by bitter rivalry between CIA and NRO, about requirements to a next-generation very large reconnaissance camera led to corporate tragedy. In February 1965, Itek discontinued its work on this new CIA's system. Never again would Itek get contracts for new space photoreconnaissance systems, and the company's role would be limited to continuation of Corona. Perkin–Elmer took over the development of the new optical camera and built the system.

After several more failures, success finally came on the 13th launch. Discoverer XIII reached the desired orbit on 10 August 1960; the satellite recovery vehicle separated on the 17th orbit and reentered the atmosphere. The descending capsule was not recovered in the air because of some miscommunications between the waiting C-119 airplanes and the control plane. The capsule splashed down and, fortunately, was quickly found and retrieved from water by helicopter and brought to a recovery ship *Haiti Victory*. It was the first ever successful recovery of a space object. And this was also the American space "first" in a competition with the Soviets. The USSR succeeded however in only nine days to safely deorbit and land a capsule with two dogs, Belka and Strelka.

Discoverer XIII

American Space "First"

Fig. 16.8. The first Corona flight (Discoverer I) being prepared for launch on the Thor Agena A vehicle at Vandenberg Air Force Base on 28 February 1959. The spacecraft did not reach the orbit because of Agena failure. Photo courtesy of U.S. Air Force.

Ironically, Discoverer XIII was a diagnostic launch with no optical equipment onboard. The satellite carried however a CIA-developed signals intelligence sensor SCOTOP to determine whether Soviet radars tracked the satellite in orbit. The recovered capsule was delivered by airplane first to the West Coast and then to Washington, D.C. The program cover story was easily preserved this time because the capsule did not have a tale-telling optical camera. Consequently, there was no need to clandestinely switch capsules, as was prearranged, and only the SCOTOP was quietly removed during an overnight stay at Sunnyvale, California. On 15 August the real, not switched capsule reached the White House where President Eisenhower was presented an American flag carried by the satellite.

Cover Preserved

The full success came a few days later with Discoverer XIV launched on 18 August 1960. The spacecraft completed the mission with all 20 lb (9 kg) of film exposed and transferred to the recovery capsule, which was captured in air by the C-119. The mission brought the first photographs of the denied Soviet territory. Ironically, the pilot of the shot-down U-2 airplane, Francis Gary Powers, was sentenced for espionage in a show trial in Moscow at the same time as the first Corona photographs were being obtained. More than 1,650,000 square miles (4,270,000 km^2) were photographed by this single mission, a larger coverage than accomplished by all prior U-2 overflights of the Soviet Union combined. The ground objects 35 ft (11 m) and larger were identifiable. The impact of this flight and the Corona missions that would follow was enormous.

Success

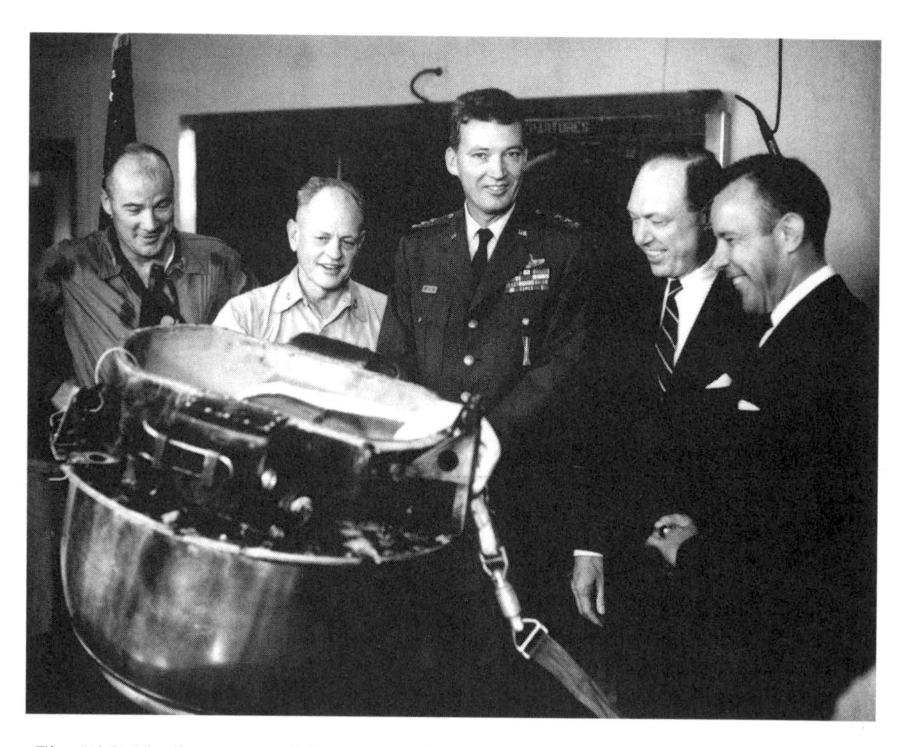

Fig. 16.9. The first successfully recovered Corona capsule (Corona flight 13 — Discoverer XIII). Left to right: Colonel Charles G. "Moose" Mathison who brought the capsule from Hawaii, Major General Osmond Ritland, Lieutenant General Bernard A. Schriever, and two civilian officials. Discoverer XIII was a developmental flight, and the capsule did not carry a photographic equipment. It had however a classified CIA-provided SCOTOP sensor detecting radar hits to determine whether the Soviet Union tracked the satellite. Photo courtesy of U.S. Air Force.

Missile Gap Re-evaluated

Similar to how U-2 overflights had definitely resolved the issue of the "bomber gap" and reduced the arms race in mid-1950s, satellite photoreconnaissance addressed fears of the "missile gap." In the late 1950s and early 1960s, the CIA and the Air Force could not agree on the estimated number of the deployed, or being deployed operationally, Soviet ICBMs. Corona's photographs showed that

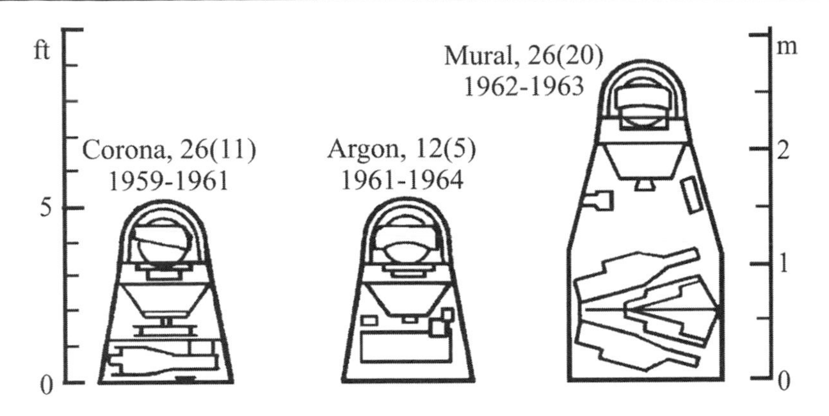

Fig. 16.10. Evolution of the family of the early Corona photoreconnaissance spacecraft, from first Corona (KH-1, KH-2, KH-3) to stereoscopic Mural (KH-4): the numbers after the system type show the total number of launches and the number (in parentheses) of successful capsule recoveries. Stereoscopic imaging with two simultaneously operating cameras was introduced first on the Mural. Argon (KH-5) was an Army's low-resolution system for mapping the terrain. Corona and Argon carried 40 lb (18.1 kg) of film each. The film load increased to 80 lb (36.3 kg) on Mural and up to 160 lb (72.6 kg) on later versions (not shown) J-1 and J-3. Based on figure courtesy of National Reconnaissance Office.

much fewer ICBMs were actually deployed than one could infer from the Soviet propaganda. Consequently, the United States could safely wait for completion of the development of the advanced, convenient, and relatively inexpensive solid-propellant ICBMs. Tremendous resources had thus been saved on avoiding a significant increase in deployment of the existing but obsolescent liquid-propellant intercontinental ballistic missiles.

Corona Modifications

The original Corona optical system KH-1 went through a number of modifications with time, such as variable image motion compensation and improved camera resolution. The Itek-designed KH-4 Mural camera provided stereoscopic capabilities and reached orbit for the first time in February 1962. The Mural consisted of two cameras, one pointed 15 deg backward and the other 15 deg forward along the spacecraft path, with two rolls of film supplied to cameras separately. Each area was thus photographed twice from different directions. Two images of the same area were then aligned in a stereo microscope and appeared three dimensional for a photointerpreter. A rule of thumb of aerial reconnaissance was that stereo coverage increased the information content of photographs by a factor of two-and-one-half.

Stereoscopic Mural

The performance characteristics of photographic film were also gradually improving. The later Corona versions (J cameras) carried two "buckets," that is, two satellite recovery vehicles, thus substantially extending mission duration and coverage. In addition, the new versions of the upper-stage Agena carried more propellant and allowed multiple restarts in orbit, providing a capability of orbit change during the mission. The lift ability of the Thor-based launcher was also increased in 1963 by strapping solid-propellant boosters. The overall reliability

Two Buckets

MISSILE GAP AND CORONA

... By 1959, the great "missile gap" controversy was very much in the forefront. The Soviets had tested ICBMs at ranges of 5,000 miles, proving they had the capability to build and fly them. What was not known was where those missiles were being deployed operationally and in what numbers. In the preparation of the National Intelligence Estimate for Guided Missiles during the fall 1959, various intelligence agencies held widely diverse views on the Soviet missile situation. In an election year (1960) the "missile gap" became a grave political issue ... .

In an episode reminiscent of the 1944 presidential election, when Thomas E. Dewey was constrained by wartime security from making potentially devastating revelations about Pearl Harbor, Richard M. Nixon in 1960 was constrained from revealing that the "missile gap," on which John F. Kennedy had earlier campaigned, was an illusion. The Discoverer XIV payload was retrieved, and its intelligence information digested, two months before the 1960 election campaign ended. Kennedy, who had been made aware of the mission results, stopped talking about the missile gap. But some of his supporters did not, and Nixon's indirect assertions that there was no missile gap had little impact, because he had been saying this much earlier, when nobody really knew, and because he had subsequently adopted the policy of promising to enlarge the US missile program in much the way Kennedy proposed. In later years, when the August 1960 findings became more widely known, there was surprisingly little discussion of the potential change in election results that might have occurred if the information had been revealed.

F.C.E. Oder, J.C. Fitzpatrick, and P.E. Worthman (1987, 61)

of the Corona system dramatically improved with time. Only 11 capsules were recovered during the first 26 launches in 1959–1961. The following 26 launches of the Corona with the Mural camera in 1961–1963 had 20 capsules recovered.

Army's Argon

In the spring of 1959, the Army Map Service proposed to ARPA a program to obtain precise geodetic data and extension of the datum planes within the Soviet Union. This Army proposal was incorporated into the Corona program, and the mapping project was code-named *Argon*. The Argon cameras, KH-5, significantly differed from "standard" Corona cameras and provided nominal resolution 460 ft (140 m) on the ground with large image frames covering areas 300×300 miles (480×480 km). Five total Argon missions were successful out of 12 launched from 1961–1964.

Policy Wisdom

The Corona photography laid bare vast territory of the Soviet Union and communist China. The Soviet missile test centers at Tyuratam, Plesetsk, Kapustin Yar, and Saryshagan were regularly visited by Corona satellites as well as many operational ICBM and IRBM sites, naval shipyards, nuclear centers, military depots and bases, and industrial and transportation facilities. During its life, the Corona program deployed 145 satellites in orbit and successfully recovered over the Pacific Ocean 167 capsules with more than 2,000,000 ft (more than 600 km) of photographic film. The success of Corona validated the wisdom of Eisenhower's policy of developing overhead reconnaissance capabilities, which played a crucial stabilizing role during the uncertainties of the Cold War.

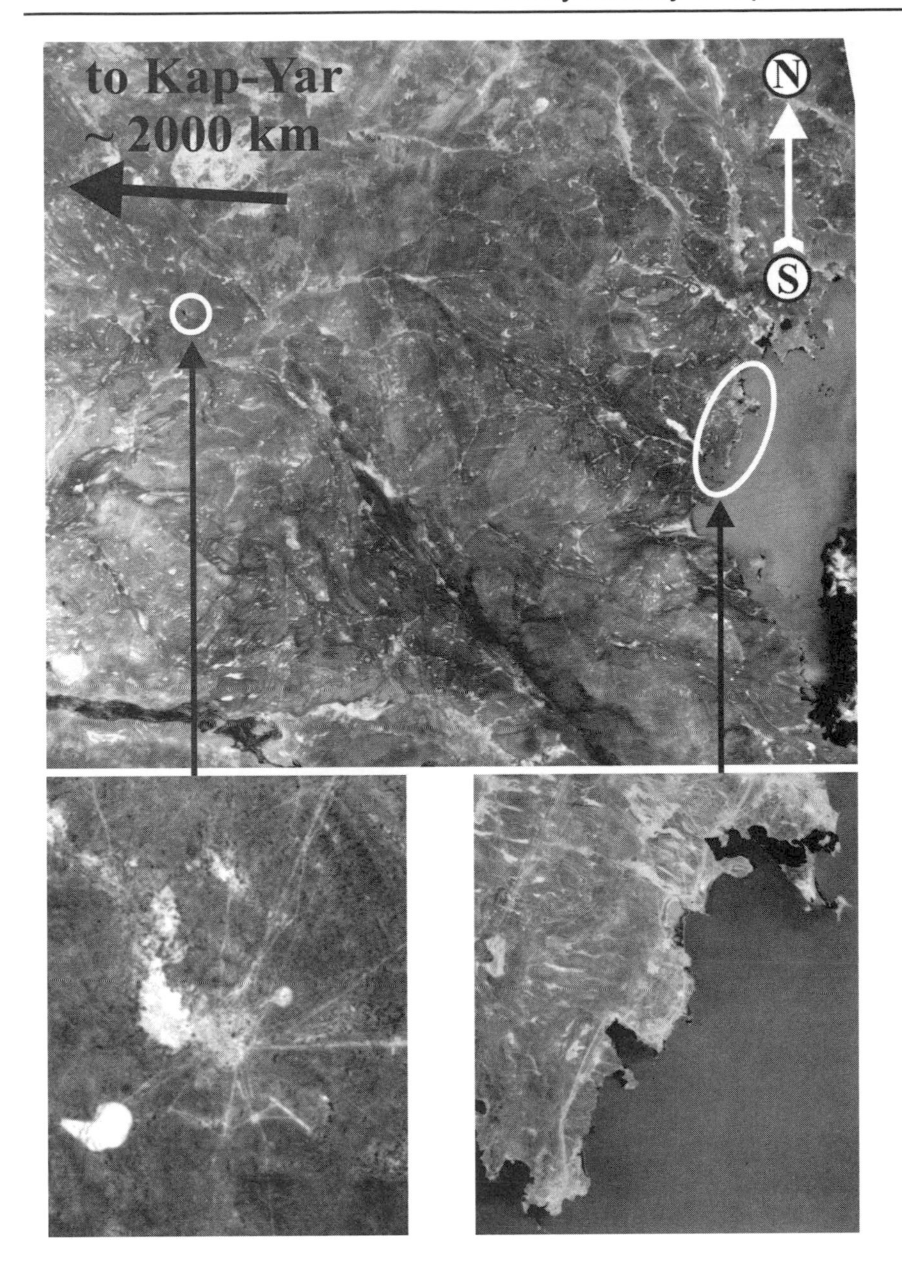

Fig. 16.11. Example of Corona satellite photographs. A low-resolution photo (top) shows the area of the Soviet Saryshagan antiballistic missile defense test range (presently in Kazakhstan) obtained by a KH-5 Argon mapping camera (Mission 9058A; 29 August 1963). Lake Balkhash is covered by fog on the right. Bottom high-resolution photos show (left) site 2 (KH-1; Mission 9009; 18 August 1960) with the guidance radar and (right) the area of long-range radars (KH-4; Mission 9035; 30 May 1962). In a missile defense "first," the long-range "Hen House" radar (*Dunai-2*) detected a Kap-Yar-launched ballistic missile R-12 (SS-4) at a distance of 975 km. The radars at site 2 and two other similarly instrumented locations precisely followed the warhead that was intercepted by a V-1000 missile on 4 March 1961 (see also Fig. 16.12 on the next page). Courtesy of Mike Gruntman.

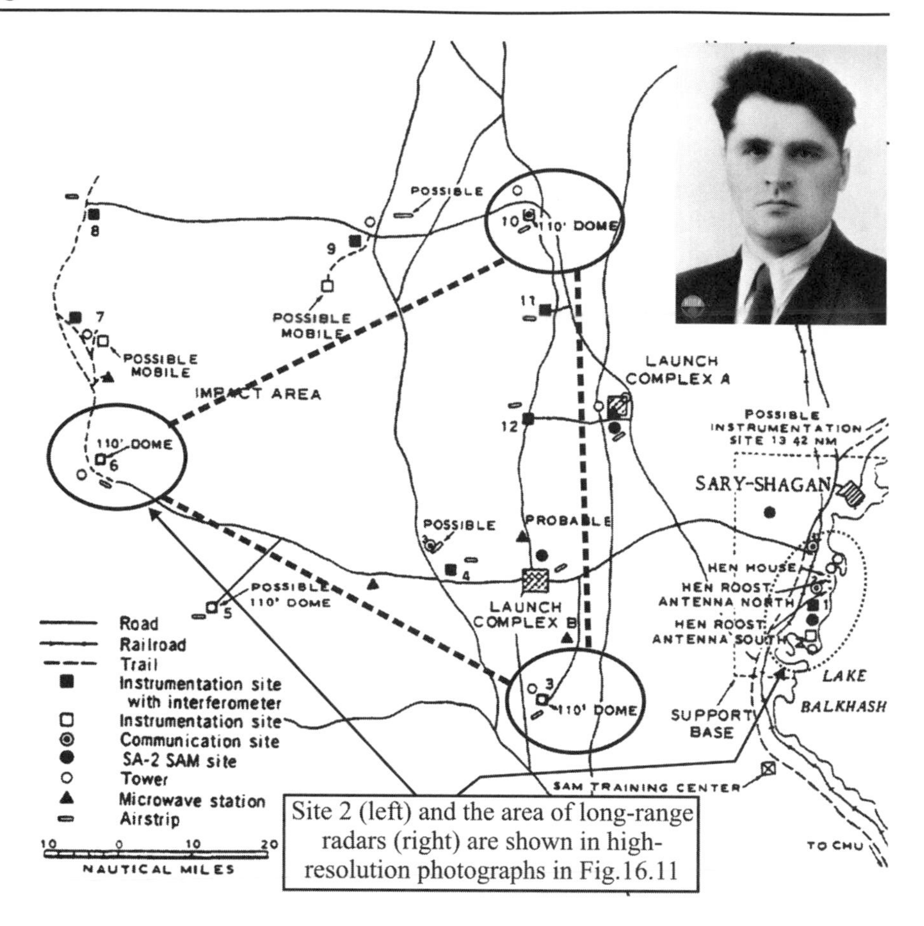

Site 2 (left) and the area of long-range radars (right) are shown in high-resolution photographs in Fig.16.11

Fig. 16.12. Example of intelligence derived from the Corona program — a map of the Saryshagan antiballistic missile defence test range, ca. 1963. Related representative Corona photographs are shown in Fig. 16.11. The figure also illustrates the first "non-nuclear" intercept of a warhead by a missile (see also Fig. 16.11) accomplished in the Soviet Union by a team led by Grigorii V. Kisun'ko on 4 March 1961. (A photograph of Kisun'ko, 1918-1998, is shown in the right top corner.) Three precise guidance radars were located at the sites (circled on the map) forming an equilateral triangle with the side 150 km (93 miles). Each radar measured distance to the incoming ballistic missile warhead with a 5-m (16-ft) error. The two-stage V-1000 interceptor missile developed by design bureau *Fakel* of Pyotr D. Grushin was launched from a site marked "Launch Complex B." The intercept with the accuracy 32 m (105 ft) was achieved at altitude 25 km (16 miles) 43.7 s after the interceptor missile launch. The interceptor was detonated 0.3 s before nominal intercept, releasing 16,000 spherical 25-g balls, each containing an explosive charge and a hard core made of carbide-tungsten-cobalt alloy. The released balls formed a uniform disk 150 m (500 ft) in diameter with high statistical probability to hit the target. On impact, the ball's explosive charge detonated and destroyed a part of the target external wall with the hard core penetrating inside and damaging the nuclear charge of the warhead. The long-range radars similar to those first observed on the shore of Lake Balkhash were later spotted near Moscow, thus revealing deployment of the antiballistic missile defense system around the Soviet capital. Original map courtesy of Central Intelligence Agency; Kisun'ko photograph courtesy of the Russian Academy of Sciences. Courtesy of Mike Gruntman.

Corona, the American image intelligence program was a large and expensive effort, but it was not the first operating American reconnaissance space system. The honor of being the first belongs to a small signals intelligence spacecraft GRAB that innocuously stood for a Galactic Radiation Background satellite. GRAB was an electronic intelligence (ELINT) satellite launched by the U.S. Navy on 22 June 1960.

The First Intelligence Satellite

The Naval Research Laboratory proposed GRAB in the spring of 1958, and President Eisenhower approved satellite launch in August 1959. The program emerged from the work of NRL's Radio Counter Measures Branch engaged in ELINT. An electronic engineer Reid Mayo had been involved in deployment of sensors designed to warn submarines under periscope when they were being detected by radar. Mayo made first estimates of the applicability of the detection technique on an Earth-circling satellite and reported favorable findings to his superiors at NRL.

Detection of Radars

The space radar-detecting sensor program was conducted under tight security. In addition to a classified ELINT payload, the satellite was designed to carry an open scientific instrumentation package SolRad (Solar Radiation). The SolRad was developed by the NRL group of Herbert Friedman and measured solar radiation in x rays and in the hydrogen Lyman-α (121.6 nm) line, particularly during solar flares. This scientific experiment would establish the importance of solar x-ray emissions for the variability of the upper atmosphere and ionosphere. Friedman's scientific package provided an excellent cover for the ELINT mission.

Scientific Cover

The GRAB intelligence payload was designed to intercept Soviet air defense radar signals when flying over the USSR. Because the spacecraft did not have a recording device, it beamed intercepted signals down to several small receiving stations established near Soviet borders. These installations actually were small huts operated by two-man Navy teams that recorded signals on magnetic tapes. The tapes were then delivered by couriers to the NRL and then to the National Security Agency and the Strategic Air Command. NSA analyzed the pattern of the Soviet air defense radars including antenna scan rates, pulse repetition frequencies, types of radar, and their locations. The information obtained by GRAB was particularly important for SAC for plotting the routes, avoiding Soviet radars and air defenses, of its bombers in case of hostilities.

GRAB Customers

The omnidirectional GRAB antenna did not allow determination of radar locations. Air defense radars, however, usually did not point upward but rather circled in search of targets. Therefore, an incoming satellite first intercepted radar signals that later disappeared when the satellite was over the radar site and then reappeared again on the outbound trajectory. This pattern helped to locate radar sites.

Locating Radar Sites

The GRAB satellite was based on a 20-in. (51-cm)-diam spacecraft originally developed by the NRL for the Vanguard program. In the first ever dual launch of spacecraft, GRAB was designed to ride to orbit together with a much larger Navy's Transit satellite built by APL. The two satellites were attached together during launch and separated in orbit. During ascent, the spacecraft were spun up to 60 rpm. After separation, Transit 2A was despun while GRAB remained spinning.

First Dual Launch

Fig. 16.13. The first American operational reconnaissance satellite known as the Galactic Radiation Background (GRAB) satellite. The satellite was based on the original 20-in. (51-cm) design of the NRL's Vanguard satellite. GRAB hitched a ride to space on 22 June 1960, with the APL's navigational satellite Transit 2A. Photo courtesy of Naval Research Laboratory.

The Thor–Able launcher successfully placed the first operational GRAB in a 66.7-deg inclination orbit on 22 June 1960. The satellite was also known as SolRad I. It was in an elliptical orbit with apogee altitude 660 miles (1061 km) and perigee altitude 382 miles (614 km). The ELINT payload was turned on for the first time on 5 July, six weeks before the first successful Corona brought its photographs. GRAB is still in orbit with the same inclination, but its apogee and perigee altitudes have decreased to 561 miles (903 km) and 362 miles (583 km), respectively.

GRAB in Orbit

The ELINT payload of GRAB operated until August 1962. Its information revealed that the strength of Soviet air defenses was greater and its radars were more powerful than expected. In addition, GRAB identified radar signals from the rapidly growing Soviet antiballistic missile defense installations. Four more GRAB launches followed, with only one being successful.

Eyes and Ears in Space

By the early 1960s, the skies had been opened giving birth to space-based image intelligence and signals intelligence. Infrared early missile-launch warning space systems were also under development. In addition, a program to detect nuclear explosions in the atmosphere and in space for verification of international treaties was rapidly advancing. This latter program was started at the initiative of technical advisors supporting negotiations in Geneva on nuclear test ban treaties.

Fig.16.14. Nuclear explosion monitoring satellites Vela 5A and 5B are assembled for flight. Photo courtesy of Los Alamos National Laboratory, Los Alamos, New Mexico.

Nuclear Explosions Detection from Space

In September 1959, ARPA was instructed to undertake a research and development program to determine the feasibility of monitoring nuclear test ban treaties. The program was named *Vela*, and its component Vela Hotel (where the phonetic word "Hotel" stood for "high-altitude") focused on detection of nuclear explosions in the upper atmosphere. (Other program components "Uniform" and "Sierra" corresponded to "underground" and "surface" nuclear explosions.) Detection of nuclear explosions in space was also desired because of its critical importance for antiballistic missile defense systems. The Vela Hotel research program became so successful that it provided interim operational detection capabilities. The program was a joint undertaking of the Department of

Vela Program

Fig.16.15. Artist rendition of satellite separation prior to injection into operational orbits of the second-generation Vela satellites. Figure courtesy of Los Alamos National Laboratory, Los Alamos, New Mexico.

Defense and the Atomic Energy Commission, with its Los Alamos Scientific Laboratory, Lawrence Radiation Laboratory, and Sandia Corporation playing major roles.

X-Ray and Visible Spectral Range Signatures

A large fraction of the energy produced in a nuclear explosion is released in the form of x rays. Therefore, nuclear detonations above the atmosphere, at altitudes higher than 40 miles (65 km), can be detected from space by x-ray sensors. The x rays produced by explosions at lower altitudes, less than 20 miles (32 km), are absorbed by the atmosphere. Such explosions could, however, be observed through emission in a spectral range where the atmosphere is transparent, namely, in the visible spectral range.

Velas in Orbit

The R-W's Space Technology Laboratory (the future TRW) was selected in November 1961 to build Vela spacecraft. The first two 300-lb (135-kg) satellites, Vela 1A and 1B, were launched by the Atlas–Agena on 17 October 1963. The spacecraft were deployed in a high-altitude nearly circular orbit with an apogee of 19.3 Earth radii (77,000 miles or 123,000 km), perigee 16.8 Earth radii (67,000 miles or 107,000 km), and orbital period of 108 h. Orbit selection was determined to a large degree by a requirement of operating outside Van Allen radiation belts in order to reduce detector background count rates.

These two intelligence satellites would be followed by two more pairs of spacecraft. The launch of three more pairs of the second generation of advanced satellites with the increasingly sophisticated instrumentation was approved in March 1965. TRW was again selected to build the spacecraft. In total, 12 Vela satellites were placed in high-altitude orbits with inclinations between 32 and 41 deg.

Second Generation of Velas

Vela satellites were built in a shape of a regular polyhedron with 20 triangular sides (icosahedron). The first satellites carried a set of instruments to measure radiation in x rays and γ rays and to detect neutrons and had 18 sides covered with solar cells, totaling 14,000 individual cells. The second set of three pairs of advanced Velas was of similar shape and carried additional instruments to measure heavy energetic particles, extreme ultraviolet emissions, and electromagnetic pulses. These satellites were correspondingly heavier and were launched on Titan IIIC launchers.

Fig. 16.16. Vela's optical sensor, also known as "bhangmeter," in the National Atomic Museum, Albuquerque, New Mexico. Bhangmeters are designed for detection of atmospheric nuclear explosions at altitudes less than 20 miles (32 km) and reveal nuclear explosions by detecting a characteristic double-hump temporal dependence of light intensity. A conical baffle protect the sensor from the direct solar light. Modern bhangmeters are carried by the constellation of GPS satellites, which allows the determination of the locations of an explosion by precisely measuring time of signal appearance on several satellites. Photo courtesy of Mike Gruntman.

Bhangmeters

Advanced Vela satellites also carried optical detectors called bhangmeters. The bhangmeter was designed to detect optical flashes from nuclear explosions in an atmosphere characterized by a double-peak temporal dependence. The time interval between peaks of emissions was indicative of the strength (yield) of the explosion. This unique temporal signal feature allowed distinguishing detection of nuclear explosions from other light flashes such as lightning.

"... Must Be Smoking ..."

The National Atomic Museum in Albuquerque, New Mexico, explains that

> the unique name [of the detector], the bhangmeter, originated with some of the early sceptics who did not believe that such sensing [of nuclear explosions in the atmosphere] was possible. 'Bhang' is an Indian hemp that is smoked for its hallucinogenic impact. Apparently some thought that anyone who believed that such an approach was feasible must have been smoking hallucinogens.

First Vela spacecraft were spin-stabilized. Each second generation Vela spacecraft carried two bhangmeters and thus required pointing spacecraft at Earth, which in turn required stabilization of the spacecraft in three axes. Correspondingly, the second-generation Velas carried reaction wheels and sensor suits to achieve the desired orientation.

Advancing Science

In addition to monitoring compliance with the test-ban treaties for more than 20 years, Velas provided monitoring of the radiation environment during Apollo missions when the astronauts left a protective shield of the Earth magnetosphere and would have been exposed to the deadly solar energetic particles had a major solar flare occurred. Vela spacecraft also significantly advanced x-ray astronomy and improved our understanding of the physics of the magnetosphere and its interaction with the solar wind.

SOVIET UNION

Being the first to achieve the ICBM and Earth-orbiting satellite, the Soviet Union also quickly initiated a program in space reconnaissance. Korolev's OKB-1 started preliminary work on a Soviet photoreconnaissance satellite in the Tikhonravov's department in 1957. At first, a new dedicated spacecraft was considered. Korolev then decided to build an experimental spaceship Vostok-1 and use it as a basis for both a reconnaissance spacecraft and a spaceship for manned flight. The reconnaissance satellite became known as *Zenit-2*. The government decree of 25 May 1959 established the schedule and priorities for the rapidly growing program. The Zenit-2 would closely resemble in size and configuration the Vostok spaceships for manned flight.

Zenit-2 Reconnaissance Spacecraft

The photographic cameras for the satellites were designed and built by an optical-mechanical plant in Krasnogorsk, 40 km (25 miles) west-northwest of downtown Moscow. The *All-Union Scientific-Research Institute of Television*, also known as NII-380, in Leningrad began designing a system that developed the exposed film onboard and transmitted images through a television channel.

The Zenit-2 spacecraft required a highly demanding attitude control system and a complex facility for operations and control from the ground. Spacecraft thermal control was another challenging problem: temperature of optical instrumentation had to be maintained within 1°C (2°F) from the nominal with the allowed rate of change of 0.1°C (0.2°F) per hour under varying spacecraft orientations with respect to the sun. The mechanically articulated solar panel provided electric power.

Thermal Control

The first test launch of the Zenit-2 failed to reach orbit on 11 December 1961, because of the malfunctioning third rocket stage. The next attempt on 26 April 1962 was successful. After completing the mission, the spacecraft fired its propulsion system to decelerate, reentered the atmosphere, deployed a parachute, and landed. The three-day flight, officially announced as *Cosmos-4*, produced photographs meeting and sometimes exceeding the requirement of a 10-m (33-ft) resolution on the ground. The obtained film was promptly delivered under guard to the Chief Intelligence Directorate GRU of the General Staff — the Soviet military intelligence.

10-m (33-ft) Resolution

During the follow-on test flights, the inefficient television system was removed from the package and replaced by two additional photographic cameras with the 1-m focal length. The film quality also improved, and the length of film carried onboard was increased. The total number of frames of a 10-m resolution optical system reached 1500 with the each frame being 30×30 cm (12×12 in.). Another camera with a short focal length had enough film for 500 frames 18×18 cm (7×7 in.) each; this camera resolution was 30–50 m (100–165 ft) on the ground, and its primary goal was mapping. In addition to photographic package, the spacecraft carried a radio system for signals intelligence.

Fig. 16.17. Soviet photoreconnaissance spacecraft *Zenit-2*, ca. 1962–1963. The left sphere is the reentry section that carries the optical system; the instrumentation section is the right part of the spacecraft; and the propulsion system for reentry is on the far right. The spacecraft diameter is 2.5 m (8.2 ft), and its length is about 5 m (16 ft). Photo courtesy of Videocosmos.

First Soviet Operational Military Space System

A total of 10 test flights had been completed by 30 October 1963, and the new system was accepted "to active duty" and declared operational on 10 March 1964. Zenit-2 became the first operational Soviet military space system. A new photoreconnaissance system, Zenit-4, would follow with the camera with the 3-m focal length. Korolev's OKB-1 had transferred the related technology and serial production of Zenit photoreconnaissance satellites to its Branch N.3 in Kuibyshev (today known as TsSKB in Samara) by the mid-1960s.

17. JOINING THE CLUB

Space Era

Two rival superpowers, the Soviet Union and the United States, achieved breakthrough into space in the late 1950s, opening the space era. Increasingly sophisticated satellites were launched to orbit the Earth and spacecraft headed to the Moon and dashed into interplanetary space beyond. Human spaceflight followed in 1961. Several other countries joined the space effort embarking on development of rocket launchers and satellites. Considerations of prestige, national pride, and demonstration of technological and military prowess encouraged these expensive undertakings.

Elite Club

With time, six more countries — France, Japan, People's Republic of China, United Kingdom, India, and Israel — joined the Soviet Union and the United States in the elite "club" of truly space-faring nations that built satellites and deployed them in orbit with their own national space launchers. In addition, a multinational organization, the European Space Agency (ESA), began operating in 1975 and became a major participant in the space enterprise, largely, but not entirely, superseding national space programs in Europe. It is expected that the ninth country, Brazil, will join space-faring nations some time in the not so distant future. First attempts of satellite launch by a national rocket failed, but it seems to be only a question of time, given the country's determination, that Brazil places a satellite in orbit.

Space-faring Nations

In 1960s, the United Kingdom, Canada, Italy, and Australia were the first to design and build satellites that were launched for them by another country, the United States. In recent years, many more countries, including those in the Third World, joined space activities by designing and building satellites, often simple and basic, to be launched by other nations. Commercial launches have become common. Although the number of countries building or procuring and operating satellites grows rapidly, the elite club of nations that possess satellite launch capabilities remains limited today to six members plus the multinational ESA. Two original club members, France and United Kingdom, no longer maintain national launch capabilities.

FIRST SPACE LAUNCHES BY SPACE-FARING NATIONS

Country	Launcher	Spacecraft	Date
Soviet Union	R-7	Sputnik 1	4 October 1957
United States	Juno 1	Explorer I	31 January 1958
France	Diamant-A	Astérix	26 November 1965
Japan	Lambda-4S	Ohsumi	11 February 1970
China (PRC)	Long March 1	DFH-1	24 April 1970
Great Britain	Black Arrow	Prospero	28 October 1971
ESA	Ariane	CAT	24 December 1979
India	SLV-3	Rohini 1	18 July 1980
Israel	Shavit	Ofeq-1	19 September 1988
Brazil	Unsuccessful launch attempts, but determined to succeed		

FRANCE

Heritage of Esnault-Pelterie and Barré

Rocketry and spaceflight in France were pioneered in the early 20th century by Robert Esnault-Pelterie. During the 1930s, the French Army sponsored some practical rocket development, particularly by Jean-Jacques Barré. The outbreak of World War II and the following defeat of France practically terminated rocket work, although Barré succeeded to continue development of his liquid-propellant EA-41 missile with the support of the Vichy government. When Barré was preparing for test firings in French Algeria in 1942, the Allied forces landed in North Africa. All experimental work thus ended. Barré resumed missile development after the liberation of France with the first EA-41 rocket launched in March 1945. The missile exploded after a few seconds of flight. This was the first flight of a liquid-propellant rocket in France.

Modest Beginnings

After the war, a group of specialists from the *Armaments Study and Construction Authority* (*Direction des Etudes et des Fabrications d'Armements — DEFA*) of the French Ministry of Defense surveyed German military rocketry. Henri Moureu led this group that included Barré. Moureu proposed initiating a French rocket program with the creation of a special institute, but his proposal lost in priority to the competing national nuclear weapons program. An initially modest missile development however gradually evolved. In late 1945, a team under engineer Lafarge started rocket and space-oriented research at the *Center for Study of Self-Propelled Projectiles* (*Centre d'Étude des Projectiles Autopropulsés*). In May 1946, a group of German rocketeers from Peenemünde and aerodynamicists from Kochel began working on liquid-propellant rockets with French engineers in the *Ballistic and Aerodynamic Research Laboratory* (*Laboratoire de Recherches Balistique et Aérodynamiques — LRBA*) at Vernon. More German scientists and engineers would join them in 1947.

LRBA

LRBA first concentrated on the development of a liquid-propellant engine based on nitric acid and kerosene as propellants with a thrust of 39 kN (8800 lbf). This program led to the French sounding rocket *Veronique*, approved for construction by the government in 1949. LRBA's Jean Corbeau would direct the

Fig. 17.1. Humble beginning of the French practical rocketry: Jean Jacques Barré, 1901–1978, with his liquid-propellant EA-41 rocket. Photo courtesy of Centre National d'Études Spatiales, France.

Veronique

effort. The Veronique-N was designed to reach altitude 70 km (44 miles), and it was launched for the first time on 20 May 1952. Eleven additional rockets were fired during the following 12 months. Fifteen of the more powerful Veronique rockets were launched during the International Geophysical Year in 1957–1958, each carrying 65 kg (143 lb) of scientific instruments to altitudes up to 210 km (130 miles).

Test Range at Hammaguir

The launches of French rockets took place at a new missile test facility, the *Joint Staff Center for Special Missile Testing* (*Centre Interarmées d'Essais d'Engines Spéciaux — CIEES*), established on 24 April 1947 in the Sahara desert of the French colony Algeria. The test range was located 450 miles (700 km) southwest of Algiers by the foothills, on the desert side, of the Atlas mountains. CIEES included two test sites, a small range B1 at Colomb-Béchar and a much larger range B2 at Hammaguir. The latter installations on the vast desert plateau would be used for testing numerous missiles and rockets and for the first French space launches of the Diamant-A rocket.

Landes Test Center

The Evian agreement of 18 March 1962 ended the French rule of Algeria but retained the right of using CIEES until 1 July 1967. The French military decided first to build an alternative missile test site in the France proper, the *Landes Test Center* (*Centre d'Essais des Landes — CEL*), in the region of Landes in southwestern France facing the Bay of Biscay. CEL became operational in 1965. Safety restrictions, however, would have allowed only highly inefficient westward space

Fig. 17.2. French space port at Hammaguir, Algeria, in 1967. Four Diamant-A rockets had been launched from this site before it was closed in 1967. The continuing launches would now be conducted from the French *Centre Spatial Guyanais*, or the *Guiana Space Center*. Photo courtesy of Centre National d'Études Spatiales, France.

launches ascending over the Atlantic Ocean into retrograde orbits. (Israel was later to experience the same disadvantage because of its geographical location.)

Kourou Space Center

Consequently a decision was made on 14 April 1964 to build a new space port, the *Guiana Space Center* (*Centre Spatial Guyanais — CSG*) near Kourou in French Guiana. Subsequently, the Kourou center has been developed into a major space port. CSG is strategically located close to the equator (latitude 5.5°N) and thus takes the full advantage of Earth rotation and allows especially efficient satellite launch into near equatorial orbits, including the geostationary orbit favored by communications satellites.

La Force de Frappe

In the 1950s, the French military undertook vigorous development of long-range ballistic missiles. Together with the nuclear weapons, they would form the future French ballistic missile "strike force" (*la force de frappe*). A comprehensive missile program included both liquid-propellant engines and solid-propellant motors. Special government and industrial organizations were formed to contribute to and support the development.

Centre National d'Études Spatiales (CNES)

The increasingly ambitious French civilian space effort was also reorganized. On 19 December 1961, President Charles de Gaulle signed a bill forming the *National Center of Space Studies* (*Centre National d'Études Spatiales — CNES*), replacing the *Committee of Space Research* (*Comité des Researches Spatiales — CRS*) that had coordinated space activities in France since 1959. An Air Force general, Robert Aubinière, became the first Director General of CNES, and he served in this position for 10 years. In addition to guiding the French national space effort, CNES was directed to develop plans for French participation in the rapidly growing European space programs.

First French Satellite

The concerted effort in ballistic missiles and space technology would make France the third country launching satellites. Although CNES was created to lead the French space effort and had a primary responsibility for the planned first satellite, designated *D-1A*, the satellite program witnessed a turf battle when — in 1963 — the *National Office of Aerospace Studies and Research* (*Office National d'Études et de Réalisations Aéronautiques — ONERA*) presented an alternative proposal to launch a satellite.

ONERA's Challenge

ONERA specialized in building solid-propellant missiles, and it proposed the launch of a small 3.5-kg (7.7-lb) spacecraft by its *Berenice* rocket in 1964. This proposed ONERA launch would have occurred one year earlier than the planned launch of the CNES satellite by a new *Diamant* (*diamond* in English) rocket. Interference of high-level government officials stopped the rivalry between two organizations and preserved the bureaucratic supremacy of CNES.

CNES conducted development of the first French space launcher, the Diamant-A, for 43 months in great secrecy. The rocket's technology relied to a large degree on the so-called "Gem Stones" (*Pierres Precieuses*) program. Under this program, a series of single-stage technological rockets including *Agate, Topaze,* and *Emeraude* (*agate*, *topaz*, and *emerald* in English) was developed for the Ministry of Defense.

Pierres Precieuses

Fig. 17.3. The first French satellite, *technological spacecraft A-1*, also named *Astérix*. The diameter of the Matra-built spacecraft was 50 cm (19.7 in.) and height 53.6 cm (21.1 in.). Nine black stripes were optimized to provide the desired passive thermal control. The spacecraft did not transmit any information because of the damaged antennas, but it was tracked by radar. Photo courtesy of Centre National d'Études Spatiales, France.

Diamant-A Launcher

The three-stage Diamant-A launcher consisted of the modified Emeraude and Agate and the specially built P-6 as the first, second, and third stages, respectively. These Diamant stages were demonstrated on two two-stage test vehicles, the *Saphir* and *Rubis* (*sapphire* and *ruby* in English). The Saphir consisted of the liquid-propellant (turpentine fuel and nitric-acid oxidizer) Emeraude and the solid-propellant Agate as the first and second stages, respectively. The rocket was fired three times during 1965. The Agate served, in turn, as the first stage of the other test vehicle, Rubis. The solid-propellant stage P-6 flew as the second stage of the Rubis. Eight test launches of the Rubis vehicles were conducted in 1964 and 1965.

Fig.17.4. Launch of the first French technological satellite A-1, *Astérix*, by the *Diamant-A* rocket on 26 November 1965 at Hammaguir. On that day, France became the third nation that placed a satellite in Earth orbit by a national space launcher. Photo courtesy of Centre National d'Études Spatiales, France.

The Diamant-A space launcher was built by an industrial *Ballistic Missiles Research and Development Company (Société pour l'Étude et la Réalisation d'Engines Balistiques — SEREB*), the developer of the gem stones. SEREB was formed in September 1959 with the participation of two major national aeronautical companies *Nord-Aviation* and *Sud-Aviation*. Total mass of the Diamant-A was 18,400 kg (40,600 lb) at liftoff, and the rocket was 18.94 m (62.1 ft) long.

Astérix in Orbit

The Diamant-A successfully launched the first French technological satellite A-1 on 26 November 1965 from Hammaguir. (The spacecraft was built by *Matra* for the French Army; therefore, its designation "A" standing for the Army; the follow on launches would carry the D-series of spacecraft, with "D" standing for the Diamant.) The A-1 spacecraft, also called *Astérix*, never transmitted any data because of the problems with the antenna, which was damaged during jettisoning of the fairing. The 38-kg (84-lb) spacecraft was however tracked by radar and its 34.3-deg-inclination orbit was confirmed with perigee 528 km (328 miles) and apogee 1769 km (1099 miles). Today, the *Astérix* orbit perigee and apogee are slightly lower at 523 km (325 miles) and 1663 km (1034 miles), respectively.

FR-1 and D-1A

Eleven days after the launch of the Astérix — on 6 December 1965 — an American Scout rocket placed in orbit another French satellite, *FR-1,* developed by the *National Center of Telecommunications Research* (*Centre National d'Études des Télécommunications — CNET*). Three months later, on 17 February 1966, another Diamant-A successfully launched the French scientific satellite *D-1A* from Hammaguir. Both FR-1 and D-1A functioned for many months. Thus, France became the third space-faring nation with the capabilities of building and launching satellites. In the process, France was unseating its eternal rival, Great Britain, as the European country leading the continent in rocketry and space technology.

WOOMERA, AUSTRALIA, GREAT BRITAIN, AND EUROPE

Test Range for British Missile Program

Although Australia never built its own space launcher, the country was among the very few nations that successfully designed and built a satellite in 1960s. And this satellite was launched from Australian territory. In addition, Australia contributed important engineering facilities and services for development of rocketry and for access to space. This early space effort was not continued, however, and Australia's role today is limited largely to housing specialized communications facilities supporting space operations. Launch installations built in the 1960s are periodically proposed by commercial enterprises as a possible location of a space port.

Weapons Research Establishment (WRE)

When Great Britain made a decision to initiate a guided missile program after World War II, it was clear that no large test range could be accommodated in the British Isles. The choices for a Commonwealth country to house such a range were between the Canadian tundra and the Australian desert. By the end of 1946, an agreement had been reached to establish test facilities on vast lands in outback South Australia. The missile work was to be supported by an engineering and technical center, the *Weapons Research Establishment* (WRE), in Salisbury 20 miles (30 km) north of Adelaide.

In mid-1947, the location for the rangehead was chosen, along with a site for a support town, which would be called Woomera, 20 miles (32 km) away. It was 300 miles (480 km) from Woomera to Adelaide. The selected plateau was unimpeded by vegetation and offered good areas for runways of the airfield. A 9-mile (14-km) spur connected the site with the Adelaide–Perth transcontinental railroad.

Missile Test Range in Outback

Fig. 17.5. Rocket establishment, Woomera, of the Australian missile testing range in the semidesert of the Arcoona Plateau. The salt lake, Island lagoon, can be seen in the distance. Rocketeers built an oasis out of the unforgiving surroundings by water and care. Photo courtesy of Defense Science and Technology Organization, Department of Defense, Australia.

Building a new test site was a major challenge. Numerous engineering facilities had to be brought to the arid outback, the living quarters had to be constructed, and hundreds and hundreds of miles of the range had to be instrumented to support testing of missiles and munitions. With an average annual rainfall of 190 mm (7.5 in.), water supply obviously presented a major problem. A special 170-km (106-mile) branch of a water pipeline was laid to provide water reliably. The conditions in this inhospitable land were not unlike those at the Soviet Tyuratam base, where water and care had also turned the desert into an oasis.

Woomera

One of the proposed names for the new site was "Red Sands," in analogy to the American White Sands Missile Range in New Mexico. An Aboriginal name *Woomera* was finally selected. A "woomera" means a spear thrower in Aboriginal languages of eastern Australia. (The new test range was located, however, in western Australia, where a different word was used with the same meaning.) The population of the new settlement steadily grew and reached 3000 in 1950. A post office, a hospital, churches, pools, and several shops were opened. The town soon boasted a movie theater and telephone exchange, and even ice cream became available.

Fig. 17.6. Radar at Central Bore near Mt. Elba at Woomera. This Stanford Research Institute-built radar and the optical site in the foreground were established in 1964–1965 for the *Dazzle* program studying plasma trails in the wakes of reentering bodies. Communications, tracking, and instrumentation sites spanned a huge territory claimed by the Woomera test range and its remote impact areas. Photo courtesy of Defense Science and Technology Organization, Department of Defense, Australia.

Security "Embargo"

Building the Woomera missile test range was not without controversy. Pacifists and communists tried to interfere with the construction, as their counterparts invariably did with defense initiatives in other countries of the free world, thus serving willingly or unwittingly as a Soviet fifth column. The security problems were further exacerbated by the tight American restrictions in 1948–1949 on providing classified information to Australia. The reasons for this "embargo" included the seemingly blind policies of the Australian left-wing Labor Party government to the menace of communism in Asia and an "unsatisfactory security situation" inside the government with leaks of classified information to Moscow linked to an uncovered Soviet spy network. The situation improved with time, especially after a new government under Robert Menzies came to power in 1949, and more efficient security procedures were introduced in defense research and development in 1950s.

First Missiles at Woomera

The first tests at the new range began in 1947 with experiments with bombs, not missiles. The first missile test took place on 22 March 1949, when a British

UP, or unrotated projectile, of the World War II heritage was fired in a trial. Since that time, numerous missiles of increasing sophistication and diversity were tested at Woomera, culminating with space launches. Most of the fired rockets were British. Subsequently, a few Australian missiles were also built as the local expertise grew beyond testing and expanded into missile design and fabrication. The Australian antitank *Malkara* and antisubmarine *Ikara* missiles were among the first fired at the range in 1950s.

Australian Missiles

The surge of interest in space in Australia coincided with the International Geophysical Year in 1957–1958. A number of British *Skylark* sounding rockets were launched at Woomera at that time. In addition, all-Australian sounding rockets appeared. The first was the *Long Tom* built by the Weapons Research Establishment. Several larger sounding rockets followed, including *Kookaburra* (reaching altitude 75 km or 47 miles) in 1968 and *Cockatoo* (reaching 130 km or 81 miles) in 1970.

Kookaburra and Cockatoo

The Weapons Research Establishment took the full advantage of a unique opportunity in the mid-1960s to build and launch the first Australian satellite. It was an exciting and pioneering event of constructing very quickly — in mere 11 months from conception to launch — a satellite that cost very little.

Taking Advantage of Opportunity

At that time, the United States conducted launches of a series of nine Redstone rockets as part of the U.S.–Australian–British project *Sparta* studying physical processes during warhead reentry. TRW was managing these launches at Woomera under contract to ARPA. When it became known that one Sparta launch vehicle, already at the test range, was not needed to complete the ongoing program, the Australians approached their ARPA and TRW counterparts at Woomera to seek its use to launch their satellite. The Americans immediately agreed to the proposal. It was necessary to win approval by the governments of both countries, and this was quickly obtained.

Sparta Project

ORIGIN OF WRESAT

Where the germ of the idea came from that the Redstone package might be used for a satellite launch in Australia is not known exactly. According to one account, the subject came up during one of those hard-drinking mess sessions that were such a feature of Woomera life. Someone from ARPA or TRW mentioned casually that a spare Sparta [project] vehicle had been sent to Australia just in case, and that the research project was going so well that probably it would not be needed. Everyone at the table was familiar with the history of the excellent Redstone. America's first satellite Explorer I had been orbited in 1958 using it as the first stage of another package called Jupiter C and since then various other combinations had taken many more satellites aloft. Since some of these were similar if not identical to the vehicles at Woomera, no one doubted that the Sparta rocket could easily be modified to orbit a small payload from Australian territory. And if a spare was going begging ... .

Peter Morton (1997, 483)

The United States provided not only the rocket but also its TRW contractor crew to prepare and launch it. In addition, NASA tracked the satellite with its network of ground stations and donated recording tapes. In total, the American contribution to the Australian satellite was at least $1,000,000 in 1967 dollars.

Fig. 17.7. Australian Weapons Research Establishment Satellite (WRESAT) being prepared at Woomera for launch by the modified American Redstone rocket, with the kangaroo symbol on its side, on 29 November 1967. The satellite was deployed in a near-polar low-Earth orbit and completed 642 orbits before reentering the atmosphere on 10 January 1968. Photo courtesy of Defense Science and Technology Organization, Department of Defense, Australia.

11 Months

Because the American program was coming to an end, it was required to design, build, and launch the satellite within 11 months. The spacecraft, appropriately named *WRESAT*, was constructed at the facilities of the Weapons Research Establishment in Salisbury, with most active roles played by Jeff Heinrich, Peter Twiss, Don Woods, Arthur Wills, Bryan Rofe, and Des Barnsley. The WRESAT carried instruments to measure solar x-ray and ultraviolet radiation as well as emissions in a spectral band important for characterization of the ozone layer. Scientific experiments were provided, in part, by the group led by John Carver at the Adelaide University.

WRESAT

The spacecraft together with the integral third stage weighed 160 lb (72.5 kg). Its body cone was made of aluminum alloy; it was over 6.5 ft (2 m) long and was painted mostly black with some silver stripes for passive thermal control. A specially produced expensive white paint was obtained to treat the spacecraft cone inside in order to evenly distribute heat. It took 15 quick-drying coats to achieve the desired surface quality. Ironically, it was later discovered that the wrong paint, no different from common household white enamel, was shipped and used for painting. Passive thermal control worked, nevertheless, within the specified requirements, offering an early lesson that the cost of spacecraft could be lowered in some applications. WRESAT was powered by batteries expected to function for several days only.

The first Australian satellite was successfully launched from Woomera on 29 November 1967. A tracking station on Guam was the first to receive the signals from the spacecraft transmitter. WRESAT was deployed in a nearly polar orbit with apogee altitude 1245 km (774 miles) and perigee altitude 169 km (105 miles). The satellite transmitted data for five days as it made 73 revolutions around the earth. Low perigee of the orbit resulted in large drag, and the spacecraft reentered the atmosphere a month later on 10 January 1968. Sadly, the success of the Australian WRESAT did not lead to a sustained national space effort.

WRESAT in Orbit

Lost Momentum

The Woomera range played an important role in Britain's national space effort and contributed to the first attempt by a united Europe to develop a European space launcher. The origins of the British space launcher could be traced back to development of independent nuclear deterrent in mid-1950s, the intermediate-range ballistic missile *Blue Streak*. The *Blue Streak* was designed as a silo-launched liquid-propellant IRBM for delivering nuclear warheads a 2740-mile (4400-km) distance, reaching from England to the heartland of the USSR. First development contracts were let in 1955 to *De Havilland Propellers* (airframe), *Rolls-Royce* (engines), *Sperry Gyroscope* (inertial navigation), and *Marconi* (radars and communications). The Blue Streak design was inspired by the American Atlas ICBM and used kerosene and liquid oxygen as propellants.

Blue Streak IRBM

Development of the missile was an expensive undertaking with a large Australian contribution. The Woomera range significantly expanded its territory and facilities in preparation for the forthcoming tests of the Blue Streak IRBM. In particular, a new launchpad was erected on an escarpment 150 ft (46 m) high above a dry salt lake. This arrangement directed much of hot exhaust gases in the flame duct and thus allowed substantial reduction of the amount of water required for cooling the pad as compared to launches from flat surfaces.

Black Knight

When the Blue Streak program was started in the 1950s, British experience in ballistic missiles was rather limited. Therefore, a new program, called *Black Knight*, was soon initiated to gain expertise through development of an inexpensive test vehicle. The *Royal Aircraft Establishment* (RAE) in Farnborough designed the Black Knight, which would become a highly successful rocket program.

Gamma Engine

The Black Knight's turbopump-fed gimballed engine *Gamma* was built by *Armstrong-Siddeley Motors* and achieved a thrust of 16,000 lbf (71 kN). (Armstrong-Siddeley was later absorbed by Rolls-Royce.) Propulsion technology for Gamma originated from a series of rocket engines that included the *Alpha* and *Beta* engines of the early high-altitude interceptor programs, produced under direction of the RAE. These engines used kerosene as fuel and high test peroxide (HTP) as oxidizer. HTP was an 86% concentration of hydrogen peroxide, with the remaining 14% water and required special care and safety precautions in storage and handling. The use of HTP in British missiles was rooted in the technology implemented in the German *Walter* rocket engine (named after the *Walter Company* of Hellmuth Walter) built for the *Me 163b* interceptor during World War II.

Hydrogen Peroxide

On 7 September 1958, the new Black Knight missile was successfully launched on the first attempt. During the next 10 years, the missiles would reach altitudes of several hundred kilometers many times carrying out various research projects including studies of warhead reentry. The final launch of the Black Knight rocket was performed on 25 November 1965.

In the meantime, the Blue Streak IRBM went through its developmental difficulties during late 1950s. It was becoming clear that the missile would be obsolescent as a weapon. Moreover, a wisdom of placing liquid-oxygen-based rockets in protected silos was questioned. Strategic consideration also pointed to vulnerability of the deployed missiles in case of a full-scale Soviet nuclear attack, thus denying the desired deterrent capabilities. After significant expenditures, the Blue Streak program was cancelled in April 1960. American-built submarine-launched solid-propellant Polaris missiles would become Britain's deterrent weapon.

Blue Streak Terminated

The cancellation of the Blue Streak called for applying already invested substantial resources and accumulated technological expertise and facilities to a new development. The situation offered an opportunity to initiate a modest space program, both national and Europe-wide. Space enterprise was expensive, so that pulling European resources together was natural and doable, as had been earlier demonstrated by creation of the *European Center for Nuclear Research* (CERN).

European Space Program

This was also the time, when diverse political forces argued for a European space effort as a balance to the dominating American and Soviet programs and in response to the anticipated emergence of presence in space by Japan and Communist China. The *European Space Research Organization (ESRO)* was promoting research in space science and a new space launch initiative, the *European Launcher Development Organization (ELDO)*, was thus formed, with Britain playing the leading role, in November 1961.

Rise and Fall of Europa

ELDO's main objective was to develop a European space launcher, the *Europa*. The program heavily relied on the Australian ground infrastructure developed at Woomera. The *Europa-1* was a three-stage vehicle with the first stage based on the cancelled British Blue Streak. France was to build the second stage and the Federal Republic of Germany the third stage. Italy was to lead construction of first satellites, Belgium focused on development of the radio guidance system, and the Netherlands was to provide the long-range telemetry link.

Coralie

The French second stage, *Coralie*, was based on the successful experience with the Veronique sounding rocket and the Emeraude propulsion stage developed for the French ballistic missile program. LRBA and Nord-Aviation built Coralie and its pressure-fed engine used nitrogen tetroxide as oxidizer and UDMH as fuel.

Astris

The German third-stage *Astris* was the first major rocket system built by German industry, marking the country's return to the space and rocket "world" after restrictions and reconstruction that followed World War II. The main engine of the Astris (the stage also carried two additional much smaller vernier thrusters) used nitrogen tetroxide as oxidizer and Aerozine-50 as fuel and was supplied by a French contractor. Italy built a satellite that was specifically designed to support testing of the launch vehicle performance, particularly its guidance and separation. Australia was responsible for rocket tests and launch at Woomera.

Europa-1 and Europa-2

The first development vehicle with the live first stage only was fired at Woomera on 5 June 1964. In total, 11 launches in support of the Europa-1 were conducted with many failures and some trials being only partially successful. In the last couple of launches, the vehicles had all stages fully active with an objective of deploying satellites in orbit. No spacecraft had ever been placed in orbit

ESRO, ELDO, AND ESA

The idea of creation of the *European Space Research Organization (ESRO)* was first brought up in the late 1959 by the physicist E. Amaldi and his colleagues in Italy. In the early 1960, informal discussions began among leading European scientists interested in promoting space research. Several informal and then more formal meetings followed, with the organizational structure being gradually developed. At first, a *Steering Committee* was established, and then the intergovernmental meeting in Geneva on 28 November 1960 formed a *preparatory commission* that included representatives of nine countries: Belgium, Denmark, France, Italy, the Netherlands, Norway, Sweden, Switzerland, and United Kingdom.

The Preparatory Commission formulated plans for space research in Europe and outlined ways of their implementation. In another important step, a *financial protocol* was signed in 1962. After contentious debates, the sites were selected for permanent laboratories that would support space research in Europe. Actually, only one site for the ESRO headquarters in Paris was unanimously approved. It was decided — with much unresolved disagreement — to locate the *Space Technology Center (ESTEC)* near Noordwijk in the Netherlands; the *Data Analysis Center (ESDAC)* in Darmstadt, Germany; and *Scientific Laboratory (ESLAB)*, later known as the *European Space Research Institute (ESRIN)*, in Frascati near Rome, Italy. The British proposals to establish ESTEC (and ESLAB) or ESDAC in Bracknell, 20 miles (36 km) from the Royal Aircraft Establishment in Farnborough and 35 miles (55 km) from London were voted down. The other European facilities included a rocket range near Kiruna in Sweden (ESRANGE) and tracking stations (ESTRACK).

The ESRO Convention was signed by all original members (with exception of Norway) of the Preparatory Commission and Austria on 14 June 1962. The ratification process brought all of the disagreements into focus of the political and public debate. Finally, a sufficient number of countries had ratified the Convention to make the organization operational by March 1964. (Italy and Norway had not ratified the Convention by that time, and Austria had requested a status of observer.)

In the meantime the *European Launcher Development Organization (ELDO)* was formed in November 1961 by six European countries — Belgium, Federal Republic of Germany, France, Italy, the Netherlands, and the United Kingdom — with participation of Australia. The formation of ELDO was initiated by Great Britain, the leading European country in missile and space technology of the 1950s and early 1960s. ELDO was to manage development of a new European space launcher, the *Europa*.

The conduct of European space programs was characterized by sharp differences among participating countries, reflecting diverging goals and political philosophies. Britain was unhappy with the growing burden of the development of the Europa launcher, which was experiencing problems with the French and German stages. A political rational of British engagement in the continent was also severely damaged when France vetoed a British application to join the Common Market. The French desire, supported by Germany, to develop independent European rocket and space technologies regardless of cost was not shared by everyone either. Italy was also unhappy with attempts to reduce its role in the development of the Europa.

The European space effort was reorganized in the early 1970s, and the new *European Space Agency (ESA)* began operations in 1975. Eventually, 11 European countries — Belgium, Denmark, France, Germany, Ireland, Italy, the Netherlands, Spain, Sweden, Switzerland, and the United Kingdom — signed the Convention forming ESA. With time, ESA subsumed all of the functions of ESRO and ELDO; it took several more years before all of the legal formalities were completed.

Fig. 17.8. ELDO's Europa-1 being prepared for launch. Australia contributed this *Launcher 6A* complex and the liquid-oxygen plant in the nearby Woomera for the European Launcher Development Organization. The launchpad was erected on an escarpment above the dry salt Lake Hart for the Blue Streak program. Photo courtesy of Defense Science and Technology Organization, Department of Defense, Australia.

though. Subsequently, the Europa program at Woomera was discontinued, with the remaining program activities transferred to a newly established launch center at Kourou.

The Europa-1 was designed to place spacecraft with mass up to one metric ton in low-Earth orbit. It was quickly recognized, however, that the capabilities of deploying communications spacecraft in geostationary orbit were of prime importance and value. Consequently, studies were initiated that led to design of the *Europa-2* space launcher. The Europa-2 had an additional fourth stage, the *perigee-apogee system* (PAS), capable of placing 150-kg (330-lb) satellites in geostationary orbit. After an unsuccessful first launch attempt on 5 November 1971, the program was terminated on 27 April 1973.

Europa-2

The European Launcher Development Organization that directed development of the space launcher was in the middle of intra-European politics, and it definitely was not the most efficiently operating organization with a clear purpose. Political disagreements among the ELDO member states, especially between France and Britain, also contributed to the problems in missile development. An editorial in the British science magazine *Nature* caustically observed in 1966 that "the first thing to be said about ELDO is that it was wrongly begotten. ELDO has also been badly organized... . Then, ELDO has been overtaken by events" (*Nature* 1966, 1085). Commenting on the political difficulties of joint European projects, *Nature* added that "there is still evidence that skill in rocketry is sometimes confused with national prestige."

"ELDO ... Was Wrongly Begotten..."

Great Britain withdrew from participation in ELDO programs in June 1971. ELDO continued its operations through 1970s and was subsumed with time by the newly formed ESA. Though Europa launch vehicles had not placed satellites in orbit, the ELDO experience provided many lessons for a future successful family of the *Ariane* space launchers, developed under the French leadership. (The original preferred choice of the name for this new European launcher was *Vega*, after a bright star in the Lyre constellation. The name had to be abandoned and deemed somehow inappropriate when it was found that it was also a brand name of a beer.) The first Ariane launcher vehicle successfully lifted off Kourou on Christmas Eve 1979.

Ariane

ARIEL SATELLITE SERIES

NASA launched first British satellites, the *Ariel* series, in 1960s, helping British scientists and engineers to gradually build up their expertise in satellite and space instrumentation technology. The first satellite, S-51 or Ariel-1, was placed in orbit by the American Thor–Delta launcher on 26 April 1962. Actually, the spacecraft bus was also built by NASA with the British scientists providing scientific instruments. The arrangements were similar for the second satellite, S-52 or Ariel-2, that was launched on 27 March 1964.

The S-53 or Ariel-3 was different. The spacecraft itself was designed and built in Great Britain. Ariel-3 was successfully launched by the American Scout rocket on 5 May 1967. In total, six Ariel satellites were placed in orbit, with the last Ariel-6 launched on 2 June 1979.

With the problems in the European ELDO initiative mounting, Great Britain made a decision in 1964 to launch a satellite by its own rocket using technological expertise acquired in missile programs. The three-stage launch vehicle was named *Black Arrow*, and it largely relied on modified and upgraded systems of the successful Black Knight missile. The first and second stages used swivelling engines with hydrogen peroxide and kerosene as propellants and provided a thrust of 51,000 lbf (227 kN) and 15,700 lbf (70 kN), respectively, for slightly more than two minutes each. Specific impulse varied form 217 s at sea level for the first stage to 265 s for the second stage. A solid-propellant motor of the third stage provided a thrust of 35,000 lbf (156 kN) with specific impulse of 276 s for 55 s. The third stage with the attached satellite was spun up to 190 rpm before firing the motor.

British National Space Launcher

Fig. 17.9. British Black Arrow rocket being prepared for launch of the Prospero satellite at Woomera on 28 October 1971. Photo courtesy of Defense Science and Technology Organization, Department of Defense, Australia.

Black Arrow

British Hovercraft Corporation integrated the entire launch vehicle at its facilities at Cowes, Isle of Wight. Rolls-Royce provided engines for the first and second stages and RAE's *Rocket Propulsion Department* at Westcott, assisted by *Bristol Aerojet, Ltd.*, built the third-stage solid-propellant motor *Waxwing*. The Black Arrow stood 29 ft (13 m) tall and was 6.5 ft (2 m) in diameter of the first stage, looking not exactly as a slim arrow as its name would suggest.

Development Launches

The first Black Arrow development vehicle, the R0, had a dummy third stage, and it was tested on 28 June 1969 at Woomera. The flight safety officer destructed the rocket after it began disintegrating at a 5-mile (8-km) altitude as a result of loss of swiveling control of engines. The next test vehicle, the R1, was a repeat of the test program of the failed R0. The R1 successfully flew to the Indian Ocean on 4 March 1970.

The next launch, the R2, was designed to test the entire launch vehicle and, if successful, to place in orbit a simple satellite, a hollow 3-ft (76-cm)-diam gold-plated aluminum sphere with a transmitter sending a continuous signal. The launch direction was northward in order to achieve a near-polar orbit. On 2 September 1970, the rocket first stage and the new third-stage Waxwing motor worked perfectly. The second stage, however, cut off 30 s earlier than scheduled, and, as a result, the satellite fell into the Gulf of Carpentaria.

First Space Launch Attempted

Finally, on 28 October 1971, the Black Arrow R3 vehicle successfully placed a British satellite, named *Prospero*, in orbit. *British Aircraft Corporation* in Bristol built the spacecraft, with electronics developed by Marconi. The satellite was placed in a near-polar orbit with inclination 82.1 deg, apogee 983 miles (1582 km), and perigee 340 miles (547 km). Today, the satellite is still in orbit with apogee 832 miles (1338 km) and perigee 329 miles (530 km).

Prospero in Orbit

Prospero's shape resembled a pumpkin and was slightly more than 1 m (3.3 ft) in diameter with 70 cm (28 in.) height and 72 kg (159 lb) mass. The spin-stabilized satellite carried only one scientific instrument for detection of micrometeoroids, which was provided by Birmingham University. Technological experiments included validation of surface coatings for passive thermal control and evaluation of performance of three types of solar cells, silica and glass slips protecting cells, and new integrated electronic components.

PROSPERO — WHAT'S IN THE NAME?

The Shakespearian names given to Britain's first satellites was a pleasantly imaginative touch, but it turned into something of a comedy of errors. Ariel, the first to be launched by US in May 1962, was named in allusion to the sprite who boasted of putting 'a girdle round the earth in forty minutes.' When it came to naming later satellites in the series, Don Hardy of Space Department [of the Royal Aircraft Establishment] RAE pointed out that the sprite was not Ariel in *The Tempest*, but Puck in *A Midsummer's Night's Dream*. This name was vetoed for fear of embarrassing the Minister with a slip of the tongue in the House [of Commons], so Hardy suggested they stick to *The Tempest* and its magical atmosphere, offering the name Prospero. When in 1974 the X4 satellite was to be launched on a Scout rocket it was named Miranda after Prospero's daughter.

Peter Morton (1997, 526)

The First and the Last

This first successful space launch of the Black Arrow on 28 October 1971 also became its last. Three months before the launch, the British Minister of Aerospace announced a cancellation of the Black Arrow program. (The second satellite built for the program was later launched by a Scout rocket in the United States.) No more space launches would be conducted by British national launch vehicles.

Great Britain thus launched its own satellite on its own rocket and joined the elite group of space-faring nations. The country did not become, however, the fourth member of the club — after the USSR, United States, and France — but the sixth. Japan and People's Republic of China had launched their satellites in the meantime.

JAPAN

Before the space era had begun, Japan's experience in rocketry was minimal. The Mongol troops of Kublai Khan used rockets in the failed invasion of Japan in the 13th century. Then a few hundreds of small rockets, the fire-arrows, were purchased and brought to the country around the year 1600. Since then, tiny rockets took part in fireworks displays for religious festivals. The first Japanese war rockets appeared during World War II. Industrial corporations *Kawasaki* and *Mitsubishi* manufactured these relatively simple short-range barrage and antiaircraft missiles. In addition, a rocket-propelled kamikaze aircraft was developed.

Hideo Itokawa

The interest in space exploration and rocketry began to grow in the 1950s, after the postwar restrictions on aircraft and missile development were rescinded in 1951. Professor Hideo Itokawa spearheaded the early missile work in Japan. During the war, Itokawa designed military aircraft at the Nakajima Aircraft Company. After the war, he joined the *University of Tokyo*. Itokawa's group at the University's *Institute of Industrial Science* became a focus point for Japanese engineers and students interested in rocketry. This small group would expand with time and grow into a major establishment, the *Institute of Space and Astronautical Science*.

Fig. 17.10. Hideo Itokawa, 1912–1999, with a "Pencil" rocket (left) and "Baby" rocket (right). Photos courtesy of Institute of Space and Astronautical Science, Japan.

Pencil Rockets

Theoretical studies in the Itokawa's group were complemented in 1954 and 1955 by designing and building small solid-propellant rockets. These first tiny rockets were just 18 mm (0.7 in.) in diameter and 23 cm (9 in.) long with mass

200 g (0.44 lb), and they were appropriately called the *Pencil* rockets. The next rocket type, *Baby*, was much larger with a diameter 8 cm (3.1 in.), and it reached altitudes up to 3 miles (5 km).

Kappa Series

Following the appeals by Itokawa and several other scientists interested in space research, the Ministry of Education agreed to support Japan's participation in the International Geophysical Year in 1957–1958. Consequently, the Itokawa's group was reorganized in 1957 into the *Sounding Rocket Group*, and it initiated a program, funded by the government, to develop sounding rockets of the series *K*, or *Kappa*. Following the order of the Greek alphabet, the successive families of rockets would be called *Lambda* and *Mu*.

Kappa rockets grew in size, and a two-stage solid-propellant *Kappa-6* was 6.5 m (21 ft) long and 25 cm (10 in.) in diameter with a mass of 260 kg (570 lb). Participation of industry in the rocket program was also increasing. The Sounding Rocket Group launched 13 rockets during the IGY. These missiles lifted 12 kg (26 lb) of scientific instruments up to 60 km (37 miles) in altitude.

ISAS

Following first successes, Japan formed the *National Space Development Center*. Then, the *Aeronautical Research Institute* and the Sounding Rocket Group were brought together in 1959 forming the *Institute of Space and Aeronautical Science*, with Professor N. Takagi as its first director. The Institute was reorganized into the *Institute of Space and Astronautical Science*, or *ISAS*, in 1984. Today, ISAS is the main national organization focused on space and astronautical science in Japan.

Fig. 17.11. Kagoshima launch site at the tip of the island of Kyushu on the southern edge of Japan. More than 300 rockets, including satellite launch vehicles, were launched from the Kagoshima Space Center since its establishment in February 1962. Photo courtesy of Institute of Space and Astronautical Science, Japan.

The Kappa series of sounding rockets reached altitudes of several hundred kilometers by the mid-1960s. The next rocket series was the *Lambda* with the first missile test fired in July 1964. In 1961, Itokawa selected a new remote site for launches of rockets that were becoming bigger and bigger. The new site, the *Kagoshima Space Center*, was located on the eastern side of the Ohsumi Peninsula in the Kagoshima Prefecture on the Island of Kyushu. This peninsula on the southern edge of Japan would also give the name, *Ohsumi*, to the first Japanese Earth-orbiting satellite. The site was ready just in time for the first launch of the Lambda rocket, *Lambda-3*, in July 1964. The three-stage 19-m (62-ft)-long missile with mass 7000 kg (15,400 lb) reached an altitude of 1000 km (620 miles). Fishermen led a local opposition to the test-site activities, and the ensuing agreement limited launches to two periods of time around February and September each year.

Kagoshima Space Center

Fig. 17.12. Launch of the first Japanese satellite by the Lambda-4S rocket from the Kagoshima Space Center on 11 February 1970. Photo courtesy of Institute of Space and Astronautical Science, Japan.

Lambda-4S

In 1965, the *National Space Activities Council* authorized development of a space launcher based on a new Mu series of rockets. In the meantime, T. Nomura built a *Lambda-4S* rocket by augmenting the existing Lambda-3 with an additional fourth stage and two solid boosters attached to the first stage. This all-solid rocket was capable of placing a small satellite in low-Earth orbit. The launcher was 16.52 m (54 ft) long and 0.74 m (29 in.) in diameter and had a mass of 9400 kg (20,700 lb). The first, second, third, and fourth stages produced thrusts of 362.6 kN (81,500 lbf), 115.7 kN (26,000 lbf), 64.5 kN (14,500 lbf), and 8030 N (1806 lb), respectively. Combined thrust of the first-stage two solid boosters was 129.0 kN (29,000 lbf).

Fig. 17.13. First Japanese satellite Ohsumi launched on 11 February 1970. Photo courtesy of Institute of Space and Astronautical Science, Japan.

Difficult Development

The first attempt to launch a satellite by the Lambda-4S failed on 16 September 1966. After two more failed launches, Hideo Itokawa resigned from ISAS in April 1967. Finally, after one more failure in 1969, the Lambda-4S rocket placed the first Japanese satellite, *Ohsumi*, into orbit on 11 February 1970. The satellite was launched in the eastern direction in a 31-deg-inclination orbit.

Ohsumi in Orbit

The spacecraft together with the integral fourth stage had mass 38 kg (84 lb); the instrumentation payload accounted for 12 kg (26 lb). In addition to a transmitter, the spacecraft carried a battery, accelerometer, and thermal sensors. The signals were received for seven revolutions around the Earth. The initial satellite orbit had 338 km (210 miles) perigee and 5150 km (3200 miles) apogee. The relatively low orbit perigee resulted in high atmospheric drag, and the satellite reentered the atmosphere 33 years later on 2 August 2003.

The launch of Ohsumi made Japan the fourth member of the elite club of the space-faring nations that placed satellites in orbit by national launchers. Only two-and-a-half months later, People's Republic of China would also join the club.

PEOPLE'S REPUBLIC OF CHINA

Ballistic Missiles a Top Priority

For eight centuries the early Chinese rockets did not change much — they remained small and inefficient. The era of modern rocketry in China began only after a communist takeover of the country in the late 1940s. After several years of consolidating power, the "great teacher, great leader, great supreme commander, and great helmsman" Chairman Mao Zedong, put nuclear weapons and rocketry on the top of priorities of the rapidly growing *People's Republic of China (PRC)*. Vice Premier and head of the *Defense Science and Technology Commission* Nie Rongzhen (Nieh Jung-Chen) would supervise the emerging defense research, development, and manufacturing complex, including nuclear and missile programs.

ROAD TO MISSILES IN CHINA

The Chinese people should have an ambitious plan for changing, in some decades, their economic, scientific, and cultural backwardness, to catch up quickly with the world advanced levels.

Chairman Mao Zedong, 1956
(*China Today* 1992, 4)

[China's missile research should be] adhering to self-reliance as our foothold, while striving to win external assistance and to use scientific achievements acquired by the capitalist countries.

Vice Premier Nie Rongzhen, 1956
(*China Today* 1992, 6)

Hsue-Shen Tsien (Qian Xuesen)

In September 1955, a prominent scientist and rocketeer Hsue-Shen Tsien (Qian Xuesen) left California for his "great ancestral socialist motherland" (Tsien 1976). In the words of Theodore von Kármán, "the United States ... gave Red China one of our [America's] most brilliant rocket experts ..." (von Kármán 1967, 308). Already by February of the next year, Tsien sent a proposal to the Central Committee of the Chinese Communist Party to establish a "defense aeronautical industry" that included missile research and development. At this time, Soviet advisors also encouraged the addition of missile development to the China's 12-year plan for science and technology.

Research Academy No.5

Fig. 17.14. Hsue-Shen Tsien (right) with Premier of communist China Zhou Enlai (left). Photo courtesy of China Astronautics Publishing House, Beijing.

Seventh Ministry of Machine Building

Soviet Help

R-2 (SS-2) in China

Jiuquan Missile Test Range

Dong Feng

Less than one year after Tsien's return to China, he was appointed president of *Research Academy No.5* of the Ministry of National Defense. The Academy was created in May 1956, and it became China's first missile research establishment; the *Second Artillery* of the *People's Liberation Army (PLA)* would become the ballistic missile force. The Academy would be reorganized into the *Seventh Ministry of Machine Building* in 1965 and absorb at that time a number of other research, development, and manufacturing facilities. The ministry would subsequently become the *Ministry of Space Industry* on 19 December 1986.

The Soviet Union offered crucial help to jump start the ballistic missile program. Two R-1 (SS-1) rockets built by Korolev's design bureau were provided to China in 1956. More advanced R-2 (SS-2) missiles reached the country in December 1957. The missiles were followed by extensive engineering documentation and equipment and a large number of Soviet specialists arriving to organize production of the R-2 in China. Simultaneously, 50 Chinese students were sent to study missile technology in a leading Soviet educational engineering institution, the Moscow Aviation Institute.

SOVIET BROTHERLY HELP

In 1950s the Soviet Union provided massive help in science and engineering to the newly established communist neighbor, People's Republic of China. Chairman Mao Zedong recognized in a speech in Moscow on 6 November 1957 that "the socialist construction of China has received the fraternal assistance of the Soviet Union in many areas" (Mao Zedong 1986, 764).

During these 10 years, the USSR Academy of Sciences trained 900 scientists from the PRC. One-thousand Chinese scientists and 2000 technical specialists visited the Soviet Union. More than 1500 Soviet scientists, engineers, educators, and doctors worked in Red China. In addition, about 3000 Soviet scientific and engineering books were translated to Chinese in 1949–1955.

A group of Soviet rocketeers that went to China to support transfer of the R-2 ballistic missile technology included 45 scientists and engineers from Korolev's design bureau and other organizations responsible for missile subsystems. Soviet specialists gave numerous lectures on rocket design, technical documentation, standards, materials, technology, and manufacturing processes and techniques. They worked closely with Chinese engineers on specific problems of rocket production, laying foundation of the Chinese missile research, development, and manufacturing base.

Establishing production of the Chinese R-2, designated model *1059*, created the foundations of the Chinese indigenous ballistic missile capabilities. More than 1400 organizations participated in this challenging undertaking and learned technology and manufacturing of various components, parts, and materials for the missile. The R-2 program was building, from scratch, the indispensable research, development, and manufacturing base for the new missile industry. At the same time PLA's 20th Corps and engineering units began construction of the *Jiuquan* missile test site in the Gobi desert. The number of students studying in the Soviet Union was also increased.

The range of the R-2 was limited to 600 km (370 miles), which was a smaller flight distance than needed to hit American military installations in Japan. In addition, the missile was not capable of carrying the projected weight of the

EAST WIND

The first series of Chinese ballistic missiles was named *Dong Feng* (*East wind*) or *DF*.

When he attended an international gathering of communist and workers' parties in the USSR, Chairman Mao Zedong talked on 17 November 1957 to Chinese students and trainees studying in Moscow. Mao emphasized that the forces of socialism had surpassed the forces of imperialism. He said that "in the struggle between the socialist camp and the capitalist camp either the West Wind prevails over the East Wind, or the East Wind prevails over the West Wind .... This is a war between two worlds .... The East Wind is bound to prevail over the West Wind" (Mao Zedong 1986, 774, 775).

Picked up by the communist propaganda machine, this characterization by Mao Zedong gave the name to the country's first ballistic missiles.

first Chinese atomic bombs. Therefore, Research Academy No.5 initiated development of the new, more capable DF series of ballistic missiles; DF standing for *Dong Feng* or *East wind* in Chinese.

SS-3 and SS-4 Denied

At first, Research Academy No. 5 sought to base its first missile design, DF-1, on the Soviet R-12 (SS-4) IRBM built by Yangel's design bureau. The Soviet Union, however, refused to provide this recently developed missile to China. Not being satisfied with "brotherly assistance" of Soviet comrades, the Chinese students studying in Moscow did their best to collect information on another ballistic missile, the Korolev's R-5 (SS-3), by copying restricted notes and talking to the instructors. But the Soviet Union considered the R-5 to be too advanced for transfer to another country.

One Sun in the Skies

There is room for only one truly Marxist sun shining in the communist skies. So, the disagreements between the USSR and PRC mounted, and the relations between two communist giants quickly deteriorated. (The late 1960s would even witness military skirmishes along the border.) So, the Soviet aid in the missile area was cut off, and Soviet specialists left Research Academy No.5 on 12 August 1960.

First R-2 Missile Rises

Chinese scientists and engineers learned diligently from their one-time Soviet brothers, as they demonstrated by a successful launch of the first R-2 from a newly established missile test site in September 1960, one month after departure of the Soviet mentors. This first fired R-2 was actually built in the USSR, but it was fueled by the Chinese-made propellants. The successful launch of the Chinese-made R-2 followed on 5 November, and two more missiles were fired in December of the same year.

DF-2

In a short time, the PRC had become capable of developing its own ballistic missiles. The new Chinese-designed and built missile *DF-2* was expected to be capable of reaching any place in Japan with a 1500-kg (3300-lb) warhead. Tsien personally initiated development of another even more advanced long-range rocket similar in performance characteristics and configuration to the first Soviet ICBM R-7 (SS-6). This task was however too challenging for Chinese rocketeers at that time, and the program was cancelled in 1963.

The ranges of the successively produced Chinese ballistic missiles were linked to specific targets. So, the first DF-2 was designed to deliver atomic bombs to Japan. (China demonstrated its atomic bomb on 16 October 1964. Then on 26 October 1966, the PLA successfully launched a ballistic missile with a live

GUIDED MISSILE — NUCLEAR WEAPON TEST

On October 27, 1966, China successfully conducted over its own territory a guided missile — nuclear weapon test. The guided missile flew normally and the nuclear warhead accurately hit the target at the appointed distance, effecting a nuclear explosion.

... The complete success of this test was ensured by the Chinese People's Liberation Army and China's scientists, technicians and broad sections of workers and functionaries, who, enthusiastically responding to the call of Comrade Lin Biao [Piao] and holding high the great red banner of Mao Zedong's thought, put politics in the forefront, creatively studied and applied Chairman Mao's works, and, propelled by the great proletarian cultural revolution, took firm hold of the revolution and stimulated production, displayed the spirit of self-reliance, hard work, collective wisdom and efforts and wholehearted cooperation. This is a great victory to Mao Zedong's thought.

Press Communique, Hsinhua (Xinhua) News Agency, 27 October 1966

Reaching Strategic Targets

atomic warhead that was detonated over the target area.) The modified and upgraded original DF-1 was renamed DF-3, and it was built to reach American bases Clark Field and Subic Bay in the Philippines. The DF-4 was capable of hitting Guam, a home of a B-52 fleet. (The shifting political priorities of 1970s forced extension of the capabilities of the DF-4 to put the Soviet capital, Moscow, within its range.) The DF-5 ICBM would cover the continental United States. Finally, the DF-6 was to be launched southward and, after flying over the Antarctic, to reach the United States from south. The Panama Canal also became one of its targets.

"... we shall develop our own artificial satellite..."

Launching of an Earth-orbiting satellite was on the minds of the Chinese leaders early. Mao Zedong stated on 17 May 1958 at a meeting of the Communist Party Central Committee, "we shall develop our own artificial satellite" (*China Today* 1992, 20). Consequently, a special program was established by the Chinese Academy of Sciences and the *Institute of Machine and Electricity* was formed in Shanghai to concentrate on the project. The focus of the effort was however shifted a year later to more realistic objectives of establishing a sounding rocket program.

"... to concentrate the forces like a clenched fist ..."

Numerous organizations participated in a large-scale cooperation established for the missile program, following a directive "to concentrate the forces like a clenched fist to make effective breakthroughs." Initially, most of rocket and space institutions were concentrated in Beijing and Shanghai. With time, a number of organizations were formed in other parts of the country. For example, development of solid-propellant motors began in the Inner Mongolia. Building installations throughout China followed an instruction of Marshal Lin Biao [Piao] to disperse facilities to be "near the mountain, well scattered, and securely covered."

Long March

The Chinese satellite program that included development of a space launcher and a spacecraft was started again in 1965 with the significant resources behind it. The space launcher would become known as the *Chang Zheng* (*Long March* in English), or *CZ-1 (LM-1)*, and the spacecraft would be called *Dong Fong Hong* (*The East is Red* in English), or *DFH-1*. This ambitious and politically important program was conducted during the bloody turmoil of the cultural revolution, which

Fig. 17.15. Long March rocket that launched the first Chinese satellite DFH-1 on 24 April 1970. Photo courtesy of China Astronautics Publishing House, Beijing.

began in 1966 and lasted for 10 years. Many functionaries (cadres), managers, scientists, and engineers were persecuted, humiliated, harassed, banished, or murdered, in a familiar story repeated again and again in advanced Marxist countries.

Cultural Revolution

Under the conditions of internal struggle in the Chinese Communist Party, Premier Zhou Enlai and Vice Premier Nie Rongzhen took steps in 1967–1968 to protect important parts of defense industry. The *Academy of Space Technology*, with Tsien as president, was established in 1968 under the Defense Science and

GREAT PROLETARIAN CULTURAL REVOLUTION

By comparison with the terrifying but almost unknown horrors of the agrarian revolution and the Great Leap Forward, the effect of the "Great Proletarian Cultural Revolution" seems almost modest. Estimates vary greatly for the number of dead: most authors cite figures between 400,000 and 1 million … .

The "Cultural Revolution Group" [formed at an extraordinary meeting of the Politburo of the Chinese Communist Party on 16 May 1966 and reporting directly to Mao Zedong] decided that the peasantry, the army, and scientific research (which for the most part centered upon nuclear weapons) should remain unaffected [by the Cultural Revolution] … .

There was never a clear aim to eliminate a whole section of the population. Even intellectuals, who were particularly affected at the outset, were not for long the prime targets of persecutions. Moreover, the persecutors often came from their own ranks....

...For the whole period of the Cultural Revolution, between 3 and 4 million of the 18 million cadres were imprisoned, as were 400,000 soldiers, despite banning of Red Guards in the PLA. Among intellectuals, 142,000 teachers, 53,000 scientists and technicians, 500 teachers of medicine, and 2,600 artists and writers were persecuted, and many of them were killed or committed suicide. These latter figures, which are not entirely reliable are those used in the trial of the Gang of Four in 1981 … .

S. Courtois et al. (1999, 513, 514, 524, 795)

Technology Commission. (The Academy would be transferred back to the Seventh Ministry of Machine Building in the early 1970s.) As part of the PLA, the Academy was exempt from the political campaign of "speaking freely, airing views fully, holding great debates, and writing big-character wall posters [*dazibao*]." Instead, the staff went through "positive education."

Military Protection

This was the time when many leading specialists were attacked as "reactionary academic authorities," and engineers and scientists were often branded "bourgeois intellectuals." Premier Zhou Enlai ordered the extension of the state's personal protection from attack and persecution to leading specialists in the missile program, including Tsien. Military control of the Seventh Ministry of Machine Building and its affiliates protected other important research, development, and industrial assets.

Long March Space Launcher

The Long March space launcher was a three-stage rocket with the first two stages using liquid-propellant engines and with a solid-propellant third stage. The vehicle was closely linked to the development of the intermediate-range ballistic missiles and multistage rockets initiated by the *Academy of Launch Vehicle Technology* in 1965. Successful demonstration of the first two-stage IRBM on 30 January 1970 paved a way for launch of the first Chinese satellite.

The *Institute of Rocket Engine Technology* designed the engines of the Long March's first two stages based on UDMH and nitric acid as propellants. The first stage YF-2A had four engines. The second stage YF-3 had one engine that was ignited at altitude 60 km (37 miles) and operated under near vacuum conditions. The *Academy of Rocket Motor Technology* developed the third-stage motor that was 4 m (13 ft) long with diameter 77 cm (30 in.) and carried 1800 kg (4000 lb) of solid propellant. The stage was designed to operate at a spin rate 180 rpm. The Long March launcher was 29.46 m (96.7 ft) long with the maximum diameter 2.25 m (7.38 ft). The rocket's mass was 81,500 kg (180,000 lb) at liftoff, and it produced a thrust of 1.02 MN (229,300 lbf).

The Chinese Communist Party attached enormous propaganda importance to the satellite program. The original plan was to launch a satellite in an 80-deg-azimuth direction. Such a near-eastward launch would have placed the satellite in orbit with inclination 42 deg. The country leaders wanted, however, the satellite to be literally "seen and heard" throughout the world. The spacecraft was designed to transmit a melody, *Dong Fong Hong*, celebrating China's love to Chairman Mao and shining worldwide an inspiration of his invincible communist thought. So, the orbit inclination was ordered to be changed to 70 deg (and launch azimuth to 160 deg) to make the satellite visible in Europe and North America. The achieved orbit inclination would be 68.4 deg.

Must Be "Seen and Heard"

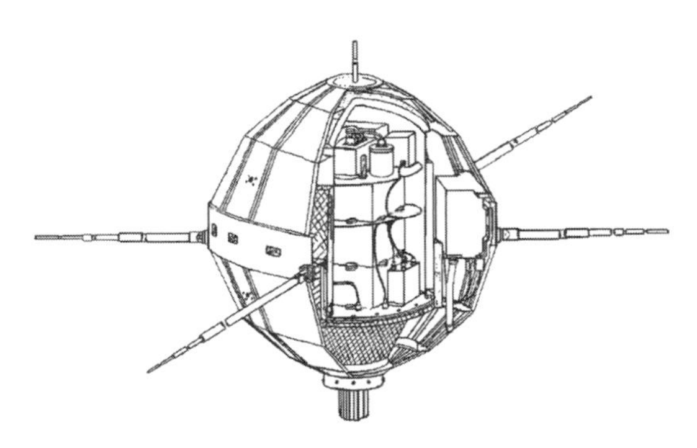

Fig. 17.16. First Chinese artificial satellite DFH-1. The satellite carried a transmitter broadcasting a Dong Fong Hong (The East is Red) melody. Photo courtesy of China Astronautics Publishing House, Beijing.

Special Skirt

After separation, the third rocket stage was to stay in orbit traveling close to the satellite. Satellite visual magnitude was estimated to be $m = +5$ to $+8$, barely visible by the naked eye at the most favorable conditions. The bigger third stage was expected to be somewhat brighter with magnitude $m = +4$ to $+7$. To make the third stage better visible to the naked eye, a special skirt was added to it, increasing its brightness by two levels of visual magnitude.

DFH-1 in Orbit

On 17 April 1970, the rocket with the attached satellite was moved to the service tower at the Jiuquan missile range. After preparation and checks completed, the first Chinese satellite, DFH-1, was successfully launched into orbit on 24 April 1970. The achieved orbit perigee was 439 km (273 miles) and apogee 2384 km (1482 miles). Today, the satellite orbit perigee is

434 km (270 miles), and apogee is 2082 km (1296 miles). The official state *Hsinhua (Xinhua) News Agency* announced the successful launch of the satellite on the afternoon of the next day, starting celebrations across the nation.

> **DONG FONG HONG**
> (The East is Red)
>
> Red is the east, rises the sun.
> China has brought forth a Mao Zedong.
> For the people's happiness he works,
> hu erh hai yo,
> He's the people's liberator.
>
> Chairman Mao loves the people.
> Chairman Mao, he is our guide.
> To build a new China,
> hu erh hai yo,
> He leads us forward.
>
> Communist Party is like the sun,
> Bringing light whereever it shines.
> Where there's the Communist party,
> hu erh hai yo,
> There the people win liberation.

The DFH-1 satellite was a spherical polyhedron 1 m (3.3 ft) in diameter with mass 173 kg (381 lb). It carried batteries and radio-transmitter with the "melody device." The spacecraft was spun up with the third stage to 180 rpm during the launch.

Spin-Stabilized Spacecraft

After separation, the centrifugal forces spread four 3-m telescopic antennas (folded during the launch) in the radial direction reducing the spinning rate to 120 rpm.

The spacecraft did not have an attitude control system, but it was equipped with sun angle and infrared horizon sensors to measure attitude. Its passive thermal control by surface coatings maintained temperature between 5 and 40°C in the instrument compartment. In addition, an electrical heater was used for thermal control of crystal oscillators. Silver-zinc batteries with the 6.165 A-h capacity provided power for 20 days of the mission.

> **OFFICIAL CELEBRATIONS**
>
> ... A revolutionary atmosphere of great joy prevailed everywhere, in the cities as well as in the countryside. The people gathered around the radio sets or under the loudspeakers in the streets to listen again and again the press communique on the successful launching of the earth satellite and the music Dong Fong Hong (The East is Red) in praise of our great leader Chairman Mao, which was transmitted by the man-made satellite and received and recorded by the Central People's Broadcasting Station.
>
> From that night through the following day, thousands upon thousands of people, holding aloft huge portraits of our great leader Chairman Mao and singing and dancing, streamed into the streets and public squares from all directions for meetings and parades.
>
> ... When the earth satellite radiant with the brilliance of Mao Zedong Thought passed over the capital of the great motherland, the whole city was thrilled. Firecrackers and fireworks were let off; gazing at the earth satellite high in the sky, the people danced with joy and cheered tumultuously "'Long live invincible Mao Zedong Thought!' ... 'A long, long life to Chairman Mao!'"
>
> *Peking Review* (1970, 6)

The 3-W transmitter sent radio signals at 20-MHz frequency. The Dong Fong Hong melody, after which the satellite was named, was broadcast for 40 s. Then after a 5-s interval, telemetry was transmitted for 10 s. Following another 5-s interval, the cycle repeated itself. The melody was generated by six onboard oscillators corresponding to the six different sound tones of the melody. In addition, the spacecraft carried a transponder and a radio beacon.

Melody Device

The launch of the first satellite DFH-1 was followed by launch of a similar sized but much more complex satellite *Practice-1* on 3 March 1971. Practice-1 carried solar cells, rechargeable batteries, and instrumentation to measure geomagnetic field, cosmic rays, and x-ray radiation. The spacecraft was designed to operate for one year but lasted for eight years.

Practice-1

Thus, the People's Republic of China became the fifth country in the world to launch its own satellite on a national space launcher.

INDIA

The glorious time of Indian war rockets was in the 18th century, followed by many decades of neglect and lack of interest in rocketry. The situation began to change only in the early 1960s when Dr. Vikram Sarabhai was appointed chairman of the *Indian National Committee for Space Research*. Sarabhai, who earned a Ph.D. in physics from the Cavendish Laboratory in Cambridge, England, set up a research group studying cosmic rays and upper-atmosphere phenomena in the late 1940s. He later reached prominence as an industrial manager and formed several Indian technology-oriented companies.

Vikram Sarabhai

Fig. 17.17. Vikram Sarabhai, 1919–1971, considered by many to be a "father" of the modern Indian space program. Photo courtesy of Indian Space Research Organization, Bangalore.

In 1962, Sarabhai made the first practical steps in organizing Indian space activities under the aegis of the Department of Atomic Energy. He began with establishing the first launch site near a fishing village Thumba in Kerala on the southwestern coast of India. The first rocket, an American Nike–Apache, was launched from the new site on 21 November 1963. In addition, Sarabhai arranged manufacturing in India of French-designed sounding rockets.

Thumba Launch Site

Design and fabrication of indigenous Indian sounding rockets, the *Rohini* family, followed. The first *Rohini 75* was launched at Thumba in 1967, with rockets growing in size to *Rohini 125* and *Rohini 560*. The latter missile was capable of lifting 200-kg (440-lb) payloads to altitudes up to 400 km (250 miles). The program led to developing expertise in propellants, pyrotechnic devices,

Rohini

testing instrumentation, structures, and materials. Japan's Hideo Itokawa spent a significant amount of time with Sarabhai in 1960s serving as an advisor to the emerging Indian rocket program.

Emergence of National Space Expertise

In 1972, India signed an agreement with the Soviet Union to launch first Indian satellites on Soviet rockets. In three years, an Indian team under Professor U.R. Rao built a satellite, named *Aryabhata* after a prominent Indian mathematician and astronomer of the fifth century. The spacecraft was successfully launched on 19 April 1975 from the Soviet Kapustin Yar missile test range. Aryabhata had mass 360 kg (794 lb) and carried scientific instrumentation to measure radiation from solar flares. The power supply failed after 60 orbits, thus prematurely terminating the mission. Two more Indian satellites, *Bhaskhara 1* and *Bhaskhara 2* (named after an Indian mathematician and astronomer of the 12th century), were launched in 1979 and 1981 from Kapustin Yar. The first Indian spacecraft were built by *Hindustan Aeronautics, Ltd.*, with several critically important components such as solar cells, batteries, and surface coatings supplied by the Soviet Union. The program was an important and valuable learning experience for Indian scientists and engineers and helped in the development of the national industrial capabilities indispensable for a space-faring nation.

Aryabhata

Fig. 17.18. The first Indian satellite *Aryabhata* launched by the Soviet Union from the Kapustin Yar missile range on 19 April 1975. Photo courtesy of Indian Space Research Organization, Bangalore.

Bhaskhara

Space Industry

Delivery of educational television programs to rural areas through the American *Applications Technology Satellite (ATS)* was another major space-related program in India. A large number of receiving stations equipped with inexpensive 3-m (9.8-ft) dishes were built throughout the country. Television programs in literacy, agriculture, family planning, and health were beamed to the American satellite that retransmitted them for four hours a day. The experience with the first Indian satellites and the ATS program led to emergence of a national space industry employing almost 10,000 people in the 1970s. In addition, more than 30 universities became involved in space research, and a number of specialists working in foreign countries decided to return home to India.

ATS Program

Design of an indigenous *Satellite Launch Vehicle (SLV)* was conceived in 1969. The launcher, SLV-3, was to place small satellites in low-Earth orbit. Building the vehicle was a major nationwide undertaking involving more than 40 industrial and government enterprises. Five-sixth of the required components were designed and fabricated in India. The SLV was a slim four-stage all-solid-propellant rocket 22.7

Fig. 17.19. First launch of a satellite by the Indian SLV-3 rocket from a launch site on Sriharicota Island on 18 July 1980. An all-solid-propellant four-stage rocket successfully placed a small spacecraft *Rohini* in low-Earth orbit. Photo courtesy of Indian Space Research Organization, Bangalore.

m (74.5 ft) tall and only 1 m (3.3 ft) in diameter. The rocket mass was 16,900 kg (37,250 lb) at liftoff. The first, second, third, and fourth stages developed thrusts of 622.7 kN (140,000 lbf), 266.7 kN (60,000 lbf), 90.2 kN (20,300 lbf), and 26.5 kN (5960 lbf), respectively. The cases of the first- and second-stage motors were made of stainless steel, whereas fiberglass was used in motors of the third and fourth stages.

SLV Development

A satellite that was built for launch by the national rocket, *Rohini*, was a small spacecraft equipped with a CCD-based camera with a 1-km resolution on the ground. It also carried Indian-made solar cells. (The USSR supplied all solar cells on the Indian satellites launched earlier from Kapustin Yar.)

Fig.17.20. *Rohini*, the first Indian satellite placed in orbit by a national space launcher, the SLV-3. The spacecraft was about 0.5 m (1.6 ft) in diameter with mass close to 40 kg (88 lb). The first flown Indian solar cells produced 3 W of electric power. Photo courtesy of Indian Space Research Organization, Bangalore.

Sriharicota Space Port

Rohini was launched from a new test site constructed on *Sriharicota Island* near the southeastern coast of India. Sriharicota, a small 44 km (27 miles) by 7 km (4.5 miles) island, is located at 13°N geographical latitude, making it an excellent location for a space port that takes advantage of the Earth's rotation and offers favorable conditions for launches into near-equatorial orbits. Only the French (and ESA) permanent launch site at Kourou is better in this respect. (Obviously, launches from an oceangoing platform by *SeaLaunch* and launches of *Pegasus* rockets from an aircraft can be conducted from any desired geographical latitude. The Italian San Marco platform off the coast of East Africa near the equator was used for launches of American rockets in early 1960s, and the facility does not exist today.)

Rohini in Orbit

After several years of delays, the first Indian space launch was ready in 1979. The attempt failed, however, because of problems with the control system of the second stage. Success came on 18 July 1980, when the Rohini spacecraft was placed in orbit with perigee 306 km (190 miles), apogee 919 km (571 miles), and inclination 45.75 deg. Rohini stayed in orbit more than 10 months and reentered the atmosphere in May 1981. Thus, India joined the elite club of space-faring nations who built and launched satellites.

ISRAEL

Ballistic Missiles and Space for Survival

Development of ballistic missiles was essential for survival of Israel as a country surrounded by hostile Arab neighbors intent on physical annihilation of the Jewish state. The achieved national ballistic missile capabilities subsequently enabled space launches. Naturally, the major objectives of the space program were also driven by the national security requirements and concentrated on space-based reconnaissance and communications. The country's space effort received important boost in 1982 with the formation of the *Israel Space Agency* (ISA). Physicist Yuval Ne'eman, 1925–2006, played a crucial role in shaping the Israel's program and became president of ISA.

First Satellite

Israel's first step into space was a launch of a simple *Ofeq-1* (*ofeq* means *horizon* in Hebrew) satellite on 19 September 1988, with the *Israel Aircraft Industries* (IAI) leading the effort as the prime contractor. The spacecraft was designed and built by IAI's *MBT Division*, and the satellite control center was established at MBT's facility in Yahud near Tel Aviv. The spin-stabilized (at 60 rpm) Ofeq-1 was deployed in a low-Earth orbit with perigee 248 km (154 miles) and apogee 1150 km (715 miles). Low orbit perigee resulted in large atmospheric drag, and the satellite reentered the atmosphere in four months on 14 January 1989.

Ofeq-1

The Ofeq-1 spacecraft was an octagon with the lower- and upper-base diameters 1.3 m (4.3 ft) and 0.7 m (2.3 ft), respectively, and height 2.3 m (7.5 ft). The spacecraft total mass was 156 kg (344 lb), with the mass breakdown among the subsystems, as follows: 33 kg (73 lb) for structures, 58 kg (128 lb) for electric power, 5 kg (11 lb) for thermal control, 12 kg (26 lb) for communications, 7 kg (15 lb) for onboard computer, 32 kg (71 lb) for spacecraft instrumentation and balancing masses, and 9 kg (20 lb) for cabling. The body-mounted solar panels supported average power consumption of 53 W, and the onboard battery had capacity 7 A-h. The S-band communications system provided a 2.5-kbit/s transmission rate.

PROGRADE AND RETROGRADE ORBITS

Typically, spacecraft are launched in the eastern direction (from left to right, if one looks at the map). Therefore, most satellites move around the Earth in the same general direction as the Earth rotates. Such orbits are called *prograde*.

The rotation of our planet is not negligible and provides significant help for launching satellites. The velocity of a point on the Earth's surface is approximately equal to $465 \times \cos(\lambda)$ m/s, where λ is geographical latitude of the launching site. Thus, one gets "for free" a significant velocity (465 m/s at the equator) when launching a satellite due east.

Possible launch directions are usually restricted by safety considerations, to prevent first rocket stages, or malfunctioning rocket, from falling on populated areas. In Israel's case, political considerations do not allow space launches in the eastern direction. Therefore, the country has to launch satellites over the Mediterranean sea in the *western* direction, against the Earth's rotation. Such orbits are called *retrograde*. Launch in low-inclination retrograde orbits is highly unfavorable and requires significantly larger rockets than for similar eastward launches.

Fig. 17.21. Israel's three-stage rocket Shavit launches a satellite from the Palmachim Air Force Base near Tel Aviv. Photo courtesy of Israel Space Agency.

Space launch is especially challenging for Israel because political constraints limit possible launch azimuth to western directions against the Earth's rotation. Therefore, Ofeq-1 (as well as the follow on satellites) was placed in retrograde orbit with inclination 143 deg (or 37 deg retrograde). The IAI's *MLM Division* built a capable *Shavit* (*comet* in Hebrew) three-stage solid-propellant launcher to deploy satellites under such adverse restrictions. The solid motors of the first two stages were built by *Israel Military Industries*; the third-stage motor was designed by *Rafael*.

Adverse Launch Conditions

Shavit launched the Israel's first satellite from the *Palmachim Air Force Base* in Yavne, 15 miles (25 km) from Tel Aviv. Israel is a small country, and this is the only place in the world where satellites are launched from a site so near to a major metropolitan area. The Shavit's second stage cut off at approximately 110 km (68 miles) altitude. After coasting, the third stage was fired at 250–260 km (160 miles) injecting the spacecraft into the desired orbit. The shroud was jettisoned just before firing of the third stage.

Shavit Launcher

Fig. 17.22. Israel's third satellite, Ofeq-3, successfully launched in April 1995. A three-axis-stabilized spacecraft (attitude accuracy 0.1 deg) was deployed in a 143-deg-inclination (or 37 deg retrograde) low-Earth orbit. Israel's first satellite, Ofeq-1, had a similar body, but without deployable solar panels. Photo courtesy of Israel Space Agency.

Focused Space Program

With the launch of Ofeq-1, Israel also became a member of the small elite group of nations with space launch capabilities. The country continued its energetic and focused space program with the successive satellites increasing in sophistication and capabilities. *Ofeq-3*, for example, was launched on 5 April 1995 and stayed in orbit more than five years. This three-axis-stabilized spacecraft had deployable solar arrays that generated 180-W power, and it carried an optical payload. A more sophisticated operational reconnaissance satellite *Ofeq-5* was launched on 28 May 2002. In addition to low-Earth-orbit Earth-observation spacecraft, Israel also built geosynchronous communication satellites *AMOS*.

BRAZIL'S TURN?

After Israel had become a space-faring nation in 1988, only one other nation, Brazil, attempted to launch a satellite. All attempts have failed so far. Given the country's determination, however, it is only a question of time until Brazil succeeds.

18. THE FIRST THOUSAND YEARS

Rockets have been known to man for almost 1000 years. A long road led us from simple fireworks to intercontinental ballistic missiles and powerful space launchers that opened a way to the cosmos in the late 1950s. By the turn of the 21st century, spacecraft had visited all of the planets of our solar system with the exception of Pluto; cosmonauts and astronauts have spent years orbiting the Earth; and 12 men have walked on the moon.

From Fire-crackers to Interstellar Flight

One operational spacecraft, Voyager 1, has already approached the solar system frontier that separates the region governed by the sun — the heliosphere — from galactic medium. We believe that this boundary region extends from 100 to 200 AU. (*AU* is astronomical unit, the distance between the sun and the Earth). The first space mission into the surrounding pristine interstellar space, the Interstellar Probe, is technically feasible for a number of years, and it is being planned by NASA. This spacecraft would reach the distance of several hundred AUs and explore in situ the interstellar medium in a first and necessarily modest step toward the truly interstellar flight of the distant future. Not bad for the first 1000 years of rocketry: from firecrackers to interstellar flight.

Difficult and Heroic Time

This breakthrough to space did not come easily. The first space launch vehicles had a low probability of success. The rockets exploded on launchpads and malfunctioned in flight. It was a difficult and heroic time of facing one technical challenge after another. But the learning curve was steep. By the mid-1960s, most of the technical problems causing failure of rocket launch, such as combustion instabilities, pogo effect, fuel sloshing, undesired structural dynamics effects, malfunctioning of unreliable components, and many others had been identified, understood, and solved. In 1960, 1961, and 1962, for example, the success rate of American space launches was 55, 71, and 88%, respectively. The situation was not any better in the rival Soviet Union. The first cosmonauts and astronauts indeed braved the dangerous unknown riding on space rockets that could turn into fireballs. Space launch remains a dangerous undertaking to this day, however, and even the highly reliable space shuttle has a probability of about 1% of deadly disaster.

Spacecraft Technologies

The practical value of rockets was realized long ago, and the military necessity and spectacular advancements of science and engineering during the last couple of centuries perfected them into efficient weapons and space launchers. As the rocket technology advanced, there was, however, very little practical interest in artificial satellites or spacecraft until the beginning of the second half of the 20th century. Although the visionaries and writers thought about spaceflight, the technology was simply not ready yet. The practical satellite, even a simple one, requires solutions to numerous challenging technical problems: How would one obtain electric power on a spacecraft? How would one navigate? How would one communicate with Earth and with other space vehicles? How would one determine and control the attitude of a spacecraft? How would one change a spacecraft's orbit? What environment would one find in space? Would a spacecraft survive in vacuum exposed to plasma, micrometeoroids, energetic particles, and harsh electromagnetic radiation?

WWII and Science and Engineering

These were serious unresolved scientific and engineering questions that usually did not deter — as many of today's technical challenges do not deter now — the enthusiastic amateurs and science-fiction fans. It took tremendous effort, time, and resources for scientists and engineers to gradually develop the technologies enabling a satellite. These technologies did not exist in the 1930s. Only rapid scientific and technological progress before, during, and immediately after World War II laid the foundation for the breakthrough to space. The role of science, particularly of physics and engineering, also fundamentally changed at that time. The Cold War confrontation that ensued produced the indispensable political will with the resulting significant investments in rocketry and spaceflight by the United States and Soviet Union.

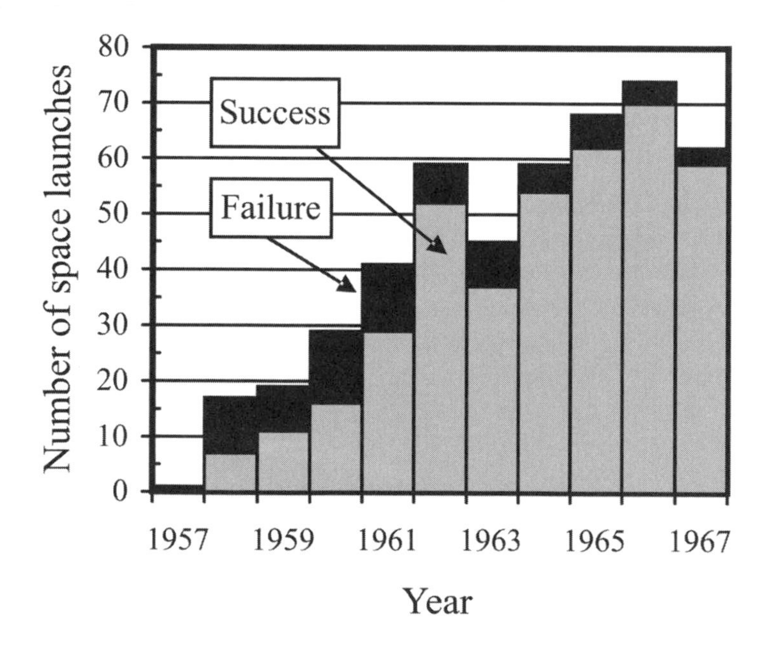

Fig. 18.1. Successes and failures of early space launches in the United States. The learning curve was steep during the first several years. Figure courtesy of Mike Gruntman.

Fig. 18.2. Breakthrough to space — a monument to the conquerors of space in Moscow. The upper 99-m (325-ft) part of the monument is made of titanium and steel; the rocket on the top is 11 m (36 ft) tall. Photo courtesy of Mike Gruntman.

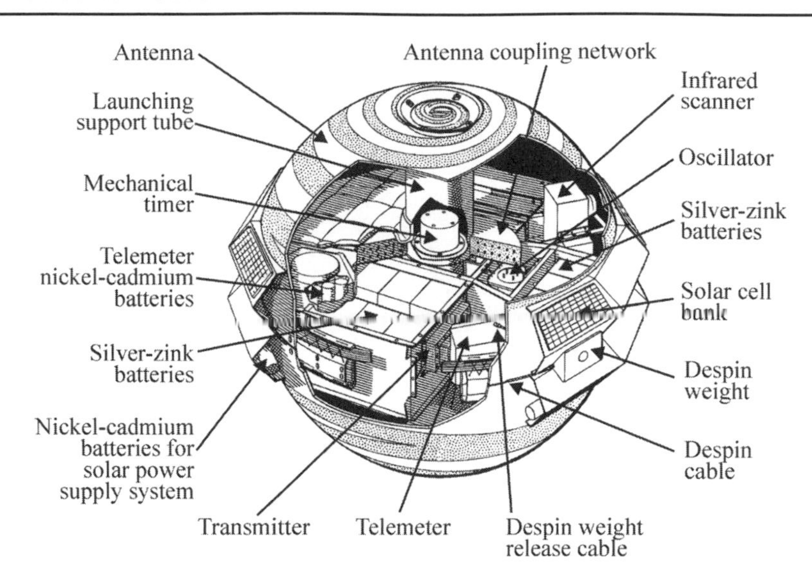

Fig. 18.3. Cutaway view of the first successfully launched (13 April 1960) navigational satellite Transit 1B. This early spacecraft demonstrates the complexity of an artificial satellite and shows the variety of various subsystems that are indispensable for functioning of such a relatively simple early spacecraft. Figure courtesy of Applied Physics Laboratory, Laurel, Maryland.

Radio-telemetry

The concept of an unmanned scientific satellite was long retarded by the slow birth of radio telemetry. Why would anybody put a satellite into orbit if one cannot communicate with it? Radio was first demonstrated in 1895, and the first experiments with transmitting technical data by radio started only in the 1930s. It was not even clear whether the radio waves could penetrate through the ionosphere. When Robert Goddard launched a rocket with an instrument payload (aneroid barometer and thermometer) on 17 July 1929, he placed a photographic camera to record the readings of the instruments. Radio telemetry did not exist at that time.

Solar Cells, SNAP, and Hydrazine

The practical silicon-based solar cells with acceptable efficiency of several percent, a key to satellite power systems, were invented in 1953 at the Bell Telephone Laboratories. This was also the time when the recently developed transistor opened the way to miniaturization of electronics and practical computers. Study of nuclear power sources for space applications began in 1950s, when the Atomic Energy Commission started the *Systems for Nuclear Auxiliary Power* (SNAP) program.

Stationkeeping and attitude control required onboard propulsion capabilities, which were at first achieved by cold-gas jets. Later, monopropellant hydrazine thrusters provided precisely measured impulses. (Hydrazine became a workhorse of small-thrust propulsion after development of the *Shell-405* catalyst in early 1960s.) Inertial navigation required for long-range ballistic missiles was also rapidly maturing in the late 1950s. In short, a number of technologies were becoming available in 1950s enabling efficient satellites. Only 12 years after the launch of the first artificial satellite, the mighty Saturn V lifted off to land men on the Moon and bring them safely back to Earth.

So, in spite of all difficulties, the human race broke away from shackles of Earth's gravity and reached to space. This feat was led mostly by two nations, the United States and Soviet Union, the ideological, military, and political rivals of the Cold War. What will happen next? Where will the next thousand years take us in space? The answer depends on our vision and determination in the pursuit of expansion in space.

Ready for Second-Stage Ignition

If one draws an analogy with space launch, we are now at the cutoff of the engine of the first rocket stage and preparing to ignite the second stage, which will take us to far corners of the solar system and beyond. Two enabling technological developments will have to happen to make it possible, dramatic reduction of cost of access to space and achieving advanced in-space power (nuclear) and propulsion (such as electric propulsion and solar sails) capabilities.

Fig. 18.4. Saturn V at Cape Canaveral. In a mere 12 years from the launch of the first artificial Earth-orbiting satellite, mighty space launchers capable of taking man to the moon became available. There is no limit to what scientists and engineers can do given the vision, determination, resources, and public support. Photo courtesy of NASA.

Cost of Access to Space

Today it takes $10,000 to place 1 kg in low-Earth orbit. The cost increases fivefold for geostationary orbit. Only a significant reduction in cost of space launch will allow us to embark on exploration of the far corners of the solar system and finally bring manufacturing, tourism, and other yet-to-be-determined activities of the commercial world, with all of its indomitable entrepreneurial spirit, to space.

Nobody and nothing will help us to achieve this enabling technological breakthrough except dogged determination and focus of scientists and engineers, supported by the public. We cannot, and should not, wait and bet on fulfilling a hope of finally meeting, one day, aliens riding flying saucers and learning from them smart ways of dashing through vastness of space.

Can Aliens Help Us?

ONE SINGLE DISCOVERY

In the near future, the first men will land on Mars. If there is a single, ancient, long-abandoned edifice [erected by ancient astronauts] there, if there is a single object indicating earlier intelligences, if there is one still recognizable rock drawing to be found, then these finds will shake foundations of our religions and throw our past into confusion. One single discovery of this kind will cause the greatest revolution and reformation in the history of humankind.

Erich von Däniken (1968, 51)

In reality, perhaps to the dismay of some, statistics tells us that more than one-half of "sightings" of UFOs in the late 1950s and most of the similar events in the 1960s could be attributed to flights of the U-2 and A-12 aircraft (SR-71 is a military variant of A-12). There is no reason to think that the sightings in the later years were fundamentally different from those earlier events.

Are We Alone in the Universe?

Our civilization is very young on the timescale of galactic history. Judging from the progress of the human race, it is reasonable to assume that our technological capabilities will continue rapidly advancing. Conceptually, if there are other worlds and intelligent life somewhere else in the universe, then some of these civilizations could be much older than we are today and thus much more technologically advanced. They might have achieved interstellar flight capabilities and perhaps even visited our solar system. There is, however, no unambiguous experimental evidence that this has ever happened. The latter fact should not preclude or discourage us from trying to understand the meaning of the mysterious geoglyphs such as Nasca

INTERSTELLAR TRAVEL

Despite the immense distance between our own solar system (including the earth) and the nearest other solar systems, a journey from one system to another is theoretically possible, once an unlimited source of power is developed.

Hermann Oberth (1954, 4)

When you want to go to the next star, which is so distant that light takes four years to get there, then you have to get rockets which really move fast; and with the best nuclear propulsion that I can imagine today, it will still take more than hundred years to get to that next star ... . Really exotic methods of propulsion ... will have to be devised to get there. How it will be done, I do not know. Whether it will be done, I am not quite certain. But I would bet it can be done ... .

Edward Teller, 1957 (*Hearings* 1958, 38)

... I am quite pessimistic about ever achieving interstellar travel.

Theodore von Kármán (1967, 347)

Fig. 18.5. Delta II rocket reaches to orbit from Vandenberg Air Force Base, 25 March 2000. Photo courtesy of Mike Gruntman.

figures and lines in Peru or searching for signatures of extraterrestrial intelligence by radioastronomical means.

Stay on This Side of Warp Drive

In the meantime, we need to stay on "this side" of the warp drive and diligently work on improving our propulsion and other space technology capabilities.

Two other issues — military space and international cooperation — were important in the past and will remain so in the future. The military traditionally drove development of rocketry and space technology, paving the way to space. The importance and value of satellites for military operations on the ground will continue to grow, as was clearly demonstrated by the campaign in Iraq in 2003.

Military Space

Air, land, and sea have witnessed conflict, and space is no different. Armed forces of various countries are routinely active and conduct operations on the high seas and in international airspace. It is logical to assume that military operations in space will become routine, driven by national security imperatives.

American economy, infrastructure, and national security depend on satellites more than those of any other nation, and consequently the United States will lead the effort to deter and to defend its space assets. It is only a question of time when the current space capabilities will be expanded to include spacecraft projecting force in, from, and through space. As navies patrol the high seas, space forces will inevitably be deployed in low-Earth orbit and then reach geostationary orbit and beyond. To try to avoid or slow down this development would invite a new Space Pearl Harbor. The unwillingness to defend oneself by force, if necessary, never brought freedom, security, and peace to anybody but blood, tears, and violent conflict that could have been otherwise averted. As Frank Loesser famously expressed it in his memorable song, "Praise the Lord and pass the ammunition and we'll all stay free!" (Loesser 2003, 92)

Space Pearl Harbor

Tremendous energy and emotions going into lengthy discussions of international cooperation in exploration of space obscures one critically important fact: most of the nations are not committed to exploration of space and avoid investing resources in space technology. In 2002, civilian government space expenditures were about $20B worldwide (Hertzfeld and Ojalehto 2003). NASA accounted for 63%, whereas industrially developed Europe spent 17 and 6% through ESA and additional national space programs, respectively. Japan's share was 12%. In addition, American programs in military space dwarf those of all other nations. Only France (and the old Soviet Union in the past) approaches the U.S. space expenditures in terms of the fraction of the gross domestic product (GDP). Most other industrialized countries spend in space, as fraction of GDP, four to six times less than the United States.

Lack of Commitment

This inequality in commitment to exploration of and expansion in space should point to many space enthusiasts, particularly in Europe, that their frustration with slow pace of space programs and limited opportunities result not from often invoked lack of or unwillingness in international cooperation, but are rather caused by disinterest of their own nations and governments. No help can be expected in the foreseeable future from the United Nations either. The United Nations was and is a forum of all nations — not a forum of free nations adhering to the same basic values of liberty, human dignity, and human rights — an organization with an abysmal record of waste, incompetence, and inefficiency and politicizing each and every issue. Nevertheless, educating the public and working with governments is absolutely essential for bringing support for our expansion in space.

Ad Astra!

So, the first 1000 years of rocketry brought us spectacular successes, and we reached the cosmos. The next 1000 years will be even more exciting.

Who said that the road to Alpha Centauri is easy? *Per Aspera ad Astra!* Through Difficulties to the Stars!

APPENDIX A: ACRONYMS AND ABBREVIATIONS

ABL — Allegany Ballistic Laboratory
ABMA — Army Ballistic Missile Agency
AEC — Atomic Energy Commission
AFB — Air Force Base
AFBMD — Air Force Ballistic Missile Division
AFCRL — Air Force Cambridge Research Laboratory
AFETR — Air Force Eastern Test Range
AIAA — American Institute of Aeronautics and Astronautics
AOMC — Army Ordnance Missile Command
APL — Applied Physics Laboratory
ARPA — Advanced Research Projects Agency
ARS — American Rocket Society
ARS — advanced reconnaissance system
ATS — advanced technology satellite
BBR — beach barrage rockets
BIS — British Interplanetary Society
BuAer — Bureau of Aeronautics
BUORL — Boston University Optical Research Laboratory
BUPRL — Boston University Physical Research Laboratory
CCD — charge-coupled device
CEFSR — Committee for Evaluating the Feasibility of Space Rocketry
CEP — circular error probable (the radius of a circle within which one-half of the warheads or projectiles is expected to land)
CIA — Central Intelligence Agency
CIEES — Center Interarmées d'Essais d'Engines Spéciaux (in French) — Joint Staff Center for Special Missile Testing
CNES — Centre National d'Études Spatiales (in French) — National Center for Space Research
CSG — Centre Spatial Guyanais (in French) — Guiana Space Center

Appendix A: Acronyms and Abbreviations

DSP	—	Defense Support Program
ELDO	—	European Launcher Development Organization
ELINT	—	Electronic Intelligence
ESA	—	European Space Agency
ESRO	—	European Space Research Organization
FBI	—	Federal Bureau of Investigations
FBM	—	fleet ballistic missile
FFRDC	—	Federally Funded Research and Development Center
GALCIT	—	Guggenheim Aeronautical Laboratory of the California Institute of Technology
GDL	—	Gasodinamicheskaya Laboratoriya (in Russian) — Gas Dynamical Laboratory
GDP	—	gross domestic product
GEO	—	geostationary orbit
GIRD	—	Gruppa Izucheniya Reaktivnogo Dvizheniya (in Russian) — Group for Study of Jet Propulsion
GPS	—	Global Positioning System
GRAB	—	Galactic Radiation and Background Satellite
GRU	—	Glavnoe Razvedyvatel'noe Upravlenuie (in Russian) — Chief Intelligence Directorate of the General Staff
GSFC	—	Goddard Space Flight Center, NASA
GTsP	—	Gosudarstvennyi Tsentral'nyi Poligon (in Russian) — State Central Test Range
HATV	—	high-altitude test vehicle
HTP	—	high test peroxide
HTPB	—	hydroxy-terminated polybutadiene
HVAR	—	high-velocity aircraft rocket
IAI	—	Israel Aircraft Industries
IBMS	—	Intercontinental Ballistic Missile System
ICBM	—	intercontinental ballistic missile
ICSU	—	International Council of Scientific Unions
IGY	—	International Geophysical Year
IMINT	—	image intelligence
IRBM	—	intermediate-range ballistic missile
IRFNA	—	inhibited red-fuming nitric acid
ISA	—	Israel Space Agency
ISAS	—	Institute of Space and Astronautical Science, Japan
JATO	—	jet-assisted takeoff
JPL	—	Jet Propulsion Laboratory
KB	—	Konstruktorskoe Byuro (in Russian) — Design Bureau
KGB	—	Komitet Gosudarstvennoi Bezopasnosti (in Russian) — Committee for State Security
LEO	—	low-Earth orbit
LOX	—	liquid oxygen
LRBA	—	Laboratoire de Recherches Balistique et Aérodynamiques (in French) — Ballistic and Aerodynamic Research Laboratory

MIDAS — Missile Detection and Alarm System
MIT — Massachusetts Institute of Technology
MMH — monomethylhydrazine
MOL — Manned Orbital Laboratory
NACA — National Advisory Committee for Aeronautics
NASA — National Aeronautics and Space Administration
NII — Nauchno-Issledovatel'skii Institut (in Russian) — Scientific-Research Institute
NIIP — Nauchno-Issledovatel'skii Ispytatel'nyi Poligon (in Russian) — Scientific-Research Test Range
NNSS — Navy Navigation Satellite System
NPO — Nauchno-Proizvodstvennoe Ob'edinenie (in Russian) — Scientific-Production Association
NRL — Naval Research Laboratory
NRO — National Reconnaissance Office
NSA — National Security Agency
NTO — nitrogen tetroxide
OKB — Osoboye Konstruktorskoe Byuro (in Russian) — Special Design Bureau
ONERA — Office National d'Études et de Réalisations Aéronautiques (in French) — National Office of Aerospace Studies and Research
OSRD — Office of Scientific Research and Development
ORDCIT — Ordnance Contract to California Institute of Technology
PLA — People's Liberation Army
PRC — People's Republic of China
PSAC — President's Scientific Advisory Council
RAE — Royal Aircraft Establishment
RAF — Royal Air Force
RCA — Radio Corporation of America
RFNA — red-fuming nitric acid
RHA — Royal Horse Artillery
RMA — Royal Marine Artillery
RMI — Reaction Motors, Inc.
RNII — Reaktivnyi Nauchno-Issledovatel'skii Institut (in Russian) — Jet Propulsion Scientific Research Institute
RP-1 — rocket propellant 1
rpm — revolutions per minute
RTV — reentry test vehicle
RUP — Radiouprableniya Punkt (in Russian) — Post of Radio Control
SAB — Science Advisory Board
SAC — Strategic Air Command
SAG — Science Advisory Group
SAMOS — Satellite and Missile Observation System
SCORE — Signal Communications by Orbiting Relay Experiment
SD — Siecherheitsdienst (in German) — Security Service (part of the Reich Security Central Office)

SINS	—	Ship Inertial Navigation System
SKB	—	Spetsial'noe Konstruktorskoe Byuro (in Russian) — Special Design Bureau
SLBM	—	submarine-launched ballistic missile
SLV	—	satellite launch vehicle
SPO	—	Special Projects Office
SRV	—	Satellite Recovery Vehicle
SS	—	Schutzstaffel (in German) — Protective Echelon (Security Organization of the National-Socialist Party)
SSME	—	space shuttle main engine
STL	—	Space Technology Laboratory
TASS	—	Telegrafnoe Agentstvo Sovestskogo Soyuza (in Russian) — Telegraph Agency of the Soviet Union
TV	—	test vehicle
UARRP	—	Upper-Atmosphere Rocket Research Panel
UDMH	—	unsymmetrical dimethylhydrazine
UFO	—	unidentified flying object
UR	—	Universal'naya Raketa (in Russian) — Universal Rocket
USAF	—	United States Air Force
VAFB	—	Vandenberg Air Force Base
WDD	—	Western Development Division
WFNA	—	white-fuming nitric acid
WRE	—	Weapons Research Establishment
WS	—	weapon system
WSMR	—	White Sands Missile Range
WSPG	—	White Sands Proving Ground

APPENDIX B: SELECTED BIBLIOGRAPHY

Adams, H., *The War of 1812*, Cooper Square Press, New York, 1999.

Aerojet: The Creative Company, Aerojet History Group, 1995.

Aircraft Division Industry Report, U.S. Strategic Bombing Survey, Washington, DC, 1945.

Albertus Magnus, *Opera Omnia*, Apud Ludovicum Vivès, Parisiis, 1890.

Allen, H.J., and Eggers, A.J., Jr., "A Study of the Motion and Aerodynamic Heating of Missiles Entering the Earth's Atmosphere at High Supersonic Speeds," NACA RM A53D28, Aug. 1953.

Antoniadi, E.M., "Considerations of the Physical Appearance of the Planet Mars," *Popular Astronomy*, Vol. 21, 1913, pp. 416–424.

Appier-Hanzelet, J., *La Pyrotechnie de Hanzelet de Lorrain, ou sont representer les plus rares & plus appreuuez secrets des machines & des feux artificiels ...*, I & Gaspard Bernard, Pont-à-Mousson, 1630.

Army Ordnance Satellite Program, Army Ballistic Missile Agency, Redstone Arsenal, AL, 1958.

Barber, J.W., *Historical Collections of the State of New York*, Clark, Austin & Co., New York, 1851.

Battisti, E., and Battisti, G.S., *Le Machine Cifrate di Giovanni Fontana*, Arcadia Edizioni, Milano, 1984.

Beatson, A., *A View of the Origin and Conduct of the War with Tippo Sultaun*, W. Bulmer, London, 1800.

Beauregard, P.G.T., *The Mexican War Reminiscences*, edited by T.H. Williams, Louisiana State Univ. Press, Baton Rouge, LA, 1956.

Benét, S.V., *A Collection of Annual Reports and Other Important Papers, Relating to the Ordnance Department*, Government Printing Office, Washington, DC, 1880.

Benson, R.L., and Warner, M. (eds.), *Venona: Soviet Espionage and American Response 1939-1957*, National Security Agency and Central Intelligence Agency, Washington, DC, 1996.

Benton, J.G., *A Course of Instruction in Ordnance and Gunnery*, Van Nostrand, New York, 1883.

Bille, M., Johnson, P., Kane, R., and Lishock, E.R., "History and Development of U.S. Small Launch Vehicles," *To Reach the High Frontier. A History of the U.S. Launch Vehicles*, edited by R.D. Launius and D.R. Jenkins, Univ. of Kentucky Press, Lexington, KY, 2002, pp. 186–228.

Biringuccio, V., *The Pirotechnia*, Basic Books, New York, 1942.

Bishop, F., *Our First War in Mexico*, Charles Scribner's Sons, New York, 1916.

Bissell, R.M., Jr., *Reflections of a Cold Warrior. From Yalta to the Bay of Pigs*, Yale Univ. Press, New Haven, CT, 1996.

Blosset, L., "Robert Esnault-Pelterie: Space Pioneer," *First Steps Toward Space*, edited by F.C. Durant III and G.S. James, AAS History Series, Vol. 6, American Astronautical Society, San Diego, 1985, pp. 5–31.

Boushey, H.A., "A Brief History of the First U.S. JATO Flight Tests of August 1941: A Memoir," *History of Rocketry and Astronautics*, AAS History Series, Vol. 14, American Astronautical Society, San Diego, 1993, pp. 117–126.

Brechtel, F.J., *Büchsenmeisterey, das ist, Kurtze doch eigentliche Erklerung deren Ding ...*, K. Gerlach, Nürnberg, 1591.

Brett-James, A., *Europe Against Napoleon*, Macmillan, London, 1970.

Brock, A., *A History of Fireworks*, George G. Harrap & Co., London, 1949.

Brugioni, D.A., "The Tyuratam Enigma," *Air Force Magazine*, March 1984, pp. 108, 109.

Bryant, W.W., *A History of Astronomy*, Methuen, London, 1907.

Buchanan, F., *A Journey from Madras through the Countries of Mysore, Canara, and Malabar*, Vol. 2, East India Company, London, 1807.

Bush, V., *Modern Arms and Free Men*, Simon and Schuster, New York, 1949.

Carlier, C., and Gilli, M., *The First Thirty Years at CNES*, La Documentation française/CNES, Paris, 1994 (English edition, 1995).

Chang, I., *Thread of the Silkworm*, Basic Books, New York, 1995.

Chapman, J.L., *Atlas: The Story of a Missile*, Harper and Brothers, New York, 1960.

Chertok, B.E., *Rakety i Lyudi (Rockets and People)*, Mashinostroenie, Moscow, 1994 (in Russian).

China Today: Space Industry, Astronautics Publishing House, Beijing, 1992.

Chulick, M.J., Meland, L.C., Thompson, F.C., and Williams, H.W., "History of the Titan Liquid Rocket Engines," *History of Liquid Rocket Engine Development in the United States: 1955-1980*, edited by S.E. Doyle, AAS History Series, Vol. 13, American Astronautical Society, San Diego, 1992, pp. 19–36.

Churchill, W., *The Second World War*, Houghton Mifflin, Boston, 1953.

Clarke, A.C., "Extra-Terrestrial Relays," *Wireless World*, October 1945, pp. 305–307.

Cochrane, T. (Earl of Dundonald), *Narratives of Services in the Liberation of Chili, Peru and Brazil*, J. Ridgway, London, 1859.

Congreve, W., *An Elementary Treatise on the Mounting of Naval Ordnance*, J. Egerton, London, 1811.

Congreve, W., *A Treatise on the General Principles, Powers, and Facility of Application of the Congreve Rocket System ...*, Longman, Rees, Orme, Brown, and Green, London, 1827.

Congreve, W. W., and Colquhoun, J.N., "Application of Rockets to the Destruction and Capture of Whales, &c.," British Patent No. 4563, 1821.

Corbeau, J., "A History of the French Sounding Rocket Veronique," AAS 89-260, *History of Rocketry and Astronautics,* edited by K.R. Lattu, AAS History Series, Vol. 8, American Astronautical Society, San Diego, 1992, pp. 147–167.

Corliss, W.R., *Scientific Satellites*, NASA SP-133, Washington, DC, 1967.

Corliss, W.R., *Spacecraft Tracking*, NASA EP-55, Washington, DC, 1969.

Courtois, S., Werth, N., Panné, J.-L., Paczkowski, A., Bartošek, K., and Margolin, J.-L., *The Black Book of Communism: Crimes, Terror, Repression*, Harvard Univ. Press, Cambridge, MA, 1999.

Cullum, G.W., *Biographical Register of the Officers and Graduates of the Military Academy at West Point*, Vols. 1 and 2, Van Nostrand, New York, 1868.

Curcio, J.A., and Sanderson, J.A., "Further Investigation of the Radiation from Rocket Motor Flames," Naval Research Lab. Rept. N-3327, Washington, DC, 26 July 1948.

Davies, M.E., and Harris, W.R., *RAND's Role in the Evolution of Balloon and Satellite Observation Systems and Related U.S. Space Technology*, R-3692-RC, RAND Corp., Santa Monica, CA, 1988.

Davis, T.L., "Fire for the Wars of China," *Ordnance*, Vol. 33, 1948, pp. 52, 53.

Davis, T.L., and Ware, J.R., "Early Chinese Military Pyrotechnics," *Journal of Chemical Education*, Vol. 24, 1947, pp. 522–537.

Day, D.A., "Listening from Above: The First Signals Intelligence Satellite," *Spaceflight*, Vol. 41, 1999, pp. 338–346.

de Clavijo, R.G., *Clavijo Embassy to Tamerlane*, George Rutledge and Sons, London, 1928.

de Joinville, J., *Chronicles of the Crusades*, H.G. Bohn, London, 1848.

de Joinville, J., *Histoire de Saint Louis*, Firmin Didot Freres, Paris, 1874.

de la Tour, M., *The History of Hyder Shah*, W. Thacker, London, 1855.

Derevyashkin, S., and Baichurin, I., "Trudnye Starty Znamenitoi 'Semyorki'" ("Difficult Starts of the Famous 'Semyorka'"), *Novosti Kosmonavtiki*, No. 10, 2000, pp. 66–69 (in Russian).

DeVorkin, D.H., *Science with a Vengeance*, Springer-Verlag, New York, 1992.

DiFrancesco, A., Boorady, F., and Sumpter, D., "The Agena Rocket Engine Story," AIAA/ASME/SAE/ASEE 25th Joint Propulsion Conference, AIAA Paper 89-2390, July 1989.

Donnelly, R.W., "Rocket Batteries in the Civil War," *Military Affairs*, Vol. 25, No. 2, 1961, pp. 69–93.

Dornberger, W., "The Lessons of Peenemunde," *Astronautics*, Vol. 3, No. 3, 1958, pp. 18–20 and 58, 59.

Dornberger, W., *V-2*, Viking Press, New York, 1954.

Draper, C.S., "Origins of Inertial Navigation," *Journal of Guidance and Control*, Vol. 4, 1981, pp. 449 463.

Draper, C.S., Wrigley, W., and Hovorka, J., *Inertial Guidance*, Pergamon, New York, 1960.

Duncan, F., *History of the Royal Regiment of Artillery*, John Murray, London, 1874.

Dyer, D., *TRW: Pioneering Technology and Innovation Since 1900*, Harvard Business School Press, Boston, MA, 1998.

Eisenhower, D.D., *Mandate for Change, 1953-1956*, Doubleday, Garden City, NY, 1963.

Esnault-Pelterie, R., *L'Astronautique*, A. Lahure, Paris, 1930.

Esnault-Pelterie, R., "Considérations sur les Résultats d'un Allégement Indéfini des Moteurs," *Journal de Physique*, Ser. 5, t. 3, 3 March 1913, pp. 218–230.

Esnault-Pelterie, R., *L'Exploration par Fusées de la Très Haut Atmosphère et la Possibilité des Voyages Interplanétaires*, Société Astronomique de France, Paris, 1928.

Falk, S.L., "Artillery for the Land Service: the Development of a System," *Military Affairs*, Vol. 28, No. 3, 1964, pp. 97–110.

Feklisov, A., and Kostin, S., *The Man Behind the Rosenbergs*, Enigma Books, New York, 2001.

The Field Manual for the Use of the Officers on Ordnance Duty, Ritchie & Dunnavant, Richmond, 1862.

Flammarion, C., *Les Terres du Ciel. Voyage Astronomique sur les Autres Mondes ...*, C. Marpon et E. Flammarion, Paris, 1884.

Flammarion, C., *Mémoires Biographique et Philosophique d'un Astronome*, E. Flammarion, Paris, 1911.

Forsyth, K.S., "Delta: the Ultimate Thor," *To Reach the High Frontier. A History of the U.S. Launch Vehicles*, edited by R.D. Launius and D.R. Jenkins, Univ. of Kentucky Press, Lexington, KY, 2002, pp. 103–146.

Fortescue, J.W., *Dundonald*, Macmillan, London, 1906.

Fraser, E., and Carr-Laughton, L.G., *The Royal Marine Artillery*, The Royal United Service Institution, London, 1930.

Fuhrman, R.A., "The Fleet Ballistic Missile System: Polaris to Trident," *Journal of Spacecraft*, Vol. 15, No. 5, 1978, pp. 265–286.

Gardner, T., "How We Fell Behind in Guided Missiles," *The Air Power Historian*, Vol. 5, No. 1, Jan. 1958, pp. 3–14.

Gentilini, E., *Il perfeto Bombardiero et Real Instrvttione di artiglieri ...*, Appresso Alessandro de'Vechi, Venetia, 1626.

Gerchik, K.V. (ed.), *Nezabyvaemyi Baikonur* (*Unforgettable Baikonur*), Moscow, 1998 (in Russian).

Gleig, G.R., *The Campaigns of the British Army at Washington and New Orleans, in the Years 1814-1815*, J. Murray, London, 1827.

Gleig, G.R., *The Subaltern: the Diaries of George Greig during the Peninsular War*, London, 1825; reprinted as Gleig, G.R., *The Subaltern: a Chronicle of the Peninsular War*, Leo Cooper, London, 1969, 2001.

Glushko, V.P., *Development of Rocketry and Space Technology in the USSR*, Novosti Press Publishing House, Moscow, 1973.

Goddard, G.W., with Copp, D.S., *Overview: A Life-Long Adventure in Aerial Photography*, Doubleday & Co., Garden City, NY, 1969.

Goddard, R.H., *The Papers of Robert H. Goddard*, edited by E.C. Goddard and G.E. Pendray, Vol. I: 1898–1924, McGraw–Hill, New York, 1970.

Gold, C., *Oriental Drawings: Sketches Between the Years 1791 and 1798*, London, 1806.

Gould, R.D., and Harlow, J., "Black Arrow: The First British Satellite Launcher," *History of Rocketry and Astronautics*, edited by J.G. Hunley, AAS History Series, Vol. 20, American Astronautical Society, San Diego, 1997, pp. 257–273.

Green, C.M., and Lomask, M., *Vanguard: A History*, Smithsonian Institution Press, Washington, DC, 1971; also NASA SP-4202, 1970.

Gruen, A.L., *Preemptive Defense*, Air Force History and Museums Program, Washington, DC, 1998.

Guier, W.H., and Weiffenbach, G.C., "Genesis of Satellite Navigation," *Johns Hopkins APL Technical Digest*, Vol. 18, No. 2, 1997, pp. 178–181.

Hale, E.E., *His Level Best and Other Stories*, Garrett Press, New York, 1968.

Hale, E.E., *The Life and Letters of Edward E. Hale, Jr.*, Little, Brown, and Co., Boston, 1917.

Haley, A.G., *Rocketry and Space Exploration*, Van Nostrand, Princeton, NJ, 1958.

Hall, R.C., "From Concept to National Policy: Strategic Reconnaissance in the Cold War," *Prologue*, Vol. 28, No. 2, 1996, pp. 107–125.

Hall, R.C., "Origins of U.S. Space Policy: Eisenhower, Open Skies, and Freedom of Space," *Exploring the Unknown*, edited by J.M. Logsdon, NASA SP-4218, Washington, DC, 1995, Chap. 2.

Hartt, J., *The Mighty Thor: Missile in Readiness*, Duell, Sloan and Pearce, New York, 1961.

Harvey, B., *The Japanese and Indian Space Programmes: Two Roads Into Space*, Springer-Verlag, New York, 2000.

Hearings Before the Preparedness Investigating Committee of the Committee on Armed Services, U.S. Senate, Eighty-Fifth Congress, First and Second Sessions, Part I and II, Government Printing Office, Washington, DC, 1958.

Heflin, W.A., "Astronautics," *American Speech*, Vol. 34, 1961, pp. 169–174.

Heitman, F.B., *Historical Register and Dictionary of the United States Army*, Government Printing Office, Washington, DC, 1903.

Hersey, I., "The Meaning of 'Sputnik'," *Astronautics*, Vol. 2, No. 4, November 1957, pp. 22–25.

Hertzfeld, H., and Ojalehto, G., "Steady Course for Civil Space in 2002," *Aerospace America*, Vol. 41, No. 10, 2003, pp. 6, 7 and 38–40.

Hohmann, W., *Die Erreichbarkeit der Himmelskörper*, R. Oldenbourg, München und Berlin, 1928.

Horne, C.F. (ed.), *Works of Jules Verne*, V. Parke and Co., New York, 1911.

Horsman, R., *The War of 1812*, Knopf, New York, 1969.

Hunley, J.D., "Minuteman and the Development of Solid-Rocket Launch Technology," *To Reach the High Frontier. A History of the U.S. Launch Vehicles*, edited by R.D. Launius and D.R. Jenkins, Univ. of Kentucky Press, Lexington, KY, 2002, pp. 229–300.

Hussein Ali, *The History of Hydur Naik*, translated by Colonel W. Miles, Oriental Translation Fund, London, 1842.

Huzel, D.K., *Peenemünde to Canaveral*, Prentice-Hall, Englewood Cliffs, NJ, 1962.

Inquiry into Satellite and Missile Programs, Hearings Before the Preparedness Investigating Subcommittee of the Committee on Armed Services, U.S. Senate, Eighty-Fifth Congress, Part I and Part II, Washington, DC, 1958.

Jackson, A.V.W., *History of India*, Vol. 5, Grolier Society, London, 1906.

Jung, P., "French Rocketry 1739-1872," *History of Rocketry and Astronautics*, edited by J.D. Hunley, AAS History Series, Vol. 20, American Astronautical Society, San Diego, 1997, pp. 3–24.

Keldysh, M.V. (ed.), *Tvorcheskoe Nasledie Akademika Sergeya Pavlovicha Koroleva* (*Creative Heritage of Academician Sergei Pavlovich Korolev)*, Nauka, Moscow, 1980 (in Russian).

Kennedy, G.P., *Vengeance Weapon 2*, Smithsonian Institution Press, Washington, DC, 1983.

Khrushchev, S.N., *Rozhdenie Sverkhderzhavy* (*Birth of Superpower)*, Vremya, Moscow, 2000 (in Russian).

Killian, J.R., *Sputnik, Scientists, and Eisenhower*, MIT Press, Cambridge, MA, 1977.

Kisun'ko, G.V., *Sekretnaya Zona* (*Secrete Zone)*, Sovremennik, Moscow, 1996 (in Russian).

Klawans, B., "The Vanguard Satellite Launching Vehicle: An Engineering Summary," NASA CR-13654, April 1960.

Konstantinov, K.I., "Boevye Rakety v Rossii v 1867 Godu" ("War Rockets in Russia in 1867"), *Artilleriiskii Zhurnal* (*Artillery Journal)*, No. 5, 1867, pp. 818–872 (in Russian).

Kyeser aus Eichstätt, C. , *Bellifortis*, VDI-Verlag, Düsseldorf, 1967.

Lamphere, R. J., *The FBI-KGB War*, Mercer Univ. Press, Macon, GA, 1995.

Lasby, C.G., *Project Paperclip*, Atheneum, NY, 1971.

Latour, A.L., *Historical Memoir of the War in West Florida and Louisiana in 1814-1815*, edited by G.A. Smith, The Historic New Orleans Collection and Univ. Press of Florida, Gainesville, 1999.

Launius, R.D., "Eisenhower, Sputnik, and the Creation of NASA," *Prologue*, Vol. 28, No. 2, 1996, pp. 127–143.

Launius, R.D., *NASA: A History of the U.S. Civil Space Program*, Krieger, Malabar, FL, 1994.

Launius, R.D., "Titan. Some Heavy Lifting Required," *To Reach the High Frontier. A History of the U.S. Launch Vehicles*, edited by R.D. Launius and D.R. Jenkins, Univ. of Kentucky Press, Lexington, KY, 2002, pp. 147–185.

Lewis, J.D., and Di, H., "China's Ballistic Missile Programs," *International Security*, Vol. 17, No. 2, 1992, pp. 5–40.

Lewis, J.E., *Spy Capitalism: Itek and the CIA*, Yale Univ. Press, New Haven, CT, 2002.

Ling, W., "On the Invention and Use of Gunpowder and Firearms in China," *Isis*, Vol. 37, 1947, pp. 160–178.

Loesser, F., *The Complete Lyrics of Frank Loesser*, edited by R. Kimball and S. Nelson, Knopf, New York, 2003.

Logsdon, J.M. (ed.), *Exploring the Unknown*, NASA SP-4407, Washington, DC, 1996.

London, J.R., III, "Vela — A Space System Success Story," *History of Rocketry and Astronautics*, edited by J.G. Hunley, AAS History Series, Vol. 20, American Astronautical Society, San Diego, 1997, pp. 215–232.

Lossing, B.J., *New History of the United States*, Gay Brothers & Co., New York, 1889.

Lossing, B.J., *The Pictorial Field-Book of the War of 1812*, Harper and Brothers, New York, 1869.

Lossing, B.J., *Pictorial History of the Civil War*, George W. Childs, Philadelphia, 1866.

Lowell, P., *Mars*, Houghton, Mifflin, and Co., Boston, 1895.

Lowell, P., *Mars and Its Canals*, MacMillan, New York, 1911.

Lowell, P., *Mars as the Abode of Life*, MacMillan, New York, 1910.

Lowell, P., "Our Solar System," *Popular Astronomy*, Vol. 24, No. 237, 1916, pp. 419–427.

MacKenzie, D., *Inventing Accuracy: an Historical Sociology of Nuclear Missile Guidance*, MIT Press, Cambridge, MA, 1993.

Malthus, F., *Traité de fevx Artificiels povr la Gverre, et povr la Recreation*, Cardin Besongne, Paris, 1640.

Maksimov, A., *Operatsiia "Turnir"* (*Operation "Turnir")*, GEIA iterum, Moscow, 1999 (in Russian).

Mao Yuan-i Chi, *Wu Pei Chih*, Hua Shih Press, Taipei, 1991.

Mao Zedong, *The Writings of Mao Zedong, 1946–1976*, Vol. 2, edited by J.K. Leung and M.Y.M. Kau, M.E. Sharpe, Armonk, New York, 1986.

Martin, R.E., "The Atlas and Centaur 'Steel Balloon' Tanks. A Legacy of Karel Bossart," 40th Congress of International Astronautics Federation, IAA Paper 89-738, 1989.

Massey, H., and Robins, M.O., *History of the British Space Science*, Cambridge University Press, Cambridge, England, 1986.

Maziukevich, M., *Zhizn' i Sluzhba Generala Ad'iutanta Karla Andreevicha Shil'dera* (*Life and Service of General Karl Andreevich Shil'der)*, Moscow, 1876 (in Russian).

McClellan, B.B., *The Mexican War Diary*, edited by W.S. Myers, Princeton Univ. Press, Princeton, NJ, 1917.

Medaris, J.B., and Gordon, A., *Countdown for Decision*, Putnam, New York, 1960.

Mercer, C., *Journal of the Waterloo Campaign*, Peter Davis, Ltd., London, 1927.

Meyer, W.E., and Anderson, R.W., "Operational Lessons Learned," *Johns Hopkins APL Technical Digest*, Vol. 3, No. 2, 1982, pp. 170–179.

Middlebrook, M., *The Peenemünde Raid*, Cassel & Co., London, 1988.

Miles, W.D., "The Polaris," *Technology and Culture*, Vol. 4, 1963, pp. 478–489.

Millard, D., *The Black Arrow Project: A History of a Satellite Launch Vehicle and Its Engines*, Science Museum, London, 2001.

Miller, J., *Memoirs of General Miller*, Longman, Rees, Orme, Brown, and Green, London, 1829.

Miller, R., *The Dream Machines*, Krieger, Malabar, FL, 1993.

Moore, T.L., "Solid Rocket Development at Allegany Ballistic Laboratory," 35-th AIAA/ASME/SAE/ASEE Joint Propulsion Conference, AIAA Paper 99-2931, June 1999.

Moore, W., *A Treatise on the Motion of Rockets*, G. and S. Robinson, London, 1813.

Mordecai, A., "The Life of Alfred Mordecai. As related by Himself," edited by J.A. Padgett, *The North Carolina Historical Review*, Vol. 22, 1945, pp. 58–108.

Mordecai, A., *Military Commission to Europe in 1855 and 1856*, George W. Bowman, Washington, DC, 1860.

Morros, B., *My Ten Years as a Counterspy*, Viking, New York, 1959.

Morton, P., *Fire Across the Desert*, Australian Government Publishing Service, Canberra, 1989, 1997.

Moulin, H., "A-1: The First French Satellite," *History of Rocketry and Astronautics,* edited by D.C. Elder and C. Rothmund, AAS History Series, Vol. 23, American Astronautical Society, San Diego, 2001, pp. 51–72.

Moynihan, M.F., "The Scientific Community and Intelligence Collection," *Physics Today*, Vol. 53, No. 12, 2000, pp. 51–56.

Mueller, F.K., "A History of Inertial Guidance," *Journal of the British Interplanetary Society*, Vol. 38, 1985, pp. 180–192.

Muratori, L.A., *Rerum Italicarum Scriptores*, Vol. 12, Mediolani, 1728.

Muratori, L.A., *Rerum Italicarum Scriptores*, Vol. 15, Mediolani, 1729.

Myers, D.D., "The Navaho Cruise Missile — A Burst of Technology," *History of Rocketry and Astronautics*, AAS History Series, Vol. 20, American Astronautical Society, San Diego, 1997, pp. 121–132.

The National Cyclopædia of American Biography, Vol. 4, James T. White and Co., New York, 1897.

The National Cyclopædia of American Biography, Vol. 5, James T. White and Co., New York, 1907.

Nature, editorial, Vol. 210, No. 5041, 11 June 1966, pp. 1085, 1086.

Neufeld, J., *The Development of Ballistic Missiles in the United States Air Force, 1945–1960*, U.S. Air Force, Washington, DC, 1990.

Neufeld, M.J., *The Rocket and the Reich*, The Free Press, New York, 1995.

Noordung, H., *The Problems of Space Travel. The Rocket Motor*, translation, NASA SP-4026, NASA, Washington, DC, 1995.

Ob Upotreblenii Boevykh Raket pri Vzyatii Ak-Mecheti i iz Del S. P. B. Raketnogo Zavedeniya (On the Use of War Rockets during Capture of Ak-Mechet' and from the Files of St. Petersburg Rocket Establishment), St. Petersburg Rocket Establishment, St. Petersburg, 1854 (in Russian).

Oberth, H., "Flying Saucers Come from a Distant World," *The American Weekly*, 24 October 1954, pp. 4, 5.

Oberth, H., "My Contributions to Astronautics," *First Steps Toward Space*, edited by F.C. Durant III and G.S. James, AAS History Series, Vol. 6, American Astronautical Society, San Diego, 1985, pp. 129–140.

Oberth, H., *Die Rakete zu den Planetenräumen*, R. Oldenbourg, Munich and Berlin, 1923.

Oder, F.C.E., Fitzpatrick, J.C., and Worthman, P.E., *The Corona Story*, National Reconnaissance Office, 1987.

Odnazhdy i Navsegda ... Dokumenty i Lyudi o Sozdatele Raketnykh Dvigatelei i Kosmicheskikh System Akademike Valentine Petroviche Glushko (Once and Forever ... Documents and People about the Developer of Rocket Engines and Space Systems Academician Valentin Petrovich Glushko), Mashinostroenie, Moscow, 1998 (in Russian).

Olejar, P.D., "Rockets in Early American Wars," *Military Affairs*, Vol. 10, 1946, pp. 16–34.

The Ordnance Manual for the Use of the Officers of the United States Army, 2nd ed., Guideon & Co., Washington, DC, 1850.

The Ordnance Manual for the Use of the Officers of the United States Army, 3rd ed., J.B. Lippincott & Co., Philadelphia, 1862.

Ordway, F.I., III, and Sharpe, M.R., *The Rocket Team*, MIT Press, Cambridge, MA, 1982.

Ordway, F.I., III, and Wakeford, R.C., *International Missile and Spacecraft Guide*, McGraw–Hill, New York, 1960.

Ordway, F.I., III, and Winter, F.H., "Reaction Motors Inc.: A Corporate History," AIAA Paper 82-277, 1982.

Pant, G.N., *Horse and Elephant Armour*, Agam Kala Prakashan, Delhi, 1997.

Partington, J.R., *A History of Greek Fire and Gunpowder*, Johns Hopkins Univ. Press, Baltimore, MD, 1999.

Pedlow, G.W., and Welzenbach, D.E., *The CIA and the U-2 Program, 1954-1974*, Central Intelligence Agency, 1998.

Peebles, C., *The Corona Project: America's First Spy Satellites*, Naval Inst. Press, Annapolis, MD, 1997.

Peking Review, "Nation's Armymen and Civilians Acclaim Successful Launching of China's First Man-Made Earth Satellite," No. 18, 30 April 1970, pp. 6–9.

Pendray, G.E., "The First Quarter Century of the American Rocket Society," *Jet Propulsion*, Vol. 25, 1955, pp. 586–593.

Perry, R.L., "The Atlas, Thor, and Titan," *Technology and Culture*, Vol. 4, 1963, pp. 466–477.

Pervov, M.A., *Sistemy Raketno-Kosmicheskoi Oborony Rossii Sozdavalis' Tak* (*Systems of Missile-Space Defense of Russia Were Created This Way)*, Aviarus-XXI, Moscow, 2003 (in Russian).

Preliminary Design of an Experimental World-Circling Spaceship, Rept. SM-11827, Douglas Aircraft Co., Santa Monica, CA, 2 May 1946.

Primakov, E.M. (chief ed.), *Ocherki Istorii Rossiiskoi Vneshnei Razvedki* (*Stories of the Russian Foreign Intelligence Service)*, Vol. 1 (published in 1996) and Vol. 3 (1997), Mezhdunarodnye Otnosheniya, Moscow (in Russian).

Raborn, W.F., "Navy within a Navy," *Aeronautics and Astronautics*, Vol. 10, No. 10, Oct. 1972, pp. 63–65.

Ramo, S., *The Business of Science*, Hill and Wang, New York, 1988.

Raschid-Eldin, *Histoire des Mongols de Perse*, trad. par E. Quatremère, Oriental Press, Amsterdam, 1968.

Raushenbakh, B.V. (ed.), *S.P. Korolev i Ego Delo: Svet i Teni v Istorii Kosmonavtiki* (*S.P. Korolev and his Cause: Light and Shadows in the History of Cosmonautics)*, compiled by G..S. Vetrov, Nauka, Moscow, 1998 (in Russian).

Remini, R.V., *The Battle of New Orleans*, Viking, New York, 1999.

Richelson, J.T., *America's Space Sentinels: DSP Satellites and National Security*, Univ. Press of Kansas, Lawrence, KS, 1999.

Robillard, G., "Explorer Rocket Research Program," *ARS Journal*, Vol. 29, No. 7, 1959, pp. 492–496.

Rosen, M.W., *The Viking Rocket Story*, Harper & Brothers, New York, 1955.

Rostow, W.W., *Open Skies*, Univ. of Texas Press, Austin, 1982.

Roys, T.W., "Improvements in Harpoon-Guns," U.S. Patent 31,190, 1861.

Rubinstein, M., "A New Sect of the Atomic Religion," *New Times*, No. 50, Moscow, 10 Dec. 1947, pp. 7–10.

Ruffner, K.C. (ed.), *Corona: America's First Satellite Program*, Central Intelligence Agency, Washington, DC, 1995.

Russo, A., "Launching Europe into Space: The Origins of the Ariane Rocket," *History of Rocketry and Astronautics*, edited by D.C. Elder, AAS History Series, Vol. 23, American Astronautical Society, San Diego, 2001, pp. 35–49.

Sakharov, A.D., *Memoirs*, Knopf, New York, 1990.

Schriever, B.A., "Breaching the Chinese Wall of 100 Approvals," *Aeronautics and Astronautics*, Vol. 10, No. 10, Oct. 1972, pp. 57–60.

Schriever, B.A., "Our Five Arsenals of Peace," *Space Age*, Vol. 2, No. 1, Nov. 1959, pp. 36–38.

Scoffern, J., *Projectile Weapons of War*, Longman, Brown, and Co., London, 1859.

Semenov, Yu.P. (chief ed.), *Raketno-Kosmicheskaya Korporatsiya "Energia" Imeni S.P. Koroleva: 1946–1996* (*Korolev Rocket-Space Corporation "Energia": 1946–1996)*, RKK Energia, Korolev, Russia, 1996 (in Russian).

Sharpe, M.R., "Nonmilitary Applications of the Rocket Between the 17th and 20th centuries," *History of Rocketry and Astronautics*, edited by R. C. Hall, AAS History Series, Vol. 7, Part 1, American Astronautical Society, San Diego, 1986, pp. 51–72.

Shea, J.T., and Baumann, R.C., "Vanguard I Satellite Structure and Separation Mechanism," NASA TN D-495, March 1961.

Sheehan, W., *The Planet Mars: A History of Observation and Discovery*, Univ. of Arizona Press, Tucson, AZ, 1996.

Skripchinskii, "Parachyut-Rakety i Rakety s Kryl'yami" ("Parachute Rockets and Winged Rockets"), *Artilleriiskii Zhurnal* (*Artillery Journal*), No. 12, Russia, 1866, pp. 615–621 (in Russian).

Sloop, J.L., *Liquid Hydrogen as a Propulsion Fuel, 1945-1959*, NASA SP-4404, 1978.

Smith, B.L.R., *The RAND Corporation*, Harvard Univ. Press, Cambridge, MA, 1966.

Smith, R.A., and Minami, H.M., "History of the Atlas Engines," *History of Liquid Rocket Engine Development in the United States: 1955-1980*, edited by S.E. Doyle, AAS History Series, Vol. 13, American Astronautical Society, San Diego, 1992, pp. 53–66.

Sokol'skii, V.N., *Russian Solid-Fuel Rockets*, NASA TT F-415, 1967.

Speer, A., *Inside the Third Reich*, Simon and Schuster, New York, 1970.

Stuhlinger, E., and Ordway, F.I., III, *Wernher von Braun. Crusader for Space*, Krieger, Malabar, FL, 1996.

Susane, L.A., *Histoire de L'Artillerie Française*, J. Heitzel et Companie, Paris, 1874.

Sutton, E.S., "From Polymers to Propellants to Rockets — A History of Thiokol," 35-th AIAA/ASME/SAE/ASEE Joint Propulsion Conference and Exhibit, Los Angeles, CA, AIAA Paper 99-2929, June 1999.

Taccola, M., *De Ingeneis*, Dr. Ludwig Reichert Verlag, Wiesbaden, 1984.

Tait, P.G. and Steele, W.J., *A Treatise on the Dynamics of a Particle with Numerous Examples*, Macmillan and Co., Cambridge, 1856.

Teller, E., "The Work of Many People," *Science*, Vol. 121, 1955, pp. 267–275.

Teller, E., with Shoolery, J.L., *Memoirs. A Twentieth-Century Journey in Science and Politics*, Perseus Publishing, Cambridge, MA, 2001.

Thybourel, F., *Recueil de Plusiers Machines Militaires et Feux Artificiels pour la Guerre & Recreation ...*, Charles Marchant, Pont-à-Mousson, 1620.

Tikhonravov, M.K., "The Creation of the First Artificial Earth Satellite: Some Historical Details," *History of Rocketry and Astronautics*, edited by K.R. Lattu, AAS History Series, Vol. 8, American Astronautical Society, San Diego, 1989, pp. 207–213.

Towards New Horizons, Commemorative Edition 1950-1992, History Office, Headquarters Air Force Systems Command, 1992.

Tsien, H.S., "Never Forget Chairman Mao's Kind teachings," *Renmin Ribao* (*People's Daily*), 16 Sept. 1976, p. 6 (in Chinese).

Tsiolkovsky, K.E., *Selected Works*, Mir Publishers, Moscow, 1968 (in English).

Tsiolkovsky, K.E., *Sobranie Sochinenii*, tom 2, *Reaktivnye Letatel'nye Apparaty* (*Collection of Works*, Vol. 2, *Jet-Propelled Flying Vehicles*), AN SSSR (Publishing House of the USSR Academy of Sciences), Moscow, 1954 (in Russian).

Tucker, J.E., "The History of the RL10 Upper-Stage Rocket Engine: 1956-1980," *History of Liquid Rocket Engine Development in the United States: 1955-1980*, edited by S.E. Doyle, AAS History Series, Vol. 13, American Astronautical Society, San Diego, 1992, pp. 123–151.

Valenzuela, R., *Cochrane: Marino y Libertador*, Marina de Chile, Valparaiso, 1961.

Van Allen, J.A., "Genesis of the International Geophysical Year," *History of Geophysics*, Vol. 1, edited by C.S. Gillmor, American Geophysical Union, Washington, DC, 1984, pp. 49, 50.

van Keuren, D.K., "Cold War Science in Black and White: US Intelligence Gathering and Its Scientific Cover at the Naval Research Laboratory, 1948-62," *Social Studies of Science*, Vol. 31, No. 2, 2001, pp. 207–229.

Verne, J., *Les Enfants du Capitain Grant*, American Book Co., New York, 1906.

Vetrov, G.S., in *Akademik S.P. Korolev. Uchenyi. Inzhener. Chelovek* (*Academician S.P. Korolev. Scientist. Engineer. Man.*), Nauka , Moscow, 1986, pp. 298- 304 (in Russian).

Villain, J., "France and the Peenemunde Legacy," *History of Rocketry and Astronautics,* edited by P. Jung, AAS History Series, Vol. 21, American Astronautical Society, San Diego, 1997, pp. 119–161.

von Braun, W., "The Explorers," *Astronautica Acta*, Vol. 5, No. 2, 1959, pp. 126–143.

von Braun, W., "Reminiscences of German Rocketry," *Journal of the British Interplanetary Society*, Vol. 15, No. 3, May–June 1956, pp. 125–145.

von Braun, W., Ordway, F.I., III, and Dooling, D., *Space Travel. A History*, Harper & Row, New York, 1975.

von Däniken, E., *Chariots of the Gods?*, Berkley Books, New York, 1968.

von Kármán, T., with Edson, L., *The Wind and Beyond*, Little, Brown and Co., Boston, 1967.

von Wrede, A., *Geschichte der K. und K. Wehrmacht*, Vol. 4, Part 1, Verlag von L. Seidel & Sons, Vienna, 1905.

Whinyates, F.A., "Captain Bogue and the Rocket Brigade," *Minutes of Proceedings of the Royal Artillery Institution*, Vol. 24, 1897, pp. 131–136.

Whitmore, W.F., "Military Operations Research — A Personal Retrospect," *Operations Research*, Vol. 9, No. 2, 1961, pp. 258–265.

Winter, F.H., *The First Golden Age of Rocketry*, Smithsonian Institution Press, Washington, DC, 1990.

Winter, F.H., *Prelude to Space Age: the Rocket Societies, 1924-1940*, Smithsonian Institution Press, Washington, DC, 1983.

Wise, J.C., *The Long Arm of Lee*, J.P. Bell Co., Lynchburg, VA, 1915.

Wrigley, W., "The History of Inertial Navigation," *The Journal of Navigation*, Vol. 30, 1977, pp. 61–68.

Yamada, N., *Ghenko. The Mongol Invasion of Japan*, E.P. Dutton and Co., New York, 1916.

York, H.F., *The Advisors. Oppenheimer, Teller, and the Superbomb*, Stanford Univ. Press, Stanford, CA, 1976.

INDEX

Numbers

A

Index

B

Index

C

Index

D

E

Index

F

G

I

J

K

L

M

N

Index

Q

R

S

T

W

X

Y

Z

SUPPORTING MATERIALS

A complete listing of AIAA publications is available at http://www.aiaa.org.